AF646318

ENCYCLOPAEDIA OF BIOLOGY

Dr. Satyavir Singh

ANMOL PUBLICATIONS PVT. LTD.
NEW DELHI - 110 002 (INDIA)

ANMOL PUBLICATIONS PVT. LTD.
H.O.: 4374/4B, Ansari Road, Daryaganj
New Delhi - 110 002
Ph.: 23278000, 23261597
B.O.: No. 1015, Ist Main Road, BSK III Stage
III Phase, III Block
Bangalore - 560 085 (India)
Visit us at: www.anmolpublications.com

Encyclopaedia of Biology

© Reserved

First Published, 2007

[Responsibility for the facts stated, opinions expressed, conclusions reached and plagiarism, if any, in these volumes is entirely that of the Editor/Author. The Publishers bear no responsibility for them, whatsoever.]

PRINTED IN INDIA

Printed at Mehra Offset Press, Delhi.

Preface

This Encyclopaedia is the perfect guide for those studying biology, either at school or university or anyone interested in Biology. It provides comprehensive coverage of Biology. The aim of this encyclopaedia is to provide teachers, students and laymen interested in Biology with clear, concise, and correct definitions and descriptions of the Biology terms, throughout. In the volume all the branches of Biology are represented.

This volume covers whole realm of Biology and includes various topics such as plant cell, animal cell, procaryotes, eucaryotes, cell structure, plastids, chloroplasts, endoplasmic reticulum, ribosomes, endosomes, vacuoles, dictyosomes, food vacuoles, pulsating vacuoles, vacuoles, golgi apparatus, microbodies, mitochondria, nucleus, cell wall, cytoskeleton, microtubules, centrioles, actin, cell functions, unicellular organisms, cell aaggregation, vegetative reproduction, alternation of generation, lignification, photosynthesis, plasmolysis, motility, protoplasmic streaming, migration of organelles, ameboid movement, muscle movement, stomatal movement, genetics, mitosis, cytokinesis and much more. The explanations of entries range from brief definitions of a few lines to expository discussions of various lengths.

The Encyclopaedia is to satisfy the immediate technical needs of the user. Concepts are presented in a clear and lucid manner. In terms of its broader objectives, the present book may be viewed as an outline or compendium of Biology ideas correlated with a large collection of specialized items. There are concern and activity in the educational field with regard to revitalizing and modernizing scientific training and creating a better understanding of this field; it is hoped that this volume will contribute to these aims in Biology.

The aim of this volume is to provide a concise repository of number of major concepts in this field. In preparation of this book, the author has freely consulted large number of books and journals so no authenticity is claimed. Author is especially thankful to M/S Anmol Publications for shaping this book in its final form.Suggestions for further improvement of this book are not only welcome but also greatly appreciated.

Author

A

A-amylase An enzyme that converts starch to sugars.

AB toxins The structure and activity of many exotoxins are based on the AB model. In this model, the B portion of the toxin is responsible for toxin binding to a cell but does not directly harm it. The A portion enters the cell and disrupts its function.

Abdomen In vertebrates, the portion of the trunk containing visceral organs other than heart and lungs; in arthropods, the posterior portion of the body, made up of similar segments and containing the reproductive organs and part of the digestive tract. Region of the body furthest from the mouth. In insects, the third body region behind the head and thorax.The hindmost of the three main body divisions of an insect. The last of the three major body divisions of an insect.The portion of a tetrapod's body between the thorax and pelvic girdle.The region of an arthropod's body behind the thorax. It contains the visceral organs.

Abductor A muscle that performs the action.Muscle taking a limb or the jaw away from the body.

Abiogenesis Early theory that held that some organisms originated from nonliving material.

Abiotic factor The physical, chemical and other non-living components of the environment that an organism lives in. These factors include all aspects of climate, geology, and atmosphere that affect ecological systems. Compare *biotic factor*. Nonliving component in the environment.

Abiotic stress Outside (nonliving) factors which can cause harmful effects to plants, such as soil conditions, drought, extreme temperatures.

Ablation experiment An experiment designed to produce an animal deficient in one or a few cell types, in order to study cell lineage or cell function. The idea is to make a transgenic mouse with a toxin gene (often diphtheria toxin) under control of a specialized promoter which activates only in the target cell type. When embryo development progresses to the point where it starts to form the target tissue, the toxin gene is activated, and that specific tissue dies. Other tissues are unaffected. Experimental removal or killing of some part of an organism; for example, in experimental embryology, used to determine what effect absence of the structure will have on development of the remaining embryo.

Abnormal hemoglobin Hemoglobin molecule with a different shape due to an altered amino acid sequence (ultimately caused by an altered DNA base sequence), such as in the inherited disease sickle-cell anemia.

Abomasum Fourth and last chamber of the stomach of ruminant mammals.

Aboral surface The surface opposite the mouth (or oral surface) in echinoderms.

Aboral The part of the body farthest from the mouth.

Abrasion The mechanical process of gradually breaking down a hard layer, as in a seed coat.

Abscess Dead cells and tissue fluid confined in a localized area, causing swelling.

Abscisic acid (ABA) A plant hormone that generally acts to inhibit growth, promote dormancy, and help the plant tolerate stressful conditions.A plant hormone associated with dormancy, abscission of organs, and water stress.A plant hormone that promotes dormancy in perennial plants and causes rapid closure of leaf stomata when a leaf begins to wilt.

Abscission In plants, the dropping of leaves, flowers, fruits, or stems at the end of a growing season, as the result of formation of a two-layered zone of specialized cells (the abscission zone) and the action of a hormone (ethylene).The detachment of leaves, flowers, or fruit from a plant, usually at a mechanically weak location, termed the abscission zone.

Absolute time One of the two types of geologic time (relative time being the other), with a definite age date established mostly by the decay of radioactive elements, although ages may also be obtained by counting tree rings, decay of a specific type of atom, or annual sedimentary layers (such as varves in lakes or layers in a glacier). The term is in some disfavor because it suggests an exactness that may not be possible to obtain.

Absorption The movement of water and dissolved substances into a cell, tissue, or organism. The process by which the products of digestion are transferred into the body's internal environment, enabling them to reach the cells. The process of taking in, as uptake by roots. The taking in of water and dissolved minerals and nutrients across cell membranes. Contrast with ingestion.

Absorption spectrum Graph of absorbance values for different wavelengths of light.The range of a pigment's ability to absorb various wavelengths of light.

Absorptive feeders Animals such as tapeworms that ingest food through the body wall.

Absorptive feeding A means of taking up dissolved foodmaterial through specialized organs or across the body wall.

Absorptive heterotroph Mode of nutrition in which organisms secrete digestive enzymes into a substrate and then absorb the products of digestion; nutritional mode of fungi.

Abyssal zone The deepest part of the oceans lying more than 2000 m beneath the surface. The main physical feature of this zone is the abyssal plain, a region of low relief beyond the continental margins. Its surface is broken in places by volcanic hills or seamounts rising out of the plain and, particularly in the Pacific Ocean, by linear trenches that reach depths in excess of 12,000 m beneath the sea surface. Sediments deposited in the upper levels of the oceans ultimately reach the abyssal zone, but little light penetrates, energy levels are low and plant or animal life is sparse. The water in the abyssal zone is considered to be in long-term storage, and its participation in the hydrological cycle has a time-scale of centuries.

Abyssopelagic zone 'Abysso' meaning 'no bottom', this zone of the ocean begins 4000 m below the surface of the ocean and extends down to the sea floor. This zone is home to a variety of unique critters that are specially adapted to the inhospitable conditions that these depths create. Visit the Abyssopelagic Zone on OceanLink to learn about life in the abyss.The 4,000 to 6000-m-depth zone, seaward of the shelf-slope break.The pelagic environment from a depth of 4,000 m (13,000 ft) to 6,000 m (20,000 ft).

Acanthella Developing acanthocephalan larva, between an acanthor and a cystacanth, in which the definitive organ systems are developed.

Acanthocephala The phylum of aschelminths commonly called the spiny-headed worms.

Acanthodians A group of the earliest known true jawed fishes from Lower Silurian to Lower Permian.

Acanthor Acanthocephalan larva (first larval stage) that hatches from the egg. The larva has a rostellum with hooks that are used in penetrating the host's tissues.

Acaricide A chemical employed to kill and control mites and ticks.

Accessory buds Those buds adjacent to a primary bud and usually smaller in size.

Accessory fruit Most or part of the fruit is derived from tissue other than the ovary of a flower.

Accessory pigment One of several nongreen photosynthetic pigments found in marine plants that absorbs light energy from the center of the visible light spectrum and transfers it to the green pigment chlorophyll.Photosynthetic pigments such as carotenoids and phycobiliproteins that aid chlorophyll in trapping light energy.

Accidental myiasis Presence within a host of a fly not normally parasitic.Also called pseudomyiasis.

Accidental parasite Parasite found in other than its normal host. Also called an incidental parasite.

Acclimatization Physiological adjustment to a change in an environmental factor.Gradual physiological adaptation in response to relatively long-lasting environmental changes.

Accommodation The adjustment of the eye for various distances.The automatic adjustment of an eye to focus on near objects.

Accrete To add terranes (small land masses or pieces of crust) to another, usually larger, land mass.

Accuracy Is the correctness of a measure when comparing to a known standard.

Acetabularia Genus of unicellular marine algae. These organisms comprise an enormous single cell consisting of a cap, stalk and rhizoid and can be up to 10 cm in size.

Acetabulum True sucker, especially in flukes and leeches; the socket in the hip bone that receives the thigh bone.

Acetospora The protozoan phylum characterized by members having multicellular spores; all parasitic in invertebrates. Examples: acetosporans (Paramyxa, Halosporidium).

Acetyl CoA An intermediate compound formed during the breakdown of glucose by adding a two-carbon fragment to a carrier molecule (Coenzyme A or CoA).The entry compound for the Krebs cycle in cellular respiration; formed from a fragment of pyruvate attached to a coenzyme.A combination of acetic acid and coenzyme A that is energy rich; it is produced by many catabolic pathways and is the substrate for the tricarboxylic acid cycle, fatty acid biosynthesis, and other pathways.

Acetylcholine One of the most common neurotransmitters; functions by binding to receptors and altering the permeability of the postsynaptic membrane to specific ions, either depolarizing or hyperpolarizing the membrane.A chemical released at neuromuscular junctions that binds to receptors on the surface of the plasma membrane of muscle cells, causing an electrical impulse to be transmitted. The impulse ultimately leads to muscle contraction.A neurotransmitter liberated by certain neurons. It is excitatory at neuromuscular junctions and inhibitory at other synapses.

Acetylcholinesterase (AChE) An enzyme that catalyzes hydrolysis of the neurotransmitter acetylcholine into acetate and choline. Expression of AChE can serve as a marker of skeletal muscle differentiation.

Achene A small, dry, one-seeded indehiscent fruit; the pericarp is easily separated from the seed coat.

Acicula Needlelike supporting bristle in parapodia of some polychaetes.

Acid H+(proton) donor; a substance that associates to release H+ and thus cause the pH of the solution to be less than 7.0.A substance that increases the hydrogen ion concentration in a solution.

Acid fast Refers to bacteria like the mycobacteria that cannot be easily decolorized with acid alcohol after being stained with dyes such as basic fuchsin.

Acid rain Acid rain is normally considered to be a by-product of modern atmospheric pollution. Even in a pure, uncontaminated world, however, it is likely that the rainfall would be acidic. Carbonic acid (H_2CO_3) is formed when carbon dioxide (CO_2) is absorbed by atmospheric water. Nitric acid (HNO_3) is created during thunderstorms and sulphuric acid (H_2SO_4) is formed from the sulphur dioxide (SO_2) released during volcanic eruptions and forest fires or from the sulphur (S) emitted by phytoplankton during their seasonal bloom period. All of these are natural processes which contribute to make normal rain acid. Current concern over acid rain is not with the naturally produced variety, but rather with that which results from modern industrial activity and the acid gases it produces. Considerable amounts of sulphur dioxide are released into the atmosphere as a by-product of metal smelting, particularly when non-ferrous ores are involved. The burning of coal and oil to provide energy for space heating or to fuel thermal electric power stations also produces sulphur dioxide. The continuing growth of transportation systems using the internal combustion engine contributes to acid rain through the release of oxides of nitrogen (NO_X) into the atmosphere.

Acid precipitation produced by human activities differs from natural acid precipitation not only in its origins, but also in its quality. Anthropogenically produced acid rain tends to be many times more acidic than the natural variety. The difference is commonly of the order of 1.0 to 1.5 points on the pH scale used to measure acidity. In North America, for example, naturally acid rain has a pH of about 5.6, whereas measurements of rain falling in southern Ontario, Canada frequently provide values in the range of 4.5 to 4.0. To put these values in perspective, vinegar has a pH of 2.7 and milk a pH of 6.6. Similar values for background levels of acidic rain are indicated by studies in Europe with the average annual pH of rain over Britain between 1978 and 1980 being within the range of 4.5 to 4.2.

The quality of the rain is determined by a series of chemical processes set in motion when acidic materials are released into the atmosphere. Some of the sulphur dioxide and oxides of nitrogen emitted will return to the surface quite quickly, and close to their source, as dry deposition. The remainder will be carried up into the atmosphere, to be converted into sulphuric and nitric acid, which will eventually return to earth as acid rain. The processes involved are fundamentally simple. Oxidation converts the gases into acids, in either a gas or liquid phase reaction, at a rate which depends upon such variables as the concentration of heavy metals in airborne particulate matter, the presence of ammonia and the intensity of sunlight. The metals and ammonia appear to act as catalysts, while the sunlight provides the energy required by the chemical reactions.

Acid rain remained mainly a local problem in the past. The introduction of the tall stacks policy in an attempt to reduce local pollution, however, increased its geographical extent. The release of pollutants at the greater altitudes possible with the taller smokestacks placed them outside the boundary layer circulation and into the larger scale atmospheric circulation system with its potential for much greater dispersal through the mechanisms involved in the Long Range Transportation of Air Pollution (LRTAP). Local problems were mitigated at the expense of the larger environment.

The main sources of acid rain are to be found in the industrialized areas of the northern hemisphere. North-eastern North America, Britain and Western Europe have received most attention, although their output of sulphur dioxide has been declining. Levels of oxides of nitrogen have not yet experienced significant decline and continue to rise in some areas. The emission of acid gases from Eastern Europe and the republics of the former USSR – Russia, Ukraine and Kazakhstan – remains high. In Asia, Japanese industries emit large quantities of sulphur dioxide, while the industrial areas of China are also major contributors. Acid emissions remain limited outside of the major industrial nations, but concern has been expressed over growing levels of air pollution, often associated with urban automobile exhaust emissions, which may already have provided a base for acid rain in some Third World countries.

Once the acid gases have been released into the atmosphere, they are at the mercy of the prevailing circulation patterns. With almost all of the areas currently producing large amounts of acidic pollution located within the mid-latitude westerly wind belt, emissions are normally carried eastwards, or perhaps north-eastwards, often for several hundred kilometres before being redeposited. In this way, pollutants originating in the US Midwest cause acid rain in Ontario, Quebec and the New England states. Emissions from the smelters and power stations of the English Midlands and the Ruhr contribute to the acidity of precipitation in Scandinavia, and acidity in the Arctic originates as far as 8000 km away to the south in North America and Eurasia. Thus the problem of acid rain transcends national boundaries, introducing political overtones to the problem and creating the need for international cooperation if a solution is to be found.

The impact of acid rain on the environment depends not only on the level of acidity in the rain, but also on the nature of the environment granitic or quartzitic bedrock, for example, are particularly susceptible to damage, lacking as they do the ability to 'buffer' or neutralize additional acidity from the precipitation. Acid levels therefore rise, the environmental balance is disturbed and serious ecological damage is the inevitable result. In contrast, areas which are geologically basic – underlain by limestone or chalk, for example – are much less sensitive, and may even benefit from the additional acidity. The highly alkaline soils and water of areas underlain by chalk or limestone ensure that the acid added to the environment by the rain is very effectively neutralized.

The impact of acid rain on the environment was first recognized in the lakes and rivers of these areas. Reduced pH values which indicated the rising acidity were accompanied by low levels of calcium (Ca) and magnesium (Mg), elevated sulphate concentrations and an increase in the amounts of potentially toxic metals such as aluminum (Al). When aquatic communities were examined in areas as far apart as New York State, Nova Scotia, Norway and Sweden there was clear evidence that increased surface water acidity had adverse effects on fish. In some cases the acidity was sufficiently high that mature fish died, but, more commonly, fish populations began to decline because of the effects of the increasing acidity on reproduction. Damage to the eggs during spawning and the inability of the young fry to survive the higher acidity, particularly during the spring flush, ensured that the older fish were not replaced as they died. As a result, fish populations in many rivers and lakes in eastern North America, Britain and Scandinavia have declined noticeably in the last two to three decades and hundreds of lakes are now completely devoid of fish. Although the impact on the fish is most obvious, all aquatic organisms will decline in number and variety during progressive acidification. Waterbodies that have lost or are in the process of losing their flora and fauna are often described as 'dead' or 'dying'. This is not strictly correct, however, for some organisms are remarkably acid-tolerant, and even the most acid lakes have some life in them. Species of protozoans are found at pH levels as low as 2.0, for example.

There is growing evidence that those areas in which the waterbodies have already succumbed to acidification must also face the effects of increasing acid stress on their forests and soils. The threat is not universally recognized, however, and there remains a great deal of controversy over the amount of damage directly attributable to acid rain. Reduction in forest growth in Sweden, physical damage to trees in West Germany, and the death of sugar maples in Quebec and Vermont have all been blamed on the increased acidity of the precipitation in these areas. Many of the impacts, such as the thinning of annual growth rings, reduction in biomass and damage to fine root systems, are only apparent after detailed examination, but others are more directly obvious and have been described as dieback. This involves the gradual wasting of the tree inwards from the outermost tips of its branches. The process is cumulative over several years until the tree dies. The symptoms of dieback have been recognized in the maple groves and red spruce forests of north-eastern North America, and across fifteen countries and some 70,000 km^2 of forest in Europe. In Germany, where it was first linked with acid rain, dieback, or Waldsterben, was particularly extensive and in the 1980s was seen as a major threat to the survival of the German forests. More recent research has indicated that forest decline in Europe is a multifaceted process in which acid precipitation may be only one of a series of contributors along with tree harvesting practices, drought and fungal attacks, although its exact role remains controversial.

Present concern over acid rain is concentrated mainly on its effect on the natural environment, but acid rain also contributes to deterioration in the built environment: it attacks limestone and marble used as building stone, for example. The faceless statues and crumbling cornices of the world's famous palaces, castles, abbeys and cathedrals, from the Parthenon in Greece to the Taj Mahal in India, attest to its power. By attacking the fabric of these buildings it not only causes physical and economic damage, but threatens the world's cultural heritage.

Acid rain in the form of aerosols or attached to smoke particles can cause respiratory problems in humans, as was the case with the infamous London Smog of 1952. There is no evidence that wet deposition is directly damaging to human health, although in its ability to mobilize metals, from lead (Pb) or copper (Cu) pipes, for example, it may have important indirect effects.

Solutions to the problem of acid rain are deceptively simple. In theory, a reduction in the emission rate of acid forming gases is all that is required to slow down and eventually stop the damage being caused by the acidification of the environment. Translating that concept into reality has proved difficult, however. One of the first approaches involved adaptation, such as the addition of lime to land or lakes to counteract the increased acidity, and to allow the recovery mechanisms to work more effectively. Experiments involving the liming of lakes in Canada and Sweden, for example, have provided encouraging results. Artificial buffering of lakes in this way is only a temporary measure which may be likened to the use of antacid to reduce acid indigestion. The neutralizing effects of the lime may last longer than those of the antacid, but they do wear off in three to five years and re-liming is necessary as long as acid loading continues. Most of the current proposals for dealing with acid rain tackle the problem at its source. They attempt to prevent, or at least reduce, emissions of acid gases into the atmosphere, and since sulphur dioxide makes the greatest contribution to acid rain in North America and Europe, it has received most attention in the development of abatement procedures. Sulphur dioxide is formed when coal and oil are burned to release energy, and the technology to control it may be applied before, during or after combustion. The exact timing will depend upon such factors as the amount of acid reduction required, the type and age of the system and the cost-effectiveness of the particular process.

Flue gas desulphurization (FGD) is the most common approach to sulphur dioxide reduction. It uses scrubbers to neutralize the acidity of the flue gases before they are released into the atmosphere. With sulphur dioxide removal rates of between 80 and 95 per cent, FGD is more effective than methods such as fuel switching or fuel desulphurization applied prior to combustion and it is technically simpler than fluidized bed combustion (FBC) and lime injection multi-stage burning (LIMB) incorporated in the combustion process itself. Although FGD systems do little to reduce emissions of oxides of nitrogen they are the preferred method of dealing with acid rain, and future requirements to reduce sulphur dioxide levels will be met in large part by the installation of FGD equipment.

The study of acid rain has declined remarkably since the 1980s, when it was viewed by many as the major environmental problem facing the northern hemisphere. It is one environmental issue in which abatement programmes have met with some success. Sulphur dioxide levels continue to decline, the rain in many areas is considerably less acid than it was a decade ago, and the transboundary disputes which absorbed large amounts of time, energy and money in the 1980s have been resolved. There are indications that lakes and forests are showing some signs of recovery in the most vulnerable areas of North America and Europe, although this is still a matter of dispute.

Developments such as these have created the perception that the acid rain problem is being solved. It would be more accurate, however, to say that it has changed in nature and in geographical extent. In the developed nations, oxides of nitrogen are gradually making a larger contribution to acidity as levels of sulphur dioxide decline. In less developed areas, from Eastern Europe to China, the full impact of acid rain may still be in the future, and the research activities currently winding down in Western Europe and North America may have to be revived to deal with it.

Acid-fast staining A staining procedure that differentiates between bacteria based on their ability to retain a dye when washed with an acid alcohol solution.

Acini The secreting units of a gland. Each acinus is a sack-like structure, lined by secreting cells. The sack opens out into a tubule.

Acinus A small lobe of a compound gland or a saclike cavity at the termination of a passage.

Acoelomate A solid-bodied animal lacking a cavity between the gut and outer body wall.Lacking a body cavity between the gut and the outer body wall musculature,as in flatworms and proboscis worms.Animals that do not have a coelom or body cavity; e.g., sponges and flatworms.

Acontium Threadlike structure bearingnematocysts located on mesentery of sea anemone.

Acquired enamel pellicle A membranous layer on the tooth enamel surface formed by selectively adsorbing glycoproteins (mucins) from saliva. This pellicle confers a net negative charge to the tooth surface.

Acquired immune deficiency syndrome (AIDS) An infectious disease syndrome caused by the human immunodeficiency virus and is characterized by the loss of a normal immune

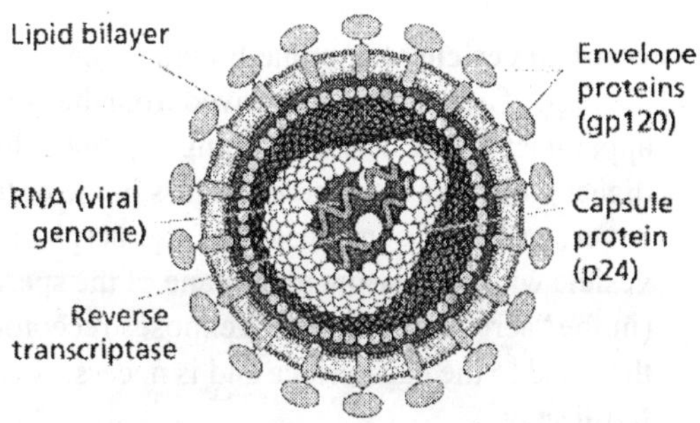

Fig. *Acquired immune deficiency syndrome (AIDS)*

response, followed by increased susceptibility to opportunistic infections and an increased risk of some cancers. A collection of disorders that develop as a result of infection by the human immunodeficiency virus (HIV), which attacks helper T cells, crippling the immune system and greatly reducing the body's ability to fight infection; results in premature death brought about by various diseases that overwhelm the compromised immune system.

Acquired immune tolerance The ability to produce antibodies against nonself antigens while "tolerating" (not producing antibodies against) self-antigens.

Acquired immunity Immunity arising from a specific immune response, stimulated by antigen in the host's body (active) or in the body of another individual with the antibodies or lymphocytes transferred to the host (passive). Refers to the type of specific immunity that develops after exposure to a suitable antigen or is produced after antibodies are transferred from one individual to another.

Acridine dyes Organic pigment molecules that are capable of causing permanent genetic changes (mutations).

Acrocentric Chromosome with centromere near the end.

Acromion A process on the scapula.

Acron Preoral region of an insect.

Acrorhagi Nematocyst-armed defensive structures of anemones.

Acrosomal vesicle Membrane-bound organelle in the head of a sperm that is derived from the golgi apparatus; the vesicle containing enzymes that digest proteins and complex sugars in the outer coverings of an egg. Fusion of the acrosomal vesicle with the plasma membrane of the sperm (in the "acrosome reaction") exposes receptors that bind to the egg surface and is necessary for fertilization.

Acrosome An organelle at the tip of a sperm cell that helps the sperm penetrate the egg.The enzyme-filled cap on the head of a sperm. Used in egg penetration.

Acrostichal bristles The two rows of hairs or bristles lying one on either side of the mid-line of the thorax of a true fly.

Acrylamide gels A polymer gel used for electrophoresis of DNA or protein to measure their sizes (in daltons for proteins, or in base pairs for DNA). Acrylamide gels are especially useful for high resolution separations of DNA in the range of tens to hundreds of nucleotides in length.

ACTH Abbreviation of adrenocorticotropic hormone.

Actin A protein of 42 kD, very abundant in eukaryotic cells (8-14% total cell protein) and one of the major components of the actomyosin motor and the cortical microfilament meshwork. First isolated from striated muscle and often referred to as one of the muscle proteins. G-Actin is the globular monomeric form of actin, 6.7 x 4.0nm: it polymerizes to form filamentous F-actin.

The distribution of filamentous actin in non-muscle cellCs is variable; sometimes it is organized into thick bundles, but more usually it forms a loose finely fibrous network.

In flattened, immobile cells in culture, actin filaments are usually organized as straight coarse bundles that radiate throughout the cell. These filamentous bundles are called 'stress' of 'sheath' fibres.

In moving cells, stress fibres lie parallel to the direction of movement and as the cell changes direction they change their allignment accordingly.

Since these filamentous bundles do not continue into the ruffled border area at the leading edge of the moving cell, it seems unlikely that they are concerned with the extended, forward phase of movement. More probably, they are responsible for 'bringing up the rear', when the extended cell contracts.

Actinobacteria A group of gram-positive bacteria containing the actinomycetes and their high G 1 C relatives.

Actinomorphic Regular or radially symmetric flowers.

Actinomycete An aerobic, gram-positive bacterium that forms branching filaments (hyphae) and asexual spores.

Actinorhizae Associations between actinomycetes and plant roots.

Actinotroch Larval form found in Phoronida.

Action potential A rapid change in the membrane potential of an excitable cell, caused by stimulus-triggered, selective opening and closing of voltage-sensitive gates in sodium and potassium ion channels.

Action threshold The pest density at which a control tactic must be implemented to avoid an economic loss.

Activated sludge Solid matter or sediment composed of actively growing microorganisms that participate in the aerobic portion of a biological sewage treatment process. The microbes readily use dissolved organic substrates and transform them into additional microbial cells and carbon dioxide.

Activation energy The energy required to bring reacting molecules together to reach the transition state in a chemical reaction.

Activational effects Effects of hormones that act as triggering influences on the expression of particular behavior patterns, response is rapid, in hours or days. Contrast with organizational effects.

Active avoidance learning A form of operant conditioning where the animal has to act in order to avoid some noxious consequence (e.g., shock).

Active carrier An individual who has an overt clinical case of a disease and who can transmit the infection to others.

Active continental margin A continental margin that is colliding with another plate and as a result is geologically active.Compare passive continental margin.

Active immunization The induction of active immunity by natural exposure to a pathogen or by vaccination.

Active ingredient (AI) The component of a pesticide formulation responsible for the toxic effect.

Active site The part of an enzyme that binds the substrate to form an enzyme-substrate complex and catalyze the reaction. Also called the catalytic site.

Active space The space within which the concentration of a pheromone or other behaviourally active substance is concentrated enough to generate the required response, remembering that like light and sound pheromones become more dilute the further they radiate out from their source.

Active transport A process that requires an expenditure of ATP energy to move molecules across a cell membrane; usually moved against the concentration gradient with the aid of specific transport proteins.The transport of solute molecules across a membrane against an electrochemical gradient; it requires a carrier protein and the input of energy.

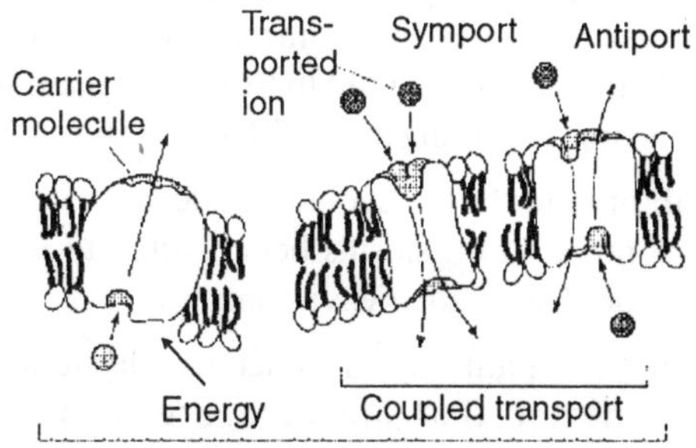

Fig. *Active transport*

Actomyosin Generally: a motor system that is thought to be based on actin and myosin . The essence of the motor system is that myosin makes transient contact with the actin filaments and undergoes a conformational change before releasing contact. The hydrolysis of ATP is coupled to movement, through the requirement for ATP to restore the configuration of myosin prior to repeating the cycle.

More specifically: a viscous solution formed when actin and myosin solutions are mixed at high salt concentrations. The viscosity diminishes if ATP is supplied and rises as the ATP is hydrolyzed. Extruded threads of actomyosin will contract in response to ATP.

Aculeate (Hymenoptera) Those members of the Hymenoptera which possess a sting.

Acute necrotising ulcerative gingivitis Abbreviated to ANUG- a painful and destructive infection of the gingiva caused by a shift in the normal balance of bacteria in the gingival sulcus, in which fusobacteria and spirochaetes become dominant.

Acute viral gastroenteritis An inflammation of the stomach and intestines, normally caused by Norwalk and Norwalklike viruses, other caliciviruses, rotaviruses, and astroviruses.

Adapt To modify in response to change. When used in regard to evolution, it means that some structure or behaviour of an organism may over time, appear to change in response toa new threat or opportunity in the environment. The bacterium which causes tuberculosis has developed certain strains which have adapted to the antibiotics used to treat the disease which is now becoming more difficult to treat.

Adaptation (1) The evolution of features that make a group of organisms better suited to live and reproduce in their environment.

(2) A peculiarity of structure, physiology, or behavior that aids the organism in its environment.Acondition or character which afford fitness to a species in a particular environment. An anatomical structure, physiological process, or behavioral trait that evolved by natural selection and improves an organism's ability to survive and leave descendants.Any change in the structure or function of an organism which makes it better suited to its environment. For example, the torpedo body shape of a tuna allows for fast swimming in the open ocean.Tendency of an organism to suit its environment; one of the major points of Charles Darwin's theory of evolution by natural selection: organisms adapt to their environment. Those organisms best adapted will have a greater chance of surviving and passing their genes on to the next generation.

Adaptive behavior Behavior patterns that make an organism more fit to survive and reproduce in comparison with other members of the same species.

Adaptive peak An equilibrium state in a population when the gene pool has allele frequencies that maximize the average fitness of a population's members.

Adaptive radiation The development of a variety of species from a single ancestral form; occurs when a new habitat becomes available to a population. Evolutionary pattern of divergence of a great many taxa from a common ancestral species as a result of novel adaptations or a recent mass extinction. Examples: mammals during the Cenozoic Era after the extinction of dinosaurs at the close of the Mesozoic Era flowering plants during the Cretaceous Period diversified because of their reproductive advantages over gymnosperm and non-seed plants that dominated the floras of the world at that time.The emergence of numerous species from a common ancestor introduced into an environment, presenting a diversity of new opportunities and problems.

Adaptive value Degree to which a characteristic helps an organism to survive and reproduce or lends greater fitness in its environment; selective advantage.

Adaptive zone A characteristic reaction and mutual relationship between environment and organism ("way of life") demonstrated by a group of evolutionarily related organisms.

Addictive drug Substance that induces an addiction by causing physiological dependence, psychological dependence, and/or tolerance.

Adductor A muscle that draws a part toward a median axis, or a muscle that draws the two valves of a mollusc shell together. Muscle bringing a limb or the jaw towards the body.

Adhesion - to form a chemical bond of attachment between two surfaces.

Adeagus The part of the male genitalia which is inserted into the female during copulation and which carries the sperm into the female. Its shape is often important in separating closely related species.

Adecticous Of pupa: referring to the state in which the pupa does not posses movable mandibles, the opposite being Decticous.

Adenine (A) A nitrogenous base, one member of the *base pair* A- T (*adenine- thymine*). A purine derivative, 6-aminopurine, found in nucleosides, nucleotides, coenzymes, and nucleic acids. One of the four nitrogen-containing bases occurring in nucleotides, the building blocks of the organic macromolecule group known as nucleic acids (DNA and RNA). Adenine is also the base in the energy carrying molecule ATP (adenosine triphosphate) which is the energy coin of the cell.

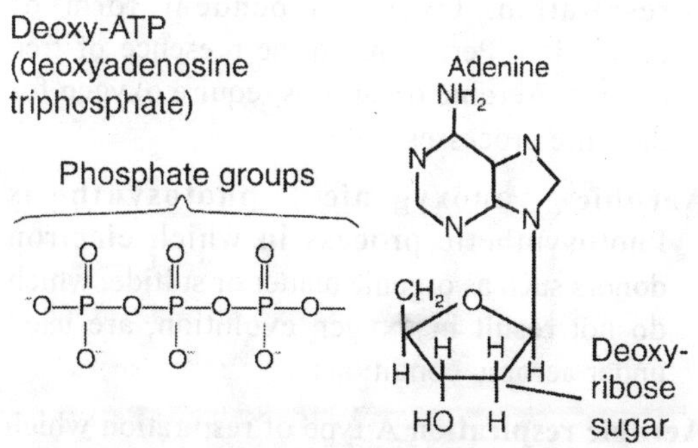

Fig. *Adenine*

Adenophorea The class of nematodes formerly called Aphasmidia. Examples: Trichinella, Trichuris.

Adenosine 5 triphosphate (ATP) The triphosphate of the nucleoside adenosine, which is a high energy molecule or has high phosphate group transfer potential and serves as the cell's major form of energy currency. An adenine-containing nucleoside triphosphate that releases free energy when its phosphate bonds are hydrolyzed. This energy is used to drive endergonic reactions in cells.

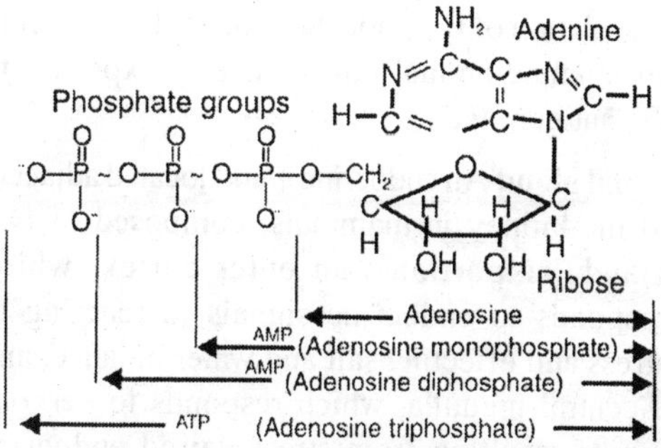

Fig. *Adenosine 5 triphosphate (ATP)*

Adenosine diphosphate (ADP) Lower energy form of ATP, having two (instead of the three in ATP) phosphhate groups attached to the adenine base and ribose sugar. The building block for ATP; by adding a terminal phosphate group and a large amount of energy, ATP can be formed.

Adenylyl cyclase An enzyme that converts ATP to cyclic AMP in response to a chemical signal.

Adhesio Interthalamica Part of the brain.

Adhesion The tendency of different kinds of molecules to stick together.Attraction between unlike molecules. The polar nature of water molecules causes them to adhere to a surface.The ability of molecules of one substance to adhere to a different substance.The attraction of unlike particles; water particles adhere to the surface of clays.

Adhesive disc Suckerlike circular organ or organelle used for attachment.

Adhesive gland Attachment glands in Turbellaria that produce a chemical that attaches part of the turbellarian to a substrate. Paired ectodermal thickenings found of the ventral surface of anuran embryos that secrete adhesive mucus. Synonyms: cement glands, mucous glands, oral suckers.

Adjuvant Material added to an antigen to increase its immunogenicity. Common examples are alum, killed Bordetella pertussis, and an oil emulsion of the antigen, either alone or with killed mycobacteria.

Adoptive immunity Immune state conferred by inoculation of lymphocytes, not antibodies, from an immune animal rather than by exposure to the antigen itself.

Adrenal gland An endocrine gland located adjacent to the kidney in mammals; composed of two glandular portions: an outer cortex, which responds to endocrine signals in reacting to stress and effecting salt and water balance, and a central medulla, which responds to nervous inputs resulting from stress.Paired endocrine glands, located next to the kidneys in the abdomen. The adrenal cortex produces steroid hormones involved in water balance, glucose metabolism, and electrolyte balance. The adrenal medulla produces adrenaline and noradrenaline, which are involved in glucose metabolism, heart rate, and blood pressure.

Adrenaline A hormone, produced by the medulla of the adrenal gland, that increases the concentration of glucose in the blood, raises blood pressure and heartbeat rate, and increases muscular power and resistance to fatigue; also a neurotransmitter across synaptic junctions. Also called epinephrine.

Adrenocorticotrophic A hormone secreted by the pituitary.

Adrenocorticotropic hormone (ACTH) A hormone produced by the anterior pituitary that stimulates the adrenal cortex to release several hormones including cortisol.

Adsorption The adhesion of molecules to solid bodies.

Adult T-cell leukemia A type of white blood cell cancer caused by the HTLV-1 virus.

Adult The mature stage of an organism, usually recognized by the organism's attaining the ability to reproduce.

Adventitious A structure arising at some location not usually expected, such as on a stem.

Adventitious roots Roots that develop from the stem following the death of the primary root. Branches from the adventitious roots form a fibrous root system in which all roots are about the same size; occur in monocots.A root produced by a stem or leaf rather than by a root.

Aeciospore Binucleate spore produced in an aecium.

Aecium Cup-shaped structure in rust fungi composed of binucleate hyphae; aeciospores are formed here.

Aedeagus Copulatory organ or penis in insects and acarines.

Aerate To supply with oxygen.

Aerial Pertaining to being in the air, such as a root projecting from an aboveground stem.

Aerobe A microorganism that grows in the presence of oxygen.

Aerobic Containing oxygen; referring to an organism, environment, or cellular process that requires oxygen. Having molecular oxygen present; an oxygen-dependent form of respiration. Oxygen-dependent form of respiration. Pertaining to the presence of free oxygen. Aerobic organisms require oxygen for their life processes.

Aerobic anoxygenic photosynthesis Photosynthetic process in which electron donors such as organic matter or sulfide, which do not result in oxygen evolution, are used under aerobic conditions.

Aerobic respiration A type of respiration which requires oxygen and in which glucose is broken down to release energy in a series of steps. The end products are carbon dioxide and water. Step 1;glucose is broken down to pyruvic acid in the cell cytoplasm with the release of 4 hydrogen atoms. Step 2; pyruvic acid is oxidised to acetylcoenzyme A (acetyl CoA), with the release of 4 further hydrogen atoms. Step 3; In the KREB cycle, 16 atoms of hydrogen are released. At all stages the hydrogen atoms are used to form the high energy molecule adenosine triphospate (ATP) via the electron transport system.

Aerosols Finely divided solid or liquid particles dispersed in the atmosphere. They include dust, soot, salt crystals, pollen grains, spores, bacteria, viruses, and a variety of other microscopic particles. Most aerosols are produced naturally by volcanic activity, forest and grass fires, evaporation, local atmospheric turbulence and biological processes, but human agricultural and industrial activities, plus increased energy consumption, have enhanced the anthropogenic contribution to the aerosol content of the atmosphere. Within the atmosphere, aerosols are redistributed by wind and pressure patterns, remaining in suspension for periods ranging from several hours to several years, depending upon particle size and altitude attained. Aerosols are responsible for atmospheric turbidity, and as a result disrupt the inward and outward flow of energy through the atmosphere. Some studies suggest that high aerosol levels contribute to global cooling through the attenuation of solar radiation, but there is also evidence that they may actually produce a slight warming through their ability to absorb and re-radiate outgoing terrestrial radiation. Resolution of this apparent contradiction will only be possible through systematic observation and monitoring of atmospheric aerosol levels and temperatures.

Aerotolerant anaerobes Microorganisms that grow equally well whether or not oxygen is present.

Aestivate To pass the summer in a certain manner or condition, often in a dormant or torpid state. It is when animals, especially desert animals, go into a state of torpor or hibernation in response to very dry conditions or hot temperatures. They usually emerge for a period of activity in the fall of the year.

Affected dentine Dentine which has been demineralised by acids in advance of invading caries bacteria. A distinction is made between affected dentine and infected dentine, because affected dentine is able to remineralise and should not be removed during cavity preparation.

Afferent Adjective meaning leading or bearing toward some organ, for example, nerves conducting impulses toward the brain or blood vessels carrying blood toward an organ; opposed to efferent.

Aflatoxin A polyketide secondary fungal metabolite that can cause cancer. A complex of four mycotoxins produced by Aspergillus flavus; often found in peanut products.

Afterbirth The placental and fetal membranes expelled from the uterus after childbirth.

Agar A gelatinous material prepared from certain red algae that is used to solidify nutrient media for growing microorganisms.Gelatinous product extracted from the walls of some red algae; used in growth media for microorganisms and in tissue culture.

Agarose gel electrophoresis A method for separating nucleic acids (DNA or RNA) within a gel made of agarose in a suitable buffer under the influence of an electrical field. Suitable for separation of large fragments of nucleic acid, separation is based primarily upon the size of the nucleic acid.

Agarose gels A polysaccharide gel used to measure the size of nucleic acids (in bases or base pairs). This is the gel of choice for DNA or RNA in the range of thousands of bases in length, or even up to 1 megabase if you are using pulsed field gel electrophoresis.

Age polyethism The regular changing of roles of colony members as they get older.

Age structure The proportion of a population that is in preproductive, reproductive, and postreproductive classes.The relative proportion of individuals in each age group in a population.

Agent orange A herbicide used as a defoliant in the Vietnam War; composed of 2,4-D and 2,4,5-T.

Agglutinates The visible aggregates or clumps formed by an agglutination reaction.

Agglutination reaction The formation of an insoluble immune complex by the cross-linking of cells or particles. The state of joining or clumping together by adhesion.

Agglutinin The antibody responsible for an agglutination reaction.

Aggregate Collection of units or particles (e.g., cells) forming a body or mass. To form such a body or mass. Clumps or collections of small particles or bacteria.Describes the concentration of most parasites of a single species in a minority of hosts.Also called overdispersed. Fairly random associations of animals with little or no internal organization; form in response to a single stimulus and disperse when the stimulus is removed; one of the three broad classes of social organization.

Aggregated spatial distribution A case where individuals in a space occur in clusters too dense to be explained by chance

Aggression Behavior that appears to be intended to inflict noxious stimulation or destruction on another organism.

Aggressive mimicry A technique for capturing prey in which the predator uses lures or other means to misinform the prey.

Agnatha A superclass of vertebrates whose members lack jaws and paired appendages and possess a cartilaginous skeleton and a persistent notochord. Lampreys and hagfishes.

Agnathan A member of a jawless class of vertebrates represented today by the lampreys and hagfishes.

Agonist The "prime mover" or muscle of reference.

Agonistic behavior A type of behavior involving a contest of some kind that determines which competitor gains access to some resource, such as food or mates.A suite of behavior patterns used during conflict with a conspecific, usually indicating whether an individual is going to submit to the other animal or fight if the other does not submit.

Agricultural infrastructure The complexities of processing, storing, packaging, and transporting farm products.

Agroecosystem A relatively artificial ecosystem in an agricultural field, pasture, or orchard.

Ahermatypic coral A coral that does not build reefs.

Ahermatypic Non-reef-building (referring to scleractinian corals)

AIDS-related complex (ARC) A collection of symptoms such as lymphadenopathy (swollen lymph glands), fever, malaise, fatigue, loss of appetite, and weight loss. It results from an HIV infection and may progress to frank AIDS.

Air sac A dilated portion of a trachea. Lateral branches of the nasal passages of smaller toothed whales; sources of echolocation sounds.

Airborne transmission The type of infectious organism transmission in which the pathogen is truly suspended in the air and travels over a meter or more from the source to the host.

Airfoil A surface, such as a wing, that provides lift by using currents of air it moves through.

Akinetes Specialized, nonmotile, dormant, thick-walled resting cells formed by some cyanobacteria.

Ala Term often applied to winglike structure on plants or animals: the lateral winglike expansions of the branchiuran carapace to form respiratory alae, cuticular winglike expansions of nematodes, and others.

Alar squama The middle of three flap-like outgrowths at the base of the wing in various flies.

Alare Usually a bone process shaped like a wing.

Alate Winged.Winged; having wings.

Albedo A description of a surface's reflective properties.

Albinism Genetic condition caused by the body's inability to manufacture pigments; an autosomal recessive trait.

Albumen The "white" of a bird's egg which provides both protein and water for the growing embryo. In reference to development, the egg white of birds eggs formed in the oviduct before the addition of the shell. It provides protein and water to the developing embryo and protection from microorganisms. Any of a large class of simple proteins that are important constituents of vertebrate blood plasma and tissue fluids and also present in milk, whites of eggs, and other animal substances.

Alcoholic fermentation A fermentation process that produces ethanol and CO2 from sugars.A form of anaerobic respiration in which sugar is degraded to alcohol and CO_2 and energy is released.

Aldehyde An organic molecule with a carbonyl group located at the end of the carbon skeleton.

Aldosterone An adrenal hormone that acts on the distal tubules of the kidney to stimulate the reabsorption of sodium (Na^+) and the passive flow of water from the filtrate.A hormone secreted by the adrenal glands that controls the reabsorption of sodium in the renal tubule of the nephron.

Aldrin (common name). A synthetic insecticide; a chlorinated hydrocarbon of not less than 95 per cent 1,2,3,4,10,10-hexachloro-1,4,4a,5,8,8a-hexahydro-1,4:5,8 dimethanonaphthalene; moderately toxic to mammals, acute oral LD,, for rats 44 mg/kg; phytotoxicity: none when properly formulated, but some crops are sensitive to solvents in certain formulations.

Aleurone layer The outermost cell layer of the endosperm of the grains (seeds) of wheat and other grasses; when acted upon by gibberellin, the aleurone layer releases enzymes that digest the stored food of the endosperm into small nutrient molecules that can be taken up by the embryo.A group of cells rich in protein granules and located as the outer layer of the endosperm of many grain seeds.

Aleurone Protein-rich, outermost layer of endosperm in a cereal grain.

Alga A common term for a series of unrelated groups of photosynthetic eucaryotic microorganisms lacking multicellular sex organs (except for the charophytes) and conducting vessels.A photosynthetic protist containing plastids. Any of several groups of autotrophs that lack the structural features (true leaves, roots, and stems) of the higher plants.

Algal bloom A proliferation of algae due to a nutrient-rich medium, usually resulting in a green scum on the water surface.

Algal grazer An animal that consumes either large algae or thin algal films.

Algal ridge A ridge of coralline algae that is found on the outer edge of some coral reefs.

Algal turf A dense growth of often filamentous algae.

Algin A polysaccharide derived from brown algae and used for many industrial processes.

Alginate component of the cell walls of many rhodophytes and kelps. Alginates have an affinity for water, and so help to slow dessication when the algae are exposed to the air; they are commercially important in the production of paper, toothpaste, beer, and frozen foods.

Alginic acid Gelatinous material extracted from the wall of certain brown algae; used in a variety of commercial and industrial products.

Alien A non-native species, especially one introduced to some part of the world through human action.

Alien species (exotic species)A species introduced by humans.

Alimentary Having to do with nutrition or nourishment.

Aliphatic A term applied to the "open chain" or fatty series of hydrocarbons.

Alitrunk Name given to the thorax and propodeum of 'wasp-waisted' hymenopterans.

Allopatric. Two or more forms of a species having essentially separate distributions.

Alkaline Denoting substances that release hydroxyl (OH^-) ions into solution. Pertaining to substances that increase the relative number of hydroxide ions (OH^-) in a solution; having a pH greater than 7; basic; opposite of acidic. Term pertaining to a highly basic, as opposed to acidic, subtance. For example, hydroxide or carbonate of sodium or potassium.A group of nitrogen-containing compounds having diverse structures; many alkaloids have medicinal, hallucinogenic, or toxic properties.

Alkaline phosphatase An enzyme which removes phosphate groups from organic compounds at an alkaline pH. It is found in high concentrations in matrix vesicles which are about to form new bone mineral. Alkaline phosphatase activity is a good indicator of bone formation.

Alkaloids Substances found in plants, many having powerful pharmacologic action, and characterized by content of nitrogen and the property of combining with acids to form 'salts'.

Alkalophile A microorganism that grows best at pHs from about 8.5 to 11.5.

Allantois An extraembryonic membrane formed by an extension of the hindgut into the extraembryonic coelom and consisting of endoderm and splanchnic mesoderm. In birds and reptiles, it fuses with the chorion to form the chorioallantoic membrane which assists in respiration, and in addition it is the site of storage of nitrogenous wastes. In eutherian mammals, it forms the vessels of the umbilical cord and fuses with the chorion to form the embryonic portion of the placenta. One of the extraembryonic membranes of the amniotes that functions in respiration and excretion in birds and reptiles and plays an important role in the development of the placenta in most mammals.

Allee effect The unfavorable consequences of undercrowding, as when populations fail to breed when numbers are below some critical density.

Allele Alternative forms of genes coding for the same trait; situated at the same locus in homologous chromosomes. One of several alternate forms of a gene occupying a given locus on a chromosome or plasmid.One of the alternative forms of a particular gene. Each gene is comprised of two alleles, one inherited from the father and one from the mother. However, within a population, many alleles may exist for one gene. Hair colour in humans is a great example! Alternative forms of a genetic *locus*; a single allele for each locus is inherited separately from each parent (e.g., at a locus for eye color the allele might result in blue or brown eyes).

Allele frequency The proportion of a particular allele in a population.

Allelopathy The release of chemicals by certain plants that inhibit the growth of competing plants.

Allergen A substance capable of inducing allergy or specific susceptibility.Substances that activate the immune system and cause an allergic response; may not be directly antigenic themselves but may make other materials antigenic.

Allergic contact dermatitis An allergic reaction caused by haptens that combine with proteins in the skin to form the allergen that produces the immune response.

Allergic reaction An inflammatory response triggered by a weak antigen (an allergen) to which most individuals do not react; involves the release of large amounts of histamine from mast cells.

Allochthonous Refers to something formed

elswhere than its present location. Antonym of autochthonous.

Allograft A piece of tissue or an organ transferred from one individual to another individual of the same species, not identical twins; homograft.

Allometric growth The variation in the relative rates of growth of various parts of the body, which helps shape the organism.Phenomenon whereby parts of the same organism grow at different rates. Contrast with isometric growth.

Allometry Relative growth of a part in relation to the whole organism.

Alloparental When individuals other than the parent assist in the caring for that parents offspring.

Allopatric In separate and mutually exclusive geographical regions.

Allopatric speciation A mode of speciation induced when the ancestral population becomes segregated by a geographical barrier.

Allopatry Populations or species with nonoverlapping geographic distributions. Contrast with sympatry.

Allopolyploid A common type of polyploid species resulting from two different species interbreeding and combining their chromosomes.

All-or-none event An action that occurs either completely or not at all, such as the generation of an action potential by a neuron.

All-or-none law The phenomenon in which a muscle fiber contracts completely when it is exposed to a stimulus of threshold strength. Also, the principle that states a neuron will "fire" at full power or not at all.

Allosteric enzyme An enzyme whose activity is altered by the binding of a small effector or modulator molecule at a regulatory site separate from the catalytic site; effector binding causes a conformational change in the enzyme and its catalytic site, which leads to enzyme activation or inhibition.

Allosteric site A specific receptor site on an enzyme molecule remote from the active site. Molecules bind to the allosteric site and change the shape of the active site, making it either more or less receptive to the substrate.

Allotype Allelic variants of antigenic determinant(s) found on antibody chains of some, but not all, members of a species, which are inherited as simple Mendelian traits.

Allozyme A variant of an enzyme type. These may be variants of a specific enzyme (e.g., cytochrome c) that are the products of a single genetic locus.Slightly different versions of the same enzyme, distinguishable via gel electrophoresis.

Alluvial fan A fan-shaped deposit of sand, mud, etc. formed by a stream where its velocity has slowed, such as at the mouth of a ravine or at the foot of a mountain.

Alluvium A deposit of sand, mud, etc., formed by flowing water.

Alpha helix A spiral shape constituting one form of the secondary structure of proteins, arising from a specific hydrogen-bonding structure. Literally the first spiral arrangement of the genetic DNA molecule; regular coiled arrangement of polypeptide chain in proteins; secondary structure of proteins.

Alpha hemolysis A greenish zone of partial clearing around a bacterial colony growing on blood agar.

Alpha-proteobacteria One of the five subgroups of proteobacteria, each with distinctive 16S rRNA sequences. This group contains most of the oligotrophic proteobacteria; some have unusual metabolic modes such as methylotrophy, chemolithotrophy, and nitrogen fixing ability. Many have distinctive morphological features.

Alternate host The alternate plant required to complete the life cycle of some microorganisms. For example, for Puccinia graminis tritice, wheat is the primary host and Berberis vulgaris is the alternate host.

Alternate Leaf arrangement in which there is only one leaf.

Alternating generations When two generations are produced within a life cycle each producing individuals of only one sex, either male first and then female or visa-versa. A life cycle in which there is both a multicellular diploid form, the sporophyte, and a multicellular haploid form, the gametophyte; characteristic of plants.A reproductive cycle in which a sexual stage alternates with an asexual one, as in the case of a gametophyte alternating with a sporophyte.

Alternative complement pathway An antibody-independent pathway of complement activation that includes the C3-C9 components of the classical pathway and several other serum protein factors (e.g., factor B and properdin).

Alternative splicing In alternative splicing, the same pre-mRNA molecule, which consists of introns and exons, is spliced in different ways to produce mature mRNAs of different lengths and different functionality.

Altitudinal zonation The sorting of plant and animal species according to elevation in response to differences in temperature and precipitation patterns.

Altricial An animal that is helpless at hatching or birth.Refers to mammals born in a fairly undeveloped state; the young require close parental care for a period of time. Rabbits are born in an altricial state, whereas, hares are precocial.

Altricial young Young that are born or hatched in a relatively immature or helpless condition. Compare with precocial.

Altruism The principle or practice of unselfish concern for or devotion to the welfare of others. 2)Altruistic - Unselfishly concerned for or devoted to the welfare of others.The principle or practice of unselfish concern for, or devotion to, the welfare of others.

Altruistic behavior The aiding of another individual at one's own risk or expense.

Altruistic preservation A philosophy of preserving nature for its own sake.

Alula A group of feathers on the wing of a bird that is supported by the bones of the medial digit. The alula reduces turbulent airflow over the upper surface of the wing.In insects (not birds) the outermost of the three flap-like outgrowths at the base of the wing in various flies: really a part of the wing membrane.

Alveolar bone Bone which develops around the roots of the teeth to hold them firmly in place. If the teeth are extracted, the alveolar bone resorbs away. Alveolar bone consists of both trabecula and cortical types of bone.

Alveolar macrophage A vigorously phagocytic macrophage located on the epithelial surface of the lung alveoli where it ingests inhaled particulate matter and microorganisms.

Alveoli Tiny, thin-walled, inflatable sacs in the lungs where oxygen and carbon dioxide are exchanged.

Alveolus *pl.* **alveoli** (1) One of the deadend, multilobed air sacs that constitute the gas exchange surface of the lungs.

(2) One of the milk-secreting sacs of epithelial tissue in the mammary glands.

Amantadine An antiviral compound used to prevent type A influenza infections.

Amber Fossilization where the organism is entrapped in resin and preserved whole.

Ambrosia The fungus cultivated by wood-boring beetles of the family Scolytidae

Ambulacra Row of tube feet of an echinoderm.

Amebiasis An infection with amoebae, often resulting in dysentery; usually it refers to an infection by Entamoeba histolytica.

Amebocyte Cell in metazoan invertebrate, often functioning in defense against invading particles.

Ameboid Ameba-like in putting forth pseudopodia.

Ameboma Granuloma containing active trophozoites, occasionally resulting from a chronic amebic ulcer; rare except in Central and South America.

Amebula Daughter cell resulting from mitosis and cytokinesis of an encysted ameba.

Ameloblasts Cells which differentiate from ectoderm and secrete enamel during tooth development.

Amensal Negatively affecting one or several species

Amensalism A relationship in which the product of one organism has a negative effect on another organism.A symbiotic relationship in which members of one population inhibit the growth of another population without being affected.

Ames test A test that uses a special Salmonella strain to test chemicals for mutagenicity and potential carcinogenicity.

Ametabola The insects which develop without metamorphosis, namely the Protura, Thysanura, and Collembola.

Ametabolous metamorphosis Development in which the number of molts is variable; immature stages resemble adults, and molting continues into adulthood.

Amictic Pertaining to female rotifers, which produce only diploid eggs that cannot be fertilized, or to the eggs produced by such females. Compare with mictic.

Amide Compound derived from carboxylic acids by replacing the hydroxyl of the -COOH by the amino group, $-NH_2-$.

Amine An organic compound containing nitrogen, derived from ammonia, NH_3, by replacing one or more hydrogen atoms by as many hydrocarbon radicals.

Amino acid An organic molecule possessing both carboxyl and amino groups. Amino acids serve as the monomers of proteins. An organic molecule including one or more amino (NH_2) and acid (COOH) groups; one of the 20 nitrogen-containing molecules that make up proteins. Any of a class of 20 molecules that are combined to form *proteins* in living things. The sequence of amino acids in a protein and hence protein function are determined by the *genetic code*. Amino acids bind by forming a peptide bond.

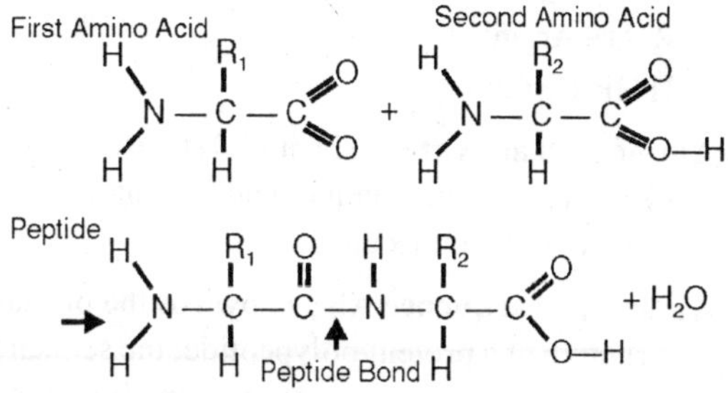

Fig. *Amino Acids*

Linear polymers of amino acids are called polypeptides.

Large polypeptides are called proteins.

A restricted set of 20 amino acids constitute the universal building blocks of proteins. The 20 basic building blocks of proteins, consisting of the basic formula NH2-CHR-COOH, where "R" is the side chain which defines the amino acid:

Nonpolar side chains (hydrophobic)

G Gly Glycine

A Ala Alanine

V Val Valine

I Ile Isoleucine

L Leu Leucine

F Phe Phenylalanine

P Pro Proline

M Met Methionine

W Trp Tryptophan

C Cys Cysteine

Noncharged polar side chains (hydrophilic)

S Ser Serine

T Thr Threonine

Y Tyr Tyrosine

N Asn Asparagine
Q Gln Glutamine
Acidic side chains (very polar, hydrophilic)
D Asp Aspartic Acid
E Glu Glutamic Acid
Basic side chains (very polar, hydrophilic)
K Lys Lysine
R Arg Arginine
H His Histidine.

Amino acid activation The initial stage of protein synthesis in which amino acids are attached to transfer RNA molecules.

Amino acid sequence Also known as the primary structure of a protein/polypeptide; the sequence of amino acids in a protein/polypeptide controlled by the sequence of DNA bases.

Amino group A functional group that consists of a nitrogen atom bonded to two hydrogen atoms; can act as a base in solution, accepting a hydrogen ion and acquiring a charge of +1.

Amino terminus Refers to the NH_2 end of a *peptide* chain (by custom drawn at the left of a protein sequence)

Aminoacyl or acceptor site (A site) The site on the ribosome that contains an aminoacyl-tRNA at the beginning of the elongation cycle during protein synthesis; the growing peptide chain is transferred to the aminoacyl-tRNA and lengthens by an amino acid.

Aminoglycoside antibiotics A group of antibiotics synthesized by Streptomyces and Micromonospora, which contain a cyclohexane ring and amino sugars; all aminoglycoside antibiotics bind to the small ribosomal subunit and inhibit protein synthesis.

Amitosis A form of cell division in which mitotic nuclear changes do not occur; cleavage without separation of daughter chromosomes.

Ammonia A colorless alkaline gas, NH_3, soluble in water.

Ammonification The process by which decomposers break down proteins and amino acids, releasing the excess nitrogen in the form of ammonia (NH_3) or ammonium ion (NH_4^+).

Ammonotelic excretion Having ammonia as the chief excretory product of nitrogen metabolism; occurs in freshwater fishes.

Amnesic shellfish poisoning The disease arising in humans and animals that eat seafood such as mussels contaminated with domoic acid from diatoms. The disease produces short-term memory loss in its victims.

Amniocentesis A technique for determining genetic abnormalities in a fetus by the presence of certain chemicals or defective fetal cells in the amniotic fluid, obtained by aspiration from a needle inserted into the uterus.A method of prenatal testing in which amniotic fluid is withdrawn from the uterus through a needle. The fluid and the fetal cells it contains are analyzed to detect biochemical or chromosomal disorders.Procedure for withdrawing a sample of fluid around the developing embryo for examination of chromosomes in the embryonic cells and other tests.

Amnion The innermost of four extraembryonic membranes; encloses a fluid-filled sac in which the embryo is suspended.An extraembryonic membrane surrounding the embryo and containing the amniotic fluid which protects the embryo from shock and desiccation. It consists of ectoderm and somatic mesoderm. The innermost membranous sac enclosing the embryo of an amniote; it becomes filled with amniotic fluid. One of the four amniote extraembryonic membranes; derived from the somatopleure (combination of ectoderm and somatic mesoderm)

Amniote A species of animal which possesses an amnion. This term is most often used in the plural (amniotes) to refer to reptiles, birds, and mammals collectively.A vertebrate possessing an amnion surrounding the embryo; reptiles, birds, and mammals are amniotes. Higher vertebrate capable of terrestrial reproduction, and having an amnion during its development. Includes reptiles, birds and mammals

Amniote egg An egg with compartmentalized sacs (a liquid-filled sac in which the embryo develops, a food sac, and a waste sac) that allowed vertebrates to reproduce on land. A shelled, water-retaining egg that enables reptiles, birds, and egg-laying mammals to complete their life cycles on dry land. The egg of reptiles, birds, and mammals. It possesses a series of extraembryonic membranes that help prevent desiccation, store wastes, and promote gas exchange. These adaptations allowed vertebrates to invade terrestrial habitats.

Amniote lineage The evolutionary lineage of vertebrates leading to modern reptiles, birds, and mammals.

Amoebocytes Amoeboid cells in sponges that occur in the matrix between the epidermal and collar cells. They transport nutrients.

Amoeboid Moving or feeding by means of pseudopodia (temporary cytoplasmic protrusions from the cell body).

Amoeboid movement A form of movement similar to that found in amoebae. Fluid endoplasm (plasmasol) flows forward inside the cell and changes state to viscous ectoplasm (plasmagel) on reaching the tip of a pseudopodium. At the opposite end of the cell, ectoplasm is converted into endoplasm.

Amorphous calcium phosphate A non crystalline form of apatite which may form as much as30% of bone mineral.

AMP Abbreviation of adenosine monophosphate.

Amphibia The class of vertebrates whose members are characterized by skin with mucoid secretions, which serves as a respiratory organ. Developmental stages are aquatic and are usually followed by metamorphosis to an amphibious adult. Frogs, toads, and salamanders.Class of terrestrial vertebrates which lay their eggs (and also mate) in water but live on land as adults following a juvenile stage where they live in water and breathe through gills. Amphibians were the first group of land vertebrates; today they are mostly restricted to moist habitats.

Amphiblastula Free-swimming larval stage of certain marine sponges; blastula-like but with only the cells of the animal pole flagellated; those of the vegetal pole unflagellated.

Amphibolic pathways Metabolic pathways that function both catabolically and anabolically.

Amphid One of a pair of chemosensory organs found on the anterior end of certain nematodes.

Amphiesma The outer covering of a dinoflagellate, consisting of several membrane layers.

Amphipathic Adjective to describe a molecule with one part soluble in water (polar) and another part insoluble in water (nonpolar). A group of small, laterally compressed crustaceans that includes beach hoppers and others.

Amphipathic molecule A molecule that has both a hydrophilic region and a hydrophobic region.

Amphistome Fluke with the ventral sucker located at the posterior end.

Amphitrichous A cell with a single flagellum at each end.

Amphotericin B An antibiotic from a strain of Streptomyces nodosus that is used to treat systemic fungal infections; it also is used topically to treat candidiasis.

Ampicillin An antibiotic derived from penicillin that prevents bacterial growth by interfering with cell wall synthesis.

Amplexus Form of sexual reproduction seen in frogs wherein the male grasps the female from behind and externally fertilizes the eggs as they are deposited. The positioning of a male amphibian dorsal to female amphibian, his forelimbs around her waist. During amplexus, the male releases sperm as the female releases eggs.

Amplification Refers to the production of additional copies of a chromosomal sequence, found either as intrachromosomal or extrachromosomal DNA. Also refers to the in vitro process in the *polymerase chain reaction*.To increase the number of copies of a DNA sequence, in vivo by inserting into a cloning vector that replicates within a host cell, or in vitro by polymerase chain reaction (PCR). Region of DNA sequence which is amplified during a *PCR* reaction and which is defined by a pair of *PCR primers* (these *primer* pairs are sometimes called amplimers).

Amplitude The maximum absolute value of a periodically varying quantity, as the peak in activity in a circadian or circannual cycle.

Ampulla Membranous vesicle; dilation at one end of each semicircular canal containing sensory epithelium; muscular vesicle above tube foot in water-vascular system of echinoderms. Upper region of the mammalian oviduct, near the ovary. Fertilization typically takes place in this region

Ampulla of Lorenzini One of several sensory structures in the head of sharks that detect weak electric fields.

Ampullary organ A receptor that can detect electrical currents. These electroreceptors are found in most fishes, some amphibians, and the platypus.

Amygdala Part of the limbic system, which seems to provide the emotional assessment of a new sensation with the memory of a similar sensation.

Amygdaloid A part of the brain.

Amylase An enzyme that breaks down starch into smaller units.

Amylopectin Component of starch consisting of highly branched chains of repeating glucose units; insoluble in water.

Amyloplast Starch-storing plastid.

Amylose Component of starch consisting of unbranched chains of repeating glucose units; soluble in water.

Anabolic Chemical reactions that synthesize and require energy.

Anabolic reactions Reactions in cells in which new chemical bonds are formed and new molecules are made; generally require energy, involve reduction, and lead to an increase in atomic order.

Anabolic steroids Synthetic chemical variants of the male sex hormone testosterone; they produce increased muscle mass but also suppress testosterone production, leading to shrinkage of the testes, growth of the breasts, and premature baldness; long-term use increases the risk of kidney and liver damage and of liver cancer.

Anabolism The synthesis of complex molecules from simpler molecules with the input of energy.Within a cell or organism, the sum of all biosynthetic reactions (that is, chemical reactions in which larger molecules are formed from smaller ones).Any constructive metabolic process by which organisms convert substances into other components of the organism's chemical architecture. System of biosynthetic reactions in a cell by which large molecules are made from smaller ones.

Anadromous fish Fish that spends most of its life feeding in the open ocean but that migrates to spawn in fresh water

Anadromous Refers to fishes that migrate up streams from the sea to spawn.

Anaerobe An organism that grows in the absence of free oxygen. Pertaining to the absence of free oxygen. Anaerobic organisms do not require oxygen for their life processes, in fact oxygen is toxic to many of them. Most anaerobic organisms are bacteria or archaeans. The phase of cellular respiration that occurs in the absence of oxygen; lacking oxygen.

Anaerobic bacteria Bacteria that do not need oxygen.

Anaerobic digestion The microbiological treatment of sewage wastes under anaerobic conditions to produce methane.

Anaerobic respiration The first step in the production of ATP is to break down glucose.

This process of glycolysis is a 10 step series of reactions leading finally to the smaller molecule pyruvate. The energy derived from this process is a hydrogen ion and an electron, which are both placed onto the carrier molecule as nicotinamide adenine dinucleotide ($NADH^+$). As the pyruvate and NAD H^+ are produced they could move into the mitochondria, for the nextstage of aerobic respiration, provided oxygen is available. If there is no oxygen, NAD H H^+ is used in a process of substrate phosphorylation to form ATP. But the pyruvate builds up. It is then converted to lactic acid and removed to the liver. If lactic acid is not removed fast enough it causes muscle weakness and pain. Anaerobic glycolysis does not produce a high yield of energy. There are still high energy bonds remaining in the pyruvate and there is no benefit from the large yield of ATP made possible by the electron transport system in the mitochondria Top.

Anagenesis A pattern of evolutionary change involving the transformation of an entire population, sometimes to a state different enough from the ancestral population to justify renaming it as a separate species; also called phyletic evolution.

Anal angle The small apical area enclosed by the inner and outer margins of the hindwing.

Anal fin Each of the last pair of ventral fins of fishes.

Anal fold A fold in the inner margin of the hindwing.

Anal Pertaining to last abdominal segment which bears the anus.

Anal veins The hindmost, or most posterior longitudinal wing veins.

Analgesic Pain-relieving remedy.

Analogous Applied to structures similar in function but different in evolutionary origin, such as the wing of a bird and the wing of an insect. Structures that have similar functions in two organisms but have not evolved from a common ancestral form.

Analogous structures Body parts that serve the same function in different organisms, but differ in structure and embryological development; e. g., the wings of insects and birds.Structures having similar function or superficial appearance, but not necessarily sharing a common evolutionary origin (contrast with homologous structures).

Analogy The similarity of structure between two species that are not closely related; attributable to convergent evolution.

Anammox process The coupled use of nitrite as an oxidant and ammonium ion as a reductant under anaerobic conditions to yield nitrogen gas.

Anamnestic response The recall, or the remembering, by the immune system of a prior response to a given antigen.Immune response to a challenge or secondary antigen inoculation, marked by more rapid and stronger manifestation of the immune reaction (specifically, antibody titer) than after the primary immunizing dose.

Anamorphic The asexual or imperfect stage in a fungus life cycle.

Anaphase The third stage of mitosis, beginning when the centromeres of duplicated chromosomes divide and sister chromotids separate from each other, and ending when a complete set of daughter chromosomes are located at each of the two poles of the cell.The stage in mitosis and meiosis, following metaphase, in which the centromeres divide and the chromatids, lined up on the mitotic spindle, begin to move apart toward the poles of the spindle to form the daughter chromosomes.

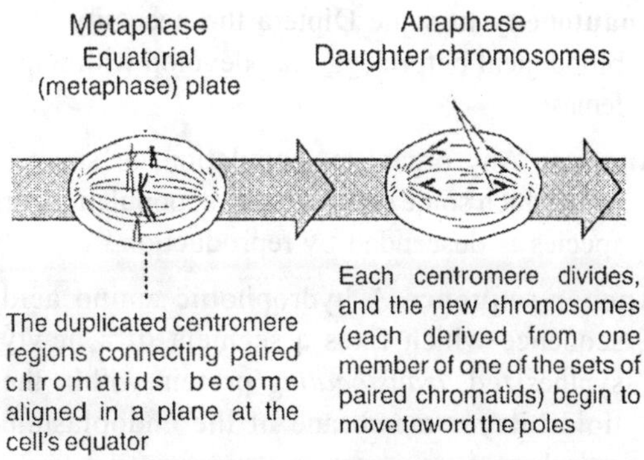

Fig. *Anaphase*

Anaphylaxis An immediate (type I) hypersensitivity reaction following exposure of a sensitized individual to the appropriate antigen. Mediated by reagin antibodies, chiefly IgE.A severe allergic reaction in which histamine is released into the circulatory system; occurs upon subsequent exposure to a particular antigen; also called anaphylactic shock.

Anaplasmosis Infection with Anaplasma, a genus of Sporozoa that infests red blood cells.

Anaplerotic reactions Reactions that replenish depleted tricarboxylic acid cycle intermediates.

Anapolysis Detachment of a senile proglottid after it has shed its eggs. Refers to reptilian skull in which there are no temporal openings. A comparative common condition in fossil forms; in living reptiles, known only in turtles.

Anapsid Amniotes in which the skull lacks temporal openings, with turtles the only living representatives. A vertebrate distinguished by a skull with no openings in the side behind the eyes, e.g. turtles.

Anasa wilt A wilt disease of cucurbits caused solely by the feeding of the squash bug, no parasitic microorganism involved.

Anastomose Coming together; usually of blood vessels. A union of two or more blood vessels, fibers, or other structures to form a branching network.

Anatomy The study of the structure of an organism and its parts.

Anautogeny In some Diptera the necessity of a blood meal before eggs can develop within the female.

Ancestor Any organism, population, or species from which some other organism, population, or species is descended by reproduction.

Anchor sequence A hydrophobic amino acid sequence which fixes a segment of a newly synthesized, *translocating* protein within the lipid bilayer membrane of the endoplasmic reticulum.

Andesite Igneous volcanic rock, less mafic than basalt, but more mafic than dacite; rough volcanic equivalent of diorite.

Androconia In male butterflies, specialised wing scales (often called scent scales) possessing special glands which produce a chemical attractive to females.

Androecium Collective term applied to all of the male (stamen) parts of the flower.

Androgen The principal male steroid hormones, such as testosterone, which stimulate the development and maintenance of the male reproductive system and secondary sex characteristics.Any male sex steroid associated with the development of male sex organs and secondary male characteristics.Any substance that contributes to masculinization, such as the hormone testosterone.

Androgenic gland Gland located near the vas deferens in many Crustacea. Its secretions are responsible for development of male secondary sexual characteristics.

Androsterone An animal hormone not synthesized by plants.

Anecdysis Ecdysis in which successive molts are separated by quite long intermolt phases; referred to as terminal anecdysis when maximal size is reached and no more ecdyses occur.

Anemia Low levels of hemoglobin due to iron deficiency or lack of red blood cells.

Anemic Deficient in blood quantity or quality.

Anemophily Seed plants which are pollinated by wind are said to be anemophilous.

Anencephaly Condition of forebrain degeneration caused by failure of anterior neuropore closure and consequent prolonged contact of the forebrain with amniotic fluid. Although infants with this neural tube defect are sometimes born alive, the condition is invariably fatal within a few hours or days.

Anergy A state of unresponsiveness to antigens.

Absence of the ability to generate a sensitivity reaction to substances that are expected to be antigenic.

Aneuploidy A chromosomal aberration in which certain chromosomes are present in extra copies or are deficient in number.Loss or gain of a chromosome, cells of the organism have one fewer than normal chromosome number, or one extra chromosome, for example, trisomy 21 (Down syndrome).Variation in chromosome number involving one or a small number of chromosomes; commonly involves the gain or loss of a single chromosome.

Angina Chest pain, especially during physical exertion or emotional stress, that is caused by gradual blockage of the coronary arteries.

Angiogenesis The development of blood vessels - a key event in embryology and healing.

Angiosperms Flowering plants. First appearing at least 110 million years ago from an unknown gymnosperm ancestor, flowering planbts have risen to dominance in most of the world's floras. The male gametophyte is 2-3 cells contained within a pollen grain; the female gametophyte is usually eight cells contained within an ovule which is retaind on the sporophyte phase of the plant's life cycle.

Angiotensin Blood protein formed from the interaction of renin and a liver protein, causing increased blood pressure and stimulating release of aldosterone and ADH.

Angstrom A unit of one ten-millionth of a millimeter (one ten-thousandth of a micrometer); it is represented by the symbol Å.

Anhydrase An enzyme involved in the removal of water from a compound. Carbonic anhydrase promotes the conversion of carbonic acid into water and carbon dioxide.

Animal behavior Activities animals perform during their lifetime.

Animal hemisphere The half of an oocyte which contains less yolk or the corresponding half of an embryo with the more actively dividing cells.

Animal pole The region of a fertilized egg where meiosis is completed. It contains less yolk and is more metabolically active than the opposite vegetal pole.

Animalia Animal Kingdom. Multicellular eukaryotic group characterized by heterotrophic nutritional mode, usually organ and tissue development, and motility sometime during the organism's life history. The kingdom of organisms whose members are multicellular, eukaryotic, and heterotrophic. The animals.

Animals Members of the kingdom Animalia, which consists of heterotrophic, eukaryotic, and multicellular organisms.

Anion A negatively charged ion.

Anisogametes Outwardly dissimilar male and female gametes. Anisogamy is the condition of having dissimilar male and female gametes.

Anisogamy The condition in which the female gamete (ovum) is larger than the male gamete (sperm).

Ankylosis Bony fusion of the two surfaces of a joint to each other, which prevents movement. Ankylosis of the tooth root to its bony socket may causes root resorption.

Anlage Rudimentary form; primordium.

Anneal The pairing of complementary DNA or RNA sequences, via hydrogen bonding, to form a double-stranded polynucleotide. Most often used to describe the binding of a short primer or probe.Formation of double-stranded molecules from two single strands of nucleic acid by base pairing of complementary sequence. Usually achieved incubation at a favourable temperature.

Annelida The phylum of triploblastic, coelomate animals whose members are metameric (segmented) and wormlike. Annelids have a complete digestive tract and a ventral nerve cord.

Annotation The process of determining the location of specific genes in a genome map after it has been produced by nucleic acid sequencing.

Annual A plant that completes its entire life cycle in a single year or growing season.A plant that lives for a single growing season.A plant with a lifespan of one year. Plants that grow and reproduce sexually during one year.

Annual ring Ring of xylem in woody stem composed of springwood and summerwood that corresponds in temperate regions to a chronological year.

Annuli Rings on the body of a parasite; not necessarily indicative of internal segmentation.

Annulus In fern sporangia, a row of cells with both thin and thickened cell walls which facilitate the opening of the sporangia and the release of spores because of the unequal expansion and tension produced as they are moistened and dried.

Anogenital condylomata Warts that are sexually transmitted and caused by types 6, 11, and 42 human papillomavirus. Usually occur around the cervix, vulva, perineum, anus, anal canal, urethra, or glans penis.

Anoxic Lacking oxygen.

Anoxygenic photosynthesis Photosynthesis that does not oxidize water to produce oxygen; the form of photosynthesis characteristic of purple and green photosynthetic bacteria.

Antagonism The condition of being an opposing principle, force, or factor, as when two hormones have opposite effects on target tissues.

Antagonist The muscle or muscles opposing the agonist.

Antagonistic muscles A pair of muscles that work to produce opposite effects&emdash;one contracts as the other relaxes: for example, the bicep and tricep muscles on opposite sides of your upper arm.

Antenna Pair of segmented appendages located on the head and usually sensory in function - the 'feelers'. A sensory appendage on the head of arthropods, or the second pair of the two such pairs of structures in crustaceans.

Antennae (second antennae of crustaceans) Second pair of appendages in Crustacea, with bases usually immediately posterior to antennules; primarily sensory but sometimes adapted for other functions; derived from appendages on primitive third preoral somite; no homologous appendage in insects.

Antennal gland The excretory organ in some crustaceans (crayfish). Called antennal glands because of their location near the base of each second antenna and their green color; also called green glands.

Antennules (first antennae) Anteriormost pair of appendages of Crustacea; primarily sensory but often adapted for additional or other functions in particular species; derived from appendages on primitive second preoral somite; homologous to antennae of insects.

Antenodal veins Small cross-veins at the front of the dragonfly or damselfly wing, between the wing base and the nodus.

Anterior Referring to the head end of a bilaterally symmetrical animal.Concerning or facing the front, towards the head.

Anterior intestinal portal The opening of the foregut in developing amniote embryos. As development progresses and the gut elongates, the opening moves posteriorly and ultimately becomes part of the yolk stalk.

Anterior station Development of a protozoan in the middle or anterior intestinal portions of its insect host, such as the section Salivaria of Trypanosomatidae.

Anther The terminal pollen sac of a stamen, inside

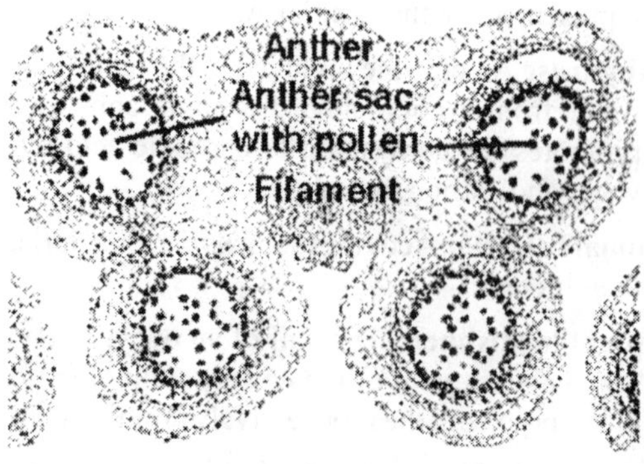

Fig. *Anther*

which pollen grains with male gametes form in the flower of an angiosperm.

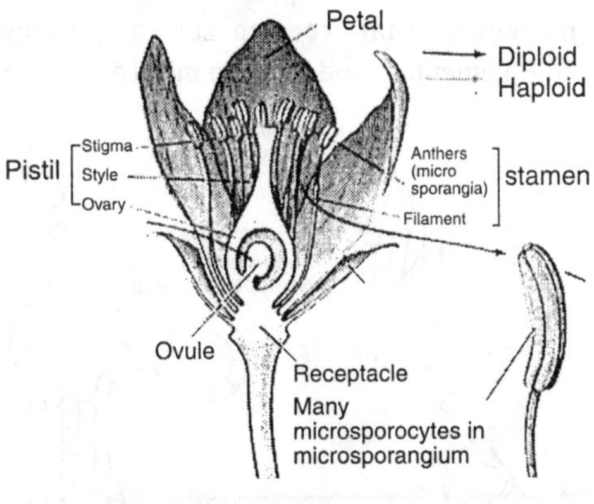

Fig. *Anther*

Antheridiophore in some liverworts, a stalk that bears antheridia embedded on an elevated and expanded tip.

Antheridium A male gamete-producing organ, which may be unicellular or multicellular.Plants, the male gametangium, a moist chamber in which gametes develop.The multicellular male sex structure of plants other than seed plants. The organ on a gametophyte plant which produces the sperm cells.

Anthocyanin Natural water-soluble pigments of blue, purple or red which are dissolved in the cell-sap vacuole of plant cells.A group of water-soluble red to blue flavonoid pigments found in certain plants; especially important pigmentation in flower petals.

Anthophyte A flowering plant, or any of its closest relatives, such as the Bennettitales, Gnetales, or Pentoxylales.

Anthozoa The class of cnidarians whose members are solitary or colonial polyps. Medusae absent; gametes originate in the gastrodermis; mesenteries divide the gastrovascular cavity. Sea anemones and corals. Cnidarians whose life cycle consists of a complex polyp and no medusa.

Anthrax An infectious disease of animals caused by ingesting Bacillus anthracis spores. Can also occur in humans and is sometimes called woolsorter's disease.

Anthropocentric The belief that humans hold a special place in nature; being centered primarily on humans and human affairs.

Anthropomorphic Regarding animals as possessing human qualities.

Anthropomorphism The attribution of human characteristics to nonhuman beings and objects.

Antibacterial Inhibiting the growth of bacteria.

Antibiosis The inhibition of growth of a microorganism by a substance produced by another microorganism.

Antibiotic A microbial product or its derivative that kills susceptible microorganisms or inhibits their growth.A chemical that kills bacteria or inhibits their growth.A class of natural and synthetic compounds that inhibit the growth of or kill other microorganisms. Substance that inhibits the growth of microorganisms, antibodiesProteins (immunoglobulins) in cell surfaces and dissolved in blood, capable of combining with the antigens that stimulated their production.

Antibiotic resistance Plasmids generally contain genes which confer on the host bacterium the ability to survive a given antibiotic. If the plasmid pBR322 is present in a host, that host will not be killed by (moderate levels of) ampicillin or tetracycline. By using plasmids containing antibiotic resistance genes, the researcher can kill off all the bacteria which have not taken up his plasmid, thus ensuring that the plasmid will be propagated as the surviving cells divide.

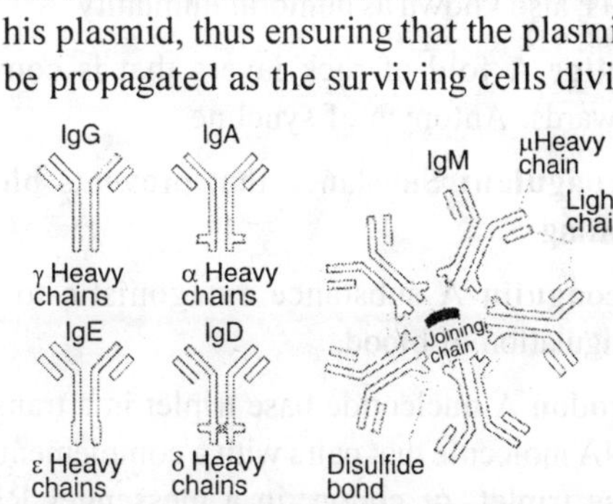

Fig. *Immunoglobulins*

Antibodies Are proteins called immunoglobulins which circulate in the blood and body fluids. They bind specifically to antigens that have induced them. Antibodies are able to inactivate bacterial toxins, viruses and help phagocytes to engulf whole bacteria. They have a vital role to play in the bodies immune response to foreign proteins.

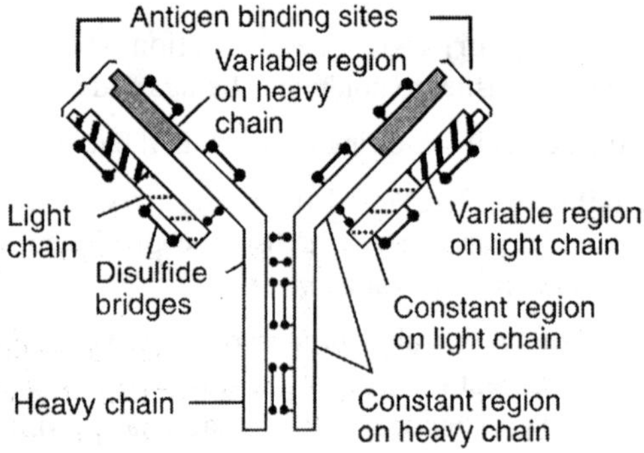

Fig. *Antibodies*

Antibody titer Measure of the amount of antibody present, usually given in units per milliliter of serum.

Antibody-dependent cell-mediated cytotoxicity (ADCC) The killing of antibody-coated target cells by cells with Fc receptors that recognize the Fc region of the bound antibody. Most ADCC is mediated by NK cells that have the Fc receptor or CD16 on their surface.

Antibody-mediated immunity Immune reaction that protects primarily against invading viruses and bacteria through antibodies produced by plasma cells; also known as humoral immunity.

Anticline A fold of rock layers that is convex upwards. Antonym of syncline.

Anticoagulant Substance that prevents blood clotting.

Anticoagulin A substance antagonistic to the coagulation of blood.

Anticodon A nucleotide base triplet in a transfer RNA molecule that pairs with a complementary base triplet, or codon, in a messenger RNA molecule. A sequence of three bases on transfer RNA that pairs with codons of messenger RNA to position amino acids during protein synthesis. A specialized base triplet on one end of a tRNA molecule that recognizes a particular complementary codon on an mRNA molecule.

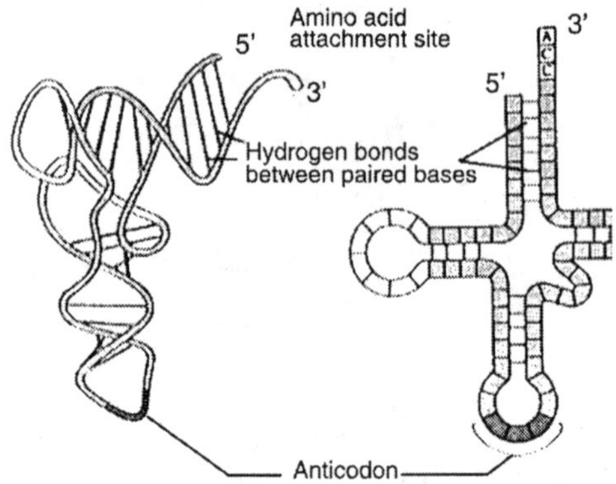

Fig. *Anticodon*

Anticodon loop The portion of a tRNA molecule responsible for the anticodon triplet, which pairs with the codon of mRNA.

Anticodon triplet The base triplet on a tRNA that is complementary to the triplet codon on mRNA.

Antidiuretic hormone (ADH) A hormone produced by the hypothalamus and released by the pituitary gland that increases the permeability of the renal tubule of the nephron and thereby increases water reabsorption; also known as vasopressin.

Antigen A foreign (nonself) substance (such as a protein, nucleoprotein, polysaccharide, or sometimes a glycolipid) to which lymphocytes respond; also known as an immunogen because it induces the immune response.A foreign macromolecule that does not belong to the host organism and that elicits an immune response.Any foreign substance, such as a virus, bacterium, or protein, that elicits an immune response by stimulating the production of antibodies. Molecules carried or produced by microorganisms that initiate antibody production; mostly proteins or proteins combined with polysaccharides.Proteins, usually foreign, which cause the bodies defense system to produce an antibody. Antigens may be food proteins, bacteria ,viruses or protozoa or cells from another individual(transplant).

Antigen challenge Dose or inoculation with an antigen given to an animal some time after primary immunization with that antigen has been achieved.

Antigen-binding fragment A monovalent antigen-binding fragment of an immunoglobulin molecule that consists of one light chain and part of one heavy chain, linked by interchain disulfide bonds.

Antigenic determinant Area on an antigen molecule that binds with antibody or specific receptor sites on the sensitized lymphocyte; it "determines" the specificity of the antibody or lymphocyte.

Antigenic drift A small change in the antigenic character of an organism that allows it to avoid attack by the immune system.A major change in the antigenic character of an organism that alters it to an antigenic strain unrecognized by host immune mechanisms.

Antigenic switching The altering of a microorganism's surface antigens through genetic rearrangement, to elude detection by the host's immune system.

Antigen-presenting cells Antigen-presenting cells (APCs) are cells that take in protein antigens, process them, and present antigen fragments to B cells and T cells in conjunction with class II MHC molecules so that the cells are activated. Macrophages, B cells, dendritic cells, and Langerhans cells may act as APCs.

Antimetabolite A compound that blocks metabolic pathway function by competitively inhibiting a key enzyme's use of a metabolite because it closely resembles the normal enzyme substrate.

Antimicrobial agent Any chemical or biological agent that harms the growth of microorganisms.

Antinutrients Chemicals produced by plants as a defense mechanism; inhibit the action of digestive enzymes in insects that attack and attempt to eat the plants.

Antiparallel Refers to opposing strands of DNA that are oriented in opposite directions.

Antipodal cell The three haploid cells at the end of the embryo sac away from the micropyle.

Antipodal One of several cells (usually three) of the embryo sac (female gametophyte) of angiosperms. They are located opposite the egg and synergids.

Antisense RNA A single-stranded RNA with a base sequence complementary to a segment of another RNA molecule that can specifically bind to the target RNA and inhibit its activity.Refers to the RNA or DNA strand of a *duplex* molecule which is *complementary* to that encoding a *polypeptide*. More specifically, the DNA strand which serves as *template* for the synthesis of RNA and which is complementary to it. "Antisense *oligonucleotides*" *hybridize* to *mRNA*, and are used to prime *cDNA* synthesis.

Antiseptic Chemical agents applied to tissue to prevent infection by killing or inhibiting pathogens.

Antiserum Serum containing induced antibodies.

Antitoxin An antibody to a microbial toxin, usually a bacterial exotoxin, that combines specifically with the toxin, in vivo and in vitro, neutralizing the toxin.

Antlers Antlers are found on members of the deer family (Cervidae). They consist of bone and are shed each year. New antlers grow each spring and are covered by skin (called velvet) which has blood vessels that carry nutrients to the growing antler. In the fall of the year the skin dies and is rubbed off on bushes and saplings. Antlers are typically found only on males, except in the caribou.

Antrum A hollow cave or SINUS, inside the maxillary bone which is lined by respiratory epithelium.

Anus End of the digestive tract, or gut, through which waste products of digestion are excreted, as distinct from the mouth.

Aorta The major artery in blood-circulating systems; the aorta sends blood to the other body tissues. Large artery carrying blood away from the heart. The artery that carries blood from the left ventricle for distribution throughout the tissues of the body. The largest diameter and thickest walled artery in the body.The front-most, non-pulsating portion of the dorsal blood vessel of an insect.

Aorta aperture An opening; the opening into the first whorl of a gastropod shell.

AP-1 site The binding site on DNA at which the transcription "factor" AP-1 binds, thereby altering the rate of transcription for the adjacent gene. AP-1 is actually a complex between c-fos protein and c-jun protein, or sometimes is just c-jun dimers. The AP-1 site consensus sequence is (C/G)TGACT(C/A)A. Also known as the TPA-response element (TRE). [TPA is a phorbol ester, tetradecanoyl phorbol acetate, which is a chemical tumor promoter]

Apatites A family of calcium phosphate salts which are found in hard tissues like bone, teeth and shells.

Aperture Small opening, for example the opening in the test of a foram.

Apex Highest or uppermost point; the lower pointed end of the heart. The tip of a structure; a leaf apex, for example, is the tip of the leaf.

Aphotic zone The part of the ocean beneath the photic zone, where light does not penetrate sufficiently for photosynthesis to occur.The portion of the ocean where the absence of sunlight prohibits plant growth.

Apiary A place where bees are kept, normally a group of hives.

Apical area Of the forewing, the area just inside of and contiguous with the apex.

Apical At or concerning the tip or furthest part of any organ: apical cells, for example are at the wing-tip.

Apical complex Dense ring and conelike structure, along with associated microtubules, micronemes, and rhoptries, at the anterior end of an apicomplexan sporozoite. (apical = at the apex.)A certain combination of organelles found in the protozoan phylum Apicomplexa.

Apical dominance The phenomenon leading to controlled growth of lateral shoots; growth occurs primarily at the top of the plant.

Apical meristem A group of cells at the tip of the stem and root that give rise by cell division to the primary tissues and are ultimately responsible for the structural organization of the entire primary plant body.

Apical organ Organ of unknown function at the apex of a cestode's scolex.

Apicomplexa The protozoan phylum characterized by members having an apical complex used for penetrating host cells; cilia and flagella lacking, except in certain reproductive stages. Examples include the gregarines (Monocystis), coccidians (Eimeria, Isospora, Sarcocystis, Toxoplasma), Pneumocystis, and Plasmodium. A sporozoan protist that lacks special locomotor organelles but has an apical complex and a spore-forming stage. It is either an intra- or extracellular parasite of animals; a member of the phylum Apicomplexa.

Aplacophora The class of molluscs whose members lack a shell, mantle, and foot. Wormlike, burrowing animals with head poorly developed. Some authors divide this group into two classes: Caudofoveata and Solengasters.

Aplanospore A nonflagellated, nonmotile spore that is involved in asexual reproduction.

Apnea A disorder in which breathing stops for periods longer than 10 seconds during sleep; can be caused by failure of the automatic respiratory center to respond to elevated blood levels of carbon dioxide.

Apneustic breathing Breathing pattern exhibited by marine mammals in which several rapid breaths alternate with a prolonged cessation of breathing.

Apocrine Applies to a type of mammalian sweat gland that produces a viscous secretion by breaking off a part of the cytoplasm of secreting cells.

Apocrine glands Sweat glands that are located primarily in the armpits and groin area; larger than the more widely distributed eccrine glands.

Apodeme Spinelike inward projection of the cuticle in arthropods on which a muscle inserts; a ridgelike projection is an apophysis.

Apodous larva Larva with no legs and with reduced head; usual in Hymenoptera, Diptera, some Coleoptera; requires maternal care or deposition in or on food source.

Apoenzyme The protein part of an enzyme that also has a nonprotein component.

Apolysis Disintegration or detachment of a gravid tapeworm segment; also, the detachment of the hypodermis from the old procuticle in arthropods before molting.

Apomorphic Adjective that refers to the form of characters, in particular characters whose form differs from that of the same character in an outgroup.

Apomorphic character derived phenotypic character, or homology, that evolved after a branch diverged from a phylogenetic tree.

Apoplast In plants, the nonliving continuum formed by the extracellular pathway provided by the continuous matrix of cell walls.Pertaining to the movement of water in the free space of tissue; free space includes cell walls and intercellular spaces.

Apoptosis Programmed cell death. The fragmentation of a cell into membrane-bound particles that are eliminated by phagocytosis. Apoptosis is a physiological suicide mechanism that preserves homeostasis and occurs during normal tissue turnover. It is responsible for cell death in pathological circumstances, such as exposure to low concentrations of xenobiotics and infections by HIV and various other viruses. Apoptotic cells display profound structural changes such as plasma membrane blebbing and nuclear collapse. DNA is cleaved into short oligonucleosomal length DNA fragments. Apoptosis usually occurs after the activation ofr calcium-dependent endogenous endonuclease.Death of a cell which is programmed by a set of specific genes. Apoptosis of chondrocytes allows osteoblasts to attach to their calcified matrix, and the epithelial cells forming webs between the fingers to die.

Apopyle In sponges, opening of the radial canal into the spongocoel.

Aporepressor An inactive form of the repressor protein, which becomes the active repressor when the corepressor binds to it.

Aposematic coloration Sharply contrasting colors of an animal that warn other animals of unpleasant or dangerous effects.

Appendage Any limb or other organ, such as an antenna, which is attached to the body by a joint

Appendicular Pertaining to appendages; pertaining to vermiform appendix.

Appendicular skeleton The bones of the appendages (wings, legs, and arms or fins) and of the pelvic and pectoral girdles that join the appendages to the rest of the skeleton; one of the two components of the skeleton of vertebrates.

Appendix Blind sac at the end of the large intestine that usually ruptures during final exams; a vestigial organ in humans. In insects, a short vein, especially a short continuation after the main vein has changed direction.

Appetitive behavior The flexible introductory phase of a behavior sequence during which the organism is searching to obtain something to meet a need, as in seeking food, a mate, stimulation of a specific type, etc.

Applied animal behavior A subdiscipline concerned with all aspects of the behavioral biology of pets, domestic animals, animals in zoos and circuses, and animals in aquaria.

Applied research Research conducted to address a specific problem for the purpose of application for productivity or commercial gain.

Apterous Without wings.

Apterygote Any member of the Apterygota - primitively wingless insects (i.e. insects which have never developed wings during their evolutionary history) in modern classifications this includes the Thysanura but not Collembola Diplura and Protura which are no longer considered insects, but are termed Hexapods instead .

Aquaporin A transport protein in the plasma membranes of a plant or animal cell that specifically facilitates the diffusion of water across the membrane (osmosis).

Aquatic Living in water.

Aqueous solution A solution in which water is the solvent.

Aquifer Underground water reservoir.Porous, water-bearing layers of sand, gravel, and rock below the earth's surface; reservoirs for groundwater.

Arachnida The class of chelicerate arthropods whose members are mostly terrestrial, possess book lungs, or tracheae, and usually have four pairs of walking legs as adults. Spiders, scorpions, ticks, mites, and harvestmen.

Arachnoid Part of the membranes covering the brain.The weblike middle covering (meninge) of the central nervous system.

Arboreal Tree-dwelling.Living in, on, or among trees. Pertaining to animals that live primarily in trees.

Arbuscular mycorrhizal (AM) fungi The mycorrhizal fungi in a symbiotic fungus-root association that penetrate the outer layer of the root, grow intracellularly, and form characteristic much-branched hyphal structures called arbuscules.

Arbuscules Branched, treelike structures formed in cells of plant roots colonized by endotrophic mycorrhizal fungi.

Arch A bridge of bone in the skull, such as zygomatic arch.

Archaea (archebacteria) Early prokaryotes, differing from bacteria in having unusual types of metabolism, membrane lipids, and amino acid and DNA and RNA base sequences. Inhabit extreme environments. Proposed, but not widely accepted, sixth taxonomic kingdom that would include the archaebacteria.The domain that contains procaryotes with isoprenoid glycerol diether or diglycerol tetraether lipids in their membranes and archaeal rRNA (among many differences).

Archaebacteria Ancient (over 3.5 billion years old) group of prokaryotes; some biologists want to place this group into a separate Kingdom, the Archaea. Most currently place it within the Kingdom Monera.

Archaeocyathids An extinct group of animals that were part of Cambrian-aged reef environments, but which were extinct by the close of the Cambrian Period.

Archaeocytes Ameboid cells of varied function in sponges.

Archean/proterozoic era The period of time beginning 4.6 billion years ago with the formation of the Earth and ending 570 million years ago.

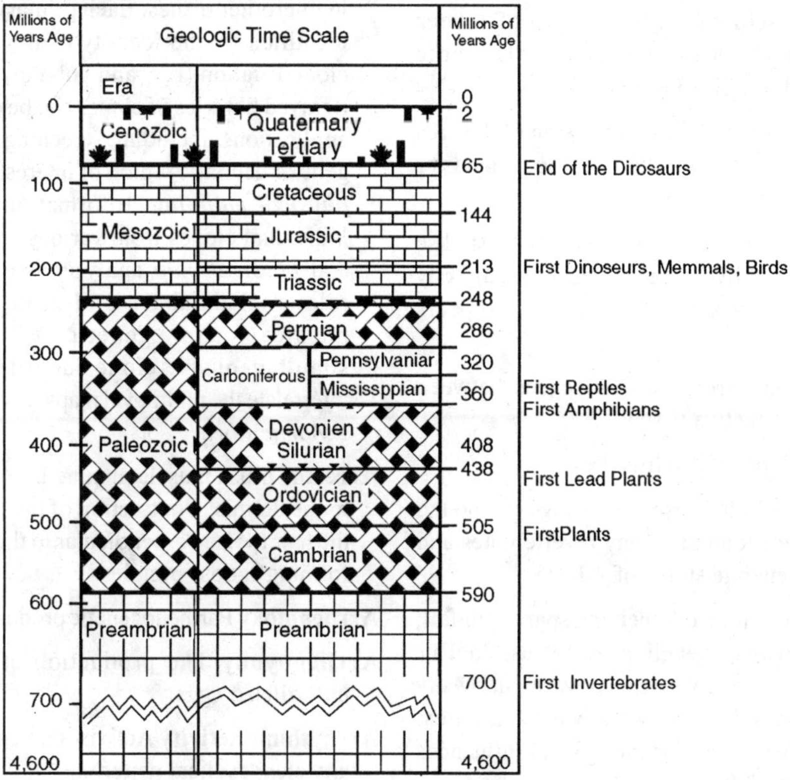

Fig. *Archean/Proterozoic Era*

Archegonium Female gametangium in which the female gamete or egg is produced and housed.The organ on a gametophyte plant which produces the egg cell, and nurtures the young sporophyte.

Archenteron The endoderm-lined cavity, formed during the gastrulation process, that develops into the digestive tract of an animal.The gut cavity formed by gastrulation in metazoan embryos; formed by infolding of part of the outer surface of the blastula and opening to exterior via blastopore. Synonyms: primitive gut, gastrocoele. Cavity formed during gastrulation by invagination of the vegetal plate cells (in sea urchin) or involution of cells at the blastoporal lip (in amphibians); it will become the interior of the primitive gut.

Archeocyte Somatic cell of sponge that can differentiate into all three other cell types of the organism.

Archezoa Primitive eukaryotic group that includes diplomonads, such as Giardia; some systematists assign kingdom status to archezoans.

Archinephros Ancestral vertebrate kidney, existing today only in the embryo of hagfishes.

Archipelago A group of islands; an expanse of water with scattered islands.

Archosaur Advanced diapsid vertebrates, a group that includes the living crocodiles and the extinct pterosaurs and dinosaurs.

Area opaca The peripheral region of the chick blastodisc surrounding the area pellucida and in direct contact with the yolk. This area develops into two regions.

Area opaca vasculosa The inner region of the area opaca where thickenings of splanchnic mesoderm form blood islands.

Area opaca vitellina The outer region of the area opaca which initially lacks any blood islands or blood vessels.

Area pellucida The central region of the blastodisc where the primitive streak forms. It lies directly over the subgerminal cavity and so appears transparent.

Areolar A small area, such as spaces between fibers of connective tissue.

Areolus A unit of diatom frustules.

Arginine phosphate Phosphate storage compound (phosphagen) found in many invertebrates and used to regenerate stores of ATP.

Aridity The condition of receiving sparse rainfall; associated with cooler climates because cool air can hold less water vapor than warm air. Many deserts occur in relatively warm climates, however, because of local or global influences that block rainfall.

Aril Thick, fleshy seed covering around some seeds.

Arista Flagellumlike appendage on the antenna of a fly of the suborder Brachycera and some members of the Nematocera.

Aristate Bearing an arista or bristle.

Aristotle's lantern The set of jaws and associated muscles used by sea urchins to bite off food.

Arithmetic growth A pattern of growth that increases at a constant amount per unit time, such as 1, 2, 3, 4 or 1, 3, 5, 7.

Arolium A small pad between the claws on an insect's foot. Usually very small, but well developed in grasshoppers and some other insects.

Arrayed library Individual primary recombinant clones (hosted in *phage, cosmid, YAC*, or other *vector*) that are placed in two-dimensional arrays in microtiter dishes. Each primary clone can be identified by the identity of the plate and the clone location (row and column) on that plate. Arrayed libraries of clones can be used for many applications, including screening for a specific gene or genomic region of interest as well as for *physical mapping*. Information gathered on individual clones from various genetic *linkage* and *physical map* analyses is entered into a relational database and used to construct physical and genetic *linkage maps* simultaneously; clone identifiers serve to interrelate the multilevel maps. Compare *library, genomic library*.

Arrector pili A muscle running from a hair follicle to the dermis. Contraction of the muscle causes the hair to rise perpendicular to the skin surface, forming "goose pimples."

Arrhenotoky Parthenogenetic production of males.

Arrhenoyoky The production of males from unfertilised eggs.

Arrhythmic activity Activity that does not exhibit any clear cyclical pattern.

Arrow worms Members of the phylum Chaetognatha, a group of planktonic carnivores

Arteries Thick-walled vessels that carry blood away from the heart. Singular=artery.

Arteriole A very small artery. The smallest arteries; usually branch into a capillary bed.

Arteriosclerosis A cardiovascular disease caused by the formation of hard plaques within the arteries.

Artery A blood vessel that carries blood away from the heart and toward a peripheral cavity.A vessel that carries blood away from the heart to organs throughout the body.

Artesian well The result of a pressurized aquifer intersecting the surface or being penetrated by a pipe or conduit, from which water gushes without being pumped; also called a spring.

Arthroconidium A thallic conidium released by the fragmentation or lysis of hypha. It is not notably larger than the parental hypha, and separation occurs at a septum.

Arthroidial arthrodia = gliding. A type of joint.

Arthropod Any of the invertebrate animals (such as insects, spiders, or crustaceans) having an exoskeleton, a segmented body and jointed limbs.A phylum of animals with segmented body, exoskeleton, and jointed legs.The phylum of animals whose members possess metamerism with tagmatization, a jointed exoskeleton, and a ventral nervous system. Insects, crustaceans, spiders, and related animals.

Arthropodization Evolutionary development of the combination of characteristics associated with the Arthropoda, including a firm cuticular exoskeleton containing chitin.

Arthrospore A spore resulting from the fragmentation of a hypha.

Articular One of the bones which together with the quadrate bones and the dentary, made/make up a reptile's jaw. In mammals the quadrate bone is incorporated into the middle ear as the malleus.A joint or union of two bones.

Artificial chromosome A vector constructed from host cell chromosomal elements such as origin of replication, telomeres and centromere (in eukaryotes). So that it replicates and is segregated during cell division in the same way as a normal chromosome. The main uses are in preparation of genomic libraries since very large inserts (in the Mbp range) can be integrated. Has also been used in expression studies. Yeast artificial chromosomes (YACs) are the most commonly used form.

Artificial mouth A laboratory device for keeping bacteria growing in a controlled environment It allows for observing bacteria and their growth under different experimental conditions.

Artificial selection The process in which breeders choose the variants to be used to produce succeeding generations.

Artificially acquired active immunity The type of immunity that results from immunizing an animal with a vaccine. The immunized animal now produces its own antibodies and activated lymphocytes.

Artificially acquired passive immunity The type of immunity that results from introducing into an animal antibodies that have been produced either in another animal or by in vitro methods. Immunity is only temporary.

Artiodactyl One of an order of mammals with two or four digits on each foot.

Arytenoid Cartilages in the larynx.

Ascaridine Protein of unknown function in the sperm of Ascaris.

Ascaroside Glycoside found in Ascaris, made of the sugar ascarylose and a series of secondary monol and diol alcohols.

Ascaroside layer The inner, thin layer of an *Ascaris* egg shell which holds the first polar body against the underside of the chitinous layer.

Aschelminth Any animal in the phyla Gastrotricha, Rotifera, Kinorhyncha, Nematoda, Nematomorpha, Acanthocephala, Loricitera, Priapulida, or Entoprocta.

Aschoff's Rule Pertaining to circadian rhythms, the rule stating that when nocturnal animals are held in constant darkness, their free-running period becomes slightly shorter each day and when a diurnal animal is held in constant darkness, the free-running period becomes slightly longer.

Ascidiacea A class of urochordates whose members are sessile as adults, and solitary or colonial.

Ascites Edema, or accumulation of tissue fluid, in the mesenteries and abdominal cavity.

Ascocarp A multicellular structure in ascomycetes lined with specialized cells called asci in which nuclear fusion and meiosis produce ascospores. An ascocarp can be open or closed and may be referred to as a fruiting body.The dense, compact mass of hyphae constituting the fruiting body of the Ascomycetes.

Ascogenous hypha A specialized hypha that gives rise to one or more asci.

AscogoniumThe receiving (female) organ in ascomycetous fungi which, after fertilization, gives rise to ascogenous hyphae and later to asci and ascospores.

Ascomycetes Division of fungi that contains the yeasts and morels; ascomycetes produce an ascus (or sac) in which ascospores are produced.

Ascon The simplest of the three sponge body forms. Asconoid sponges are vaselike, with choanocytes directly lining the spongocoel.Simplest form of sponges, with canals leading directly from the outside to the interior.

Ascorbic acid Or Vitamin C is a dietary requirement for the proper formation of collagen. Deficiency causes scurvy.

Ascospore Sexual spore of an ascomycete fungus that is produced by meiosis, followed by mitosis.

Ascus A specialized cell, characteristic of the ascomycetes, in which two haploid nuclei fuse to produce a zygote, which immediately divides by meiosis; at maturity an ascus will contain ascospores.A saclike spore capsule located at the tip of the ascocarp in dikaryotic hyphae; defining feature of the Ascomycota division of fungi.Saclike reproductive structure of ascomycetes in which meiosis, followed by mitosis, produces eight haploid cells called ascospores.

Asexual (vegetative) reproduction The type of reproduction that takes place without the formation of gametes. Compare sexual reproduction.A method of reproduction in which genetically identical offspring are produced from a single parent; occurs by many mechanisms, including fission, budding, and fragmentation. A type of reproduction involving only one parent that usually produces genetically identical offspring. Asexual reproduction occurs without **meiosis** or **syngamy**, and may happen though budding, by the division of a single cell, or the breakup of an entire organism into two or more parts.

Aspartate Four-carbon amino acid found with malate to be the first products of the C4.

Aspect ratio Index of propulsive efficiency obtained by dividing the square of a fish's fin height by the fin area.

Aspergillosis A fungal disease caused by species of Aspergillus.

Asphalt A dark bituminous substance found in natural beds. Residue from petroleum distillation.

Asphyxiants Chemicals that exclude oxygen or actively interfere with oxygen uptake and distribution; includes inert chemicals, such as nitrogen gas or halothane, that can displace oxygen and fill enclosed spaces.

Assimilation Absorption and building up of digested nutriments into complex organic protoplasmic materials. The energy-requiring process by which plant cells convert nitrate ions (NO_3^-) taken up by the roots of plants into ammonium ions (NH_4^+), which can then be used in the synthesis of amino acids and other nitrogenous compounds.

Assimilation efficiency The fraction of ingested food that is absorbed and used in metabolism

Assimilatory reduction The reduction of an inorganic molecule to incorporate it into organic material. No energy is made available during this process.

Associative learning The acquired ability to associate one stimulus with another; also called classical conditioning.

Associative nitrogen fixation Nitrogen fixation by bacteria in the plant root zone (rhizsphere).

Assortative mating A type of nonrandom mating in which mating partners resemble each other in certain phenotypic characters.The mating of a given genotype mates with another genotype at a frequency disproportionate to that expected from random

Assortment A way in which meiosis produces new combinations of genetic information. Paternal and maternal chromosomes line up randomly during synapsis, so each daughter cell is likely to receive an assortment of maternal and paternal chromosomes rather than a complete set from either.

Astelocyttarus Pertaining to nests, normally those of social wasps, in which the come is attached directly to the support.

Aster Radiating formation of microtubules at each pole of a spindle apparatus, formed during mitosis; comes from word meaning "star." The star-shaped structure seen in a cell during the prophase of mitosis; composed of a system of microtubules arranged in astral rays around the centrosome; may emanate from a centrosome or from a pole of a mitotic spindle.

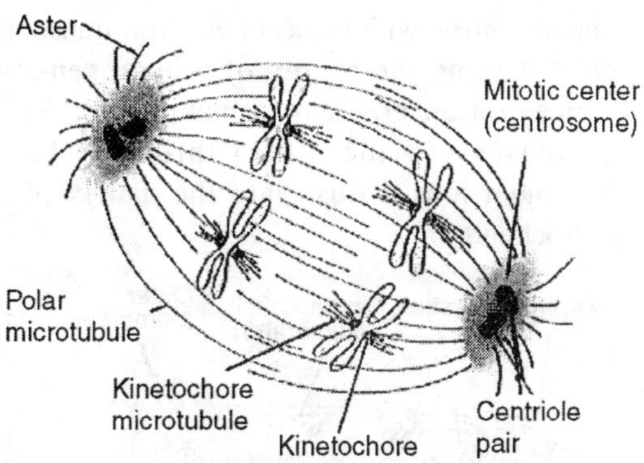

Fig. *Aster*

Aster yellows A virus disease of many kinds of plants transmitted by the six spotted leaf hopper and characterized by stunting of plants, sterility, and chlorosis in foliage.

Asteroid impacts Hypothesis that links certain mass extinction events with the impact of a comet or asteroid, most notably the mass extinction 65 million years that caused the disappearance of dinosaurs and many other reptilian groups. Asteroid impacts early in earth history also contributed to the formation of the atmosphere and oceans.

Asteroidea The class of echinoderms whose members have rays that are not sharply set off from the central disk; ambulacral grooves with tube feet; and suction disks on tube feet. A respiratory disorder caused by allergies that constrict the bronchioles by inducing spasms in the muscles surrounding the lungs, by causing the bronchioles to swell, or by clogging the bronchioles with mucus.

Asymmetric carbon A carbon atom covalently bonded to four different atoms or groups of atoms.

Asymmetrical In animals, a term referring to organisms that lack a general body plan or axis of symmetry that divides the body into mirror-image halves.

Asymmetry Without a balanced arrangement of similar parts on either side of a point or axis.

ATG The codon for methionine; the translation initiation codon. Usually, protein translation can only start at a methionine codon (although this codon may be found elsewhere within the protein sequence as well). In eukaryotic DNA, the sequence is ATG; in RNA it is AUG. Usually, the first AUG in the mRNA is the point at which translation starts, and an open reading frame follows - i.e. the nucleotides taken three at a time will code for the amino acids of the protein, and a stop codon will be found only when the protein coding region is complete.

Atherosclerosis Disease characterized by fatty plaques forming in the inner lining of arteries.

Athlete's foot A disease caused by imperfect fungi that flourish under warm, wet conditions. Usually dissappears if the feet are kept dry.

Atmosphere A unit of pressure equal to 14.7 lbs/in.2 and equivalent to the pressure created by a 10 m column of water.The envelope of gases that surrounds the Earth; consists largely of nitrogen (78%) and oxygen (21%).

Atmospheric deposition Sedimentation of solids, liquids, or gaseous materials from the air. Snow, rain, and dust are the most familiar examples, but acids, metals, and toxic organic chemicals also can be transported by winds and deposited far from the source of origin.

Atmospheric pressure Ambient pressure created near the earth's surface by large air cells that circulate around the globe.

Atoke Anterior, nonreproductive part of a marine polychaete, as distinct from the posterior, reproductive part (epitoke) during the breeding season.

Atoll A horseshoe or circular array of islands, capping a coral reef system perched around an oceanic volcanic seamount

Atom The basic unit of matter; the smallest complete unit of the elements, consisting of protons, neutrons, and electrons.The smallest indivisible particle of matter that can have an independent existence.

Atomic force microscope A type of scanning probe microscope that images a surface by moving a sharp probe over the surface at a constant distance; a very small amount of force is exerted on the tip and probe movement is followed with a laser.

Atomic mass A mass unit determined by arbitrarily assigning the carbon-12 isotope a mass of 12 atomic mass units.

Atomic nucleus Most of the mass of an atom; composed of protons and neutrons.

Atomic number Number of protons in the atomic nucleus; there is a characteristic number for each type of element.

Atomic weight The sum of the weights of an atom's protons and neutrons, the atomic weight differs between isotopes of the same element.The weight of an atom determined by adding the number of protons and neutrons (the mass of electrons is usually considered to be negligible).

ATP - adenosine triphosphate ATP is a convenient packet of energy used by both animals and plant cells. The energy in ATP is stored in its three negatively charged phosphate groups which are held close together, in spite of their repulsion for each other. This energy, multiplied many hundreds of thousand of times, for each cell is able to move our muscles, transport molecules across membranes and power all the cells other energy requirements. Once the energy has been used the ATP molecule now only has two phosphate groups. It needs energy now from either aerobic or anaerobic respiration to charge it up again, a process known as phosphorylation. Large stores of ATP are not kept as it is highly reactive. The long term storage of energy in animals is in carbon rich molecules, such as glycogen or fatty acids. In plants energy is stored as starch.

ATP synthase A cluster of several membrane proteins found in the mitochondrial cristae (and bacterial plasma membrane) that function in chemiosmosis with adjacent electron transport chains, using the energy of a hydrogen-ion concentration gradient to make ATP. ATP synthases provide a port through which hydrogen ions diffuse into the matrix of a mitrochondrion.

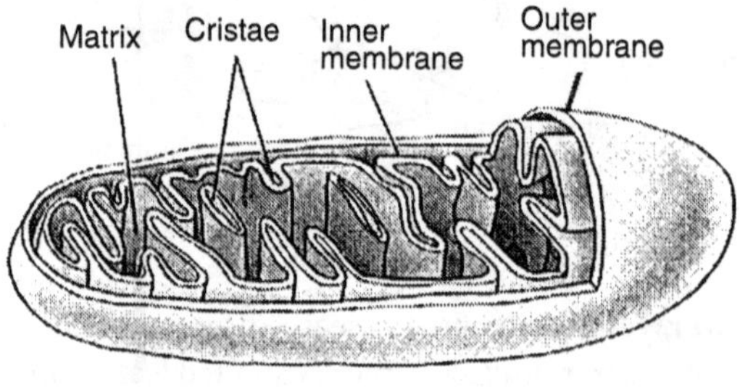

Fig. *Cristae*

ATP-binding cassette transporters (ABC transporters) Membrane protein complexes that use ATP energy to move substances across membranes without modifying the compound being transported. They require an extracytoplasmic substrate-binding protein for proper function.

Atretic follicle The resulting end-product of a degenerated ovarian follicle. The developing follicle may stop maturing at any time during maturation causing atresia, or degeneration of the follicle, its components and the ovum.

Atria Chambers of the heart.

Atrioventricular (AV) node Tissue in the right ventricle of the heart that receives the impulse from the atria and transmits it through the ventricles by way of the bundles of His and the Purkinje fibers. A group of slow-conducting fibers in the atrium of the vertebrate heart that are stimulated by impulses originating in the sinoatrial node (the pacemaker) and that conduct impulses to the bundle of His, a group of fibers that stimulate contraction of the ventricles.

Atrioventricular valve A valve in the heart between each atrium and ventricle that prevents a backflow of blood when the ventricles contract.

Atrium In the embryo, it is the chamber of the heart that receives blood from the sinus venosus and delivers blood to the ventricle.One of the chambers of the heart; also, the tympanic cavity of the ear; also, the large cavity containing the pharynx in tunicates and cephalochordates.

Attenuation 1. A mechanism for the regulation of transcription of some bacterial operons by aminoacyl-tRNAs.

2. A procedure that reduces or abolishes the virulence of a pathogen without altering its immunogenicity. A rho-independent termination site in the leader sequence that is involved in attenuation.

Attenuation (of light) Diminution of light intensity; explained, in the ocean, in terms of absorption and scattering.

Attractants Substances which elicit a positive directional response; chemicals having positive attraction for animals such as insects, usually in low concentration and at considerable distances.

Augmentation Biological control practices intended to increase the number or effectiveness of existing natural enemies.

Auricle The portion of the external ear not connected within the head. Also used to designate an atrium of a heart. In the class Turbellaria, the sensory lobes that project from the side of the head.

Auricularia A type of larva found in Holothuroidea.

Autochthonous Refers to something formed in its present location. Antonym of allochthonous.

Autoclave An apparatus for sterilizing objects by the use of steam under pressure. Its development tremendously stimulated the growth of microbiology.

Autocrine Cell messengers which areproduced by the cell itself and regulate the expression of genes.

Autogamy Condition in which the gametic nuclei produced by meiosis fuse within the same organism that produced them to restore the diploid number.

Autogenesis model According to this model, eukaryotic cells evolved by the specialization of internal membranes originally derived from prokaryotic plasma membranes.

Autogenous infection An infection that results from a patient's own microbiota, regardless of whether the infecting organism became part of the patient's microbiota subsequent to admission to a clinical care facility.

Autoimmune disease A disease produced by the immune system attacking self-antigens. Autoimmune disease results from the activation of self-reactive T and B cells that damage tissues after stimulation by genetic or environmental triggers.

Autoimmunity Autoimmunity is a condition characterized by the presence of serum autoantibodies and self-reactive lymphocytes. It may be benign or pathogenic. Autoimmunity is a normal consequence of aging; is readily inducible by infectious agents, organisms, or drugs; and is potentially reversible in that it disappears when the offending "agent" is removed or eradicated.Immune response to ones' own proteins or other antigens.

Autoinfection Reinfection by a parasite juvenile without its leaving the host.

Autolysins Enzymes that partially digest peptidoglycan in growing bacteria so that the peptidoglycan can be enlarged.

Autonomic nervous system Controls routine body functions such as gut activity, respiration, blood pressure and heart rate. There are two main divisions the sympathetic nervous system and the parasympathetic system.

Autonomic system The portion of the peripheral nervous system that stimulates smooth muscle, cardiac muscle, and glands; consists of the parasympathetic and sympathetic systems.

Autonomous specification Determination of cell fate by cytoplasm acquired during cleavage, independent of interactions with neighboring cells.

Autopolyploid A type of polyploid species resulting from one species doubling its chromosome number to become tetraploid, which may self-fertilize or mate with other tetraploids.

Autoradiography A process to detect radioactively labeled molecules (which usually have been separated in an *SDS-PAGE* or *agarose gel*) based on their ability to create an image on photographic or X-ray film. This process does not result in a linear relationship between the intensity of the signal and the amount of radioactivity unless special steps are taken. There is now increasing use of phosphorimagers and other modern devices to detect and quantitate radioactive molecules which have been separated in gels.A technique that uses X- ray film to visualize radioactively labeled molecules or fragments of molecules; used in analyzing length and number of DNA fragments after they are separated by gel *electrophoresis*.

Autosome A *chromosome* not involved in sex determination. The *diploid* human *genome* consists of 46 chromosomes, 22 pairs of autosomes, and 1 pair of *sex chromosomes* (the X and Y chromosomes). A type of chromosome, found in all cells, that is NOT concerned with sex determination. In humans there are 44 autosomes and two sex chromosomes in all body cells. The chromosomes other than the sex chromosomes. Each member of an autosome pair (in diploid organisms) is of similar length and in the genes it carries.

Autotomy plane The zone or septum of soft tissue which passes through a caudal vertebrae, along which breakage takes place in autotomy.

Autotomy The ability of certain lower animals, such as lizards and starfish, to cast off injured body parts, such as the tail and, usually, to regenerate new ones.The self amputation of an appendage. For example, the casting off of a section of a lizard's tail caught in the grasp of a predator. The autotomized appendage is usually regenerated.

Autotroph An organism that obtains organic food molecules without eating other organisms. Autotrophs use energy from the sun or from the oxidation of inorganic substances to make organic molecules from inorganic ones.An organism that makes its organic nutrients from inorganic raw materials by using an external energy source, such as light energy.An organism that synthesizes organic molecules from inorganic starting materials through photosynthesis or chemosynthesis. Autotrophs are ecologically important as primary producers as they ultimately provide energy for all heterotrophic organisms. Having the ability to synthesize food from inorganic compounds.

Autotrophic algae Algae capable of photosynthesis and growth using only dissolved inorganic nutrients

Autotrophic nutrition Nutrition characterized by the ability to use simple inorganic substances for the synthesis of more complex organic compounds, as in green plants and some bacteria.

Autotrophic Refers to organisms that synthesize their nutrients and obtain their energy from inorganic raw materials.Organisms that synthesize their own nutrients; include some bacteria that are able to synthesize organic molecules from simpler inorganic compounds.

Autotrophy Any organism that is able to manufacture its own food. Most plants are autotrophs, as are many protists and bacteria. Contrast with consumer. Autotrophs may be photoautotrophic, using light energy to manufacture food, or chemoautotrophic, using chemical energy.

Auxin A class of plant hormones, including indoleacetic acid (IAA), having a variety of effects, such as phototropic response through the stimulation of cell elongation, stimulation of secondary growth, and the development of leaf traces and fruit.A group of hormones involved in controlling plant growth and other functions; once thought responsible for phototropism by causing the cells on the shaded side of a plant to elongate, thereby causing the plant to bend toward the light.

Auxospore The resistant stage of diatoms that restores the maximum size characteristic to the species.

Auxotroph A mutated prototroph that lacks the ability to synthesize an essential nutrient and therefore must obtain it or a precursor from its surroundings.A nutritional mutant that is unable to synthesize and that cannot grow on media lacking certain essential molecules normally synthesized by wild-type strains of the same species.

Auxotrophic mutant A bacterial strain which has a mutation in at least one of the enzymes in a biochemical pathway responsible for synthesising an essential substance, for example an amino acid. The mutant strain will be unable to grow on minimal salts medium with glucose as sole carbon source unless this is supplemented with the substance which is deficient in synthesis. The mutant phenotype can therefore be easily detected.

Avalonia A separate plate in the Early Paleozoic consisting of much of Northern Europe, Newfoundland, Nova Scotia, and some coastal parts of New England.

Aves The vertebrate class of birds, characterized by feathers and other flight adaptations.

Avicularium Modified zooid that is attached to the surface of the major zooid in Ectoprocta and resembles a bird's beak.

Avidin A glycoprotein which binds to *biotin* with very high affinity.

Awn A slender bristle.

Axial cells Central cells of a dicyemid mesozoan.

Axial filament The organ of motility in spirochetes. It is made of axial fibrils or periplasmic flagella that extend from each end of the protoplasmic cylinder and overlap in the middle of the cell. The outer sheath lies outside the axial filament.

Axial skeleton Portion of the skeleton that supports and protects the organs of the head, neck, and trunk.

Axil The angle formed between a leaf stalk and the stem to which it is attached. In flowering plants, buds develop in the axils of leaves.

Axilla Axillae = armpit; beneath the axis.

Axillary bud An embryonic shoot present in the angle formed by a leaf and stem.

Axis An imaginary line passing through a body or organ around which parts are symmetrically aligned.The second cervical vertebra; any axial part.

Axocoel The most anterior of three coelomic spaces that appear during larval echinoderm development.

Axolotl Larval stage of any of several species of the genus Ambystoma (such as Ambystoma tigrinum) exhibiting neotenic reproduction.

Axon A typically long extension, or process, from a neuron that carries nerve impulses away from the cell body toward target cells.A fiber that conducts a nerve impulse away from a neuron cell body. The extension of a nerve cell, as a thin tube which may be as long a metre or a few short microns in length. The axon, like the cell body is able to depolarise and carry impulses along its length. The impulses from one axon to another nerve cell are transmitted at a synapse. Axons may be myelinated or unmeyelinated, and they may vary in diameter. Thicker, myelinated axons transmit impulses faster than thins unmeyelinated axons.

Axoneme The axial thread of the chromosome in which is located the axial combination of genes. The central core of a cilium or flagellum, consisting of a central pair of filaments surrounded by nine other pairs; also called axial filament.The motor section of the flagellum which is a construction of microtubules emanating from the centriole at the base of the flagellum.

Axopodium Fine, needlelike pseudopodium that contains a central bundle of microtubules. Also called axopod. Found in certain sarcodine protozoa.

Axostyle Tubelike organelle in some flagellate protozoa, extending from the area of the kinetosomes to the posterior end, where it often protrudes.

Azygous A blood vessel.

B

B cell A type of lymphocyte derived from bone marrow stem cells that matures into an immunologically competent cell under the influence of the bursa of Fabricius in the chicken, and the bone marrow in nonavian species; following interaction with antigen, it becomes a plasma cell, which synthesizes and secretes antibody molecules involved in humoral immunity; B lymphocyte.Type of lymphocyte responsible for antibody-mediated immunity; mature in the bone marrow and circulate in the circulatory and lymph systems where they transform into antibody-producing plasma cells when exposed to antigens.

B memory cells Long-lived B cells that are produced after an initial exposure to an antigen and play an important role in secondary immunity. They remain in the body and facilitate a more rapid responce if the antigen is encountered again.

BAC: bacterial artificial chromosome A cloning vector capable of carrying between 100 and 300 kilobases of target sequence. They are propagated as a mini-chromosome in a bacterial host. The size of the typical BAC is ideal for use as an intermediate in large-scale genome sequencing projects. Entire genomes can be cloned into BAC libraries, and entire BAC clones can be shotgun-sequenced fairly rapidly.

Bacillary bands Lateral zones in the body wall of some nematodes, consisting of glandular and nonglandular cells of unknown function.

Bacillus thuringiensis A bacterium that kills insects; a major component of the microbial pesticide industry.

Back corals Colonial anthozoans that secrete a black protein skeleton.

Back mutation Reverse the effect of a point or frame-shift mutation that had altered a gene; thus it restores the wild-type phenotype.

Back reef The inner part of a barrier reefor atoll.

Back translation Use of the experimentally determined amino acid sequence of part or all of a polypeptide to determine the theoretical nucleic acid base sequence(s) which could code for it. This is normally done using a computer programme.

Bacteria The domain that contains procaryotic cells with primarily diacyl glycerol diesters in their membranes and with bacterial rRNA. Bacteria also is a general term for organisms that are composed of procaryotic cells and are not multicellular.Spherical or rod-shaped procaryotic cells with a linear dimension of several micrometers.

Consists of a *cell wall* beneath which a plasma membrane encloses a single cytoplasmic compartment containing DNA, RNA, proteins and small molecules in an apparently unorganized internal sturcture.

They replicate quickly by dividing in two by binary fission.

Bacterial artificial chromosome (BAC) A cloning vector constructed from the E. coli F-factor plasmid that is used to clone foreign DNA fragments in E. coli.

Bacterial vaginosis Bacterial vaginosis is a sexually transmitted disease caused by Gardnerella vaginalis, Mobiluncus spp., Mycoplasma hominis, and various anaerobic bacteria. Although a mild disease it is a risk factor for obstetric infections and pelvic inflammatory disease.

Bactericide An agent that kills bacteria.

Bacteriochlorophyll A modified chlorophyll that serves as the primary light-trapping pigment in purple and green photosynthetic bacteria.A type of chlorophyll found only in bacterial systems.

Bacteriocide A class of antibiotics that kills bacterial cells.

Bacteriocin A protein produced by a bacterial strain that kills other closely related strains.

Bacteriophage Bacterial virus. Lambda bacteriophage is the basis of many *E. coli* vectors which are used for cDNA and genomic DNA libraries. Filamentous bacteriophages are used to produce single-stranded DNA for sequencing or *in vitro* mutagenesis (eg. M13), and as vectors for Phage display expression screening systems.

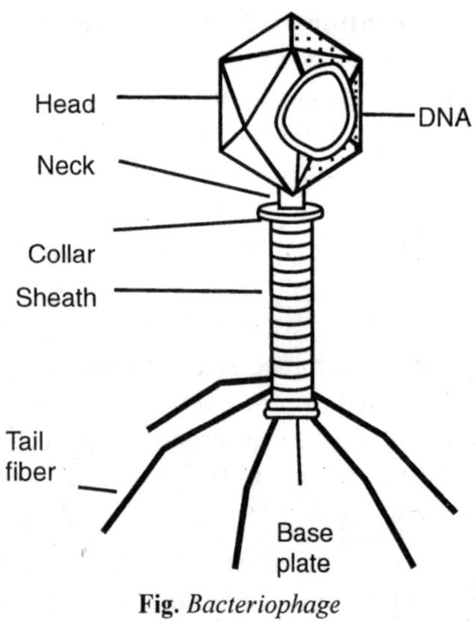

Fig. *Bacteriophage*

Bacteriophage lambda A virus which infects *E. coli* , and which is often used in molecular genetics experiments as a vector, or cloning vehicle. Recombinant phages can be made in which certain non-essential ? DNA is removed and replaced with the DNA of interest. The phage can accommodate a DNA "insert" of about 15-20 kb. Replication of that virus will thus replicate the investigator's DNA. One would use phage ? rather than a plasmid if the desired piece of DNA is rather large.

Bacterioplankton The component of the plankton that consists of bacteria.

Bacteriostat A class of antibiotics that prevents growth of bacterial cells.

Bacteria Unicellular, microscopic organisms that multiply by simple division. They take various forms, but are generally classified into four different types based on their shape; some are spherical (cocci), some are rod-shaped (bacilli), some are spiral-shaped (spirelli) and some are filamentous (actinomycetes). Bacteria can also be subdivided according to their requirement for free oxygen (O), into aerobic and anaerobic groups or, depending upon their energy source, into autotrophic and heterotrophic groups. Autotrophic bacteria oxidize inorganic material to obtain energy, whereas heterotrophic bacteria depend upon the decomposition of organic material for their energy. Some bacteria are sensitive to acidity in the environment, and the destruction of soil bacteria is common in areas where the terrestrial environment is subject to acid rain. Soil bacteria help to maintain soil fertility by breaking down organic matter and converting nitrogen into a form usable by plants. Bacteria also have an important role in pollution control, being able to decompose a great variety of organic wastes – from sewage to crude oil – into simpler compounds that can be reincorporated into the natural environment. Pollution problems arise, in part, because of society's ability to produce waste in amounts that exceed the capacity of natural decomposers such as bacteria. Despite performing such essential functions, bacteria are also responsible for a great variety of diseases including anthrax, tetanus, plague and tuberculosis. Most of these respond to treatment with antibiotics, but strains of bacteria that are resistant to antibiotics have evolved, raising fears of a resurgence of diseases such as tuberculosis, once thought to have been conquered. The apparent increased frequency of uncommon diseases such as the so-called 'flesh-eating disease' caused by a streptococal bacterium may also be indicative of the ability of new strains of bacteria to resist

Bacteroid A modified, often pleomorphic, bacterial cell within the root nodule cells of legumes; after transformation into a symbiosome it carries out nitrogen fixation.

Baculum baculum = rod. Penis bone.

Baeocytes Small, spherical, reproductive cells produced by pleurocapsalean cyanobacteria through multiple fission.

Balanced growth Microbial growth in which all cellular constituents are synthesized at constant rates relative to each other.

Balanced polymorphism A type of polymorphism in which the frequencies of the coexisting forms do not change noticeably over many generations.

Balancers Lateral appendages on the heads of some larval salamanders.

Balanitis Inflammation of the glans penis usually associated with Candida fungi; a sexually transmitted disease.

Baleen plate A keratined growth in toothless whales.

Baleen The filtering plates that hang from the upper jaws of baleen whales.

Baleen whales The filter-feeding whales.

Ballonets Four inflated areas within the "head" of nematodes of the family Gnathostomatidae; each is connected to an internal cervical sac of unknown function.

Baltica A separate continental plate of the Early Paleozoic composed of the United Kingdom, Scandinavia, European Russia and Central Europe; named for the Baltic Sea.

Banded Iron formation Rock consisting of alternating light and dark layers of iron-rich chert (the dark layers have more iron minerals) formed from 3.8 to 1.7 billion years ago.

Bar-built estuary An estuary that is formed when a barrier island or sand bar separates a section of the coast where fresh water enters.

Bark Those portions of a woody plant stem or trunk exterior to the vascular cambium.

Baroreceptor A specialized nerve ending that is stimulated by changes in pressure.

Barotolerant Organisms that can grow and reproduce at high pressures but do not require them.

Barr body Inactivated X-chromosome in mammalian females. Although inactivated, the Barr body is replicated prior to cell division and thus is passed on to all descendant cells of the embryonic cell that had one of its X-chromosomes inactivated.

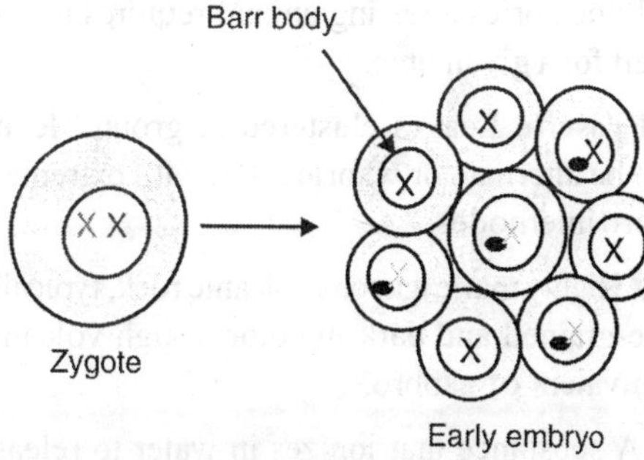

Fig. *Barr body*

Barrier reef A coral reef that runs approximately parallel to the shore and is separated from the shore by a lagoon.

Barriers to gene flow Factors, such as geographic, mechanical, and behavioral isolating mechanisms that restrict gene flow between populations, leading to populations with differing allele frequencies.

Basal Concerning the base of a structure - that part nearest the body. Basal cells in Diptera are generally small cells near the base of the wing.

Basal body A centriole that has given rise to the microtubular system of a cilium or flagellum, and is located just beneath the plasma membrane. Serves as a nucleation site for the growth of the axoneme.

Basal group The earliest diverging group within a clade; for instance, to hypothesize that sponges are basal animals is to suggest that the lineage(s) leading to sponges diverged from the lineage that gave rise to all other animals.

Basal lamina A term used to describe the different layers which make up the basement membrane. These different layers of the basement membrane, the lamina lucida and lamina densa are only visible with electron microscopy. Into the lamina dense collagen fibres of the lamina propria are anchored. And on the epithelial side are anchored bundles of tonofilaments from the hemi-desmosomes which anchor the basal epithelial cells to the basement membrane.

Basal metabolic rate (BMR) The minimal number of kilocalories a resting animal requires to fuel itself for a given time.

Basal rosette Leaves clustered at ground level; can be alternate or whorled, but with extremely short internodes.

Basalt Highly mafic igneous volcanic rock, typically fine-grained and dark in color; rough volcanic equivalent of gabbro.

Base A substance that ionizes in water to release hydroxyl ions (OH^-) or other ions that combine with hydrogen ions. The *purine* or *pyrimidine* component of a *nucleotide*; often used to refer to a *nucleotide* residue within a nucleic acid chain.

Base pair Two bases, each containing the element nitrogen, held together by relatively weak chemical bonds. Two base pairs make up DNA molecules: adenine and thymine, and guanine and cytosine.

Base sequence analysis A method, sometimes automated, for determining the *base sequence.*

Basement membrane A thin sheet of proteins held together by type IV collagen. On this sheet epithelial cells attach with hemi-desmosomes. All epithelia, whether simple cuboidal cells such as found in the salivary glands, or endothelial cells lining capillaries or thick stratified squamous epithelia of the skin, are all anchored to a basement membrane.

Basement rock The oldest rocks in a given area; a complex of metamorphic and igneous rocks that underlies the sedimentary deposits. Usually Precambrian or Paleozoic in age.

Base-pair substitution A point mutation; the replacement of one nucleotide and its partner from the complementary DNA strand by another pair of nucleotides.

Base-pairing principle In the formation of nucleic acids, the requirement that adenine must always pair with thymine (or uracil) and guanine with cytosine.

Basic dyes Dyes that are cationic, or have positively charged groups, and bind to negatively charged cell structures. Usually sold as chloride salts.

Basic Possessing a large number of hydroxyl (OH^-) ions; a pH of more than 7.0.

Basic research Research conducted to elucidate basic concepts alone with no direct application in mind.

Basidia Specialized club-shaped structures on the underside of club fungi (Basidiomycetes) within which spores form (sing.: basidium).

Basidiocarp The densely packed hyphae of the fruiting body in the Basidiomycetes.

Basidiomycetes The club fungi, a major group of

fungi that all produce a structure (basidium) on which basidiospores are produced. Includes mushrooms and toadstools.

Basidiospores The spores formed on the basidia of club fungi (Basidiomycetes).

Basidium A structure that bears on its surface a definite number of basidiospores (typically four) that are formed following karyogamy and meiosis. Basidia are found in the basidiomycetes and are usually club-shaped.

Basilar membrane A flat band of tissue that bears the auditory hair cells in the cochlea of the vertebrate ear; the membrane vibrates in response to pressure waves.

Basin Any large depression in which sediments are deposited.

Basin and Range topography The surface features typical of the Basin and Range Province.

Basipetal Movement down the stem as with auxin moving from the shoot apex downward.

Basis (basipodite) Joint of a crustacean appendage from which the exopod and the endopod originate; that is, the joint between the coxa and the exopod and endopod.

Basitarsus The Ist segment of the tarsus - usually the largest.

Basking Resting in the direct rays of the sun.

Basophil White blood cell characterized by the presence of cytoplasmic granules that become stained by a basophilic dye.

Bast fiber Fibers located in phloem.

Batch culture A culture of microorganisms produced by inoculating a closed culture vessel containing a single batch of medium.

Bateman effect The large variation in the reproductive success of males compared to females in relation to the number of mates they have.

Batesian mimicry Evolution by one species to resemble the coloration, body shape, or behavior of a related species that is protected from predators by a venomous stinger, bad taste, or some other adaptation.

Bathyal zone The bottom between the shelf break and a depth of approximately 4,000 m (13,000 ft).

Bathypelagic Relating to or inhabiting the deep sea.

Bathypelagic zone The zone of the ocean that extends from 1000m to 4000m below the surface of the ocean.

Batumen A protective layer of propilis or hard cerumen that encloses the nest cavity of a stingless bee colony.

Bronchus One of a pair of respiratory tubes branching into either lung at the lower end of the trachea; it subdivides into progressively finer passageways, the bronchioles, culminating in the alveoli.

B-carotene An important plant carotenoid and precursor of vitamin A.

B-cell antigen receptor (BCR) A transmembrane immunoglobulin complex on the surface of a B cell that binds an antigen and stimulates the B cell. It is composed of a membrane-bound immunoglobulin, usually IgD or a modified IgM, complexed with another membrane protein (the Ig-a/Ig-b heterodimer).

Bdelloidea A class of rotifers containing members where there are no males; anterior end retractile and bearing two disks; mastax adapted for grinding; paired ovaries; cylindrical body. Example: Rotaria.

Beak Colloquial expression for the protruding mouthpart structures of a sucking insect (= proboscis).

Beard worms (pogonophorans; phylum Pogonophora)Tube-dwelling invertebrates that lack a digestive system.

Bedrock The general term referring to the rock underlying other unconsolidated material, i.e. soil.

Behavior All of the acts an organism performs, as in, for example, seeking a suitable habitat, obtaining food, avoiding predators, and seeking a mate and reproducing.

Behavior genetics The study of the role that genes play in controlling behavior.

Behavioral ecology A subdiscipline within animal behavior that deals with the ways in which animals interact with their environment and the survival value of behavior as well as its contribution to reproductive success.

Behavioral isolating mechanisms Differences in behavior (usually courtship) that prevent genetic exchange between members of different populations or species.

Behaviorism A view of the actions of animals that postulates that behavior can be analyzed functionally in terms of stimulus and response combinations. Thus, most of behavior is a function of the experience of the organism in this view.

Benign Not harmful. In the sense of tumours, not malignant. When referring to parasites, quite harmless.

Benthic Refers to organisms that live on or in the ocean bed. Benthic epifauna are organisms that live on the ocean floor or upon bottom objects sch as sea anemones and barnacles, whereas benthic infauna are organisms that live within the surface sediments such as clams and worms. Compare *pelagic*.

Benthic zone One of the two basic subdivisions of the marine biome; includes the sea floor and bottom-dwelling organisms.

Benthic-pelagic coupling The cycling of nutrients between the bottom sediments and overyling water column

Benthos Organisms that live associated with the sea bottom. Examples include burrowing clams, sea grasses, sea urchins, acorn barnacles.

Benzene hexachloride (chemical name) or BHC. (common name). A synthetic insecticide, a chlorinated hydrocarbon, 1,2,3,4,5,6-hexachlorocvclohexane of mixed isomers; slightly more toxic to mammals than DDT, acute oral LD51 for rats about 200 mg/kg; phytotoxicity: more toxic than DDT, interferes with germination, suppresses growth and reduces yields except at low concentration; certain crop plants, as potato absorb crude BHC with consequent tainting of tubers.

Bering land bridge The vast tundra plain that was exposed between Asia and North America during the Last Glacial Maximum, about 21,000 years ago; it served as a migration route for people, animals, and plants. Also known as Beringia.

Berm A broad area of low relief in the upper part of a beach

Berry A fleshy, two- or multiple-carpeled ovary, each carpel having many seeds.

Best practical control technology (BPT) The best technology for pollution control available at reasonable cost and operable under normal conditions.

Beta hemolysis A zone of complete clearing around a bacterial colony growing on blood agar. The zone does not change significantly in color.

Beta particles High-energy electrons released by radioactive decay.

Beta-carotene Yellow to red pigment in plants; one of the most important of the carotenoids; converts to vitamin A in the body.

Beta-DNA The normal form of DNA found in biological systems, which exists as a right-handed helix.

Beta-Lactamase Ampicillin resistance gene.

Beta-proteobacteria One of the five subgroups of proteobacteria, each with distinctive 16S rRNA sequences. Members of this subgroup are similar to the alpha-proteobacteria metabolically, but tend to use substances that diffuse from organic matter decomposition in anaerobic zones.

Between-habitat comparison A contrast of

diversity in two localities of differing habitat type (e.g., sand versus mud bottoms)

B-galactosidase The enzyme responsible for the splitting of lactose into glucose and galactose.

Biantitropical distribution A pattern of a species' geographical distribution that extends across comparable bands of latitude in both the Northern Hemisphere and the Southern Hemisphere.

Bib A patch of colored feathers under the chin of a bird

Bicarbonate ions A weak base present in saliva that helps to neutralize acids in food.

Biennial A plant with a lifespan of two years. Often only flowers and sets seed during the second year.

Big bang theory A model for the evolution of the universe that holds that all matter and energy in the universe were concentrated in one point, which suddenly exploded. Subsequently, matter condensed to form atoms, elements, and eventually galaxies and stars.

Bilateral body symmetry Animal body plan with mirror image left and right sides.

Bilateral symmetry A form of symmetry in which only the midsagittal plane will divide an organism into mirror images; bilateral symmetry is characteristic of actively moving organisms that have definite anterior (head) and posterior (tail) ends.

Bilateria Group containing all multicellular animals with bilateral symmetry.

Bile A fluid secreted by the liver and poured into the small intestine via the bile duct; emulsifies fats. A yellow secretion of the vertebrate liver, temporarily stored in the gallbladder and composed of organic salts that emulsify fats in the small intestine.

Bilharziasis Disease caused by Schistosoma spp. Also called schistosomiasis.

Biliary system The bile-producing system consisting of the liver, gallbladder, and associated ducts.

Bilirubin A breakdown product of the heme group of hemoglobin, excreted in the bile.

Bimodal breathing The ability of an organism to exchange respiratory gases simultaneously with both air and water; usually using gills for water breathing and lungs for air breathing.

Bimodal foraging Having two modes of foraging

Binal symmetry The symmetry of some virus capsids (e.g., those of complex phages) that is a combination of icosahedral and helical symmetry.

Binary fission The method by which bacteria reproduce. The circular DNA molecule is replicated; then the cell splits into two identical cells, each containing an exact copy of the original cell's DNA.

Bindin Acrosomal protein (of sea urchin sperm) mediating species-specific recognition and binding between the sperm and the egg once the sperm has penetrated the egg jelly.

Binding site A place on cellular DNA to which a protein (such as a transcription factor) can bind. Typically, binding sites might be found in the vicinity of genes, and would be involved in activating transcription of that gene (promoter elements), in enhancing the transcription of that gene (enhancer elements), or in reducing the transcription of that gene (silencers). Note that whether the protein in fact performs these functions may depend on some condition, such as the presence of a hormone, or the tissue in which the gene is being examined. Binding sites could also be involved in the regulation of chromosome structure or of DNA replication.

Binocular vision Having to do with or the use of both eye simultaneously.

Binomial The two-part Latinized name of a species, consisting of genus and specific epithet.

Binomial system of nomenclature A system of taxonomy developed by Linnaeus in the early eighteenth century. Each species of plant and animal receives a two-term name; the first term is the genus, and the second is the species.

Bioaccumulation Increasingly concentrated accumulation of substances, especially pollutants, at successively higher trophic levels in food chains.

Bioassay A quantitative assay of a particular substance using a portion of or an entire living organism.

Bioaugmentation Increasing the activity of bacteria that decompose pollutants; a technique used in bioremediation.

Biocentric preservation A philosophy that emphasizes the fundamental right of living organisms to exist and to pursue their own goods.

Biocentrism The belief that all creatures have rights and values; being centered on nature rather than humans.

Biochemical cycle The flow of an element through the living tissue and physical environment of an ecosystem; e. g., the carbon, hydrogen, oxygen, nitrogen, sulfur, and phosphorus cycles.

Biochemical oxygen demand (BOD) The amount of oxygen used by organisms in water under certain standard conditions; it provides an index of the amount of microbially oxidizable organic matter present.

Biochemical pathway An ordered series of chemical reactions in a living cell, in which each step is catalyzed by a specific enzyme; different biochemical pathways serve different functions in the life of the cell.

Biochemical reactions Specific chemical processes that occur in living things.

Biochemical taxonomy A biochemical approach to defining evolutionary relationships of organisms by determining molecular similarities and differences among organisms.

Biochemistry The chemistry of living organisms and of vital processes; also known as physiological or biological chemistry. The study of the molecular basis of life.

Biocide A broad-spectrum poison that kills a wide range of organisms.

Biodegradable A chemical that can be broken down by bacteria or other organisms.

Biodegradable plastics Plastics that can be decomposed by microorganisms.

Biodegradation The breakdown of a complex chemical through biological processes that can result in minor loss of functional groups, fragmentation into larger constitutents, or complete breakdown to carbon dioxide and minerals. Often the term refers to the undesired microbial-mediated destruction of materials such as paper, paint, and textiles.

Biodiversity Biological diversity; can be measured in terms of genetic, species, or ecosystem diversity. The total variation in life, including the number of species, the degree of genetic variation within species, the different types of ecosystems, and the all ecosystem functions.

Biodiversity hotspot A relatively small area with an exceptional concentration of

Bioenergetics The study of how organisms manage their energy resources.

Bioenrichment Adding nutrients or oxygen to increase microbial breakdown of pollutants.

Biofilms Organized microbial systems consisting of layers of microbial cells associated with surfaces, often with complex structural and functional characteristics. Biofilms have physical/chemical gradients that influence microbial metabolic processes. They can form on inanimate devices (catheters, medical prosthetic devices) and also cause fouling (e.g., of ships' hulls, water pipes, cooling towers).

Biogenesis The doctrine that life originates only from preexisting life.

Biogenic graded bedding A regular change of sediment median grain size with depth below the sediment-water interface caused by the activities of burrowing organisms

Biogenically reworked zone The depth zone, within a sediment, that is actively burrowed by benthic organisms

Biogenous sediment The type of sediment that is made up of the skeletons and shells of marine organisms.

Biogeochemical cycling The oxidation and reduction of substances carried out by living organisms and/or abiotic processes that results in the cycling of elements within and between different parts of the ecosystem (the soil, aquatic environment, and atmosphere).

Biogeographical area An entire self-contained natural ecosystem and its associated land, water, air, and wildlife resources.

Biogeography The study of the distribution of plants and animals across the Earth.

Bioinsecticide A pathogen that is used to kill or disable unwanted insect pests. Bacteria, fungi, or viruses are used, either directly or after manipulation, to control insect populations.

Biologic transmission A type of vector-borne transmission in which a pathogen goes through some morphological or physiological change within the vector.

Biological clock Proposed internal factor(s) in organisms that governs functions that occur rhythmically in the absence of external stimuli.

Biological communication An action by one organism (or cell) that causes a reaction from another organism (or cell) in a fashion adaptive to either one or both of the participants.

Biological community The populations of plants, animals, and microorganisms living and interacting in a certain area at a given time.

Biological control The use of living organisms, such as predators, parasitoids, and pathogens, to control pest insects, weeds, or diseases. Typically involves some human activity.

Biological magnification A trophic process in which retained substances become more concentrated with each link in the food chain. The increased concentration of nonbiodegradable chemicals in the higher levels of the food chain.

Biological or biotic factors Organisms and products of organisms that are part of the environment and potentially affect the life of other organisms.

Biological oxygen demand (BOD) A standard test for measuring the amount of dissolved oxygen utilized by aquatic microorganisms over a five-day period.

Biological pests Organisms that reduce the availability, quality, or value of resources useful to humans.

Biological resources The earth's organisms.

Biological rhythm A cyclical pattern of behavior, occurring at some regular period (e.g., diurnal).

Biological species A population or group of populations whose members have the potential to interbreed.

Biological species concept A reproductive community of populations (reproductively isolated from others) that occupies a specific niche in nature.

Biological vector Vector in which a disease organism lives or develops. Contrast with mechanical vector.

Biologics Agents, such as vaccines, that give immunity to diseases or harmful biotic stresses.

Biology The study of all living organisms.

Bioluminescence Meaning living (bio) light (luminescence) is the light produced by living organisms and the emission of such biologically produced light. Also commonly referred to as 'phosphorescence'.

Biomagnification Increase in concentration of certain stable chemicals (for example, heavy metals or fat-soluble pesticides) in successively higher trophic levels of a food chain or web.

Biomass The dry weight of organic matter comprising a group of organisms in a particular habitat.

Biomass fuel Organic material produced by plants, animals, or microorganisms that can be burned directly as a heat source or converted into a gaseous or liquid fuel.

Biomass pyramid A metaphor or diagram that explains the relationship between the amounts of biomass at different trophic levels.

Biome One of the largest recognizably distinct ecosystems on earth; the plant and animal communities and associated soils that are characteristic of a given regional climate type.

Bionomics The study of the habits, breeding, and adaptations of living forms.

Biopesticide The use of a microorganism or another biological agent to control a specific pest.

Biorational Having a minimal disruptive influence upon the environment and its inhabitants (e.g., a biorational insecticide). Broad-spectrum (insecticide): Active against a wide range of insects.

Bioregionalism Organization of human activities according to natural geographic or ecological boundaries and associations. This philosophy emphasizes a sense of place and living within the resources of one's local ecosystem.

Bioremediation The use of biologically mediated processes to remove or degrade pollutants from specific environments. Bioremediation can be carried out by modification of the environment to accelerate biological processes, either with or without the addition of specific microorganisms.

Biosensor The coupling of a biological process with production of an electrical signal or light to detect the presence of particular substances.

Biosphere All ecosystems on Earth as well as the Earth's crust, waters, and atmosphere on and in which organisms exist; also, the sum of all living matter on Earth.

Biosphere reserves World heritage sites identified by the IUCN as worthy for national park or wildlife refuge status because of high biological diversity or unique ecological features.

Biostratigraphy The study of rock layers (e.g., distribution, environment of deposition, age) based on their fossils; biostratigraphic- adj.

Biostratinomy The study of what happens between the death of an organism and burial. Part of taphonomy.

Biosurfactants Products of bacteria which increase the hydrophilic nature of a surface so as to allow for better adhesion.

Biosynthesis Formation by living organisms of organic compounds from elements or simple compounds.

Biotechnology A set of biological techniques developed through basic research and now applied to research and product development. In particular, the use by industry of *recombinant DNA*, cell fusion, and new bioprocessing techniques.

Bioterrorism The intentional or threatened use of viruses, bacteria, fungi, or toxins from living organisms to produce death or disease in humans, animals, and plants.

Biotic Pertaining to the living organisms in the environment.

Biotic Factor A living component of the environment which arises from and affects living organisms (distinct from physical factors). For example, the interaction between predators and prey is a biotic interaction. Compare *abiotic factor*.

Biotic potential The maximum reproductive rate of an organism, given unlimited resources and ideal environmental conditions. Compare with environmental resistance.

Biotic stress Living organisms which can harm plants ,

Such as viruses, fungi, and bacteria, and harmful insects.

Biotin A coenzyme which is essential for carboxylation reactions.

Biotransformation or microbial transformation The use of living organisms to modify substances that are not normally used for growth.

Bioturbation The disturbance of sediment by organisms, e.g. burrows, trails, or complete mixing.

Bipectinate Feathery, with branches growing out oil both sides of the main axis: applied mainly to antennae.

Bipedal Describes an animal that walks on two legs.

Bipinnaria larva In the class Asteroidea, the free-swimming, ciliated stage of larval development.

Bipinnate Describing a pinnate leaf in which the leaflets themselves are further subdivided in a pinnate fashion.

Biramous Arthropod appendages that are biramous have two branches, an outer branch and an inner branch. These branches may have separate functions; in crustaceans, for instance, the inner branch of a leg is used for walking, while the outer branch may be paddle-shaped or feathery and often functions as a gill. Contrast with uniramous.

Birds Taxonomic class of terrestrial vertebrates that are characterized by endothermy and feathers; descended from some group of reptiles (or possibly dinosaurs).

Birth control Any method used to reduce births, including celibacy, delayed marriage, contraception; devices or medication that prevent implantation of fertilized zygotes, and induced abortions.

Birth rate The ratio between births and individuals in a specified population at a particular time.

Bisexual Having two sexes distinct and separate;i.e. a species with males and females.

Bisporangiate When a flower or cone produces both megaspores and microspores, it is said to be bisporangiate. Most flowers are bisporangiate.

Bivalent A pair of homologous chromosomes closely adjacent to each other formed during the first meiotic division within a primary spermatocyte or oocyte.

Bivalvia The class of molluscs whose members are enclosed in a shell consisting of two dorsally hinged valves, lack a radula, and possess a wedge-shaped foot. Clams, mussels, oysters.

Bivoltine Having two generations per year.

Bivouac The mass of army ant workers within which the queen and brood, live while the colony is not on the move.

Black fly fever Combination of symptoms resulting from sensitization to bites by black flies (Simuliidae).

Black lung disease Inflammation and fibrosis caused by accumulation of coal dust in the lungs or airways.

Black peidra A fungal infection caused by Piedraia hortae that forms hard black nodules on the hairs of the scalp.

Black smoker A chimney-like accumulation of mineral deposits that is found at hydrothermal vents.

Blackhead Disease of turkeys caused by the protozoan Histomonas meleagridis. Also called histomoniasis or infectious enterohepatitis.

Bladder A hollow, distensible organ with muscular walls that stores urine and expels it through the urethra.

Bladder worm The unilocular hydatid cyst of a tapeworm.

Blade (1) The broad, expanded part of a leaf.

(2) The broad, expanded photosynthetic part of the thallus of a multicellular alga or a simple plant.

Blastocoel The large fluid-filled cavity of the blastula formed by cleavage of the zygote. It may eventually be replaced by the growing archenteron.

Blastocyst The developmental stage of the fertilized ovum by the time it is ready to implant; formed from the morula and consists of an inner cell mass, an internal cavity, and an outer layer of cells (the trophoblast).

Blastoderm "Primary epithelium" formed in early embryonic development of many arthropods when the nuclei migrate to the periphery and undergo superficial cleavage; usually encloses the central yolk mass.

Blastodisc Disklike area on the surface of a large, yolky egg that undergoes cleavage and gives rise to the embryo.

Blastogenesis The origination of different castes, within a species, from the egg by means other than genetic.

Blastomeres Any of the cells produced by cleavage of a zygote.

Blastomycosis A systemic fungal infection caused by Blastomyces dermatitidis and marked by suppurating tumors in the skin or by lesions in the lungs.

Blastopore Site of gastrulation initiation and later the opening of the archenteron at the vegetal region of certain embryos (e.g., echinoderm and amphibian); in deuterostome embryos it is the future anus of the organism.

Blastospore A spore formed by budding from a hypha.

Blastula A cleavage stage embryo, typically a hollow ball of cells surrounding a cavity called the blastocoel; this term is used for (among others) echinoderm and amphibian embryos.

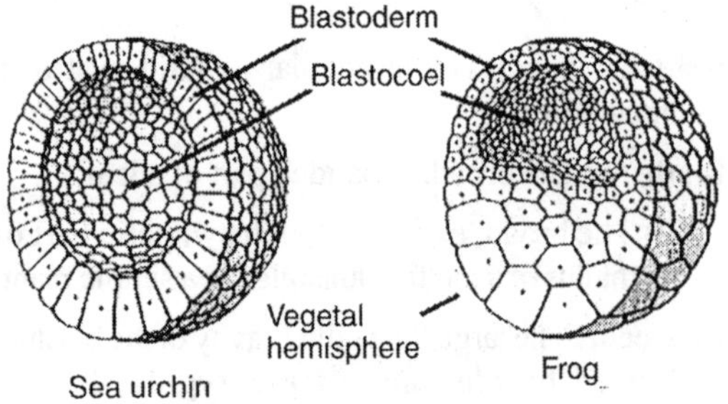

Fig. *Blastula*

Bleaching The expulsion of zooxanthellae by reef corals in response to stress.

Blending Term applied to 19th century belief that parental traits "blended" in their offspring; disproven by Mendel's work.

Blood A type of connective tissue with a fluid matrix called plasma in which blood cells are suspended. The fluid that circulates through the heart, arteries, capillaries, and veins.

Blood clotting One of three key processes in haemostasis, the prevention of blood loss. After three minutes of rupture of a small blood vessel, the entire cut is filled with a blood clot. After an hour, the clot has retracted inside the vessel making the plug even more effective. Within a few more hours, fibroblasts have moved into the clot, followed by capillary -forming endothelial cells. Within 10 days the clot is replaced by fibrous scar tissue. Clotting takes place in three steps. 1. In response to damage to the blood vessel prothrombin activator is formed.

2 this activator converts prothrombin into thrombin.

3.The thrombin acts as an enzyme converting fibrinogen into fibrin threads which adhere to the damaged walls of the blood vessel, trap platelets, blood cells and plasma to form a clot.

Blood groups Blood cells from different people do not always have the same cell surface antigens. A transfusions of blood whose antigens do not match the recipient evokes an immune response and the donor cells are broken down. Two of the common blood groups are the A,B,O group and the Rh group. The blood group antigens are glycoproteins or glycolipids.

Blood islands Also known as angiogenetic clusters; masses of splanchnic mesodermal cells found in the yolk sac of amniotes. The first blood forming tissue of the embryo, responsible for red blood cells and vitelline blood vessels.

Blood pigment A molecule used by an organism to

transport oxygen efficiently, usually in a circulatory system (e.g., hemoglobin)

Blood plasma The liquid, noncellular fraction of blood, including dissolved substances.

Blood pressure The force (energy) with which blood is pushed against the walls of blood vessels and circulated throughout the body when the heart contracts.

Blood-brain barrier A specialized capillary arrangement in the brain that restricts the passage of most substances into the brain, thereby preventing dramatic fluctuations in the brain's environment.

Bloom A dense concentration of phytoplankton that occurs in response to optimum growth conditions.

Blotting A technique for detecting one RNA within a mixture of RNAs (a Northern blot) or one type of DNA within a mixture of DNAs (a Southern blot). A blot can prove whether that one species of RNA or DNA is present, how much is there, and its approximate size. Basically, blotting involves gel electrophoresis, transfer to a blotting membrane (typically nitrocellulose or activated nylon), and incubating with a radioactive probe. Exposing the membrane to X-ray film produces darkening at a spot correlating with the position of the DNA or RNA of interest. The darker the spot, the more nucleic acid was present there.

Blowhole The nostrils, or nasal openings, of cetaceans.

Blubber The fat found between the skin and muscle of whales and other cetaceans, from which oil is made.

Blue revolution New techniques of fish farming that may contribute as much to human nutrition as miracle cereal grains but also may create social and environmental problems. Commercial fish farming already is a $5 billion per year industry, and domestic fish ponds supply as much as two-thirds of the protein consumed by subsistence farmers in many countries.

Blue-green algae Prokaryote organisms with photosynthetic phycobilin pigments.

Blueschist Metamorphic rock formed under great pressures, but not so great temperatures.

Bluetongue Virus disease of ruminants transmitted by biting midges (Ceratopogonidae).

Blunt end End of a DNA fragment produced by a restriction enzyme which cuts both strands of DNA at the same point, leaving no single-stranded sections. DNA with this type of end is more difficult to ligate.

B-lymphocytes Cells that manufacture antibodies involved in immunity.

Body fossil The actual remains (however permineralized, compressed or otherwise post-mortem altered) of an organism; includes bones, shells, and teeth.

Bog An area of waterlogged soil that tends to be peaty; fed mainly by precipitation; low productivity; some bogs are acidic.

Bohr effect When blood pH decreases, the ability of hemoglobin to bind to oxygen decreases. An adaptation to release oxygen in the oxygen starved tissues in capillaries where respiratory carbon dioxide lowers blood pH

Bolus A mass of chewed food mixed with salivary secretions that is propelled into the espohagus during the swallowing phase of digestion.

Bomb calorimeter An instrument used for measuring the caloric content of any organic material; measures energy release in combusting organic materials to CO_2 and H_2O.

Bond energy The quantity of energy that must be absorbed to break a particular kind of chemical bond; equal to the quantity of energy the bond releases when it forms.

Bond strength The strength with which a chemical bond holds two atoms together; conventionally measured in terms of the amount of energy, in kilocalories per mole, required to break the bond.

Bone cells The hard, rigid form of connective tissue constituting most of the skeleton of vertebrates;

composed chiefly of calcium salts. Also called bone (osseous) tissue.

Bone membrane A theoretical membrane separating the fluid surrounding bone crystals from the fluid of the surrounding connective tissue. The membrane would be formed by the endosteum.

Bone morphogenic proteins (BMP) Part of the cytokine family of transforming growth factors. BMP have a powerful ability to cause differentiation of stem cells into osteoblasts and to initiate bone formation.

Bony fish A term applied collectively to all groups of fish with bony (as opposed to cartilaginous) skeletons.

Book gill Modifications of a horseshoe crab's exoskeleton into a series of leaflike plates that serve as a surface for gas exchange between the arthropod and the water (phylum Arthropoda, class Merostomata).

Book lung A set of soft overlapping flaps, covered up by a plate on the abdomen, through which oxygen is taken up and carbon dioxide given off. Characteristic of many terrestrial arachnids such as scorpions and spiders.

Boreal forest A broad band of mixed coniferous and deciduous trees that stretches across northern North America (and also Europe and Asia); its northernmost edge, the taiga, intergrades with the arctic tundra.

Boring sponges Sponges that bore through calcareous skeletons and shells.

Bot The larva of certain flies that are parasitic in the body of mammals.

Botany The study of plants.

Bothridium Muscular lappet on the dorsal or ventral side of the scolex of a tapeworm. Bothridia are often highly specialized, with many types of adaptations for adhesion.

Bottle cells Epithelial cells found at the initial site of gastrulation, lining the initial archenteron, that temporarily become bottle-shaped; they maintain contact with the outer surface of the embryo, but the majority of the cell is inside the embryo. Also known as flask cells.

Bottleneck effect A period in the history of a population during which the population number is severely depleted, perhaps by disease or changed environmental conditions. During this time, the gene pool may also become depleted, leading to reduced genetic variability within the surviving population.

Bottom-up control Refers to food webs. A control of a population that comes from change lower in a food web (e.g., control of a population of mussels by abundance of phytoplankton food).

Botulism Food poisoning by the neurotoxin botulin, characterized by vomiting, abdominal pain, difficulty of vision, central nervous systems, disturbances of secretion, dyspepsia, a barking cough, and ptsosis.

Boundary layer A layer of fluid near a surface, where flow is affected by viscous properties of the fluid. At the surface, fluid velocity must be zero, and the boundary layer is a thin film that depends on surface texture, fluid velocity in the "mainstream of flow," and fluid mass properties such as salinity.

Bowman's capsule A cup-shaped receptacle in the vertebrate kidney that is the initial, expanded segment of the nephron where filtrate enters from the blood.

Box Refers to a short nucleic acid *consensus sequence* or *motif* that is universal within kingdoms of organisms. Examples of DNA boxes are the Pribow box (TATAAT) for *RNA polymerase*, the Hogness box (TATA) that has a similar function in eukaryotic organisms, and the homeo box. RNA boxes have also been described, such as Pilipenko's Box-A motif that may be involved in ribosome binding in some viral RNAs.

B-oxidation pathway The major pathway of fatty

acid oxidation to produce NADH, FADH2, and acetyl coenzyme A.

BP Abbreviation for base pair(s). Double stranded DNA is usually measured in bp rather than nucleotides (nt).

Brachiolaria This asteroid larva develops from the bipinnaria larva and has three preoral holdfast processes.

Brachiopoda A phylum of marine animals whose members possess a bivalved calcareous and/or chitinous shell that is secreted by a mantle and encloses nearly all of the body. Unlike the molluscs, the valves are dorsal and ventral. Possess a lophophore. Lampshells.

Brachium The arm from shoulder to elbow.

Brachydactly Human genetic disorder that causes production of an extra digit; an autosomal dominant trait. Sometimes referred to as polydactly.

Brachypterous With short wings that do not cover the abdomen, used of individuals of a species which otherwise has longer wings.

Brackish Describing water that is slightly salty; water whose salinity is between that of normal fresh water and normal sea water.

Brackish sea Semienclosed water body of large extent in which tidal stirring and seaward flow of freshwater do not exert enough of a mixing effect to prevent the body of water from having its own internal circulation pattern

Bracts 1. a small modified leaf with a relative undeveloped blade, in the axil of which grows an inflorescence or flower.

2. a leaf or leaflike structure associated with a sporangium.

Bradycardia Marked slowing of the heartbeat rate during a breath-hold dive.

Bradykinin One of several substances, all known as kinins, which cause vasodilation and increased capillary permeability, both events associated with inflammation.

Bradyzoite Small stage in various coccidia of the Isospora group that develops in a zoitocyst; similar to a merozoite.

Brain Collection of nerve cells usually located at the anterior end of an animal, when present at all. The nerves coordinate information gathered by sense organs, locomotion, and most internal body activities.

Brainstem The hindbrain and midbrain of the vertebrate central nervous system. In humans, it forms a cap on the anterior end of the spinal cord, extending to about the middle of the brain.

Bran The husk of a cereal grain; including the pericarp and seed coat of a cereal grain. The bran is removed in the processing of refined grains.

Branch roots Lateral roots arising from the pericycle of a

Breakbone fever Another name for dengue, a virus disease transmitted by mosquitoes.

Breeder reactor A nuclear reactor that produces fuel by bombarding isotopes of uranium and thorium with high-energy neutrons that convert inert atoms to fissionable ones. These reactors are inherently less stable and more dangerous than normal nuclear reactors.

Brevitoxin Neurotoxin produced by the dinoflagellate *Ptychodiscus brevis*.

Bright-field microscope A microscope that illuminates the specimen directly with bright light and forms a dark image on a brighter background.

Brittle stars Echinoderms with five flexible arms that radiate from a conspicuous central disk and tube feet that are used in feeding.

Broadcast spawners Marine animals that reproduce by releasing eggs and sperm into the water.

Broad-spectrum drugs Chemotherapeutic agents that are effective against many different kinds of pathogens.

Bronchi The paired ventilatory tubes of a vertebrate that branch into each lung at the lower end of the trachea.

Bronchial asthma An example of an atopic allergy involving the lower respiratory tract.

Bronchial-associated lymphoid tissue (BALT) The type of defensive tissue found in the lungs. Part of the nonspecific immune system.

Bronchioles Small tubes in the lungs that are formed by the branching of the bronchi; terminate in the alveoli.

Bronchitis A respiratory disorder characterized by excess mucus production and swelling of the bronchioles; caused by long-term exposure to irritants such as cigarette smoke and air pollutants.

Bronchus Either of two primary divisions of the trachea that lead to the right and left lung.

Brood In insects, a group of individuals of a given species which have hatched into young or which have become adult at approximately the same time and which live together in a defined and limited area. Often referring to the immature stages of ants bees and wasps.

Brood patch The patch of feathers used by birds to incubate eggs; also known as incubation patch.

Brown algae Multicellular protistans placed in the Division Phaeophyta, includes kelp.

Brown fat Mitochondria-rich, heat-generating adipose tissue of endothermic vertebrates.

Browse 1. to eat; nibble at; eat from.

2. to graze; feed on; pasture on. Browser - Mammals that feed predominantly on the growing shoots and buds of certain woody shrubs and trees. This includes many members of the deer family.

Browsers Organisms that feed by scraping thin layers of living organisms from the surface of the substratum (eg., periwinkles feeding on rock-surface diatom films; urchins scraping a thin, filmy sponge colony from a rock)

Bruce effect In mice, the effect of a strange male, or his odor, that causes females to abort and become receptive.

Brush border The collection of microvilli forming a border on the intestinal side of the epithelial cells of the small intestine.

Bryophyte Plants in which the gametophyte generation is the larger, persistent phase; they generally lack conducting tissues. Bryophytes include the Hepaticophyta (liverworts), Anthocerotophyta (hornworts), and Bryophyta (mosses). The nonvascular plants, characterized by life cycles dominated by the gametophyte phase. This group includes the mosses, liverworts, and hornworts, which lack lignified conducting tissues.

Bryozoans Colonial invertebrates in which the anus is located outside the edge of a lophophore.

Bt The bacterium, Bacillus thuringiensis.

Bubo Swollen lymph node.

Bubonic plague The most common form of plague, characterized by an abrupt onset of fever, chills, weakness, and headache followed by pain, tenderness, and buboes associated with a marked hemorrahagic tendency, the development of of disseminated coagulation, necrotic purpura, and extensive symmetrical gangrene; severe complications include pneumonia and septicemia. The notorious Black Plague that devastated the population of Europe and Asia in the 1300s was a form of bubonic plague.

Buccal cone Portion of the mouthparts of acarines composed of hypostome and labrum.

Buccal pump The mechanism by which lung ventilation occurs in amphibians; muscles of the mouth and pharynx create positive pressure to force air into the lungs.

Buccopharyngeal respiration The diffusion of gases across moist linings of the mouth and pharynx of amphibians.

Bud (1) In plants, an embryonic shoot, including rudimentary leaves, often protected by special bud scales.

(2) In animals, an asexually produced outgrowth that develops into a new individual.

Bud scale scar A scar or impression encircling the twig caused by the abscission of bud scales.

Bud sports Buds that produce fruit that is different from the rest of the fruit on the tree; vegetatively propagated by grafting cuttings onto another plant.

Budding A form of asexual reproduction in which new individuals develop from a portion of thc parent, as in all bryozoans and in many protozoans, cnidarians, and polychaetes.

Buff Yellowish-brown; medium or dark tan.

Buffers Chemical which are able to keep the pH of a solution within a normal range, neither acid nor alkali. Salivary buffers are important in reducing the progress of caries by neutralising plaque acids.

Bulb An underground storage organ characterized by fleshy leaves attached to a stem base.

Bulb of krause A sensory receptor in the skin believed to be the sensor for touch-pressure; also called bulbous corpuscle.

Bulbourethral gland One of a pair of glands near the base of the penis in the human male that secrete fluid that lubricates and neutralizes acids in the urethra during sexual arousal.

Bulbus arteriosus The anterior chamber of the early embryo's heart which functions to connect the ventricle to the ventral aorta. Synonyms: bulbus cordis, conus arteriosus, truncus arteriosus.

Bulk flow The movement of water due to a difference in pressure between two locations.

Bulking sludge Sludges produced in sewage treatment that do not settle properly, usually due to the development of filamentous microorganisms.

Bulla Nonliving structure serving as an anchor to which the maxillae are permanently attached; secreted by head and maxillary glands of female copepods in the family Lernaeopodidae.

Bulliform cell A large epidermal cell found on the upper surface of many grass leaves; turgor pressure in these cells controls the lateral rolling of the leaves during water stress.

Bundle of His In the vertebrate heart, a group of muscle fibers that carry impulses from the atrioventricular node to the walls of the ventricles; the only electrical bridge between the atria and the ventricles.

Bundle sheath cell Cells surrounding the vascular bundle in C_4 plants.

Bundle sheath Sheath of parenchyma or sclerenchyma cells surrounding the vascular bundles in leaves.

Bursa A sac-like cavity. In ophiuroid echinoderms, pouches opening at bases of arms and functioning in respiration and reproduction (genitorespiratory bursae). Small sacs lined with synovial membrane and filled with synovial fluid; act as cushions to reduce friction between tendons and bones.

Bursa copulatrix That part of the female genitalia which receives the aedeagus and sperm during copulation. Its structure is often important in separating closely related species.

Bursa of Fabricius Found in birds; the blind saclike structure located on the posterior wall of the cloaca; it performs a thymuslike function. A primary lymphoid organ where B-cell maturation occurs. Bone marrow is the equivalent in mammals.

Burst size The number of phages released by a host cell during the lytic life cycle.

Butanediol fermentation A type of fermentation most often found in the family Enterobacteriaceae in which 2,3-butanediol is a major product; acetoin is an intermediate in the pathway and may be detected by the Voges-Proskauer test.

Buttocks Prominence formed by the gluteal muscles.

Buttress Seaward face of a coral reef, extending from a depth of about 20 m to just below the low tide line.

By-catch The less valuable catch that is captured while fishing for a more valuable catch.

Byssal thread a strong elastic fiber used by mussels to attach themselves to a solid substrate

C

C_3 plant A plant that uses the Calvin cycle for the initial steps that incorporate CO_2 into organic material, forming a three-carbon compound as the first stable intermediate.

C_4 plant A plant that prefaces the Calvin cycle with reactions that incorporate CO_2 into four-carbon compounds, the end-product of which supplies CO_2 for the Calvin cycle.

C_4 pathway The set of reactions by which some plants initially fix carbon in the four-carbon compound oxaloacetic acid; the carbon dioxide is later released in the interior of the leaf and enters the Calvin cycle.

Cache A collection of items kept in one location.

Cadre Sclerotized mouth lining of a pentastomid.

Caecum Refers to a sack or pouch connected to the intestine of mammals. A caecum is usually found in herbivorous mammals and they house bacteria, which help digest the plant material that the mammal eats.

Calabar swelling Transient subcutaneous nodule, provoked by the filarial nematode Loa loa.

Calcar 1. a spur or spurlike process projecting from the leg of a bird.

2. a bony or cartilaginous process on the heel bone of bats, which helps to support the portion of the wing membrane lying between the legs.

Calcarea The class of sponges whose members are small and possess monaxon, triaxon, or tetraaxon calcium carbonate spicules.

Calcareous Term used to describe a structure, secreted by an organism, that consists of or contains calcium carbonate ($CaCO_3$), e.g., the shell of a bivalve.

Calcareous ooze A type of biogenous sediment that is made of the calcium carbonate shells and skeletons of marine organisms.

Calciferous glands Glands in an earthworm that secrete calcium ions into the gut.

Calcite A common crystalline form of natural calcium carbonate, CaCO3, that is the basic constituent of limestone, marble, and chalk. Also called calcspar.

Calcitonin A hormone produced by the thyroid that plays a role in regulating calcium levels.

Calcium carbonate ($CaCO_3$) A mineral that is the major component of the shell, skeleton, and other parts of many organisms.

Calcium-binding proteins proteins which have the ability to store calcium ions and to bind onto calcium in the hydroxyapatite of the enamel surface.

Calculus A hard deposit of calcified plaque which is found around the neck of the tooth. When it is above the free gingival margin (supra-gingival) it is white and chalky. When it is below (sub-gingival) it is dark and hard.

Caldera A large circular volcanic depression, often originating due to collapse.

Callose A complex carbohydrate found in the sieve areas of sieve-tube elements; particularly abundant at the time of injury.

Callow Newly eclosed workers in social insect colonies whose exoskeletons are still soft and whose colour has not fully matured.

Callus A mass of immature plant cells growing in laboratory tissue culture. Can be stimulated to develop into an entire plant with appropriate treatment with plant hormones.

Calmodulin An intracellular protein to which calcium binds in its function as a second messenger in hormone action.

Calorie The amount of heat energy required to raise the temperature of 1 g of water 1°C; the amount of heat energy that 1 g of water releases when it cools by 1°C. The Calorie (with a capital C), usually used to indicate the energy content of food, is a kilocalorie.

Calotte "Head" end of a dicyemid mesozoan.

Calvin cycle Series of biochemical, enzyme-mediated reactions during which atmospheric carbon dioxide is reduced and incorporated into organic molecules, eventually some of this forms sugars. In eukaryotes, this occurs in the stroma of the chloroplast.

Calypter Innermost of the three flap-like outgrowths at the base of the wing in various flies. Also known as the thoracic squama, it generally conceals the haltere.

Calyptodomous Of the nests of wasps, referring to those which are surrounded by an envelope.

Calyx 1. A boat-shaped or cuplike central body of an entoproct or crinoid. The body and tentacles of an entoproct.

2. A cuplike set of ossicles that support the crown of a sea lily or feather star (class Crinoidea, phylum Echinodermata).

CAM plant A plant that uses crassulacean acid metabolism, an adaptation for photosynthesis in arid conditions, first discovered in the family Crassulaceae. Carbon dioxide entering open stomata during the night is converted into organic acids, which release CO_2 for the Calvin cycle during the day, when stomata are closed.

Cambium A lateral meristem in plants. Types of cambiums include vascular, cork, and intercalary.

Cambium In vascular plants, a layer of meristematic tissue that gives rise to the xylem, phloem and (in woody plants) bark.

Cambrian A geological period of the Paleozoic beginning about 590 million years ago and lasting about 85 million years, during which time many divisions of protists occurred in the oceans.

Cambrian explosion A burst of evolutionary origins when most of the major body plans of animals appeared in a relatively brief time in geological history; recorded in the fossil record about 545 to 525 million years ago.

Camerostome Ventral groove in the propodosoma of soft ticks wherein lies the capitulum.

Camp cyclic 3',5' - adenosine monophosphate, an important intracellular second messenger molecule formed from ATP by the enzyme adenylate cyclase. Also used by the slime mold Dictyostelium for extracellular cell-to-cell communication to promote chemotaxis during aggregation and to promote spore cell differentiation during migration and culmination.

Campestral Characteristic of rural locations, especially open country and grasslands.

Campodactyly A dominant trait in which a muscle is improperly attached to bones in the little finger, causing the finger to be permanently bent.

Campodeiform. (applied to a larva) Grub-like, flattened and elongated with well-developed legs and antennae. Many beetle larvae are of this type, and so are those of the lacewings.

Canaliculus A small canal.

Cancellous Having a spongy or porous structure.

Cancer Invasive, out-of-control cell growth that results in malignant tumors.

Candidal vaginitis Vaginitis caused by Candida species.

Candidiasis An infection caused by Candida albicans, a normal commensal of the mouth; also called "thrush";

Cannulation The introduction of a substance to a tissue (e.g., an area of the brain) via hollow electrode, fine tube (cannula), or needle.

Canopy Layer of vegetation elevated above the ground, usually of tree braches and epiphytes. In tropical forests, the canopy may be more than 100 feet above the ground.

Cap All eukaryotes have at the 5' end of their messages a structure called a "cap", consisting of a 7-methylguanosine in 5'-5' triphosphate linkage with the first nucleotide of the mRNA. It is added post-transcriptionally, and is not encoded in the DNA.

Cap site Two usages: In eukaryotes, the cap site is the position in the gene at which transcription starts, and really should be called the "transcription initiation site". The first nucleotide is transcribed from this site to start the nascent RNA chain. That nucleotide becomes the 5' end of the chain, and thus the nucleotide to which the cap structure is attached. In bacteria, the CAP site (note the capital letters) is a site on the DNA to which a protein factor (the Catabolite Activated Protein) binds.

Capacitation Change in mammalian sperm that occurs after exposure to female genital tract making the sperm competent to undergo the acrosome reaction; this change is necessary for penetration of the cumulus matrix and for fertilization. Numerous molecular changes in the sperm are associated with capacitation, but the extent to which each event causes sperm capacitation is uncertain.

Capillaries Small, thin-walled blood vessels that allow oxygen to diffuse from the blood into the cells and carbon dioxide to diffuse from the cells into the blood.

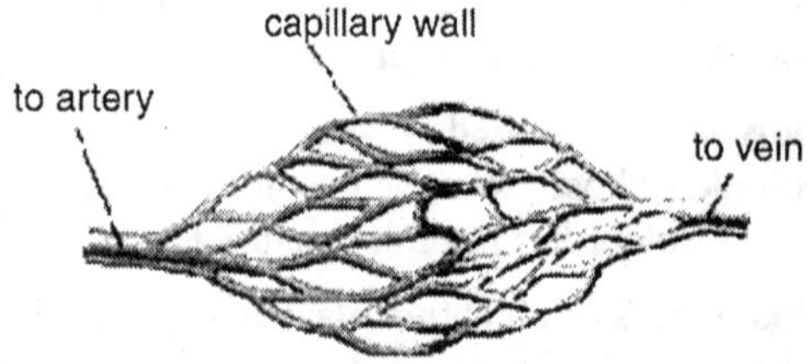

Fig. *Capillaries*

Capillary action The movement of water or any liquid along a surface; results from the combined effect of cohesion and adhesion.

Capillary bed A branching network of capillaries supplied by arterioles and drained by venules.

Capillary pores Small spaces in the soil that become filled with a fluid (such as water) because of the adhesion of particles to the matrix (solid substrate) and the cohesion of the water molecules to themselves.

Capitate. With an apical knob like enlargement.

Capitulum Anterior of two basic body regions of a mite or tick. Also called a gnathosoma.

Capsid The protein shell that encloses the viral genome; rod-shaped, polyhedral, or more completely shaped.

Capsomer The ring-shaped morphological unit of which icosahedral capsids are constructed.

Capsule A fibrous casing surrounding an organ or gland; also a coating for some bacteria which protects them, from the bodies immune system. It is only the variety of Pneumococcus sp. which has a capsule which is able to pass the immune barrier and cause pneumonia

Captacula Tentacles extending from head of scaphopod molluscs, used in feeding.

Captive breeding Raising plants or animals in zoos

or other controlled conditions to produce stock for subsequent release into the wild.

Carabiform larva A larva shaped like the larva of a carabid beetle, that is etiolate, flattened, and with well-developed legs; with no filaments on the end of the abdomen.

Carageenan A polysaccharide extracted from red algae and used for many industrial products.

Carapace Structure formed by posterior and lateral extension of dorsal sclerites of the head in many Crustacea, usually covering and/or fusing with one or more thoracic somites; considered as arising from a fold of head exoskeleton. Also a dorsal sclerotized plate often covering the idiosoma of acarines. The dorsal portion of the shell of a turtle. Formed from a fusion of vertebrae, ribs, and dermal bone.

Carbaminohemoglobin Compound formed by the union of carbon dioxide and hemoglobin.

Carbohydrate Compounds of carbon, hydrogen, and oxygen having the generalized formula $(CH_2O)n$; aldehyde or ketone derivatives of polyhydric alcohols, with hydrogen and oxygen atoms attached in a 2:1 ratio.

Carbon An element that is an essential constituent of all organic compounds.

Carbon cycle Worldwide circulation and reutilization of carbon atoms, chiefly due to metabolic processes of living organisms. Inorganic carbon, in the form of carbon dioxide, is incorporated into organic compounds by photosynthetic organisms; when the organic compounds are broken down in respiration, carbon dioxide is released. Large quantities of carbon are "stored" in the seas and the atmosphere, as well as in fossil fuel deposits.

Carbon dating A procedure used to determine the age of fossils.

Carbon dioxide (CO_2) A colorless gas that is required in the process of photosynthesis.

Carbon film Thin layer of carbon remains of past life found in sedimentary rocks.

Carbon fixation The incorporation of carbon from CO_2 into an organic compound by an autotrophic organism (a plant, another photosynthetic organism, or a chemo-autotrophic bacterium).

Carbon monoxide (CO) Colorless, odorless, nonirritating but highly toxic gas produced by incomplete combustion of fuel, incineration of biomass or solid waste, or partially anaerobic decomposition of organic material.

Carbon sink Places of carbon accumulation, such as in large forests (organic compounds) or ocean sediments (calcium carbonate); carbon is thus removed from the carbon cycle for moderately long to very long periods of time.

Carbon skeleton The central portion of any organic molecule consisting only of the carbon molecule.

Carbon source Originating point of carbon that reenters the carbon cycle; cellular respiration and combustion.

Carbonate bank A narrow (10s of meters), fairly flat, shallow, submarine plateau of carbonate rock, more common from the middle-late Paleozoic to the present, e.g., the Bahama Banks.

Carbonate A mineral composed mainly of calcium (Ca) and carbonate (CO_3) ions, may also include magnesium (Mg), iron (Fe) and others; rock or sediments derived from debris of organic materials composed mainly of calcium and carbonate (e.g., shells, corals, etc.) or from the inorganic precipitation of calcium (and other ions) and carbonate from solution (seawater). For example, limestone or dolomite.

Carbonate platform A broad (100s of meters), flat, shallow submarine expanse of carbonate rock, more common in the early-middle Paleozoic.

Carbonic acid Atmospheric CO_2 dissolved in water to yield H_2CO_3, a weak acid.

Carboniferous A geological period of the Paleozoic beginning approximately 360 million years ago and lasting about 85 million years. During this time forests appeared and became dominant, composed of arborescent club mosses, horsetails, and ferns. Their extinction resulted in the formation of coal deposits.

Carbonyl group A functional group present in aldehydes and ketones, consisting of a carbon atom double-bonded to an oxygen atom.

Carboxyl The acid group of organic molecules _COOH.

Carboxyl group A functional group present in organic acids, consisting of a single carbon atom double-bonded to an oxygen atom and also bonded to a hydroxyl group.

Carboxyl terminus Refers to the COOH end of a *peptide* chain (by custom drawn at the right of a protein sequence)

Carboxysomes Polyhedral inclusion bodies that contain the CO2 fixation enzyme ribulose 1,5-bisphosphate carboxylase; found in cyanobacteria, nitrifying bacteria, and thiobacilli.

Carcinogens Substances that cause cancer.

Cardiac Belonging or relating to the heart.

Cardiac cycle One heartbeat; consists of atrial contraction and relaxation, ventricular contraction and relaxation, and a short pause.

Cardiac muscle A type of muscle that forms the contractile wall of the heart; its cells are joined by intercalated discs that relay each heartbeat.

Cardiac output The volume of blood pumped per minute by the left ventricle of the heart.

Cardioactive glycoside Sugar-containing molecules with an active compound that affects the heartbeat.

Cardiovascular system The human circulatory system consisting of the heart and the vessels that transport blood to and from the heart.

Cardo The basal segment of the maxilla or secondary jaw.

Caries The demineralisation, and breakdown of tooth structure by plaque acids.

Carina A ridge or keel.

Carinate Having a keel, in particular the flying birds with a keeled sternum for the insertion of flight muscles.

Cariogenic Likely to cause caries. Sugar is cariogenic because it supports the growth of plaque.

Carious plaque Types of plaque which are associated with caries .

Carnauba wax A high quality, hard industrial wax extracted from the carnauba palm tree.

Carnivores/ Carnivorous Term applied to a heterotroph, usually an animal, that eats other animals. Carnivores function as secondary, tertiary, or top consumers in food chains and food webs.

Carotenoids Accessory pigments, yellow and orange, in the chloroplasts of plants; by absorbing wavelengths of light that chlorophyll cannot, they broaden the spectrum of colors that can drive photosynthesis.

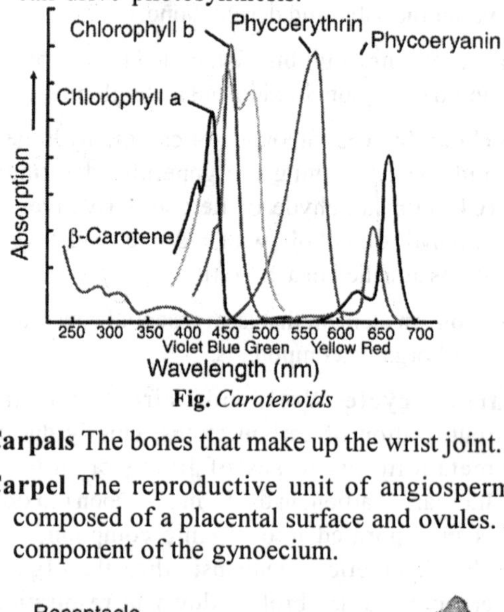

Fig. *Carotenoids*

Carpals The bones that make up the wrist joint.

Carpel The reproductive unit of angiosperms composed of a placental surface and ovules. A component of the gynoecium.

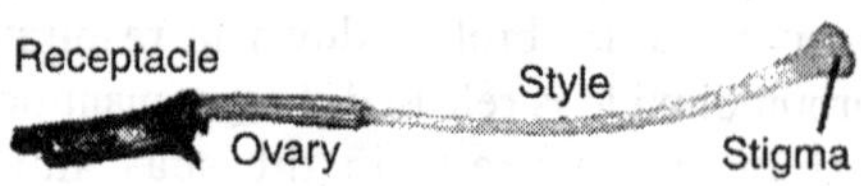

Fig. *Carpel*

Carpellate A unisexual flower having carpels but no stamens.

Carpospore spores produced by the carposporophyte form of red algae

Carposporophyte A diploid generation found in the red algae. It produces non-motile carpospores.

Carrageenan Chemical extracted from red algae that is added to commercial ice creams as an emulsifying agent.

Carrier An infected individual who is a potential source of infection for others and plays an important role in the epidemiology of a disease.

Carrion's disease Bacterial disease transmitted by sand flies.

Carrying capacity The maximum population size that can be regularly sustained by an environment; the point where the population size levels off in the logistic growth model.

Cartilage Type of connective tissue in which cells are located within lacunae and are separated by a semisolid matrix. Provides a site for muscle attachment, aids in movement of joints, and provides support.

Cartilaginous fishes Fishes with a skeleton made of cartilage: sharks, rays, skates, and ratfishes (or chimaeras).

Caryopsis Simple, dry, indehiscent fruit with a single seed that is fused to the ovary wall. Also called a grain. Example: wheat.

Caseous lesion A lesion resembling cheese or curd; cheesy. Most caseous lesions are caused by M. tuberculosis.

Cash crops Crops that are sold rather than consumed or bartered.

Casparian strip In plants, an impermeable waxy

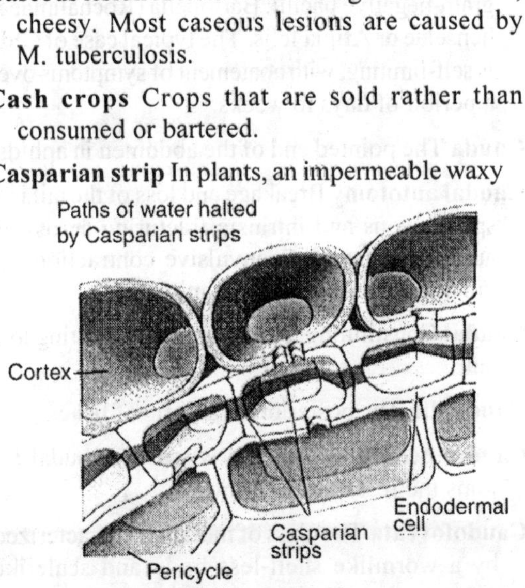

Fig. *Casparian strip*

layer between the cells of the endodermis that stops water and solutes from entering the xylem, except by passing through the cytoplasm of adjacent cells.

Castes Groups of individuals that become irreversibly behaviorally distinct at some point prior to reproductive maturity. One of three or more distinct forms which make up the population among social insects. The usual three castes are queen, drone (male), and worker. The termites and some of the ants have one or more soldier castes as well.

Castor Scent from the gland of a beaver

Casts Fossils formed when water containing minerals leaks into a mold. The minerals harden to form a copy of the original structure or organism.

Casual carrier An individual who harbors an infectious organism for only a short period.

CAT assay An enzyme assay CAT stands for chloramphenicol acetyl transferase, a bacterial enzyme which inactivates chloramphenicol by acetylating it. CAT assays are often performed to test the function of a promoter. The gene coding for CAT is linked onto a promoter (transcription control region) from another gene, and the construct is "transfected" into cultured cells. The amount of CAT enzyme produced is taken to indicate the transcriptional activity of the promoter (relative to other promoters which must be tested in parallel). It is easier to perform a CAT assay than it is to do a Northern blot, so CAT assays were a common method for testing the effects of sequence changes on promoter function. Largely supplanted by the reporter gene luciferase.

Catabolic Chemical reactions that release energy by degrading complex compounds into simpler ones.

Catabolic pathway A metabolic pathway that releases energy by breaking down complex molecules into simpler compounds.

Catabolic reactions Reactions in cells in which existing chemical bonds are broken and molecules are broken down; generally produce energy, involve oxidation, and lead to a decrease in atomic order.

Catabolism That part of metabolism in which larger, more complex molecules are broken down into smaller, simpler molecules with the release of energy.

Catabolite activator protein (CAP) In E. coli, a helper protein that stimulates gene expression by binding within the promoter region of an operon and enhancing the promoter's ability to associate with RNA polymerase.

Catabolite repression Inhibition of the synthesis of several catabolic enzymes by a metabolite such as glucose.

Catadromous fish Fish that spawns in seawater but feed and spends most of its life in estuarine or fresh water

Catalysis An increase in the velocity of a chemical reaction or process produced by the presence of a substance that is not consumed in the net chemical reaction or process.

Catalyst A substance that lowers the activation energy of a chemical reaction by forming a temporary association with the reacting molecules; as a result, the rate of the reaction is accelerated. Enzymes are catalysts.

Cataphyll In cycads, a scale-like modified leaf which protects the developing true leaves.

Catastrophic systems Dynamic systems that jump abruptly from one seemingly steady state to another without any intermediate stages.

Catastrophism Once-popular belief that events in earth history had occurred in the past a sudden events and by processes unlike those operating today. Periods of catastrophic change were followed by long periods of little change. A subgroup, the Diluvialists, contended that Noah's Flood was the last of many floods which had occurred throughout earth history.

Caterpillar The larva of a moth, butterfly, or saw-fly.

Catfacing The injury caused by the feeding of such insects as plant bugs and stink bugs on developing fruit which results in uneven growth and a deformed mature fruit.

Cathaysian terranes A set of small landmasses that developed in tropical to subtropical latitudes on the eastern side of Pangea during the Permian and Triassic, includes modern North China (Sino-Korea), South China (Yangtze), Eastern Qiangtang, Tarim, and Indochina.

Catheter A tubular surgical instrument for withdrawing fluids from a cavity of the body, especially one for introduction into the bladder through the urethra for the withdrawal of urine.

Cation An ion with a positive charge, produced by the loss of one or more electrons.

Cation exchange A process in which positively charged minerals are made available to a plant when hydrogen ions in the soil displace mineral ions from the clay particles.

Catkin Inflorescence of unisexual flowers.

Cat-scratch disease (CSD) A loosely defined syndrome caused by either of the following gram-negative bacilli: Bartonella (Rochalimaea) henselae or Afipia felis. The typical case of CSD is self-limiting, with abatement of symptoms over a period of days to weeks.

Cauda The pointed end of the abdomen in aphids.

Caudal autotomy Breakage and loss of the tail in a spontaneous and intrinsic fracturing across an autotomy plane by convulsive contractions of tail muscle initiated by the animal.

Caudal Constituting, belonging to, or relating to a tail.

Caudal fin The posterior, or tail, fin of fishes.

Caudal peduncle The area where the caudal fin joins the rest of a fish's body.

Caudofoveata The class of molluscs characterized by a wormlike shell-less body and scalelike calcareous spicules; lack eyes, tentacles, statocysts, crystalline style, foot, and nephridia. Deep-water marine burrowers. Chaetoderma.

Caveolae The invaginated vesicles and pits in potocytosis.

Cavum cavus = hollow. Cavity.

CCAAT box (CAT box, CAAT box, other variants) A sequence found in the 5' flanking region of certain genes which is necessary for efficient expression. A transcription factor (CCAAT-binding protein, CBP) binds to this site.

CD95 pathway The CD95 receptor is found on many nucleated eucaryotic cells. When the receptor is bound to a specific ligand (CD95L), the CD95-CD95L complex activates several cytoplasmic proteins that initiate a cellular suicide cascade leading to apoptosis.

CDNA clone "complementary DNA"; a piece of DNA copied from an mRNA. The term "clone" indicates that this cDNA has been spliced into a plasmid or other vector in order to propagate it. A cDNA clone may contain DNA copies of such typical mRNA regions as coding sequence, 5'-untranslated region, 3' untranslated region or poly(A) tail. No introns will be present, nor any promoter sequences (or other 5' or 3' flanking regions). A "full-length" cDNA clone is one which contains all of the mRNA sequence from nucleotide #1 through to the poly(A) tail.

CDNA complementary DNA A DNA molecule which was originally copied from an RNA molecule by *reverse transcription*. The term "cDNA" is commonly used to describe double-stranded DNA which originated from a single-stranded RNA molecule, even though only one strand of the DNA is truly complementary to the RNA.

CDNA library A collection of *cDNA* fragments, each of which has been cloned into a separate *vector* molecule.

Cecum 1. Each arm of the blind-ending Y-shaped digestive tract of trematodes (phylum Platyhelminthes).

2. A region of the vertebrate digestive tract where fermentation can occur. It is located at the proximal end of the large intestine.

Cell An area of the wing bounded by a number of veins. A cell is closed if it is completely surrounded by veins and open if it is bounded partly by the wing margin.

Cell body Portion of a nerve cell that includes a cytoplasmic mass and a nucleolus, and from which the nerve fibers extend.

Cell center A region in the cytoplasm near the nucleus from which microtubules originate and radiate.

Cell culture (tissue culture)The culture of single cells or tissues to form callus and then to develop whole plants asexually.

Cell cycle Complete sequence of steps which must be performed by a cell in order to replicate itself, as seen from mitotic event to mitotic event. Most of the cycle consists of a growth period in which the cell takes on mass and replicates its DNA. Arrest of the cell cycle is an important feature in the reproduction of many organisms, including humans.

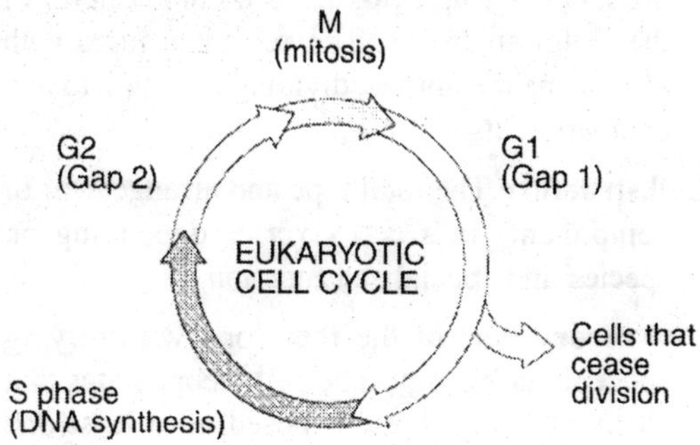

Fig. *Cell cycle*

Cell differentiation The, generally irreversible, process by which a cell or group of cells adopt a specialised function - controlled at the level of genes by regulatory proteins.

Cell enlargement The expansion in volume of a cell typically as a result of increasing turgor pressure.

Cell fractionation The disruption of a cell and separation of its organelles by centrifugation.

Cell junctions Sites on the cell membrane where

cells attach to neighbouring cells. There are three main types. 1. adhering junctions, which anchor cells to each other to resist separation. They may form a belt of adhesions between cells (as between muscle cells) or spot attachments like desmosomes which hold epithelial cells together.

2. tight junctions have no space between the membranes and allow no leakage between cells. They are found between cells of a secreting glands and between endothelial cells of blood vessels to prevent fluid leaking out.

3. gap junctions are channels which allow transfer of small molecules like ions, sugars and amino acids, between cells.

Cell membrane Structure forming outer cell covering and found inside cells, a lipid bilayer with proteins on or within the bilayer, refer to the diagram in notes for cell structure and function.

Cell plate In plants, a membrane-bound space produced during cytokinesis by the vesicles of the Golgi apparatus. The cell plate fuses with the plasma membrane, dividing the cell into two compartments.

Cell structure The total type and arrangement of components in a cell - varies depending on species and specialised function.

Cell theory One of the four (or five) unifying concepts in biology. The cell theory states that all living things are composed of at least one cell and that the cell is the fundamental unit of function in all organisms. Corollaries: the chemical composition of all cells is fundamentally alike; all cells arise from preexisting cells through cell division.

Cell wall A protective layer external to the plasma membrane in plant cells, bacteria, fungi, and some protists. In the case of plant cells, the wall is formed of cellulose fibers embedded in a polysaccharide-protein matrix. The primary cell wall is thin and flexible, whereas the secondary cell wall is stronger and more rigid, and is the primary constituent of wood.

Cell-cycle control system A cyclically operating set of proteins that triggers and coordinates events in the eukaryotic cell cycle.

Cell-mediated immune response Immune response involving cell surfaces only, not antibody production, specifically the TH1 arm of the immune response. Contrast humoral immune response.

Cell-mediated immunity (CMI) Immunity in which antigen is bound to receptor sites on the surface of sensitized T lymphocytes that have been produced in response to prior immunizing experience with that antigen and in which manifestation is through macrophage response with no intervention of antibody.

Cellular differentiation The structural and functional divergence of cells as they become specialized during a multicellular organism's development; dependent on the control of gene expression.

Cellular immune response Binding of antigen with receptor sites on sensitized T lymphocytes to cause release of lymphokines that affect macrophages, a direct response with no intervention of antibody. Also, the entire process by which the body responds to an antigen, resulting in a condition of cell-mediated immunity.

Cellular respiration The most prevalent and efficient catabolic pathway for the production of ATP, in which oxygen is consumed as a reactant along with the organic fuel.

Cellular slime molds Slime molds with a vegetative phase consisting of amoeboid cells that aggregate to form a multicellular pseudoplasmodium; they belong to the division Acrasiomycota.

Cellulase An enzyme that breaks down cellulose.

Cellulitis A diffuse spreading infection of subcutaneous skin tissue caused by streptococci, staphylococci, or other organisms. The tissue is inflamed with edema, redness, pain, and interference with function.

Cellulose Carbohydrate polymer of the simple sugar glucose. It is found in the cell walls of plants and green algae, as well as dinoflagellates. Cellulose is the most abundant compound on earth that is manufactured by living things.

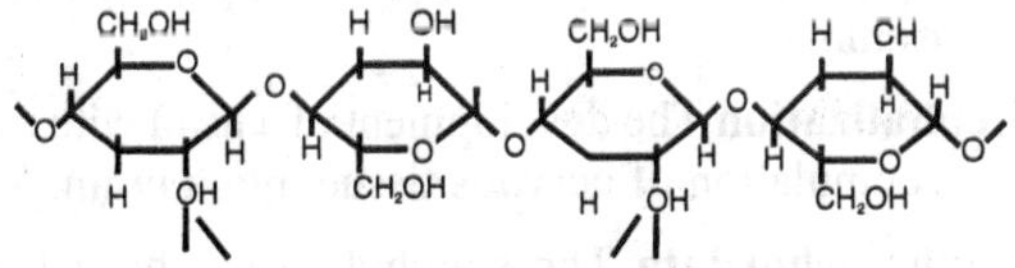

Hydrogen bonding to other cellulose molecules can occur at these points

Fig. *Cellulose*

Celsius scale A temperature scale (°C) equal to 5/9 (°F – 32) that measures the freezing point of water at 0°C and the boiling point of water at 100°C.

Cement glands Glands in a male acanthocephalan that produce secretions sealing the female reproductive tract after copulation.

Cement layer A thin laver on the surface of insect cuticles formed by the hardened secretion of the dermal glands.

Cementation Compacting and natural "gluing together" of sedimentary deposits to form sedimentary rocks.

Cementoblasts Cells of mesenchyme origin, induced by proteins from cells of ectodermal origin, to form a layer of cementum around the roots of teeth.

Cemento-enamel junction The junction between the enamel covering the crown of the tooth and the cementum covering its root. Often referred to as the CEJ.

Cementum A thin layer of bone-like material covering the roots of teeth and sometimes the enamel surface, containing both extrinsic and intrinsic fibres.

Cenozoic A geological era beginning 65 million years ago and extending to the present, characterized by major adaptive radiation of specialized life forms and formation of biomes.

Cenozoic era The period of geologic time beginning after the end of the Mesozoic Era 65 million years ago and encompassing the present. Commonly referred to as the age of mammals.

Centimorgan (CM) A unit of measure of *recombination* frequency. One centimorgan is equal to a 1% chance that a marker at one genetic *locus* will be separated from a marker at a second locus due to *crossing over* in a single generation. In human beings, 1 centimorgan is equivalent, on average, to 1 million *base pairs*.

Central bud The main bud in a cluster located at one position.

Central cell Also known as the polar cell, this is the binucleated cell in the center of an embryo sac containing the two haploid polar nuclei.

Central dogma The relationship among the steps from DNA to the production of a protein. The synthesis of messenger RNA from DNA, and the movement mRNA out of the nucleus (transcription). Carries the genetic code into the cytoplasm. Within the cytoplasm, ribosomes and transfer RNA convert this genetic code in messenger RNA into a protein (translation).

Central nervous system The brain and spinal cord. The nerves which leave the spinal cord and brain comprise the peripheral nervous system.

Central place foragers Animals that collect more than one food item at a time and bring the items to a central location for storage or feeding to offspring.

Central rift valley A depression in the mid-ocean ridge.

Central vacuole A membrane-enclosed sac that takes up most of the volume of a mature plant cell.

Centrifugal force The force that tends to push a body away from the center of rotation.

Centriole Microtubule-based structure that divides prior to mitosis; with the pericentriolar material constitutes the centrosome it is associated with the poles of the spindle apparatus during karyokinesis.

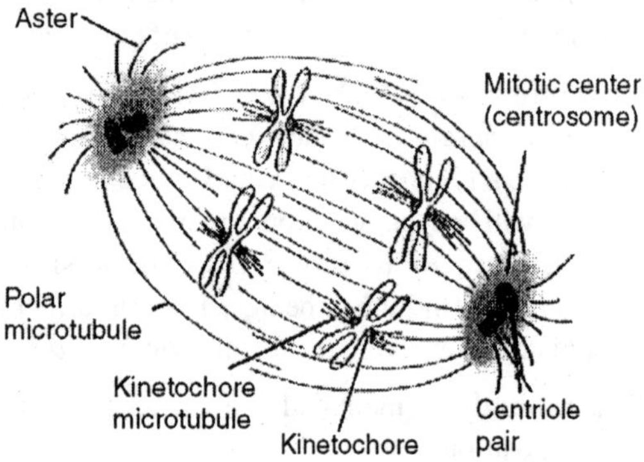

Fig. *Centriole*

Centrolecithal egg Type of egg found in many arthropods, in which the nucleus is located centrally in a small amount of nonyolky cytoplasm, surrounded by a large mass of yolk. After fertilization and some nuclear divisions, the nuclei migrate to the periphery to proceed with superficial cleavage, the yolk remaining central.

Centrolecithal Pertaining to an insect egg with the yolk concentrated in the center.

Centromere Constricted region of a mitotic chromosome that holds sister chromatids together; also the site on the DNA where the kinetochore forms and then captures microtubules from the mitotic spindle.

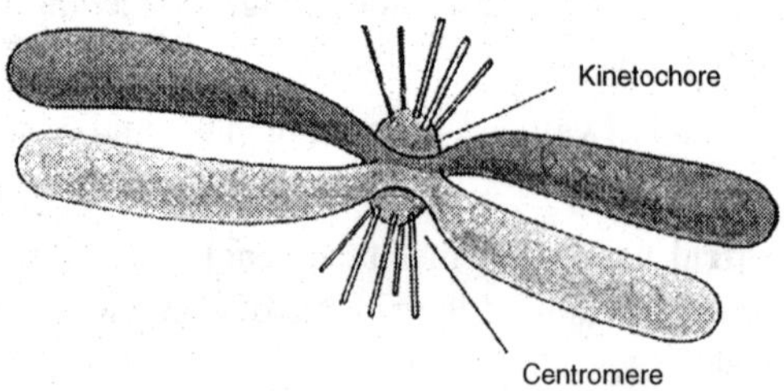

Fig. *Centromere*

Centrosome Minute body in cytoplasm of many plant and animal cells that contains one or two centrioles and is the center of dynamic activity in mitosis.

Cephalaspidomorphi The class of vertebrates characterized by the absence of paired appendages and the presence of sucking mouthparts with teeth and a rasping tongue. Lampreys.

Cephalic Having to do with, or toward, the head of an animal.

Cephalization The development of a head with an accumulation of nervous tissue into a brain.

Cephalochordata The subphylum of chordates whose members possess a laterally compressed, transparent body. They are fishlike and possess all four chordate characteristics throughout life. Amphioxus.

Cephalogaster Contractile organ in adult epicaridean isopods that functions in sucking blood and perhaps in respiration.

Cephalon In trilobites, the head shield bearing the eyes, antennae, and mouth.

Cephalosporin A group of b-lactam antibiotics derived from the fungus Cephalosporium, which share the 7-aminocephalosporanic acid nucleus.

Cephalothorax A body division found in many Arachnida and higher Crustacea, in which the head is fused with some or all of the thoracic segments.

Cerata Fingerlike projections along the dorsal sides of some nudibranches.

Cercaria Juvenile digenetic trematode, produced by asexual reproduction within a sporocyst or redia. Cercaria are freeswimming and have a digestive tract, suckers, and a tail. They develop into a metacercaria.

Cerci The paired appendages, often very long, which spring from the tip of the abdomen in many insects.

Cercomer Posterior, knoblike attachment on a procercoid or cysticercoid. It usually bears the hooks of the oncosphere.

Cereals Members of the grass family that yield a large amount of grain rich in carbohydrates, fats, proteins, and vitamins.

Cerebellum That part of the brain concerned with fine motor coordination and body movement, posture, and balance; is part of the hindbrain and is attached to the rear portion of the brain stem.

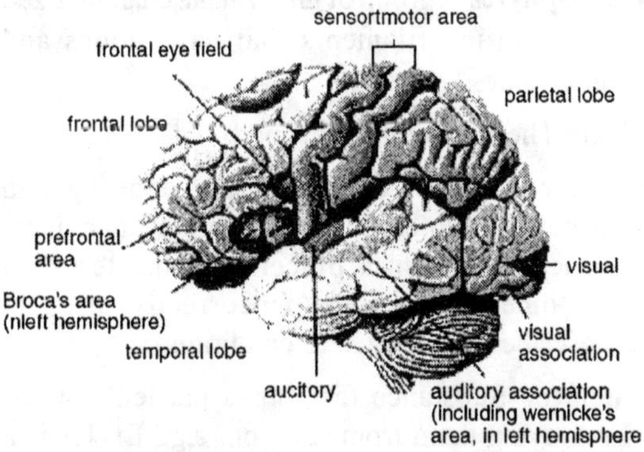

Cerebral cortex The outer layer of gray matter in the cerebrum; consists mainly of neuronal cell bodies and dendrites in humans; associated with higher functions, including language and abstract thought.

Cerebrum The main portion of the brain, occupying the upper part of the cranial cavity; its two hemispheres are united by the corpus callosum; forms the largest part of the central nervous system in mammals.

Cerumen A mixture of wax and propolis used by social bees in nest construction.

Cervical Concerning the neck region, just behind the head.

Cervix The lower neck of the uterus that opens into the vagina.

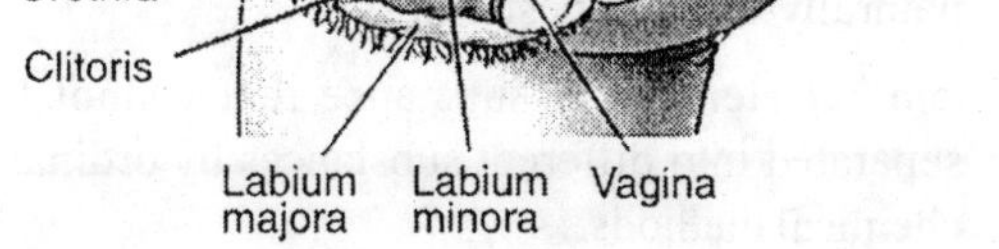

Fig. *Cervix*

Cestoidea The class of platyhelminthes that has members that are all parasitic with no digestive tract; have great reproductive potentials. Tapeworms.

CetaceansMarine mammals with anterior flippers, no posterior limbs, and a dorsal fin: whales, dolphins, and porpoises.

Chaetae Stiff bristles characteristic of annelids.

Chaetotaxy Taxonomic study of the location and arrangement of bristles on an insect. Especially important in the order Diptera.

Chaff Bracts surrounding a cereal grain; removed during threshing.

Chagas' disease Disease of humans and other mammals caused by Trypanosoma cruzi.

Chagoma Reddish nodule that forms at the site of entrance of Trypanosoma cruzi into the skin.

Chain reaction A self-sustaining reaction in which the fission of nuclei produces subatomic particles that cause the fission of other nuclei.

Chalaza One of a pair of twisted, cords of albumin found at each end of an egg, joining the shell membrane to the yolk and supporting the yolk centrally within the shell.Ovule end opposite the micropyle.

Chalimus Specialized, parasitic copepodid, found in the copepod order Siphonostomatoida; attached to its host by an anterior "frontal filament" that is secreted by the frontal gland.

Chalk A soft compact calcite, CaCO3, with varying amounts of silica, quartz, feldspar, or other mineral impurities, generally gray-white or yellow-white and derived chiefly from fossil seashells.

Chancre The primary lesion of syphilis, occurring at the site of entry of the infection.

Chancroid A sexually transmitted disease caused by the gram-negative bacterium Haemophilus ducreyi. Worldwide, chancroid is an important cofactor in the transmission of the AIDS virus. Also known as genital ulcer disease due to the painful circumscribed ulcers that form on the penis or entrance to the vagina.

Channels Transport proteins that act as gates to control the movement of sodium and potassium ions across the plasma membrane of a nerve cell.

Chaotic systems Systems that exhibit variability, which may not be necessarily random, yet whose complex patterns are not discernible over a normal human timescale.

Chaparral A type of vegetation characterized by low, thickly growing evergreen shrubs or bushes with flat, broad leaves and interlacing branches; the typical natural growth of many areas with a climate of cool moist winters and long dry summers, as in much of the western United States.

Chaperone proteins A series of proteins present in the endoplasmic reticulum which guide the proper folding of secreted proteins through a complex series of binding and release reactions.

Character A component of phenotype (including specific molecular, morphological, behavioral or other features) used by systematists to diagnose species or higher taxa, or to evaluate phylogenetic relationships among different species or higher taxa, or relationships among populations within a species.

Character displacement A pattern in which two species with overlapping ecological requirements differ more when they co-occur than when they do not. The difference is usually in a morphological feature related to resource exploitation, as in the case of head size, which may be related to prey size

Chargaff's rules Named for biochemist Erwin Chargaff, these rules describe the relationship of the amounts of the four nitrogenous bases which are found in DNA molecules. Specifically, Chargaff's rules state that the amount of adenine equals the amount of thymine, and that the amount of guanine equals the amount of cytosine in any DNA molecule.

Charging. In protein synthesis, a reaction catalyzed by tRNA synthetase, in which an amino acid is attached to its particular tRNA molecule.

Charophytes A group of green algae characterized by calcified filaments that have nodes and internodes.

Chela The claw of an arthropod.

Chelate Condition of an arthropod appendage in which the subterminal podomere bears a distal process to form a pincer with the terminal podomere; sometimes (incorrectly) used to describe the subchelate condition.

Chelator Substance that binds particular ions, removing them from solution, e.g., EDTA is a chelator of divalent cations such as Mg^{++}.

Chelicera The first pair of appendages of a chelicerate arthropod. Originally a short clawed appendage, the chelicerae of many arachnids are highly modified for feeding; in spiders, for instance, they are modified into poisonous fangs.

Chelicerata The subphylum of arthropods whose members have a body that is divided into prosoma and opisthoma. The first pair of appendages are feeding appendages called chelicerae. Spiders, scorpions, mites, and ticks.

Chelipeds Pincerlike first pair of legs in most decapod crustaceans; specialized for seizing and crushing.

Chemical bond An attraction between two atoms resulting from a sharing of outer-shell elctrons or the presence of opposite charges on the atoms; the bonded atoms gain complete outer electron shells.

Chemical control Pest management practices which rely upon the application of synthetic or naturally-derived pesticides.

Chemical element A substance that cannot be separated into different substances by ordinary chemical methods.

Chemical equilibrium In a reversible chemical reaction, the point at which the rate of the forward reaction equals the rate of the reverse reaction.

Chemical oxygen demand (COD) The amount of chemical oxidation required to convert organic matter in water and wastewater to CO2.

Chemical reaction A process leading to chemical changes in matter; involves the making and/or breaking of chemical bonds.

Chemical synapse A synapse at which neurotransmitters released by one neuron diffuse across an extracellular gap to influence a second neuron's activity.

Chemiosmosis The process whereby a proton gradient and an electrochemical gradient are generated by electron transport and then used to drive ATP synthesis by oxidative phosphorylation.

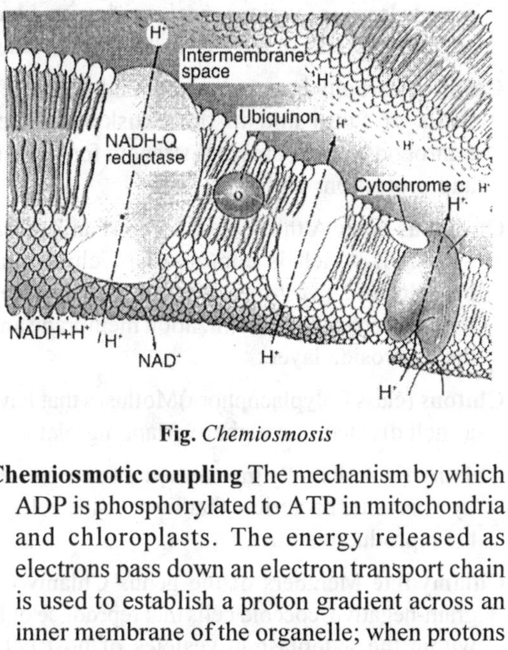

Fig. *Chemiosmosis*

Chemiosmotic coupling The mechanism by which ADP is phosphorylated to ATP in mitochondria and chloroplasts. The energy released as electrons pass down an electron transport chain is used to establish a proton gradient across an inner membrane of the organelle; when protons subsequently flow down this electrochemical gradient, the potential energy released is captured in the terminal phosphate bonds of ATP.

Chemiosmotic hypothesis The hypothesis that a proton gradient and an electrochemical gradient are generated by electron transport and then used to drive ATP synthesis by oxidative phosphorylation.

Chemistry The science dealing with the elements and atomic relations of matter, and various elemental compounds. The study of the properties of substances and how substances react with one another.

Chemoautotroph An organism that needs only carbon dioxide as a carbon source but that obtains energy by oxidizing inorganic substances.

Chemolithoautotroph An organism that synthesizes organic matter by oxidizing inorganic compounds.

Chemolithotrophic autotrophs Microorganisms that oxidize reduced inorganic compounds to derive both energy and electrons; CO2 is their carbon source. Also called chemolithoautotrophs.

Chemoorganotrophic heterotrophs Organisms that use organic compounds as sources of energy, hydrogen, electrons, and carbon for biosynthesis.

Chemoreceptors Special protein receptors in the plasma membrane or periplasmic space that bind chemicals and trigger the appropriate chemotaxic response.

Chemostat A continuous culture apparatus that feeds medium into the culture vessel at the same rate as medium containing microorganisms is removed; the medium in a chemostat contains one essential nutrient in a limiting quantity.

Chemosynthesis The process whereby chemical energy is used to make organic compounds from inorganic compounds. One example is the oxidation of ammonia to nitrite by nitrifying bacteria. Compare *photosynthesis*.

Chemosynthetic Applied to autotrophic bacteria that use the energy released by specific inorganic reactions to power their life processes, including the synthesis of organic molecules.

Chemotaxis The movement of cells in response to chemical messengers. The movement of neutrophils and macrophages into damaged tissues is brought about by signals released by damaged tissues, and bacterial products.. The term applies to the movement of any organism attracted by a specific chemical, which may be a suitable nutrient.

Chemotaxonomy Using the identification of groups of chemical compounds as genetic indicators in the establishment of taxonomic relationships.

Chemotherapeutic agents Compounds used in the treatment of disease that destroy pathogens or inhibit their growth at concentrations low enough to avoid doing undesirable damage to the host.

Chemotherapy The use of chemical substances to treat people suffering from malignant cancers. The chenotherapeutic agents are usually designed to interfere with mitosis in some way, thus slowing the rate of division of the malignant cell mass.

Chemotroph An organism that derives nourishment from inorganic substances without using chlorophyll.

Chemotrophs Organisms (usually bacteria) that derive energy from inorganic reactions; also known as chemosynthetic.

Chert Hard, dense sedimentary rock, composed of interlocking quartz crystals and possibly amorphous silica (opal). The origin of the silica is normally biological, from diatoms, radiolaria or sponge spicules. Synonymous with flint.

Chiasma A decussation or X-shaped crossing; the places where pairs of homologous chromatids remain in contact during late prophase to anaphase of the first meiotic division. The chiasma indicates where an exchange of homologous segments has taken place between nonsister chromatids by crossing over.

Chickenpox A highly contagious skin disease, usually affecting 2- to 7-year-old children; it is caused by the varicella-zoster virus, which is acquired by droplet inhalation into the respiratory system.

Chief cel l Cell of a gastric gland that secretes various digestive enzymes, including pepsinogen.

Chigger Mite of the family Trombiculidae. Also, sometimes applied to Tunga penetrans, the chigoe flea.

Chilopoda The class of uniramous arthropods whose members have one pair of legs per segment and whose body is oval in cross section. Centipedes.

Chimera An organism, usually a plant, with a genotype containing genes from two different species.

Chinampas Long, narrow strips of land bordered on three sides by irrigation canals which can produce several crops per year.

Chironomids Minute, long-legged nonbiting two-winged flies with piercing mouthparts; the aquatic larvaş of various species are green, blue, yellow, colorless, or red type called bloodworms.

Chitin A carbohydrate polymer found in the cell walls of fungi and in the exoskeletons of arthropods, which provides strength for support and protection; chitinous-

Chitinous layer A thick, clear layer of an *Ascaris* egg shell which is composed of chitin and protein and provides a protective coating located between the fertilization membrane and the ascaroside layer.

Chitons (class Polyplacophora)Molluscs that have a shell divided into eight overlapping plates.

Chlamydia A sexually transmitted disease caused by a parasitic bacterium that lives inside cells of the reproductive tract.

Chlamydiae Members of the genus Chlamydia: gram-negative, coccoid cells that reproduce only within the cytoplasmic vesicles of host cells using a life cycle that alternates between elementary bodies and reticulate bodies.

Chlamydial pneumonia A pneumonia caused by Chlamydia pneumoniae. Clinically, infections are mild and 50% of adults have antibodies to the chlamydiae.

Chlamydospore An asexually produced, thick-walled resting spore formed by some fungi.

Chloragogen tissue Cells covering the dorsal blood vessel and digestive tract of annelids; function in glycogen and fat synthesis and urea formation.

Chloramphenicol A broad-spectrum antibiotic that is produced by Streptomyces venezuelae or synthetically; it binds to the large ribosomal subunit and inhibits the peptidyl transferase reaction.

Chlorenchyma Parenchyma tissue containing chloroplasts as in the leaf mesophyll.

Chloride cells Cells in the gills of fishes that are involved in the excretion of excess salts.

Chlorinated hydrocarbons (or organochlorides)Nonbiodegradable and synthetic chemical pollutants used as pesticides.

Chlorinity Grams of chloride ions per 1000 grams of seawater

Chlorocruorin A greenish iron-containing respiratory pigment dissolved in the blood plasma of certain marine polychaetes.

Chlorofluorocarbons (CFCs) Chemical substances used in refrigerators, air conditioners, and solvents that drift to the upper stratosphere and dissociate. Chlorine released by CFCs reacts with ozone, eroding the ozone layer.

Chlorogogen cells Modified peritoneal cells, greenish or brownish, clustered around the digestive tract of certain annelids; apparently they aid in elimination of nitrogenous wastes and in food transport.

Chlorophyll A green pigment located within the chloroplasts of plants; chlorophyll a can participate directly in the light reactions, which convert solar energy to chemical energy. The green-colored pigment that absorbs light during photosynthesis, often found in plants, algae, and some bacteria; it includes a porphyrin ring, and often has a long hydrophobic tail. The pigment in green plants that absorbs solar energy.

Chlorophyta A division of protists commonly referred to as the green algae. May be unicellular, multicellular, or coenocytic.

Chloroplasts Disk-like organelles with a double membrane found in eukaryotic plant cells; contain thylakoids and are the site of photosynthesis. ATP is generated during photosynthesis by chemiosmosis.

Choanocyte One of the flagellate collar cells that line cavities and canals of sponges.

Choanomastigote Like a promastigote but with the flagellum emerging from a collarlike process, as in Crithidia spp.

Cholecystokinin A hormone secreted in the duodenum that causes the gallbladder to release bile and the pancreas to secrete lipase.

Cholera An acute infectious enteritis, endemic and epidemic in Asia, which periodically spreading to the Middle East, Africa, Southern Europe, and South America; caused by Vibrio cholerae.

Choleragen The cholera toxin; an extremely potent protein molecule elaborated by strains of Vibrio cholerae in the small intestine after ingestion of feces-contaminated water or food. It acts on epithelial cells to cause hypersecretion of chloride and bicarbonate and an outpouring of large quantities of fluid from the mucosal surface.

Cholesterol A steroid that forms an essential component of animal cell membranes and acts as a precursor molecule for the synthesis of other biologically important steroids.

Cholinergice Cell receptors specific for the neurotransmitter acetyl choline. Cholinergic receptors are found at neuromuscular junctions of muscle fibres and at all the synaptic junctions of the parasympathetic nervous system. They are also found at the pre- ganglionic synapse of the sympathetic nervous system.

Chondrichthyes The class of vertebrates whose members are fishlike, possess paired appendages and a cartilaginous skeleton, and lack a swim bladder. Skates, rays, and sharks.

Chondrin A protein-carbohydrate complex secreted by chondrocytes; chondrin and collagen fibers form cartilage.

Chondroblast A cell which gives rise to cartilage.

Chondrocyte A mature chondroblast located in a lacunae and surrounded by a matrix. These cells make up the cellular component of cartilage.

Chondroitin sulphate The major glycosaminoglycan of cartilage , the other being keratan sulphate.

Chorda A cord or string.

Chordamesoderm Tissue in the amphibian gastrula that forms between ectoderm and endoderm in the dorsal lip region of the blastopore; develops into the mesoderm and notochord.

Chordate An animal with a notochord (a cartilaginous rod that extends the length of the body), dorsal hollow nerve cord (a fluid-filled tube that runs the length of the body), gill slits or pouches, and a tail at some stage in its life cycle.

Chorion One of the four extraembryonic membranes of amniotes; it forms from the somatopleure (ectoderm and somatic mesoderm). In birds and reptiles, the membrane adheres to the shell and is highly vascularized to serve in gas exchange. In mammals, it forms the fetal contribution to the placenta, made by trophoblastic tissue and extraembryonic mesoderm, containing blood vessels that allow exchange of materials with maternal circulation.

Chorionic somatomammotropin Aka placental lactogen, a hormone that promotes maternal breast development during pregnancy.

Chorionic villus sampling (CVS) A technique for diagnosing genetic and congenital defects while the fetus is in the uterus. A small sample of the fetal portion of the placenta is removed and analyzed.

Chorioptic mange Disease caused by mites of the genus Chorioptes.

Choroid Delicate, highly vascular membrane; in vertebrate eye; the layer between the retina and sclera.

Chromaffin tissue Specialized endocrine cells located near the kidneys.

Chromatid A replicated chromosome joined to its sister chromatid by the centromere; separates and becomes daughter chromosome at anaphase of mitosis or anaphase of the second meiotic division.

Chromatin A complex of DNA and protein in eukaryotic cells that is dispersed throughout the nucleus during interphase and condensed into chromosomes during meiosis and mitosis.

Chromatoid bar Masses of RNA, visible with light microscopy, in young cysts of Entamoeba spp.

Chromatophore Pigment cell, usually in the dermis, in which usually the pigment can be dispersed or concentrated.

Chromista In some classification systems, a kingdom consisting of brown algae, golden algae, and diatoms.

Chromoblastomycosis A chronic fungal infection of the skin, producing wartlike nodules that may ulcerate. It is caused by the black molds Phialophora verrucosa or Fonsecaea pedrosoi.

Chromogen A coloriess substrate that is acted on by an enzyme to produce a colored end product.

Chromomere One of the chromatin granules of characteristic size on the chromosome; may be identical with a gene or a cluster of genes.A convoluted thread in prophase of mitosis or the central thread in a chromosome.

Chromophore group A chemical group with double bonds that absorbs visible light and gives a dye its color.

Chromoplast Membrane-bound organelle containing pigments other than chlorophyll; the pigments may be yellow, orange, or red.

Chromosomal mutations Changes to the chromosome number or structure. Chromosomal mutations may result in whole chromosomes being absent or extra chromosomes being present (as in Trisomy 21), or to sizeable portions of a chromosome being absent or in the wrong location.

Chromosomal puff Expanded region of a polytene chromosome indicative of active messenger RNA synthesis.

Chromosome crossover During meiosis, when DNA replication has occurred and homologous chromosomes are paired, the chromatids may intertwine and break. This process is known as crossing over. Enzymes repair these breaks, but not always in the correct manner to restore each chromatid to its previous location. Chromatids, therefore, can 'swap' between different DNA strands. The breaking and repair of chromosomes in this way can lead to chromosomal mutations.

Chromosome map A diagram of the linear order of the genes on a chromosome.

Chromosome Rodlike structure that appears in the nucleus of a cell during mitosis; contains the genes responsible for heredity. Structure composed of a very long DNA molecule and associated proteins that carries part (or all) of the hereditary information of an organism.

Chromosome theory of inheritance Holds that chromosomes are the cellular components that physically contain genes; proposed in 1903 by Walter Sutton and Theodore Boveri.

Chromosome walking A technique for cloning everything in the genome around a known piece of DNA (the starting probe). You screen a genomic library for all clones hybridizing with the probe, and then figure out which one extends furthest into the surrounding DNA. The most distal piece of this most distal clone is then used as a probe, so that ever more distal regions can be cloned. This has been used to move as much as 200 kb away from a given starting point (an immense undertaking). Typically used to "walk" from a starting point towards some nearby gene in order to clone that gene. Also used to obtain the remainder of a gene when you have isolated a part of it.

Chronic carrier An individual who harbors a pathogen for a long time.

Chronic effects Long-lasting results of exposure to a toxin; can be a permanent change caused by a single, acute exposure or a continuous, low-level exposure.

Chronic food shortages Long- term undernutrition and malnutrition; usually caused by people's lack of money to buy food or lack of opportunity to grow it themselves.

Chrysalis The pupal case of a butterfly that forms from the exoskeleton of the last larval instar.

Chrysolaminarin The polysaccharide storage product of the chrysophytes and diatoms.

Chrysophyta A division of protists consisting of golden-brown and yellow-green algae as well as the diatoms.

Chrysophytes Protistan division that is referred to as the golden brown algae; includes the diatoms.

Chylomicron A particle of the class of lipoproteins responsible for the transport of cholesterol and triglycerides from the small intestine to tissues after meals.

Chyluria Lymph in the urine, characterized by a milky color.

Chyme Semifluid mass of food materials that pass from the stomach to the small intestine.

Chytrids A group of chytridiomycetes, which are simple terrestrial and aquatic fungi that produce motile zoospores with single, posterior, whiplash flagella. Also considered protists.

Ciguatera A type of poisoning found in tropical fishes. It may result from a toxin produced by a dinoflagellate.

Cilia Cilia are hair-like structures that can beat in synchrony causing the movement of unicellular paramaecium. Cilia are also found in specialize linings in eukaryotes. For example, cilia sweep fluids past stationary cells in the lining of trachea and tubes of female oviduct.

Ciliary comb One of eight bands of cilia fused at the base that is found in comb jellies.

Ciliary creeping The principal means of nemertine locomotion.

Ciliary organelles Organelles of specialized function formed by the fusion of cilia.

Ciliated Bearing minute hairs (cilia).

Ciliates A group of protozoans that have cilia.

Ciliophora The protozoan phylum characterized by members with simple or compound cilia at some stage in their life history; heterotrophs with a well-developed cytostome and feeding organelles; at least one macronucleus and micronucleus present. Examples: Paramecium, Stentor, Vorticella, Balantidium.

Cilium A short, hair-like flagellum that is found in large numbers and used in movement, for pushing food particles, and in other functions.

In moving particles along the cell surface or, in ciliate protozoans, for locomotion.

Cimmerian terranes An archipelago of small landmasses that developed in tropical and subtropical latitudes on the eastern side of Pangea during the Triassic, blocks that comprised it include modern Turkey, Iran, Afghanistan, Tibet, and Malaysia; also called Cimmeria.

Cinclides Small pores in the external body wall of sea anemones for extrusion of acontia.

Circadian Refers to a 24 hour period. A "circadian cycle" would be a 24 hour period and a "circadian rhythm" would refer to the rhythm of activity of a specific mammal.

Circle of poisons Importation of food contaminated with pesticides banned for use in this country but made here and sold abroad.

Circulation cells Air movement in a circular pattern from the earth's surface up into the outer atmosphere and then back down again; produce high and low pressure systems and dictate precipitation patterns at different latitudes.

Circulatory system One of eleven major body organ systems in animals; transports oxygen, carbon dioxide, nutrients, and waste products between cells and the respiratory system and carries chemical signals from the endocrine system; consists of the blood, heart, and blood vessels.

Circulatory system, closed A system that uses a continuous series of vessels of different sizes to deliver blood to body cells and return it to the heart; found in echinoderms and vertebrates.

Circulatory system, open A system in which the circulating fluid is not enclosed in vessels at all times; found in insects, crayfish, some mollusks, and other invertebrates.

Circumcision The removal of all or part of the prepuce or foreskin.

Cirri Any of various slender or filamentous, usually flexible appendages, such as one of the compound organelles composed of groups of fused cilia seen in certain peritrichious ciliate protozoa that are used for locomotion; an eversible penis in flatworms; a fingerlike projection of a polychete parapodium.

Cirripedia The class of crustaceans whose members are sessile and highly modified as adults. Enclosed by calcium carbonate valves. Barnacles.

Cirrus A hairlike tuft on an insect appendage; locomotor organelle of fused cilia; male copulatory organ of some invertebrates.

CIS As used in molecular biology, an interaction between two sites which are located within the same molecule. However, a *cis*-acting protein can either be one which acts only on the molecule of DNA from which it was expressed, *or* a protein which acts on itself (e.g., self-proteolysis).

Cisternae Space between membranes of the endoplasmic reticulum within cells.

Cistron A series of codons in DNA that code for an entire polypeptide chain.

Citric acid cycle A series of chemical reactions in the mitochondrion by which various molecules are oxidized and energy is released from them; Krebs cycle, TCA or tricarboxylic acid cycle.

City A differentiated community with a sufficient

population and resource base to allow residents to specialize in arts, crafts, services, and professional occupations.

Clade A monophyletic taxon; a group of organisms which includes the most recent common ancestor of all of its members and all of the descendants of that most recent common ancestor. From the Greek word "klados", meaning branch or twig.

Cladistics A methodology for reconstructing evolutionary relationships of taxa, both living and extinct, by using the distribution of shared derived characters.

Cladogenesis The development of a new clade; the splitting of a single lineage into two distinct lineages; speciation.

Cladogram A diagram, resulting from a cladistic analysis, which depicts a hypothetical branching sequence of lineages leading to the taxa under consideration. The points of branching within a cladogram are called nodes. All taxa occur at the endpoints of the cladogram.

Cladophyll A stem or branch that resembles a leaf.

Clamp Complex set of sclerotized bars, forming a "pinching" organ on the opisthaptor of a monogenetic trematode.

Clasper A copulatory organ along the inner edge of each pelvic fin in male sharks and other cartilaginous fishes.

Class A category of the classification scheme of living organisms ranking below a phylum and above an order (e.g., Insecta).

Classical (Pavlovian) conditioning A type of learning whereby an unconditioned stimulus (US) that elicits a specific response (unconditioned response [UCR]) is paired with a neutral stimulus (becomes the conditioned stimulus [CS]) so that the response becomes conditioned (CR).

Classical biological control The importation of foreign natural enemies to control previously introduced, or native, pests.

Classical complement pathway The antibody-dependent pathway of complement activation; it leads to the lysis of pathogens and stimulates phagocytosis and other host defenses.

Classical conditioning A type of associative learning; the association of a normally irrelevant stimulus with a fixed behavioral response.

Classification The arrangement of organisms into groups based on mutual similarity or evolutionary relatedness.

Clast An individual grain or constituent of a rock.

Clastic Describes a rock or sediment composed mainly of fragments of preexisting rocks or minerals that have been transported some distance from their place of origin, e.g., sandstone, shale.

Clathrin A protein forming a lattice structure lining the invaginated pits during receptor-mediated endocytosis.

Claustral foundation A way of setting up of a new colony by a queen, or king and queen in the termites, which involves her/them being sealing her/themselves a way in a small chamber and raising the first group of workers entirely (or almost so) on stored body reserves (fat and often the flight muscles).

Clavate Club-shaped, with the distal end swollen: most often applied to antennae.

Clavus Posterior part of the forewing of of heteropteran bugs.

Claw The sharp, usually curved, nail on the foot of an animal or insect. The pincerlike extremity of specific limbs of certain arthropods (e.g., lobster claws).

Clay The smallest soil particle size at less than .002 mm

Cleaning symbiosis The type of symbiosis in which the smaller partner regularly removes parasites from fishes.

Clearance The removal by swallowing, of substance in the mouth. Clearance is dependent on the completeness of swallowing and the rate of flow of saliva.

Clear-cut Cutting every tree in a given area, regardless of species or size; an appropriate harvest method for some species; can be destructive if not carefully controlled.

Cleavage The process of cell division in the zygote to form a multicellular embryo. There is usually no growth of cells (blastomeres) at this time so the cells decrease in volume by about 50% with each subsequent division. The embryo is referred to as a blastula at this time.

Cleavage furrow A constriction of the cell membrane at the equator of the cell that marks the beginning of cytokinesis in animal cells. The cell divides as the furrow deepens.

Cleptoparasitism Where one female uses the resources and nest of another individual (of either the same or a different species) to provide for her young thus usurping the owners efforts and preventing her from using them.

Climacteric rise A point during the ripening process of certain fruit when the respiration rates rise to very high levels.

Climate A description of the long-term pattern of weather in a particular area.

Climatology The study of climates and the factors influencing them.

Climax community A self-perpetuating, more-or-less stable community of organisms that continues as long as environmental conditions under which it developed prevail. The final stage in an ecological succession.

Climber A soft- or non-woody-stemmed vine that clasps the stems or branches of trees and shrubs to raise its foliage and flowers above the ground; a type of growthform.

Cline A pattern of directional, often gradual, change in phenotype or genotype of populations across a geographic transect.

Clitellum In annelids, a swelling of the body towards the head of the animal, where the gonads are located. Both oligochaetes and leeches have a clitellum.

Clitoris A short shaft with a sensitive tip located where the labia minora meet; consists of erectile tissue and is important in female sexual arousal.

Cloaca Posterior chamber of digestive tract in many vertebrates, receiving feces and urogenital products. In certain invertebrates, a terminal portion of digestive tract that serves also as respiratory, excretory, or reproductive tract.

Clonal selection The mechanism that determines specificity and accounts for antigen memory in the immune system; occurs because an antigen introduced into the body selectively activates only a tiny fraction of inactive lymphocytes, which proliferate to form a clone of effector cells specific for the stimulating antigen.

Clone The term "clone" can refer either to a bacterium carrying a cloned DNA, or to the cloned DNA itself.

Cloning site A restriction enzyme site or group of sites (multiple cloning site) located in a vector at the best position for insertion of DNA.

Cloning vector Intentionally designed artificial DNA construct used by molecular biologists to amplify selected pieces of DNA inserted into the construct; examples include plasmid, phage, phagemid, cosmid, fosmid, yeast artificial chromosome (YAC) and bacterial artificial chromosome (BAC). Cloning vectors minimally contain an origin of replication, selectable marker gene (e.g., ampicillin resistance gene), and multiple cloning site containing unique restriction enzyme sites; other useful features may also be present.

Closed canopy A forest where tree crowns spread over 20 percent of the ground; has the potential for commercial timber harvests.

Closed circulatory system A circulatory system in an animal in which blood is confined to vessels throughout its circuit.

closed community A community in which

populations have similar range boundaries and density peaks; forms a discrete unit with sharp boundaries.

Clostridial myonecrosis Death of individual muscle cells caused by clostridia. Also called gas gangrene.

Cloud forests High mountain forests where temperatures are uniformly cool and fog or mist keeps vegetation wet all the time.

Clouds of electrons The distribution of electrons in space around the atomic nucleus.

Club The thickened terminal (farthest from the head) end of the antennae.

Cluster of differentiation molecules (CDs) Functional cell surface proteins or receptors that can be measured in situ from peripheral blood, biopsy samples, or other body fluids. They can be used to identify leukocyte subpopulations. Some examples include interleukin-2 receptor (IL-2R), CD4, CD8, CD25, and intercellular adhesion molecule-1 (ICAM-1).

Clutch A nest of eggs or a brood of newly hatched birds.

Clypeus Lowest part of the insect face, just above the labrum.

Cnidaria The phylum of animals whose members are characterized by radial or biradial symmetry, diploblastic organization, a gastrovascular cavity, and nematocysts. Jellyfish, sea anemones, and their relatives.

Cnidocil Modified cilium on nematocyst-bearing cnidocytes in cnidarians; triggers nematocyst.

Cnidocyst The "stinging cell" of a cnidarian.

Cnidocyte Modified interstitial cell that holds the nematocyst; during development of the nematocyst, the cnidocyte is a cnidoblast.

Coacervate An aggregate of colloidal droplets held together by electrostatic forces.

Coaggregation The collection of a variety of bacteria on a surface such as a tooth surface because of cell-to-cell recognition of genetically distinct bacterial types. Many of these interactions appear to be mediated by a lectin on one bacterium that interacts with a complementary carbohydrate receptor on another bacterium.

Coagulase An enzyme that induces blood clotting; it is characteristically produced by pathogenic staphylococci.

Coagulation Process in which a series of enzymes are activated, resulting in clotting of blood.

Coal A black or brown combustible material composed of carbon (C), various carbon compounds and other minerals such as sulphur (S). The most abundant of the fossil fuels, it was formed through the accumulation of vegetable matter over millions of years in environments (e.g. swamps, deltas) which reduced the rate of decay of the organic material and allowed the preservation of the solar energy absorbed by it when it was growing. When coal is burned it is that energy which is released. Most coal was formed from plant material – the humic coals – but some resulted from the accumulation of finely divided vegetable matter such as spore cases, algae and fungal material – the sapropelic coals. The compression and temperature increase brought about by the deposition of many metres of sediment on top of such organic matter caused the progressive elimination of moisture and volatile constituents and a subsequent increase in the proportion of carbon in the deposit. The classification of coal is based on its carbon content, with peat having least carbon and the proportion increasing through lignite and bituminous coal to anthracite. The increased use of coal as a fuel and an industrial raw material during the Industrial Revolution laid the foundations for modern society. Coal use peaked initially at the time of the First World War, after which it suffered in competition with petroleum and electricity. In the mid-1970s, its position improved again as a result of the 1973 oil crisis, and its use in developing nations such as China and India has continued to grow. At the present rate of use it is estimated that there is enough coal to last for another 300 to 500 years. The environmental costs of coal use have been high, both in extraction and use. Underground mining is costly in human terms and surface mining has despoiled the landscape in many areas. Acid mine drainage is common to both. The use of coal has contributed to such major environmental issues as acid rain, atmospheric turbidity and global warming. Although some improvement has been made in dealing with the environmental problems of coal, near-term solutions are unlikely to be found as long as modern industrial societies continue to depend on coal as an energy source and a raw material.

Coal gasification The heating and partial combustion of coal to release volatile gases, such as methane and carbon monoxide; after pollutants are washed out, these gases become efficient, clean-burning fuel.

Coal washing Coal technology that involves crushing coal and washing out soluble sulfur compounds with water or other solvents.

Coarctate (applied to pupae) Enclosed within the last larval skin, which therefore acts as a cocoon and protects the pupa. Such pupae are found in the flies (Diptera, of the sub-order Cyclorrhapha.).

Coarctate pupa Pupa in which the last larval cuticle is retained as a puparium.

Coastal management The use of coastal resources with the intention of preserving them.

Coastal plain estuary An estuary created by flooding a coastal river valley with seawater.

Coastal reef A coral reef occurring near and parallel to a coastline

Coastal shelf The shallow region of the ocean surrounding a large landmass.

Coat In reference to the eggshell of many cestodes, the portion contributed by the outer envelope, derived from embryonic blastomeres.

Coccidioidomycosis A fungal disease caused by Coccidioides immitis that exists in dry, highly alkaline soils. Also known as valley fever, San Joaquin fever, or desert rheumatism.

Coccolith A small calcareous plate imbedded in the cell wall of coccolithophores, a type of phytoplankton.

Coccolithophorids Unicellular, eukaryotic members of the phytoplankton that have calcareous, button-like structures, or coccoliths.

Coccus Any of various spherical-shaped bacteria.

Coccyx kokkyx = cuckoo. Anthropoid caudal vertebrae.

Cochlea A tubular cavity of the inner ear containing the essential organs of hearing; occurs in crocodiles, birds, and mammals; spirally coiled in mammals.

Co-composting Microbial decomposition of organic materials in solid waste into useful soil additives and fertilizer; often, extra organic material in the form of sewer sludge, animal manure, leaves, and grass clippings are added to solid waste to speed the process and make the product more useful.

Cocoon A case, made partly or completely of silk, which protects the pupa in many insects, especially the moths. The cocoon is made by the larva before it pupates.

Code degeneracy The genetic code is organized in such a way that often there is more than one codon for each amino acid.

Code The code of nucleotides is written in "words" of three letters using an"alphabet" of four "letters". These four components of the code are adenine, thymine, cytosine, and guanine.

Coding sequence The portion of a gene or an mRNA which actually codes for a protein. Introns are not coding sequences; nor are the 5' or 3' untranslated regions (or the flanking regions, for that matter - they are not even transcribed into mRNA). The coding sequence in a cDNA or mature mRNA includes everything from the AUG (or ATG) initiation codon through to the stop codon, inclusive.

Coding strand An ambiguous term intended to refer to one specific strand in a double-stranded gene.

Codominance A condition in which both alleles of a heterozygous pair are expressed independently.

Codon In an mRNA, a codon is a sequence of three nucleotides which codes for the incorporation of a specific amino acid into the growing protein. The sequence of codons in the mRNA unambiguously defines the primary structure of the final protein. Of course, the codons in the mRNA were also present in the genomic DNA, but the sequence may be interrupted by introns.

Codon bias The tendency for an organism or virus to use certain codons more than others to encode a particular amino acid. An important detrminant of codon bias is the guanosine-cytosine (GC) content of the genome. An organism that has a relatively low G+C content of 30% will be less likely to have a G or C at the third position of a codon (wobble position) than a A or T to specify an amino acid that can be represented by more than one codon.

Coefficient of relationship The fraction of genes identical by common descent shared between two individuals.

Coelacanths A group of lobed-fin fossil fishes. Latimeria was discovered alive in 1952.

Coelenterates Cnidaria; a group of aquatic animals represented by hydras, sea anemones, jellyfish, and corals.

Coelenteron Internal cavity of a cnidarian; gastrovascular cavity; archenteron.

Coelom Fluid-filled cavity within the body of an animal; usually refers to a cavity lined with specialized tissue peritoneum in which the gut is suspended. The structure and development of the coelom is an important character for recognizing major groups of animals.

Coelomate An animal whose body cavity is completely lined by mesoderm, the layers of which connect dorsally and ventrally to form mesenteries.

Coelomic fluid The fluid that fills the coelom of echinoderms and other invertebrates.

Coelomocyte Another name for amebocyte; primitive or undifferentiated cell of the coelom and the water-vascular system.

Coelomoduct A duct that carries gametes or excretory products (or both) from the coelom to the exterior.

Coelozoic Living in the lumen of a hollow organ, such as the intestine.

Coenecium, coenoecium The common secreted investment of an ectoproct colony; may be chitinous, gelatinous, or calcareous.

Coenenchyme Extensive mesogleal tissue between the polyps of an alcyonarian (phylum Cnidaria) colony.

Coenocyte Organisms possessing a large mass of cytoplasm with many nuclei.

Coenocytic Condition in which an organism consists of filamentous cells with large central vacuoles, and whose nuclei are not partitioned into separate compartments. The result is a long tube containing many nuclei, with all the cytoplasm at the periphery.

Coenosarc The inner, living part of hydrocauli in hydroids.

Coenurus Tapeworm metacestode in the family Taeniidae, in which several scolices bud from an internal germinative membrane; not enclosed in an internal secondary cyst.

Coenzyme An organic nonprotein molecule, frequently a phosphorylated derivative of a water-soluble vitamin, that binds with the protein molecule (apoenzyme) to form the active enzyme (holoenzyme). Examples include biotin, NAD+, and coenzyme A.

Coevolution The mutual influence on the evolution

of two different species interacting with each other and reciprocally influencing each other's adaptations.

Cofactor Any nonprotein molecule or ion that is required for the proper functioning of an enzyme. Cofactors can be permanently bound to the active site or may bind loosely with the substrate during catalysis.

Cogeneration The simultaneous production of electricity and steam or hot water in the same plant.

Cognition The processes in the minds of animals or their general mental functions, including perception, representation, and memory.

Cohesion-adhesion theory Describes the properties of water that help move it through a plant. Cohesion is the ability of water molecules to stick together (held by hydrogen bonds), forming a column of water extending from the roots to the leaves; adhesion is the ability of water molecules to stick to the cellulose in plant cell walls, counteracting the force of gravity and helping to lift the column of water.

Cohesion-tension theory A theory accounting for the upward movement of water in plants. According to this theory, transpiration of a water molecule results in a negative (below 1 atmosphere) pressure in the leaf cells, inducing the entrance from the vascular tissue of another water molecule, which, because of the cohesive property of water, pulls with it a chain of water molecules extending up from the cells of the root tip.

Cold aclimation response The process by which plants increase their tolerance to freezing by exposure to low, nonfreezing temperatures.

Cold front A moving boundary of cooler air displacing warmer air.

Cold sore A lesion caused by the herpes simplex virus; usually occurs on the border of the lips or nares. Also known as a fever blister or herpes labialis.

Cole crops Crops such as cabbage, broccoli, Brussels sprouts, and other crucifers.

Coleoptile The sheath enclosing the apical meristem and leaf primordia of a germinating monocot.

Coleorhiza Sheath surrounding the radicle (embryonic root) of monocotyledons.

Colicin A plasmid-encoded protein that is produced by enteric bacteria and binds to specific receptors on the cell envelope of sensitive target bacteria, where it may cause lysis or attack specific intracellular sites such as ribosomes.

Coliform bacteria Bacteria that live in the intestines (including the colon) of humans and other animals; used as a measure of the presence of feces in water or soil.

Collagen The most common protein found in the body. It has a fibrous structure and makes up the main organic component of bone and dentine, and the fibres of tendons and ligaments.

Collagenase An enzyme produced by fibroblasts which breaks down collagen fibres. The fibroblast recycles the component amino acids, and secretes new collagen fibres. This process of remodelling occurs throughout life. Osteoclasts also secrete collagenases in order to remove bone matrix. Several bacteria are able to secrete collagenases and are thus able to break down and penetrate through collagen fibres in the periodontal ligament.

Collar cell (or choanocyte)A flagellated, food-trapping cell of sponges.

Collecting duct The location in the kidney where filtrate from renal tubules is collected; the filtrate is now called urine.

Collenchyma One of the three major cell types in plants; are elongated and have thicker walls than parenchyma cells and are usually arranged in strands; provide support and are generally in a region that is growing.

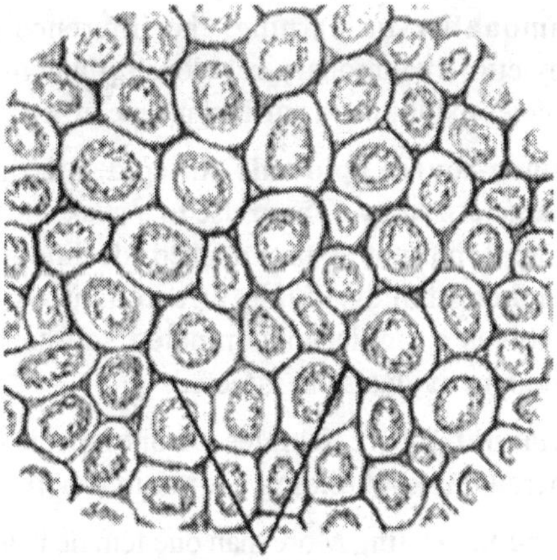

Fig. *Collenchyma*

Collencyte A type of cell in sponges that is star shaped and apparently contractile.

Colleterial glands Female accessory glands in insects that produce a substance to cement eggs together or material for an ootheca.

Colloblast A glue-secreting cell on the tentacles of ctenophores.

Colloid A two-phase system in which particles of one phase are suspended in the second phase.

Colonial 1. Level of organization intermediate between unicellular and multicellular - organisms are composed of multiple cells but fail to exhibit specialization of those cells. Examples: *Volvox*, a colonial alga.

2. Term applied to organisms that occur in a fixed location, with one generation growing atop previous generations, as in coral reefs.

Colonial hypothesis A hypothesis formulated to explain the origin of multicellularity from protist ancestors; animals may have been derived when protists associated together and cells became specialized and interdependent.

Colonies Communities of organisms which have taken up residence in a habitat.

Colonization The establishment of a population in a place formerly unoccupied by that species. Colonization implies successful reproduction in the new area, not simply the presence of a species there.

Colonizer An organism that initiates the biological "conquest" of soil or rock.

Colony A cluster of genetically identical individuals formed asexually from a single colonizing individual.

Colony A cluster or assemblage of microorganisms growing on a solid surface such as the surface of an agar culture medium; the assemblage often is directly visible, but also may be seen only microscopically.

Colony forming units (CFU) The number of microorganisms that can form colonies when cultured using spread plates or pour plates, an indication of the number of viable microorganisms in a sample.

Colony lift Blot taken from colonies (normally bacterial) growing on an agar plate. Normally used for detection of a colony containing a plasmid with a specific inserted sequence.

Colorado tick fever A disease that occurs in the mountainous regions of the western United States. It is caused by an RNA virus of the genus Coltivirus that is spread from ground squirrels, rabbits, and deer to humans by the tick, Dermacentor andersoni. Complications are rare.

Colorless sulfur bacteria A diverse group of nonphotosynthetic proteobacteria that can oxidize reduced sulfur compounds such as hydrogen sulfide. Many are lithotrophs and derive energy from sulfur oxidation. Some are unicellular, whereas others are filamentous gliding bacteria.

Colostrum The first secretion of the mammary glands following the birth of an infant.

Columella A small column of tissue which runs up through the center of a spore capsule. It is present in hornworts, mosses, and some rhyniophytes.

Comb A group of spines on the leg of an insect

specifically used for cleaning other parts of the insects body.

Comb jellies The invertebrates with a gelatinous body, radial symmetry, and eight rows of ciliary combs.

Comb plate One of the plates of fused cilia that are arranged in rows for ctenophore locomotion.

Comb rows Rows of cilia that serve as the locomotor organs of ctenophorans.

Combinatorial biology Introduction of genes from one microorganism into another microorganism to synthesize a new product or a modified product, especially in relation to antibiotic synthesis.

Cometabolism The modification of a compound not used for growth by a microorganism, which occurs in the presence of another organic material that serves as a carbon and energy source.

Commensal Having benefit for one member of a two-species association but neither positive nor negative effect on the other

Commensalisms A symbiotic relationship in which one species benefits and the other is not affected.

Commissure A bridge connecting any two bodies or structures on a body.

Common ancestor An ancestor, often only known from fossil remains, from which two or more different species or lines of species are thought to have evolved.

Common cold An acute, self-limiting, and highly contagious virus infection of the upper respiratory tract that produces inflammation, profuse discharge, and other symptoms.

Common name A regional name for well-known plants; in the language of the region, rather than in Latin, and not necessarily paralleling any scientific name.

Common vehicle transmission The transmission of a pathogen to a host by means of an inanimate medium or vehicle.

Commonality In learning, the inference of a phyletic, evolutionary relationship among the species exhibiting a particular type of learning.

Common-source epidemic An epidemic that is characterized by a sharp rise to a peak and then a rapid, but not as pronounced, decline in the number of individuals infected; it usually involves a single contaminated source from which individuals are infected.

Communal Where females of one species co-operate in nest building but not in brood care.

Communal nesting More than one female in a nest raising the young of more than one female.

Communal resource management systems Resources managed by a community for long-term sustainability.

Communicable disease A disease associated with a pathogen that can be transmitted from one host to another.

Communication Act on the part of one organism (or cell) that alters the probability of patterns of behavior in another organism (or cell) in an adaptive fashion.

Community A group of populations of different species occupying a given place at a given time that are viewed as interdependent. An aggregation of interacting species. Sometimes used to refer to only the assemblage of populations of a particular class of organisms, such as the bird community, the herb community, and so forth.

Community age One of the factors that helps cause the latitudinal diversity gradient. Tropical communities have had more time to evolve because they have been less disrupted by advancing ice sheets and other relatively recent climatic changes.

Community diversity The number of different kinds of organisms living in an area.

Community ecology The study of interactions of all populations living in the ecosystem of a given area.

Community simplification The reduction of overall species diversity in a community; generally caused by human activity.

Community succession The sequential replacement of species in a community by immigration of new species and by local extinction of old ones.

Compact bone The outer dense layer that forms the shaft of the long bones; made up of concentric layers of mineral deposits surrounding a central opening.

Compaction Event in early cleavage-stage mammalian embryo during which blastomeres become tightly joined, forming gap junctions enabling the exchange of ions and small molecules to pass from one cell to the next.

Companion cell A type of plant cell that is connected to a sieve-tube member by many plasmodesmata and whose nucleus and ribosomes may serve one or more adjacent sieve-tube members.

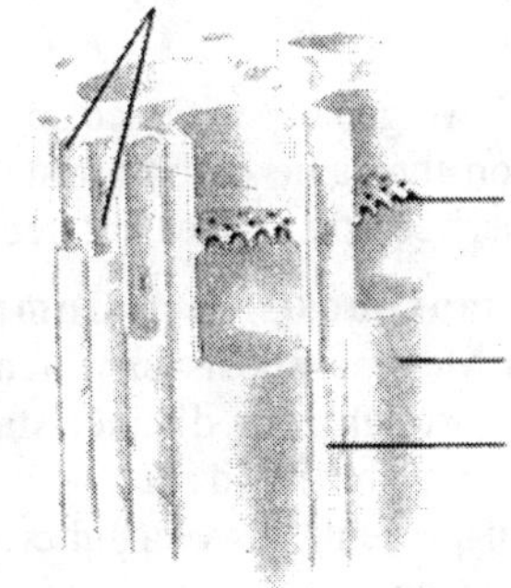

Fig. *Companion cells*

Comparative anatomy The study of animal structure in an attempt to deduce evolutionary pathways in particular animal groups.

Comparative embryology The study of animal development in an attempt to deduce evolutionary pathways in particular animal groups.

Comparative method A comparison of the behavior of two or more species for the purpose of either elucidating some common aspects of the ecology and evolution of behavior or exploring the mechanisms underlying behavior.

Comparative psychologist An ethologist who studies the genetic, neural, and hormonal bases of animal behavior.

Comparative psychology A branch of psychology involving the study of animals. Some comparative psychologists are more concerned with learning, cognition, and intelligence in human and nonhuman animals, while others are indistinguishable from other animal behaviorists in that they explore the causation, development, evolution, and functional aspects of behavior in a broad range of species.

Compartmentalization The division of labor in living cells such that enzymes related to a particular function are packaged and separated from the other cell contents, usually by a membrane.

Compensation depth The depth of the compensation light intensity

Compensation light intensity That light intensity at which oxygen evolved from a photosynthesizing organism equals that consumed in its respiration

Compensation point The condition in a living system in which the uptake of CO_2 equals the release of CO_2; that is, photosynthesis equals respiration.

Competency The ability of a cell to respond to messengers which could cause it to differentiate into a more specialised cell. Some cells, like pericytes remain competent throughout life, whereas others, such as the oral epithelium, are only able to form an tooth bud during the 12th to 16th week of foetal development.

Competent Bacterial cells which are capable of accepting foreign extra-chromosomal DNA. There are a variety of processes by which cells may be made competent.

Competition Some degree of overlap in ecological niches of two populations in the same community, such that both depend on the same food source, shelter, or other resources, and negatively affect each other's survival.

Competitive exclusion A theory that no two populations of different species will occupy the same niche and compete for exactly the same resources in the same habitat for very long; disputed by some ecologists who see biological communities as highly individualistic and variable.

Competitive inhibitor A substance that reduces the activity of an enzyme by entering the active site in place of the substrate whose structure it mimics.

Competitive release Occurs when one of two competing species is removed from an area, thereby releasing the remaining species from one of the factors that limited its population size.

Complement Collective name for a series of proteins that bind in a complex series of reactions to antibody (either IgM or IgG) when the antibody is itself bound to an antigen; produces lysis of cells if the antibody is bound to antigens on the cell surface.

Complement fixation An immune response in which antigen-antibody complexes activate complement proteins.

Complement fixation test Immunological method used to detect presence of antibodies that bind (or fix) complement; a standard diagnostic test for many infections.

Complement system A group of at least 20 blood proteins that cooperate with other defense mechanisms; may amplify the inflammatory response, enhance phagocytosis, or directly lyse pathogens; activated by the onset of the immune response or by surface antigens on microorganisms or other foreign cells.

Complementary DNA (cDNA) A DNA molecule made in vitro using mRNA as a template and the enzyme reverse transcriptase. A cDNA molecule therefore corresponds to a gene, but lacks the introns present in the DNA of the genome.

Complementary nucleotides The bonding preferences of nucleotides, Adenine with Thymine, and Cytosine with Guanine. Also referred to as complementary base pairing.

Complementary sequences *Nucleic acid base sequences* that can form a double- stranded structure by matching *base pairs*; the complementary sequence to G- T- A- C is C- A- T- G.

Complementary strand Two polynucleotide chains in which the pairing of adenine is always with thymine (in DNA) or uracil (in RNA), and guanine is always paired with cytosine.

Complete digestive tract A digestive tube that runs between a mouth and an anus; also called alimentary canal. An incomplete digestive tract has only one opening.

Complete dominance The type of inheritance in which both heterozygotes and dominant homozygotes have the same phenotype.

Complete flower A flower that has sepals, petals, stamens, and carpels.

Complete linkage Two genes positioned so close to one another on the same chromosome that recombination between them does not occur.

Complete metamorphosis or Complex metamorphosis Metamorphosis in which the insect develops through four distinct stages, e.g.., ova or egg, larva, pupa, and adult or imago; the wings (when present) develop internally during the larval stage.

Complete protein A protein that has all of the essential amino acids and in the correct proportions.

Complex camera eye The type of image-forming eye found in squids and octopuses.

Complex life cycle A life cycle that consists of several distinct stages (e.g., larva and adult)

Complex medium Culture medium that contains some ingredients of unknown chemical composition.

Complex viruses Viruses with capsids having a complex symmetry that is neither icosahedral nor helical.

Complexity (ecological)The number of species at each trophic level and the number of trophic levels in a community.

Composite A material made from two or more different types of material which contribute different properties. For example bone is a combination of a resilient fibres of collagen in a brittle matrix (hydroxyapatite).

Composite signal A signal formed by combining two or more simpler signals.

Compost Partially decayed organic matter used in gardening and farming to enrich the soil and increase water holding capacity.

Composting The microbial processing of fresh organic matter under moist, aerobic conditions, resulting in the accumulation of a stable humified product, which is suitable for soil improvement and stimulation of plant growth.

Compound A substance composed of atoms of two or more elements joined by chemical bonds and chemically united in fixed proportions.

Compound eye A type of multifaceted eye in insects and crustaceans consisting of up to several thousand light-detecting, focusing ommatidia; especially good at detecting movement.

Compound fruit A fruit consisting of several individual fruits held together (a multiple fruit) or in which separate carpels of a flower stay together (an aggregate fruit).

Compound leaf A leaf in which the blade forms small leaflets. Compound leaves that have several small leaflets originating from a central axis are termed pinnately compound; example: rose. Compound leaves that have their leaflets originating from a common point are termed palmately compound; example: palm.

Compound pistil A pistil composed of more than one carpel.

Compression Fossil formed when an organism is flattened (compressed) and a thin film of organic material from its body is left in the rock.

Compression wood The reaction wood produced along the lower side of leaning trees, straightening the trunk by expanding and pushing the tree upright.

Compressive strength Ability to withstand a crushing force.

Compromised host A host with lowered resistance to infection and disease for any of several reasons. The host may be seriously debilitated (due to malnutrition, cancer, diabetes, leukemia, or another infectious disease), traumatized (from surgery or injury), immunosuppressed, or have an altered microbiota due to prolonged use of antibiotics.

Concatemer A long DNA molecule consisting of several genomes linked together in a row.

Concentration gradient A regular increase of decrease in the intensity or density of a chemical substance. Cells often maintain concentration gradients of H^+ ions across their membranes. When a gradient exists, the ions or other chemical substances involved tend to move from where they are more concentrated to where they are less concentrated.

Concentricycloidea The class of echinoderms whose members are characterized by two concentric water-vascular rings encircling a disklike body; no digestive system; and internal brood pouches. Sea daisies.

Conchae Thin bones in the nasal chamber.

Concomitant immunity Premunition.

Concretion A hard, rounded mass, commonly of silica, calcite, dolomite, iron oxide, pyrite, or gypsum, that formed within a rock from the precipitation of these minerals around a nucleus, such as a leaf, bone, shell, or fossil, and ranging in diameter from centimeters to meters.

Condensation The aggregation of water molecules from vapor to liquid or solid when the saturation concentration is exceeded.

Condensation nuclei Tiny particles that float in the air and facilitate the condensation process.

Condensation reaction A chemical reaction in which reactant molecules are combined by the removal of a water molecule (a hydrogen from one and a hydroxyl from the other reactant).

Conditional mutations Mutations that are expressed only under certain environmental conditions.

Conditioned response The response to a stimulus that occurs when an animal has learned to associate the stimulus with a certain positive or negative effect.

Conditioned stimulus The previously neutral stimulus that, through classical conditioning, now elicits the conditioned response.

Conduction The conveyance of energy, such as heat, sound, or electricity. The direct transfer of thermal motion (heat) between molecules of the environment and those on the body surface of an animal.

Condylar process The vertical extension of the mandible which ends in the condyle head, the moveable part of the temporomandibular joint.

Cone A vertebrate photoreceptor cell with a tapered outer segment; unlike rod cells, cone cells are usually sensitive

Cone cell (1) In plants, the reproductive structure of a conifer.

(2) In vertebrates, a type of photoreceptor cell in the retina, concerned with the perception of color and with the most acute discrimination of detail.

Confocal scanning laser microscope (CSLM) A light microscope in which monochromatic laser-derived light scans across the specimen at a specific level and illuminates one spot at a time to form an image. Stray light from other parts of the specimen is blocked out to give an image with excellent contrast and resolution.

Conformational epitope An epitope which is dependent upon folding of a protein; amino acid residues present in the antibody binding site are often located at sites in the primary sequence of the protein which are at some distance from each other. The vast majority of B-cell (antibody binding) epitopes are conformational.

Conformer An organism whose physiological state (e.g., body temperature) is identical to, and varies identically with, that of the external environment

Confusion effect The reduced capture efficiency experienced by a predator caused by high densities or swarms of prey.

Congener An organism that is a member of the same genus as another animal.

Congenital Existing at or before birth; acquired at or during birth.

Congenital rubella syndrome A wide array of congenital defects affecting the heart, eyes, and ears of a fetus during the first trimester of pregnancy, and caused by the rubella virus.

Congenital syphilis Syphilis that is acquired in utero from the mother.

Conglomerate A coarse-grained sedimentary rock, with clasts larger than 2 mm.

Congo craton A separate continental plate that rifted from the supercontinent Rodinia in the Late Precambrian; contained a large part of north-central Africa.

Conidiophore Hyphae on which one or many conidia are produced.

Conidiosporangium A structure that produces conidiospores.

Conidiospore An asexual, thin-walled spore borne on hyphae and not contained within a sporangium; it may be produced singly or in chains.

Conidium Asexual reproductive spores of some ascomycetes and imperfect fungi.

Conifer A gymnosperm whose reproductive structure is the cone. Conifers include pines, firs, redwoods, and other large trees.

Coniferous forest A community type dominated by cone-bearing gymnosperms mostly in the northern hemisphere.

Conjugants Complementary mating types that participate in a form of protozoan sexual reproduction called conjugation.

Conjugation Temporary union of two ciliate protozoa while they are exchanging chromatin material and undergoing nuclear phenomena resulting in binary fission. Also, formation of cytoplasmic bridges between bacteria for transfer of plasmids.

Conjugative plasmid A plasmid that carries the genes for sex pili and can transfer copies of itself to other bacteria during conjugation.

Connective A longitudinal cord of nerve fibers connecting successive ganglia.

Connective tissue One of the four main types of collections of cells (tissues) which consists of cells in a matrix of ground substance and fibres. Some connective tissues support structures like blood vessels and glands. Others are more structural, like bone, tendons and cartilage.

Conoid Truncated cone of spiral fibrils located within the polar rings of the suborder Eimeriina.

Consensus sequence A 'nominal' sequence inferred from multiple, imperfect examples. Multiple lanes of shotgun sequence can be merged to show a consensus sequence. The optimal sequence of nucleotides recognized by some factor. A DNA binding site for a protein may vary substantially, but one can infer the consensus sequence for the binding site by comparing numerous examples. For example, the (fictitious) transcription factor ZQ1 usually binds to the sequences AAAGTT, AAGGTT or AAGATT. The consensus sequence for that factor is said to be AARRTT (where R is any purine, i.e. A or G). ZQ1 may also be able to weakly bind to ACAGTT (which differs by one base from the consensus).

Conservation Any biological control practice designed to protect and maintain populations of existing natural enemies.

Conservation biology A field of science that deals with threats to biodiversity. The goals of conservation biology are to investigate human impacts of biodiversity and to develop approaches to prevent extinction through stewardship of entire biological communities.

Conservation of matter In any chemical reaction, matter changes form; it is neither created nor destroyed.

Conservative substitution A nucleotide mutation which alters the amino acid sequence of the protein, but which causes the substitution of one amino acid with another which has a side chain with similar charge/polarity characteristics. The size of the side chain may also be an important consideration. Conservative mutations are generally considered unlikely to profoundly alter the structure or function of a protein, but there are many exceptions.

Conserved sequence A *base sequence* in a DNA molecule (or an *amino acid* sequence in a *protein*) that has remained essentially unchanged throughout evolution.

Consortium A physical association of two different organisms, usually beneficial to both organisms.

Constant proportions, rule of A principle that states that the relative amounts of ions in seawater are always the same.

Constant region (CL and CH) The part of an antibody molecule that does not vary greatly in amino acid sequence among molecules of the same class, subclass, or type.

Constitutive mutant A strain that produces an inducible enzyme continually, regardless of need, because of a mutation in either the operator or regulator gene.

Constraints In optimality models, the limits placed on an animal because of its inherent abilities.

Constructed wetlands Intentional creation of marshland plant communities and their associated microorganisms for environmental restoration or to purify water by the removal of bacteria, organic matter, and chemicals as the water passes through the aquatic plant communities.

Construction gland A gland of wasps producing a size-like substance which enables them to make paper out of wood-pulp.

Constructs New arrangements of genes produced by genetically modifying DNA.

Consumers The higher levels in a food pyramid; consist of primary consumers, which feed on the producers, and secondary consumers, which feed on the primary consumers.

Consumer, in ecological systems A heterotroph that derives its energy from living or freshly killed organisms or parts thereof. Primary consumers are herbivores; higher-level consumers are carnivores.

Consummatory behavior Actions of an animal completing a behavior sequence, as in consuming food, mating, etc.

Consumption The fraction of withdrawn water that is lost in transmission or that is evaporated, absorbed, chemically transformed, or otherwise made unavailable for other purposes as a result of human use.

Contact poison A pesticide that is absorbed through the body wall, as opposed to one that must be ingested.

Contact transmission Transmission of the pathogen by contact of the source or reservoir of the pathogen with the host.

Contagious Capable of being transmitted through direct contact. Also used to describe population distributions that are aggregated, such as in an area.

Contaminative antigen Antigen borne by the parasite that is common to both the host and the parasite but that genetically is of host origin.

Context In communication, stimuli other than the signal that are impinging on the receiver that might alter the meaning of a signal.

Contig A series of two or more individual DNA sequence determinations that overlap. In a sequencing project the contigs get larger and larger until the gaps between the contigs are filled in.

Contig map A map depicting the relative order of a linked *library* of small overlapping clones representing a complete chromosomal segment.

Contiguity The association of events in time with particular reference here to classical conditioning.

Continental boundary current A surface ocean current flowing generally north or south along a continental edge.

Continental crust The Earth's crust that includes both the continents and the continental shelves.

Continental drift The breakup and movement of land masses of the earth. The earth had a single landmass about 250 million years ago. This mass broke apart into continents, which have moved slowly to their present positions.

Continental margin The edge of a continent; the zone between a continent and the deep-sea floor.

Continental rise The gently sloping area at the base of the continental slope.

Continental shelf The part of the continental margin from the coastal shore to the continental slope; usually extending to a depth of about 200 meters and with a very slight slope, roughly 0.1 degrees; includes conetinental and oceanic sediments down to the ocean floor.

Continental slope Part of the continental margin; the ocean floor from the continental shelf to the continental rise or oceanic trench. Usually to a depth of about 200 meters. The continental slope typically has a relatively steep grade, from 3 to 6 degrees.

Continuous culture system A culture system with constant environmental conditions maintained through continual provision of nutrients and removal of wastes.

Continuous feeder Usually slow-moving sessile animals that feed all of the time.

Continuous variable One that can assume any value whatsoever between certain limits, versus a discrete variable that can assume only specific values (usually integers).

Continuous variation Occurs when the phenotypes of traits controlled by a single gene cannot be sorted into two distinct phenotypic classes, but rather fall into a series of overlapping classes.

Contour feathers Feathers that cover the body, wings, and tail of a bird. Contour feathers provide flight surfaces and are responsible for plumage colors.

Contour plowing Plowing along hill contours; reduces erosion.

Contraception The prevention of pregnancy.

Contractile ring The mechanical agent of cytokinesis, made of actin microfilaments. It is located in the cortical cytoplasm and provides the necessary force to split a single cell into two cells.

Contractile root A specialized root, common in bulbs, that contracts, pulling the stem deeper into the ground.

Contractile vacuole In many protists, a specialized vacuole with associated channels designed to collect excess water in the cell. Microtubules periodically contract to force this excess water out of the cell, regulating the cell's osmotic balance.

Contralateral The opposite side as distinct from ipsilateral. Often used to refer to the teeth, joint or muscles on the opposite side from the chewing side.

Contrast In relation to microscopes, the ability to distinguish different densities of structures.

Control group An observed, unmanipulated set of test subjects. In an experimental manipulation, we must also perform the actions involved in that manipulation on a second test group of organisms in order to ascertain whether any effects of the experimental treatment are due to that treatment and not to the actions involved in providing the treatment.

Control rods Neutron-absorbing material inserted into spaces between fuel assemblies in nuclear reactors to regulate fission reaction.

Conurbation An area with more than one million people that is formed by the merger of two or more population centers and that meets other specified requirements.

Convalescent carrier An individual who has recovered from an infectious disease but continues to harbor large numbers of the pathogen.

Convection The act of conveying or transmission. The movement of air (or a liquid) over the surface of a body and contributes to heat loss (if the air is cooler than the body) or heat gain (if air is warmer than the body).

Convection currents Rising or sinking air currents that stir the atmosphere and transport heat from one area to another. Convection currents also occur in water.

Convective mixing The vertical mixing of water masses driven by wind stresses or density changes at the sea surface.

Conventional pollutants The seven substances (sulfur dioxide, carbon monoxide, particulates, hydrocarbons, nitrogen oxides, photochemical oxidants, and lead) that make up the largest volume of air quality degradation.

Convergence Penotypic similarity in distantly related (or unrelated) forms, presumably in response to similar selective pressures.

Convergent evolution The development of similar structures in distantly related organisms as a result of adapting to similar environments and/or strategies of life. Example: wings of birds and insects, the body shape of dolphins, sharks, and the extinct marine reptiles known as ichthyosaurs.

Convergent extension Cell movement resulting in tissue elongation via intercalation of adjacent cells in an epithelial sheet to form a narrower, longer strip of tissue.

Convergent plate boundary The boundary between two plates that are moving toward one another.

Convex Curving outward; the opposite of concave.

Cool deserts Deserts such as the American Great Basin characterized by cold winters and sagebrush.

Cooperation Behavior patterns that are mutually beneficial to the cooperators.

Cooperative breeding A breeding system in which nonparents share in rearing the young.

Cooperativity An interaction of the constituent subunits of a protein causing a conformational change in one subunit to be transmitted to all the others.

Copal Brittle aromatic yellow to red resins of recent or fossil origin, obtained from tropical trees.

Copepod Order of crustaceans found often in the plankton

Copepodid Juvenile stage that succeeds the naupliar stages in copepods, often quite similar in body form to the adult.

Copepods Small, mostly planktonic crustaceans.

Copiotrophic Having a high nutrient level.

Coprolite Fossilized fecal material.

Coprophagy Feeding on dung or excrement as a normal behavior among animals; reinjestion of feces.

Copularium The first chamber built by a newly mated pair of sexual termites.

Copulation The sexual act of a male and female, of the same species, coming together for the purpose of delivering and receiving sperm respectively. The term can also refer to sexual behavior that enhances the possibility of fertilization.

Coracidium Larva with a ciliated epithelium hatching from the egg of certain cestodes; a ciliated free-swimming oncosphere.

Coral knoll (pinnacle)A column of coral within the lagoon of an atoll.

Coral reef A wave-resistant structure resulting from cementation processes and the skeletal construction of hermatypic corals, calcareous algae, and other calcium carbonate-secreting organisms

Coral rubble Coral fragments.

Coralline algae Any red alga that is impregnated with calcium carbonate. Coralline algae often contribute to coral reefs.

Corallite The calcareous skeletal cup in which a coral polyp sits.

Corbicula The pollen basket on the hind leg of many bees, formed by stout hairs on the borders of the tibia.

Cord grasses Salt-tolerant grasses, species of Zostera, that inhabit salt marshes.

Core That portion of the interior of the Earth that lies beneath the mantle, and goes all of the way to the center. The Earth's core is very dense, rich in iron and the source of the magnetic field.

Core region The primary industrial region of a country; usually located around the capital or largest port; has both the greatest population density and the greatest economic activity of the country.

Corepressor A substance that inhibits production of a particular enzyme.

Corer Tubular benthic sampling device that is plunged into the bottom in order to obtain a vertically oriented cylindrical sample.

Coriolis effect The apparent change in direction of a moving object (to the left in the Southern Hemisphere and to the right in the Northern Hemisphere) due to the rotation of the earth.

Corium The main part of the forewing of a heteropteran bug.

Cork A secondary tissue that is a major constituent of bark in woody and some herbaceous plants; made up of flattened cells, dead at maturity; restricts gas and water exchange and protects the vascular tissues from injury.

Cork cambium Meristematic tissue that produces cork cells on its outer surface and phelloderm on its inner surface.

Corm Underground, enlarged, food-storing stem covered by papery leaves.

Cornea The outer transparent coat of the eye.

Corneum Epithelial layer of dead, keratinized cells. Stratum corneum.

Cornicle One of the pair of small tubular outgrowths on the hind end of the aphid abdomen.

Cornified Adjective for conversion of epithelial cells into nonliving, keratinized cells.

Cornucopian fallacy The belief that nature is limitless in its abundance and that perpetual growth is not only possible but essential.

Corolla Petals, collectively; usually the conspicuously colored flower parts.

Corona A crown; an encircling structure. The ciliated organ at the anterior end of rotifers used for swimming or feeding.

Corona radiata A layer of protective follicle cells derived from the cumulus oophorus surrounding the zona pellucida of an oocyte after ovulation.

Coronary Encircling as a crown. 1. Vessels encircling the heart.

2. A condition caused by a decreased blood flow to the heart muscle.]

3. Specifically, a blocked coronary artery or a branch of a coronary artery.

Coronary arteries Arteries that supply the heart's muscle fibers with nutrients and oxygen.

Coronoid process The vertical extension of the mandible anterior to the condyle to which the temporal muscle attaches.

Corpora allata Paired invertebrate endocrine glands, located in the thoracic region, which secrete juvenile hormone that prevents maturation.

Corpora cardiaca Paired invertebrate endocrine glands, located near the heart, which are involved in regulation of metabolism. They receive neurosecretions from the brain and release endocrine products into the hemocoel.

Corpus albicans The fibrous scar tissue formed as a result of the degradation of the corpus luteum in the ovary.

Corpus callosum Tightly bundled nerve fibers that connect the right and left hemispheres of the cerebrum.

Corpus luteum A temporary gland formed from a ruptured Graafian follicle and the surrounding thecal cells after ovulation. If fertilization occurs, the corpus luteum persists and continues to secrete progesterone. If fertilization does not occur, the corpus luteum degenerates.

Corridor A connection between adjacent land areas that allows the passage of fauna form one area to the other.

Cortex 1) The outer part of an organ, e.g., the adrenal cortex, which produces several steroid hormones;

2) in plants, the region of the stem or root between the epidermis and the vascular bundle(s).

Cortical granules/vesicles Membrane-bound structures in the egg, derived from the Golgi apparatus, and found just beneath the plasma membrane. The cortical reaction is the release of their contents (proteases, mucopolysaccharides, and peroxidases) by fusion with the plasma membrane; the changes caused by the cortical reaction help to prevent polyspermy.

Cortical-bone The outer layer of bone which is dense and made up of lamellae.

Cortisol A steroid hormone, produced by the adrenal cortex, that promotes the formation of glucose from protein and fat; also suppresses the inflammatory and immune responses.

Cosmid An artificial hybrid vector into which large (~40kbp) DNA fragments can be inserted; this makes it useful in genomic library preparation. Cosmids replicate as plasmids in the host cell, but are inserted into the bacterium from lambda phage infectious particles, since transformation would be extremely inefficient with plasmids of this size. The cosmids carry cos sequences from lambda phage so that they can be packaged into phage heads (in vitro), and circularise once in the cell.

Cosmopolitan Occurring throughout most of the world.

Cost of meiosis The reduction in numbers of alleles passed on to subsequent generations as a result of reproducing sexually compared to asexually.

Cost/benefit analysis An evaluation of large-scale public projects by comparing the costs and benefits that accrue from them.

Costa Prominent striated rod in some flagellate protozoa that courses from one of the kinetosomes along the cell surface beneath the recurrent flagellum and undulating membrane.

Costal cell The cell between the costa and the sub-costal vein.

Costal fold A narrow, thin membrane folded back on the upper surface of the costa of the forewing of butterflies, it contains androconia

Costal grooves The deep, vertical groove on the side of a salamander's body, indicating the position of a rib.

Cotransport The coupling of the "downhill" diffusion of one substance to the "uphill" transport of another against its own concentration gradient.

Cotyledon The "seed leaves" produced by the embryo of a seed plant that serve to absorb nutrients packaged in the seed, until the seedling is able to produce its first true leaves and begin photosynthesis; the number of cotyledons is a key feature for the identification of the two major groups of flowering plants.

Cotylocidia Larva of Aspidobothria.

Counteracting osmolyte strategy An osmolyte (ion) that counteracts another ion.

Countercurrent An ocean current that flows directly back into another current; also, in some animals, paired blood vessels containing blood flowing in opposite directions.

Countercurrent exchange mechanism Mechanism by which two vessels are set side by side, with fluid flowing in opposite directions, allowing efficient uptake and retention of heat, oxygen, or gas, depending upon the type of exchanger

Countercurrent flow An arrangement by which fish obtain oxygen from the water that flows through their gills. Water flows across the respiratory surface of the gill in one direction while blood flows in the other direction through the blood vessels on the other side of the surface.

Counter-illumination Having bioluminescent organs that are concentrated on the ventral surface so as to increase the effect of countershading.

Countershading Condition of organisms in the water column that are dark-colored on top but light-colored on the bottom

Coupled reactions In cells, the linking of endergonic (energy-requiring) reactions to exergonic (energy-releasing) reactions that provide enough energy to drive the endergonic reactions forward.

Coursers Predators that chase their prey over long distances.

Courtship behavior Behavior related to the attraction of the opposite sex and mating.

Covalent bond A strong bond between atoms formed by sharing outer electrons. When an atom has 8 outer electrons it is stable. Those which naturally have 8, like neon and argon gasses are quite unreactive. The carbon atom has 4 outer electrons and therefore needs 4 extra electrons to be stable. Four hydrogen atoms make a good partnership for carbon,(CH_4, C_2H_6 ... etc) hence hydrocarbons,(saturated with hydrogen atoms) are quite stable, insoluble and unreactive. One oxygen atom (outer shell has 6 electrons) and two hydrogen atoms (H_2O) also makes a stable arrangement, although not as stable as the hydrocarbon, family as the water molecule is a little unbalanced, providing hydrogen bonds and other unusual properties of biological importance, such as its ability to hold other molecules in a solution.

Cover crops Plants, such as rye, alfalfa, or clover, that can be planted immediately after harvest to hold and protect the soil.

Coxa The basal segment of the insect leg, often immovably attached to the body.

Coxal glands Excretory organs of arachnids, consisting of a sac, tubule, and opening on the coxa.

Crack Form of purified cocaine with widespread street use.

Cranium The braincase; composed of several bones fitted together at immovable joints.

Crassulacean acid metabolism A process by which some species of plants in hot, dry climates take in carbon dioxide during the night, fixing it in organic acids; the carbon dioxide is released during the day and used immediately in the Calvin cycle.

Craton A part of the Earth's crust that has attained stability and has been little deformed for a long period of time, refers only to continents; cratonic- adj.

Crawler The active first instar of a scale insect.

Creatine phosphate. High-energy phosphate compound found in the muscle of vertebrates and some invertebrates, used to regenerate stores of ATP.

Creeping eruption Skin condition caused by hookworm larvae not able to mature in a given host.

Cremaster The cluster of minute hooks (sometimes just one larger hook) at the hind end of a lepidopterous pupa: used to grip the pupal support.

Crepuscular Active or primarily active at the time around dawn or dusk.

Cretaceous western interior seaway The epicontinental sea that formed as marine waters from the north spread over North America from around 130 to 70 million years ago (Ma), at its peak in the Middle Cretaceous (~ 90 Ma) it extended from present-day Utah to the Appalachians and from the Arctic to the Gulf of Mexico; also refered to as the Western Interior Seaway

Cretin A human with severe mental, somatic, and sexual retardation resulting from hypothyroidism during early stages of development.

Crinoidea The class of echinoderms whose members are attached by a stalk of ossicles or are free living. Possess a reduced central disk. Sea lilies and feather stars.

Crista A crest or ridge on a body organ or organelle; a platelike projection formed by the inner membrane of mitochondrion.

Critical depth That depth above which total integrated photosynthetic rate equals total integrated respiration of photosynthesizers

Critical factor The single environmental factor closest to a tolerance limit for a given species at a given time.

Critical period A discrete portion of a sensitive period during development, during which the probability for forming and reinforcing the behavior is greatest.

Critical salinity A salinity of approximately 5 to 8% that marks a minimum of species richness in an estuarine system

Crochets Hooked spines at tip of the prolegs of lepidopterous larvae.

Crop The dilated section of the foregut just behind the esophagus.

Croplands Lands used to grow crops.

Cropper a deep-sea animal in which the roles of predator and deposit feeder have merged

Cross-bedding The arrangement of sedimentary beds tilted at different angles to each other, indicating that the beds were deposited by flowing wind or water.

Cross-fertilization Fusion of gametes formed by different individuals; as opposed to self-fertilization.

Crossing over The breaking during *meiosis* of one maternal and one paternal *chromosome*, the exchange of corresponding sections of DNA, and the rejoining of the chromosomes. This process can result in an exchange of alleles between chromosomes. Compare *recombination.*

Crossopterygians A type of lobe-finned fish with lungs that were ancestral to amphibians.

Cross-pollination Transfer of pollen from the stamen to the stigma of a flower of another plant.

Cross-vein A short vein joining any two neighboring longitudinal veins.

Crown The topmost portion of a plant.

Crown group All the taxa descended from a major cladogenesis event, recognized by possessing the clade's synapomorphy. See: stem group.

Crown-of-thorns sea star (Acanthaster planci)A predator of reef corals.

Crude birth rate The number of births in a year divided by the midyear population.

Crude death rate The number of deaths per thousand persons in a given year; also called crude mortality rate.

Crura Branches of the intestine of a flatworm.

Crust The outermost layer of the Earth, varying in thickness from about 10 kilometers (6 miles) below the oceans, to 65 kilometers (about 40 miles) below the continents; represents less than 1 percent of the Earth's volume.

Crustacea The subphylum of mandibulate arthropods whose members are characterized by having two pairs of antennae, one pair of mandibles, two pairs of maxillae, and biramous appendages. Crabs, crayfish, lobsters.

Cryptic Colouring and or pattern adapted for the purpose of protection from predators or prey by concealment.

Cryptic colouration To protect themselves against predators, many animals acquire colouring and markings to match and conceal them in their usual surroundings. For example the nudibranch (sea slug), *Rostanga pulchra*, is practically indiscernible from the red sponges on which it lays its eggs.

Cryptins Peptides produced by Paneth cells in the intestines. Cryptins are toxic for some bacteria, although their mode of action is not known.

Cryptobiotic Living in concealment; refers to insects and other animals that live in secluded situations, such as underground or in wood; also tardigrades and some nematodes, rotifers, and others that survive harsh environmental conditions by assuming for a time a state of very low metabolism.

Cryptococcosis An infection caused by the basidiomycete, Cryptococcus neoformans, which may involve the skin, lungs, brain, or meninges.

Cryptogams Nonvascular plants such as lichens and mosses that make up the ground or surface layer of vegetation.

Cryptogonochorism Separate sexes joined or associated to form the appearance of hermaphroditism.

Cryptomonads Unicellular, eukaryotic members of the phytoplankton that have two flagella and no skeleton.

Cryptoniscus Intermediate, free-swimming larval stage of the isopod suborder Epicaridea, developing after microniscus; attaches to definitive host.

Cryptosporidiosis Infection with protozoa of the genus Cryptosporidium. The most common symptoms are prolonged diarrhea, weight loss, fever, and abdominal pain.

Cryptozoite Preerythrocytic schizont of Plasmodium spp.

Crystal A solid that consists of a regular pattern of molecules.

Crystalline style A proteinaceous, rodlike structure in the digestive tract of a bivalve (Mollusca) that rotates against a gastric shield and releases digestive enzymes.

Crystallizable fragment The stem of the Y portion of an antibody molecule. Cells such as macrophages bind to the Fc region, and it also is involved in complement activation.

Crystallized song The final stage in song development, when the song variation (repertoire) has decreased and become more fixed.

Ctene Bands of cilia found on the body surfaces of ctenophores.

Ctenidia Comb-like structures, especially gills of molluscs; also applied to comb plates of Ctenophora.

Ctenoid scales Thin, overlapping dermal scales of the more advanced fishes; exposed posterior margins have fine, toothlike spines.

Ctenophora The phylum of animals whose members are characterized by biradial symmetry, diploblastic organization, colloblasts, and meridionally arranged comb rows. Comb jellies.

Cubitus One of the major longitudinal veins, situated in the rear half of the wing and usually with 2 or 3 branches: abbreviated to Cu.

Cubozoa The class of cnidarians whose members have prominent cuboidal medusae with tentacles that hang from the corner of the medusa. Small polyp, gametes gastrodermal in origin. Chironex.

Cucurbits Vine crops such as cucumbers, melons, squash, and pumpkins.

Culm Hollow, jointed stem of grasses.

Cultivar Abbreviation for cultivated variety of plant.

Cultural control Pest management practices that rely upon manipulation of the cropping environment (e.g., cultivation of weeds harboring insect pests).

Cultural eutrophication An increase in biological productivity and ecosystem succession caused by human activities.

Cumulus oophorus A thickening of follicular cells that encloses the ovum and projects from the stratum granulosum into the antrum.

Cuneus A more or less triangular region of the forewing of certain heteropteran bugs, separated from the corium by a groove or suture.

Cupula Small inverted cup-like structure housing another structure; gelatinous matrix covering hair cells in lateral line and equilibrium organs.

Curare A variety of poisonous plant extracts from the bark, roots, stems, and tendrils of several woody lianas, including Chondodendron tomentosum.

Currency The resource that is being maximized in an optimality model; usually assumed to be energy in an optimal foraging model.

Current A horizontal movement of water.

Cursorial Adapted for running.

Cushion plant Low growing mat formed by tightly massed individuals of the same species of plant. Generally associated with tundra or high alpine communities.

Cuspidal Two segments of curved lines meeting and terminating at a sharp point.

Cusps Peaks or raised areas of a tooth which usually fit into a fossa on the opposing tooth.

Cutaneous Pertaining to the skin (e.g., a skin infection).

Cutaneous anthrax A form of anthrax involving the skin.

Cutaneous diphtheria A skin disease caused by Corynebacterium diphtheriae that infects wound or skin lesions, causing a slow-healing ulceration.

Cutaneous respiration Exchange of gases across thin, moist surfaces of the skin. Also cutaneous exchange or integumentary exchange.

Cuticle (1) A waxy covering on the surface of stems and leaves that acts as an adaptation to prevent desiccation in terrestrial plants.

(2) The exoskeleton of an arthropod, consisting of layers of protein and chitin that are variously modified for different functions.

Cuticulin Protein component of arthropod exoskeletons.

Cutin Waxy material secreted by epidermal cells.

Cuttings Portions of stems, usually consisting of two or three nodes, used for propagation by the production adventitious roots.

Cyanobacteria Blue-green bacteria; unicellular or filamentous chains of cells that carry out photosynthesis.

Cyanogenic glycoside Glycoside that releases cyanide.

Cyanophyte A cyanobacterium, blue-green alga.

Cycads Members of the Cycadophyta, which have unbranched stems and a terminal grouping of leathery, compound leaves.

Cycle A recurring sequence of events; e. g., the secretion of certain hormones at regular intervals.

Cyclic AMP Cyclic adenosine monophosphate, a ring-shaped molecule made from ATP that is a common intracellular signaling molecule (second messenger) in eukaryotic cells, for example, in vertebrate endocrine cells. It is also a regulator of some bacterial operons.

Cyclic electron flow A route of electron flow during the light reactions of photosynthesis that involves only photosystem I and produces ATP but not NADPH or oxygen.

Cyclic photophosphorylation The formation of ATP when light energy is used to move electrons cyclically through an electron transport chain during photosynthesis; only photosystem I participates.

Cyclin Protein active in regulating the cell cycle, typically synthesized and degraded during the cell cycle to regulate the activity of a cyclin-dependent kinase. Cyclin B is the large protein subunit of mitosis promoting factor (MPF); cyclin B is synthesized and degraded during the cell cycle to regulate MPF activity. Cyclin B plus cyclin dependent kinase 1 (cdk1, aka cdc2, aka p34 kinase) together form an active MPF protein. There are many other cyclins that regulate the cell cycle.

Cyclin-dependent kinase (Cdk) A protein kinase that is active only when attached to a particular cyclin.

Cycloid scales Thin, overlapping dermal scales of the more primitive fishes; posterior margins are smooth.

Cyclorrhaphous diptera The group of flies which emerge from the puparium through a circular opening at one end of the puparium. These flies belong to the more advanced families.

Cyclosis The spontaneous movement of cytoplasm and some organelles within the cell.

Cydippid larva Free-swimming larva of most ctenophores; superficially similar to the adult.

Cynodonts A group of mammal-like carnivorous synapsids of the Upper Permian and Triassic.

Cypris Postnaupliar larva of barnacles (crustacean subclass Cirripedia) in which the carapace largely envelops the body; so called because of its resemblance to the ostracod genus Cypris.

Cyrtocyte A protonephridial cell with a single flagellum enclosed in a cylinder of cytoplasmic rods.

Cyst A general term used for a specialized microbial cell enclosed in a wall. Cysts are formed by protozoa and a few bacteria. They may be dormant, resistant structures formed in response to adverse conditions or reproductive cysts that are a normal stage in the life cycle.

Cystacanth Juvenile stage of an acanthocephalan that is infective to the definitive host.

cystic fibrosis An autosomal recessive genetic disorder that causes the production of mucus that clogs the airways of the lungs and the ducts of the pancreas and other secretory glands.

Cysticercoid A type of juvenile tapeworm composed of a solid-bodied cyst containing an invaginated scolex; contrast with cysticercus.

Cystid In an ectoproct, the dead secreted outer parts plus the adherent underlying living layers.

Cystogenic cells Secretory cells in a cercaria that produce a metacercarial cyst.

Cytochrome Several iron-containing pigments that serve as electron carriers in aerobic respiration.

Cytokines Chemical messengers that allow neighbouring cells to communicate with each other. They are paracrine messages as distinct from endochrine or hormonal messengers. There are several main families of cytokines including growth factors , neurotransmitters, , < interleukins,>lymphokines and many others. The method of communication involves detection of the chemical message, a ligand , by receptor proteins (for example an integrin on the cell membrane of the receiving cell. The result of the message is a shift in the level of gene expression or the expression of new genes and altered cell behaviour. Cytokines are complex as they sometimes inhibit and/or facilitate the actions of each other.

Cytokinesis Process in which the cytoplasm of a cell is divided after nuclear division (mitosis) is complete.

Cytokinins A group of hormones that promote cell division and inhibit aging of green tissues in plants.

Cytomegalovirus inclusion disease An infection caused by the cytomegalovirus and marked by nuclear inclusion bodies in enlarged infected cells.

Cyton Cell body; contains nucleus and some other organelles but excludes processes extending from cell. For example, the neurocyton is the nerve cell body, excluding the axon and dendrites.

Cytopathic effect The observable change that occurs in cells as a result of viral replication. Examples include ballooning, binding together, clustering, or even death of the cultured cells.

Cytophaneres Fibers radiating out from a zoitocyst into surrounding muscle; found in some species of Sarcocystidae.

Cytopharynx A region of the plasma membrane and cytoplasm of some ciliated and flagellated protists specialized for endocytosis. A permanent oral canal.

Cytoplasm Substance contained within the plasma membrane excluding, in eukaryotes, the nucleus.

Sometimes, the cellular region within the plasma membrane, including the cytosol and the organelles but excluding the nucleus (in eukaryotes) is also refered to as the cytoplasm.

Cytoplasmic determinants In animal development, substances deposited by the mother in the eggs she produces that regulate the expression of genes affecting the early development of the embryo.

Cytoplasmic inclusion The basic food material or stored product of a cell's metabolic activities.

Cytoplasmic localization A phenomenon wherein the morphogenetic determinants are found in a specific region of the oocyte and/or are later segregated to specific blastomeres. Associated with autonomous specification as a mechanism of cell specification.

Cytoplasmic matrix The protoplasm of a cell that lies within the plasma membrane and outside any other organelles. In bacteria it is the substance between the cell membrane and the nucleoid.

Cytoplasmic streaming A circular flow of cytoplasm, involving myosin and actin filaments, that speeds the distribution of materials within cells.

Cytoproct Site on a protozoan where undigestible matter is expelled.

Cytopyge A region of the plasma membrane and cytoplasm of some ciliated protists specialized for exocytosis of undigested wastes.

Cytosine (C) A *nitrogenous base*, one member of the *base pair* G- C (*guanine* and *cytosine*).

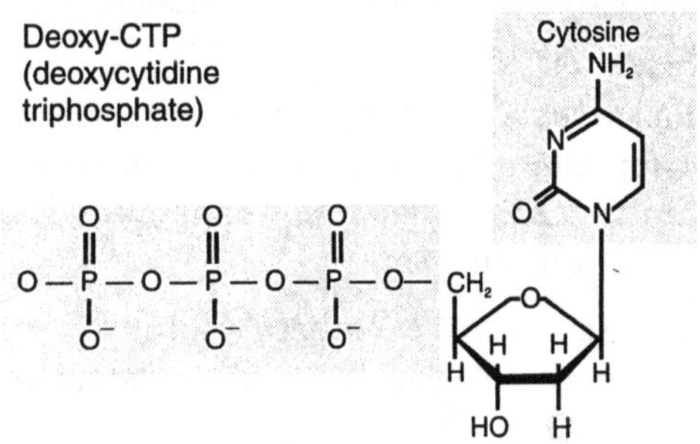

Fig. *Cytosine*

Cytoskeleton The cytoskeleton is unique to eukaryotic cells. It is a dynamic three-dimensional structure that fills the cytoplasm. This structure acts as both muscle and skeleton, for movement and stability. The long fibers of the cytoskeleton are polymers of subunits. The primary types of fibers comprising the cytoskeleton are microfilaments, microtubules, and intermediate filaments.

The cytoskeleton acts as a "track" on which cells can move organelles, chromosomes and other things.

Cytosol Contents of the main compartment of the cytoplasm, excluding membrane-bound organelles.

Cytosome The cell body inside the plasma membrane.

Cytotoxic T cells A special T cell activated during cell-mediated immune responses that recognizes and destroys virus-infected cells.

Cytotoxin A toxin or antibody that has a specific toxic action upon cells; cytotoxins are named according to the cell for which they are specific (e.g., nephrotoxin).

Cytotrophoblast Inner cellular layer of the trophectoderm (trophoblast), between the syncitiotrophoblast, and chorionic villus capillaries; part of the mammalian placenta.

Cytoxic T cells T cells that destroy body cells infected by viruses or bacteria; also attack bacteria, fungi, parasites, and cancer cells and will kill cells of transplanted organs if they are recognized as foreign; also known as killer T cells.

D

D.N.A. An abbreviation for Dioxyribonucleic Acid a large molecule which stores the data in our genes in the form of a 3 character code. D.N.A. is a self replicating molecule.

Dacite Igneous volcanic rock, less mafic than andesite, typically fine-grained and light in color; rough volcanic equivalent of granodiorite.

Dactylozooid A polyp of a colonial hydroid specialized for defense or killing food.

Daily estuary An estuary in which tidal movements cause substantial changes in salinity at any one location on a daily basis

Daily torpor Daily sluggishness that some animals experience. A period of inactivity that is normally induced by cold; observed in numerous ectotherms and also endotherms; often used to describe the specific physiological state of endotherms with a circadian cycle of lowered body temperature and depressed metabolic rate.

Dalton Measure of molecular weight or mass. One hydrogen atom has mass of 1 Da. Proteins and other macromolecule molecular weights are usually measured in kDa or kD (kilodaltons) - 1000 Da.

Dane particle A 42 nm spherical particle that is one of three that are seen in hepatitis B virus infections. The Dane particle is the complete virion.

Dark reactions The photosynthetic process in which food (sugar) molecules are formed from carbon dioxide from the atmosphere with the use of ATP; can occur in the dark as long as ATP is present.

Dark-field microscopy Microscopy in which the specimen is brightly illuminated while the background is dark.

Darwinian fitness A measure of the relative contribution of an individual to the gene pool of the next generation.

Darwinian medicine An approach to the treatment of infective diseases which takes account of the co-evolution between the host and its parasite.

Darwinism Theory of evolution emphasizing common descent of all living organisms, gradual change, multiplication of species and natural selection.

Data The results in a scientific experiment, or descriptive observations, upon which a conclusion is based.

Database search Once an open *reading frame* or a partial amino acid sequence has been determined, the investigator compares the sequence with others in the databases using a computer and a search algorithm. This is usually done in a protein database such as PIR or Swiss-

Prot. Nucleic acid sequences are in GenBank and EMBL databases. The search algorithms most commonly used are BLAST and FASTA.

Dauer juvenile Nematode juvenile in which development is arrested during unsuitable conditions and resumed when conditions improve.

Dauerblastula Permanent "blastula" of ciliated epidermal cells formed by experimental isolation of animal pole cells of the sea urchin embryo. ["dauer" is from the German dauern - "to endure"]

Daughter cell A cell that is the offspring of a cell that has undergone mitosis or meiosis. The term "daughter" does not indicate the sex of the cell.

Daughter sporocysts In digenetic trematodes the embryonic cells that develop from sporocysts and give rise to rediae.

Day length The number of hours that sunlight illuminates a given area on earth; dependent on the angle of the earth relative to the rays of the sun.

Day-neutral plant A plant whose flowering is not affected by photoperiod.

DDT (common name). A widely used synthetic insecticide; a chlorinated hydrocarbon, dichloro diphenyl trichloroethane.

Dealate Wingless as a result of the insect casting or breaking off its own wings, as in newly mated queen ants and termites.

Deamination The removal of amino groups from amino acids.

Deamination reaction A reaction in which an amino group, _NH2, is enzymatically removed from a compound.

Death phase The decrease in viable microorganisms that occurs after the completion of growth in a batch culture.

Death rate The ratio between deaths and individuals in a specified population at a particular time.

Debt-for-nature swap Forgiveness of international debt in exchange for nature protection in developing countries. Environmental groups and nongovernmental organizations often pay banks to write off uncollectable debts of developing countries at a steep discount in exchange for a promise by the debtor country to establish nature preserves.

Decacanth Ten-hooked larva that hatches from the egg of a cestodarian tapeworm. Also called a lycophora.

Decapods Crustaceans with five pairs of walking legs and a well-developed carapace. The group includes the shrimps, lobsters, hermit crabs, and crabs.

Decarboxylation The removal of a single carbon atom as CO_2 from an organic molecule.

Decidua The material of the endometrium of the mammalian uterus during pregnancy.

Decidua basalis Maternal portion of the decidua that comes in contact with the embryonic chorionic villi and underlies the conceptus.

Decidua capsularis Portion of the decidua surrounding the embryo conceptus where it projects into the uterine lumen. It will eventually fuse with the decidua parietalis.

Deciduate placenta Type of placenta where the lining of the uterus is shed along with the placenta, after birth has taken place. This is due to the extensive penetration of the chorionic villi into the endometrium. There are two types of deciduate placenta which reflect the arrangement of the chorionic villi: zonary and discoidal.

Deciduous Falling off or being shed during a certain season or at a partial stage of growth such as leaves or antlers.

Deciduous forest Forest biome in which the dominant trees are deciduous.

Decimal reduction time (D or D value) The time required to kill 90% of the microorganisms or spores in a sample at a specified temperature.

Decline spiral A catastrophic deterioration of a species, community, or whole ecosystem; accelerates as functions are disrupted or lost in a downward cascade.

Decomposer Mostly heterotrophic bacteria and fungi that obtain organic nutrients by breaking down the remains or products of other organic compounds; their activities help cycle the simple compounds back to the autotrophs.

Decomposition The biochemical process where biological materials are broken down into smaller particles and eventually into basic chemical compounds and elements.

Decongestant An agent that relieves nasal or respiratory congestion.

Decticous Of pupa: referring to the state in which the pupa possesses movable mandibles which can be used for biting, the opposite being Adecticous.

Deduction Reasoning from the general to the particular, that is, from given premises to their necessary conclusion.

Deep (bottom) layer The deepest and coldest of the three layers of the ocean.

Deep ecology A philosophy that calls for a profound shift in our attitudes and behavior based on voluntary simplicity; rejection of anthropocentric attitudes; intimate contact with nature; decentralization of power; support for cultural and biological diversity; a belief in the sacredness of nature; and direct personal action to protect nature, improve the environment, and bring about fundamental societal change.

Deep layer The layer extending from the lowest part of the thermocline to the bottom

Deep scattering layer (DSL) A sound-reflecting layer of many types of organisms that migrates daily from the mesopelagic zone to the epipelagic zone.

Deep sea The dark waters below the mesopelagic zone.

Deep-scattering layer Well-defined horizon in the ocean that reflects sonar; indicates a layer usually consisting of fishes, squid, or other larger zooplankton

Deep-sea fan A fan-like accumulation of sediment at the base of a submarine canyon.

Defecate To extrude feces from the bowel.

Defensin Specific peptides produced by neutrophils that permeabilize the outer and inner membranes of certain microorganisms, thus killing them.

Deficit irrigation The concept of low levels of irrigation to achieve a moderate level of productivity; even though yields are not maximized, limited water supplies are conserved.

Defined medium Culture medium made with components of known composition.

Definitive host Host in which a parasite achieves sexual maturity. If there is no sexual reproduction in the life of the parasite, the host most important to humans is the definitive host.

Deforestation The clearing of forested areas as part of a commercial forestry enterprise or for some other economic purpose such as the expansion of settlement or the development of agriculture. Deforestation has traditionally been associated with economic development, but it is no longer characteristic of developed nations. In such areas, all the trees that are likely to be removed have already been removed. There are exceptions, however. In Canada, for example, large areas are cleared every year in the commercial logging of the boreal forest. There and in other areas of the developed world, where forestry continues to be an important element in the economy, good forest management requires that deforested areas be regenerated, but the process is not always effective. Currently, deforestation is most extensive in the world's less-developed nations, particularly those in the tropics. Clear cutting and burning of the tropical rainforest in South and Central America, Africa and South East Asia is proceeding rapidly, although the exact rate is difficult to assess accurately. Estimates for the Amazon region in Brazil, for example, have ranged from 13,800 to 21,000 sq km per annum. Whatever the true value, given the demographic and economic pressures on these areas, the regeneration of such a large area of rainforest is unlikely.

The impact of deforestation extends beyond the destruction of the trees. It destroys the habitat for other plants and for forest-dwelling animals. By opening up the environment, and altering such elements as radiation, temperature, wind and moisture, deforestation also produces significant microclimatological changes. The higher levels of solar radiation reaching the surface cause temperatures to rise, accelerating the decomposition of organic matter. Temperature ranges and wind speeds increase following deforestation and the intensity of the rainfall reaching the surface is higher. The latter causes more rapid leaching of the nutrients released from the organic matter. Runoff also increases, creating a greater potential for soil erosion, particularly in hilly areas. On a larger scale, deforestation may contribute to the enhancement of the greenhouse effect. Much of the deforestation in tropical regions is achieved through burning which releases carbon dioxide (CO_2) directly into the atmosphere. In addition, growing forests are significant carbon (C) sinks, and once they are removed, the recycling of carbon dioxide is reduced, allowing it to remain in the atmosphere to enhance global warming.

Degeneracy Backtranslation of amino acid sequences usually leads to a collection of possible base sequences which can code for the amino acid sequence, due to the degeneracy of the genetic code. In practice, when synthesising an oligonucleotide primer or probe based on this information a mixture of all the possible sequences is made, and referred to as a degenerate oligonucleotide.

Degeneration The death or functional impairment of cells connected to destroyed or severely damaged neurons.

Degradation Deterioration in water quality due to contamination or pollution; makes water unsuitable for other desirable purposes.

Dehiscent fruit Fruit that splits open at maturity, facilitating seed dispersal.

Dehydration reaction A chemical reaction in which two molecules covalently bond to one another with the removal of a water molecule.

Deirid Sensory papilla on each side near the anterior end of some nematodes.

Delamination Splitting of one cellular sheet or layer into two parallel layers.

Delayed fertilization Occurs when fertilization of an egg does not occur immediately following coitus, but may be delayed for weeks or months.

Delayed implantation Usually, the fertilized egg implants in the wall of the female's uterus shortly after mating and fertilization. However, in delayed implantation, the fertilized egg does not implant in the female's uterus until much later, thus implantation is delayed. Once implantation occurs the fertilized egg quickly becomes a developing embryo. This process is found in family Mustelidae (weasel family) and family Ursidae (bear family).

Delayed response problem An operant conditioning procedure in which the test animal sees the experimenter set up the test situation, including what will be the correct response, and then must wait until the experimenter provides the animal with an opportunity to respond.

Delayed type hypersensitivity (DTH) Manifestation of cell-mediated immunity, distinguished from immediate hypersensitivity in that maximal response is reached about 24 hours or more after intradermal injection of the antigen. The lesion site is infiltrated primarily by monocytes and macrophages.

Deletion A type of mutation where a section of a gene or of a chromosome is missing. Deletions may or may not have a phenotypic consequence, depending on the normal role of the DNA which is absent.

Delta agent A defective RNA virus that is transmitted as an infectious agent, but cannot cause disease unless the individual is also infected with the hepatitis B virus.

Delta A low, nearly flat accumulation of sediment deposited at the mouth of a river or stream, commonly triangular or fan-shaped; deltaic– adj.

Delta-proteobacteria One of the five subgroups of proteobacteria, each with distinctive 16S rRNA sequences. Chemoorganotrophic bacteria that usually are either predators on other bacteria or anaerobes that generate sulfide from sulfate and sulfite.

Demand The amount of a product that consumers are willing and able to buy at various possible prices, assuming they are free to express their preferences.

Deme A small, local subpopulation. Isolated subpopulations sometimes display genetic changes that may contribute to evolutionary change in the subpopulation.

Demersal fish A bottom-dwelling fish.

Demineralisation Reduction in amount of mineral in tissue. This reduction occurs when the crystals of apatite are dissolved, usually in an acid environment.

Demographic Referring to numerical characteristics of a population (e.g., population size, age structure)

Demographic transition A pattern of falling death and birth rates in response to improved living conditions; could be reversed in deteriorating conditions.

Demography Vital statistics about people: births, marriages, deaths, etc.; the statistical study of human populations relating to growth rate, age structure, geographic distribution, etc., and their effects on social, economic, and environmental conditions.

Demospongiae The class of poriferans whose members have monaxon or tetraaxon siliceous spicules or spongin. Leuconoid body forms are present and vary in size from a few centimeters to 1 m in height.

Denaturation Separation of two complementary strands of nucleic acid by breakage of the hydrogen bonds involved in base pairing. This is necessary prior to probe hybridisation and most methods involving enzymic DNA synthesis on a DNA template e.g. sequencing, PCR, some labelling methods. Denaturation may be achieved by heating or treatment with NaOH.

Denaturing gel An *agarose* or *acrylamide gel* run under conditions which destroy *secondary* or *tertiary* protein or RNA structure. For protein, this usually means the inclusion of 2-ME (which reduces disulfide bonds between cysteine residues) and SDS and/or urea in an acrylamide gel. For RNA, this usually means the inclusion of formaldehyde or glyoxal to destroy higher ordered RNA structures. In DNA sequencing gels, urea is included to denature dsDNA to ssDNA strands. In denaturing gels, macromolecules tend to be separated on the basis of size and (to some extent) charge, while shape and oligomerization of molecules are not important. Contrast with NATIVE GEL.

Dendrites Small branching patterns on rocks made of iron and manganese oxides that show the passage of fluids through the rock.

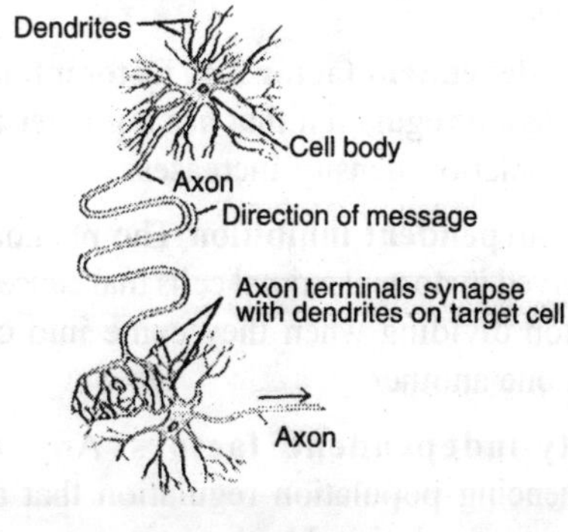

Fig. *Dendrites*

Dendritic cell An antigen-presenting cell that has long membrane extensions resembling the dendrites of neurons. These cells are found in the lymph nodes, spleen, and thymus (interdigitating dendritic cells); skin (Langerhans cells); and other tissues (interstitial dendritic cells). Express MHC class II and B7 costimulatory molecules and thus are efficient presenters of antigens to T-helper cells.

Dendrochronology The process of determining the age of a tree or wood used in structures by counting the number of annual growth rings.

Dendroclimatology Study of the annual rings of trees in order to interpret climatic changes in the past.

Dendrogram A treelike diagram that is used to graphically summarize mutual similarities and relationships between organisms.

Dengue A virus disease of man marked by severe pains in head, eyes, muscles, and joints and transmitted by certain mosquitoes.

Denitrification The process by which certain bacteria living in poorly aerated soils break down nitrates, using the oxygen for their own respiration and releasing nitrogen back into the atmosphere.

Denitrifying bacteria Free-living soil bacteria that convert nitrates to gaseous nitrogen and nitrous oxide.

Density (insect populations) The number of insects per unit of measure (e.g., beetles per square meter).

Density-dependent factor Any factor influencing population regulation that has a greater impact as population density increases.

Density-dependent inhibition The phenomenon observed in normal animal cells that causes them to stop dividing when they come into contact with one another.

Density-independent factors Any factor influencing population regulation that acts to reduce population by the same percentage, regardless of size.

Dental abscess An abscess around the apex of a tooth due to spread of infection from the pulp .

Dental formula A notation that indicates the number of incisors, canines, premolars, and molars in the upper and lower jaw of a mammal.

Dental papilla The condensation of dental mesenchyme which provides the stem cells from which ondontoblasts, cementoblasts and osteoblasts will form the pulp-dentine, cementum and alveolar bone of the tooth socket.

Dental plaque A thin film on the surface of teeth consisting of bacteria embedded in a matrix of bacterial polysaccharides, salivary glycoproteins, and other substances.

Dentary One of several bone which together made up the lower jaw in early reptiles. During evolution the other bones, the articular and quadrate bones, became part of the inner ear, and the dentary became the single the mandible of mammals.

Dentate Toothed, possessing teeth or teeth like structures.

Denticle A small toothlike process.

Dentine A hard material like bone which forms the root and inner core of the crown of teeth. Unlike bone, dentine has fine tubules which contain the elongated process of odontoblasts, the dentine forming cells.

Dentine-pulp A term used to describe the unity between dentine and pulp, and to view it as one integrated tissue.

Dentition 1. the kind, number, and arrangement of the teeth of man and animals.

2. the development and cutting or eruption of teeth.

Deoxynucleotides Components of DNA, containing the phosphate, sugar and organic base; when in the triphosphate form, they are the precursors required by DNA polymerase for DNA synthesis (i.e., ATP, CTP, GTP, TTP).

Deoxyribonucleic acid Also known as "DNA." The molecule which contains the hereditary information of most living organisms. This information is encoded in the sequence of bases arrayed along the length of the DNA molecule. DNA is a double helix, coiled around a central axis. The DNA backbones are composed of alternating sugar and phosphate subunits, while the interior portion of the molecule is composed of pairs of nitrogenous bases. The helix has a radius of 1 nm, and completes one 360 degree turn every 3.4 nm. The base pairs are 0.34 nm apart from each other. The two strands which make up the DNA molecule are anti-parallel—that is they run in opposite directions from each other.

Deoxyribonucleotide N*ucleotides* which are the building blocks of DNA and which lack the 2' hydroxyl moiety present in the ribonucleotides of RNA.

Deoxyribose A 5-carbon sugar having 1 oxygen atom less than ribose; a component of deoxyribonucleic acid (DNA).

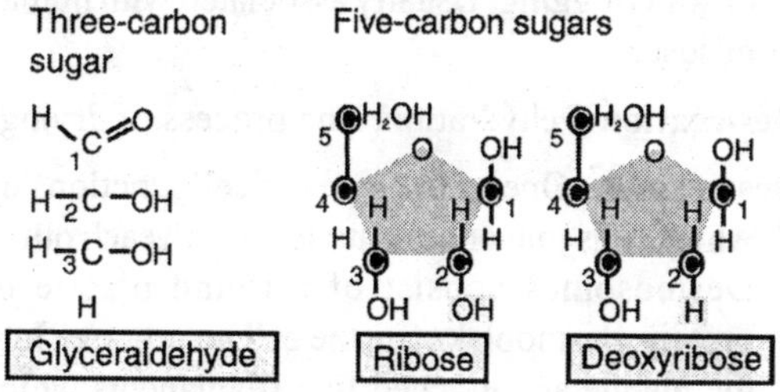

Fig. *Deoxyribose*

Dependency ratio The number of nonworking members compared to working members for a given population.

Dependent variable What may change (e.g., aggression) in response to manipulation of an independent variable (e.g., food supply) by the experimenter.

Deplasmolysis Water flowing into a plasmolized cell to correct the process that would cause cellular death from dehydration.

Depolarisation All cells have a slight difference in electrical potential between the inside and outside of the cell membrane. This difference is called a membrane potential and is due to a greater number of sodium ions (positively charged) outside the cell than inside. This imbalance is maintained by a membrane pump which pushes sodium ions out of the cell. Another pump also pushes potassium ions into the cell so there should be no difference in the balance of positive ions. But the potassium ions leak back out again, so there is always a potential difference across the membrane. Nerve cells have the ability to depolarise or reverse the membrane potential so that the inside is positive and outside negative. This reversal is short lived and is soon corrected, but it is long enough to influence the adjacent parts of the membrane and to be carried, like a wave, all the way along a nerve axon to the next nerve where it reaches a synapse The reversal is caused by a sudden opening of cell membrane gates which allow a flood of sodium ions into the cell. This flood causes the inside to become positive, but the gates are soon shut and potassium gates opened, which allows potassium ions to flood out and restore the membrane potential. This can all happen several times in one second, but after a while there is no flood, and the sodium pump has to get to work to build up enough pressure for the depolarisation to work again.

Depolarized The reduction in the difference in charge (potential) between the outside and inside of a membrane.

Deposit feeder An animal that feeds on organic matter that settles on the bottom. Compare suspension feeder.

Deposition Any accumulation of material, by mechanical settling from water or air, chemical precipitation, evaporation from solution, etc.

Depressant Psychoactive drug that has a sedative effect on the central nervous system; actions include dulling mental awareness and inducing sleep.

Deprivation experiments Experimental manipulations involving removal of particular types of stimuli (e.g., social, sensory, motor) to ascertain the effects later in development.

Depth diversity gradient The increase in species richness with increasing water depth until about 2000 meters below the surface, where species richness begins to decline.

Derived Describes a character state that is present in one or more subclades, but not all, of a clade under consideration. A derived character state is inferred to be a modified version of the primitive condition of that character, and to have arisen later in the evolution of the clade. For example, "presence of hair" is a primitive character state for all mammals, whereas the "hairlessness" of whales is a derived state for one subclade within the Mammalia.

Dermal branchiae Thin folds of the body wall of a sea star that extend between ossicles and function in gas exchange and other exchange processes.

Dermal Pertaining to the skin; cutaneous.

Dermal system Plant organ system that provides the covering for the plant.

Dermal tissue system The protective covering of plants; generally a single layer of tightly packed epidermal cells covering young plant organs formed by primary growth.

Dermatan sulphate A glycosaminoglycan found in skin, tendon, blood vessel and heart valves.

Dermatitis Infection or inflammation of the skin.

Dermatomycosis A fungal infection of the skin; the term is a general term that comprises the various forms of tinea, and it is sometimes used to specifically refer to athlete's foot (tinea pedis).

Dermatophytes Certain imperfect fungi that cause skin diseases in humans; includes the organisms causing ringworm and athlete's foot.

Dermis The layer of the skin deep to the epidermis, consisting of a dense bed of vascular connective tissue.

Desensitization To make a sensitized or hypersensitive individual insensitive or nonreactive to a sensitizing agent.

Desert A type of biome characterized by low moisture levels (below 25 cm per year), and infrequent and unpredictable precipitation. Daily and seasonal temperatures fluctuate widely. Deserts are characterized by cacti and desert shrubs or sagebrush. Animals include birds, rodents, reptiles, and numerous species of arthropods.

Desert biome Characterized by dry conditions and plants and animals that have adapted to those conditions; found in areas where local or global influences block rainfall.

Desert crust A crust formed by microbial binding of sand grains in the surface zone of desert soil; crust formation primarily involves cyanobacteria.

Desertification A process by which fragile, semiarid ecosystems lose productivity because of loss of plant cover, soil erosion, salinization, or waterlogging. Usually associated with human misuse.

Desiccation Dehydration. The process of drying.

Desmosomes One of the types of cell junctions by which cells join or communicate with each other. Desmosomes consist of a round plaque of protein, desmoplakin on the cell membrane. Into the plaque are attached fine filamanents which are part of the cell's cytoskeleton. So the plaque is attached to the skeleton of the cell. Where the filaments enter the plaque the are so dense as to be visible with a light microscope. They are then called tonofilaments. The plaque of one cell adheres to the plaque of another. This system of joining cells is designed to resist mechanical separation, so we see desmosomes joining epithelial cells which hold tightly to each other. If epithelium is processed for histology, some shrinkage occurs and the epithelial cells separate from each other, except where the desmosomes hold them together. The pulled out tags of cell membrane give these cells a star-like shape, and so they are called the stellate cells.

Despot A dominant individual that controls resources.

Desquamation The detachment of cells from the surface of an epithelium.

Dessication Mummification.

Detergent An organic molecule, other than a soap, that serves as a wetting agent and emulsifier; it is normally used as cleanser, but some may be used as antimicrobial agents.

Determinate cleavage A type of embryonic development in protostomes that rigidly casts the developmental fate of each embryonic cell very early.

Determinate growth A type of growth characteristic of animals, in which the organism stops growing after it reaches a certain size.

Determination The progressive restriction of developmental potential, causing the possible fate of each cell to become more limited as the embryo develops.

Detritivores Organisms that live on dead and discarded organic matter; include large scavengers, smaller animals such as earthworms and some insects, as well as decomposers (fungi and bacteria).

Detritus Accumulated organic debris from dead organisms, often an important source of nutrients in a food web.

Detrivore An organism that feeds on large bits of dead and decaying organic matter. What detrivores leave behind is used by decomposers. Crabs and seabirds are examples of detrivores. Compare *decomposer*.

Deuteromycetes A miscellaneous assemblage of fungi also termed the imperfect fungi because the sexual reproductive features are either not known, not used, or have been reduced

Deuterostomes Broad classification of triploblastic animals including echinoderms and chordates that tend to share certain embryological traits; among these the formation of the "mouth second" (hence the name) during gastrulation, after the future anus, which is comes from the blastopore, the site of gastrulation initiation.

Deuterostomia A group of higher phyla in which cleavage is indeterminate (regulative) and primitively radial. The endomesoderm is enterocoelous, and the mouth is derived away from the blastopore. Includes Echinodermata, Chordata, and a number of minor phyla. Compare with Protostomia.

Deuterotoky Type of parthenogenesis in which all individuals are uniparental but in which both males and females occur.

Deutomerite Posterior half of a cephaline gregarine protozoan.

Deutonymph In the life cycle of some mesostigmatid mites, a nonfeeding stage that molts into the adult.

Deutovum Incompletely developed larva that hatches from the egg of a chigger mite.

Development The summation of all activities leading to changes in cells, tissues, organs, or organisms; a genetically controlled sequence.

Developmental biology The study of how organisms, their

Devonian A geological period of the Paleozoic era beginning about 400 million years ago. During this period major diversification of the early land plants occurred.

Dew point The temperature at which condensation occurs for a given concentration of water vapor in the air.

Dewlap A fold of skin hanging from the neck of some bovines, reptiles, and birds.

Dextral Pertaining to the right; in gastropods, shell is dextral if opening is to right of columella when held with spire up and facing observer.

Dextrans polysaccharides made by bacteria. They have a slimy consistency and contribute to the sticky nature of plaque.

Diabetes Condition characterized by a high blood glucose level and the appearance of glucose in the urine due to a deficiency of insulin or the inability of body cells to respond to insulin; diabetes mellitus.

Diabetes mellitus, Types I and II A disorder associated with defects in insulin action. Type I diabetes is characterized by inadequate insulin secretion; Type II diabetes is characterized by impaired insulin secretion in response to elevated blood glucose levels or by loss of sensitivity to insulin by .

Diagenesis All of the changes that occur to a fossil (or more generally any sediment) after initial burial; includes changes that result from chemical, physical as well as biological processes. The study of diagenesis is part of taphonomy.

Diapause A period of arrested development in the life cycle of insects and certain other animals in which physiological activity is very low and the animal is highly resistant to unfavorable external conditions.

Diapause phase A period of dormancy, common in insect species, which occurs during the more rigorous portions of the annual climatic cycle.

Diaphragm The domed respiratory muscle between thoracic and abdominal compartments of mammals.

Diapolar cells Ciliated somatodermal cells located between the parapolar and uropolar cells of a mesozoan.

Diapsid A vertebrate distinguished by a skull with two pairs of openings in the side behind the eyes, e.g., lizards, snakes, crocodiles, dinosaurs, and pterosaurs.

Diastema A space, gap.

Diastole The filling of the ventricle of the heart with blood.

Diastolic pressure The pressure in an artery during the ventricular relaxation phase of the heart cycle.

Diatom The common name for the Bacillariophyceae, a class of unicellular microscopic algae with a symmetrical siliceous exoskeleton.

Diatomaceous earth A powdery, soil-like material formed by the glass cell walls of dead diatoms deposited on the marine floor.

Diatomaceous ooze A biogenous sediment that consists mostly of the siliceous frustules of diatoms. It is known as diatomaceous earth when found inland.

Diatomite Diatomite, or diatomaceous earth, is a siliceous sedimaentary rock formed from the accumulations of diatoms or other nanoplankton.

Diatoms Unicellular and eukaryotic autotrophs with a siliceous frustule; mostly planktonic.

Diauxic growth A biphasic growth pattern or response in which a microorganism, when exposed to two nutrients, initially uses one of them for growth and then alters its metabolism to make use of the second.

Dichotomous keyA device for identifying organisms based on the answers to a series of questions, with each question involving alternate choices.

Dichotomous Pertaining to the division or forking of a single axis into two branches.

Dicots One of the two main types of flowering plants; characterized by having two cotyledons, floral organs arranged in cycles of four or five, and leaves with reticulate veins; include trees (except conifers) and most ornamental and crop plants.

Dicotyledon A class of angiosperms in which the seedlings typically possess two cotyledons; commonly abbreviated to dicot.

Dictyosomes Organelles in plant cells composed of a series of flattened membrane sacs that sort, chemically modify, and package proteins produced on the rough endoplasmic reticulum. Also known as the Golgi Apparatus.

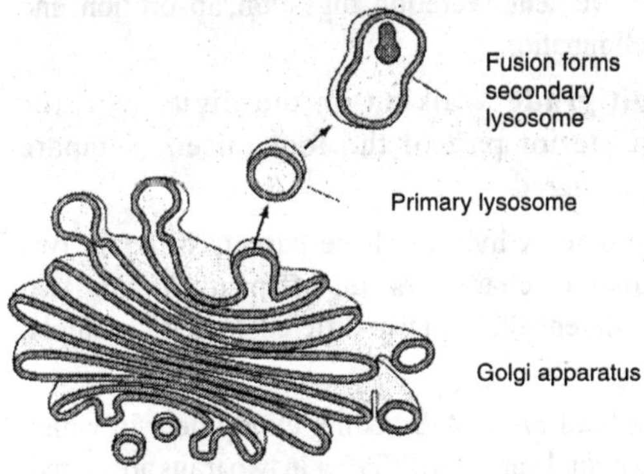

Fig. *Dictyosomes*

Dideoxy sequencing Enzymatic determination of DNA or RNA sequence by the method of Sanger and colleagues, based on the incorporation of chain terminating *dideoxynucleotides* in a growing nucleic acid strand copied by *DNA polymerase* or *reverse transcriptase* from a DNA or RNA *template*. Separate reactions include dideoxynucleotides containing A,C, G, or T bases. The reaction products represent a collection of new. labeled DNA strands of varying lengths, all terminating with a dideoxynucleotide at the 3' end (at the site of a complementary base in the template nucleic acid), and are separated in a polyacrylamide/urea gel to generate a sequence "ladder". This method is more commonly used than "Maxam-Gilbert" (chemical) sequencing.

Dideoxynucleotides Chain-terminating precursors of DNA synthesis that block further polymerization when added to the end of the DNA strand by DNA polymerase. These nucleotides lack a 3'-OH (hydroxyl) group necessary for continued 5'-to-3' DNA synthesis. Dideoxynucleotides are used in molecular biology for Sanger-type DNA sequencing, and in medicine as anti-retroviral drugs for the treatment of HIV infection (e.g., ddI, ddC, and AZT).

Dideoxyribonucleotide A *nucleotide* which lacks both 3' and 2' hydroxyl groups. Such dideoxynucleotides can be added to a growing nucleic acid chain, but do not then present a 3'-OH group which can support further propagation of the nucleic acid chain. Thus such compounds are also called "chain terminators", and are useful in DNA and RNA sequencing reactions.

Dieback A sudden population decline; also called a population crash.

Diecdysis Condition in which ecdysis processes are going on continuously and one ecdysis cycle grades rapidly into another.

Diencephalons Part of the forebrain; consists of the thalamus and hypothalamus.

Differential display A form of RT-PCR in which primers are used to select a subset of the total mRNA population. This allows comparison of mRNAs from different cells. leading to identification of those mRNAs which are expressed only in certain situations e.g. after stimulation.

Differential interference contrast (DIC) microscope A light microscope that employs two beams of plane polarized light. The beams are combined after passing through the specimen and their intereference is used to create the image.

Differential media Culture media that distinguish between groups of microorganisms based on differences in their growth and metabolic products.

Differential RNA synthesis Concept that states that cells are different from each another because they turn on (transcribe mRNAs from) different sets of genes; thus each cell makes different proteins appropriate to its particular function.

Differential staining procedures Staining procedures that divide bacteria into separate groups based on staining properties.

Differentially permeable membrane A membrane that permits the passage of certain types of particles and inhibits the passage of others; also termed selectively permeable membrane.

Differentiate Change in the pattern of genes expressed by a cell resulting in altered function, from a more primitive parent cell to a more specialised group of daughter cells.

Differentiation Process whereby cells acquire their mature morphological and biochemical characteristics. Differentiation is often considered a 'final step' of development in which cells take on their mature function.

Differentiation-inducing factor (DIF) A low molecular weight lipid that induces posterior cells of a Dictyostelium slug to differentiate as stalk cells as opposed to spore cells.

Diffusely adhering E. coli (DAEC) DAEC strains of E. coli adhere over the entire surface of epithelial cells and usually cause diarrheal disease in immunologically naive and malnourished children.

Diffuse-porous wood Secondary xylem characterized by the same-sized vessels and tracheids throughout the growing season so that growth rings are difficult or impossible to detect.

Diffusion The random movement of molecules from one location to another because of random thermal molecular motion; net diffusion always occurs from a region of higher concentration to a region of lower concentration.

Digestion The process of breaking down food into molecules small enough for the body to absorb.

Digestion efficiency The fraction of living food that does not survive passage through a predator's gut

Digestion The process of breaking down food into its molecular and chemical components so that these nutrient molecules can cross plasma membranes.

Digestive gland An enzyme-producing gland in several groups of invertebrates where digestion and absorption take place.

Digestive system One of eleven major body organ systems in animals; converts food from the external environment into nutrient molecules that can be used and stored by the body and eliminates solid wastes; involves five functions: movement, secretion, digestion, absorption, and elimination.

Digitigrade Walking on the digits with the posterior part of the foot raised; compare plantigrade.

Dihybrid A hybrid whose parents differ in two distinct characters; an offspring having two different alleles at two different loci, for example, A/a B/b.

Dihybrid cross A breeding experiment in which parental varieties differing in two traits are mated.

Dikaryon Mycelium of some fungi that have two separate haploid nuclei in each cell.Having two different and distinct nuclei per cell; found in the fungi. A dikaryotic individual is called a dikaryon.In fungi, having pairs of nuclei within cells or compartments. Each cell contains two separate haploid nuclei, one from each parent.

Diminishing returns A condition in which unrestrained population growth causes the standard of living to decrease to a subsistence level where poverty, misery, vice, and starvation makes life permanently drab and miserable. This dreary prophecy has led economics to be called "the dismal science."

Dimorphic Having more than one form, size, or appearance; usually referring to the difference between males and females of a species.

Dimorphism The existence of distinct genetically determined forms of the same species, such as distinct male and female forms or distinct young and mature forms.

Dinoflagellates Single-celled to colonial protistans characterized by two flagella, one girdling the cell and the other trailing the cell.

Dinosaurs Any of the Mesozoic reptiles belonging to the groups designated as ornithischians and saurischians.

Dinosteranes/dinosteroids chemicals found in dinoflagellates, which have been useful in documenting their existence early in the fossil record.

Dioecious Characterized by having separate sexes; that is, an individual is either male or female, but never both. "Gonochoristic" means the same thing.

Diorite Igneous plutonic rock, less mafic than gabbro, but more mafic than granite and granodiorite; rough plutonic equivalent of andesite.

Dip The angle that a bedding plane or fault makes with the horizontal when measured perpendicular to the strike of the bedding plane or fault.

Diphtheria An acute, highly contagious childhood disease that generally affects the membranes of the throat and less frequently the nose. It is caused by Corynebacterium diphtheriae.

Diphycercal A tail that tapers to a point, as in lungfishes; vertebral column extends to tip without upturning.

Diphyodont Only two sets of teeth, one deciduous and one permanent (from "di" = two,"phyo+ = generation and "dont" =teeth).

Dipicolinic acid A substance present at high concentrations in the bacterial endospore. It is thought to contribute to the endospore's heat resistance.

Diploblastic Animals whose body parts are organized into layers that are derived embryologically from two tissue layers: ectoderm and endoderm. Animals in the phyla Cnidaria and Ctenophora are diploblastic.

Diploid A full set of genetic material, consisting of paired *chromosomes* one chromosome from each parental set. Most animal cells except the *gametes* have a diploid set of chromosomes. The diploid human *genome* has 46 chromosomes. Compare *haploid.*

Diploid cell A cell containing two sets of chromosomes (2n), one set inherited from each parent.

Diploid life cycle Occurs when the only multicellular stage in an organism's life cycle is diploid.

Diplostomulum Strigeoid metacercaria in the family Diplostomatidae.

Diplotene The fourth stage of prophase 1 of meiosis, following the pachytene stage, in which there is continued shortening and thickening of the chromosomes. Final gene transcription is completed, after which diakinesis occurs.

Diporpa Larval stage in the life cycle of the monogenean Diplozoon.

Diptera The true flies, an order of insects with sucking mouth parts; nearly all posses a hind pair of wings modified to serve as stabilizers during flight.

Direct action Civil disobedience, guerrilla street theater, picketing, protest marches, road blockades, demonstrations, and other techniques borrowed from the civil rights movement and applied to environmental protection.

Direct development In arthropods refers to development in which a juvenile hatches from the egg, and the juvenile is not distinctly different from adult except in size and maturity.

Direct fitness A measure of an individual's potential to contribute genes to future generations via personal reproduction.

Direct flight Insect flight that is accomplished by flight muscles acting on wing bases and in which a single nerve impulse results in a single wing cycle; also called synchronous flight.

Direct repeats Identical or related sequences present in two or more copies in the same orientation in the same molecule of DNA; they are not necessarily adjacent.

Directed molecular evolution A laboratory version of evolution at the molecular level that can produce "designer molecules." A large starting population of molecules (typically nucleic acids) that varies randomly in base sequence and shape is subjected to replication with variation, followed by selection. After several cycles of replication and selection, the population of molecules will evolve toward one containing a high proportion of molecules well adapted to the selection criterion applied.

Directed- or adaptive mutation A mutation that seems to be chosen so the organism can better adapt to its surroundings.

Directional selection Natural selection that occurs when individuals at one phenotypic extreme have an advantage over individuals with more common phenotypes.

Dirt Displaced soil.

Disaccharides 1. Sugars made up of two monosaccharides held together by a covalent bond; e.g., sucrose and lactose.

2. Type of sugar (saccharide) composed of two sugar molecules bonded together with an ester (covalent) bond examples include sucrose, maltose, and lactose.

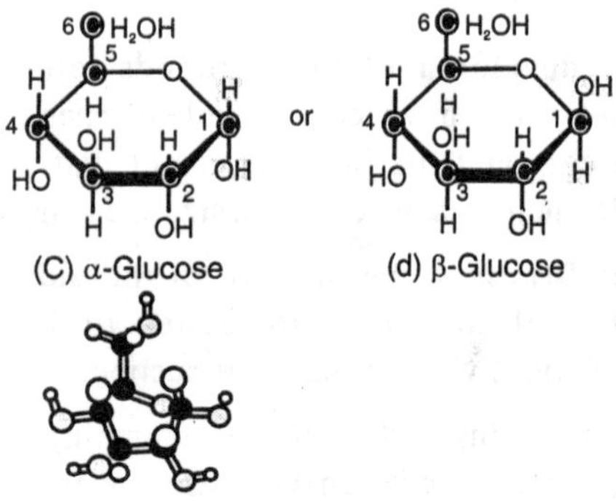

Fig. *Monosaccharides*

Discal cell Name given to a prominent and often quite large cell near the middle of the wing. The discal cell of one insect group may not be bounded by the same veins as that of another group.

Discharge The amount of water that passes a fixed point in a given amount of time; usually expressed as liters or cubic feet of water per second.

Discoidal cleavage Incomplete division of the blastodisc, a region of yolk-free, active cytoplasm; characteristic of birds, fishes and reptiles.

Discontinuous feeder An animal that does not feed all the time; instead it generally eats large meals sporadically and does not spend time in the continuous pursuit of prey.

Discontinuous variation Occurs when the phenotypes of traits controlled by a single gene can be sorted into two distinct phenotypic classes.

Discrete signal A form of communication that is digital, or all or none; usually given at the same intensity each time, such as an alarm call.

Discrete variable A variable that can take on only certain values (usually integers), as opposed to a continuous variable.

Discrimination The ability to detect differences between two or more stimuli and make choices based on those differences in a learning situation.

Disease Organisms suffer from disease when their normal function is impaired by some genetic disorder, or more often from the activity of a parasite or other organism living within them. Many diseases are caused by viruses, bacteria, or fungi.

Disease syndrome A set of signs and symptoms that are characteristic of the disease.

Disinfection The killing, inhibition, or removal of microorganisms that may cause disease. It usually refers to the treatment of inanimate objects with chemicals.

Disinfection by-products (DBPs) Chlorinated organic compounds formed during chlorine use for water disinfection. Many are carcinogens.

Dispersal The scattering of organisms of a species, often following a major reproductive event. Spores and larvae are commonly dispersed into the environment. Pollen or gametes may also be dispersed, but in this case the intent is to target another individual so that reproduction may occur. Organisms may disperse as spores, seeds, eggs, larvae, or adults.

Displacement activity An innate or stereotypic response to a stimulus that seems inappropriate or irrelevant to the situation.

Displacement Of a tooth refers to its movement within the confines of the tooth socket. A tooth can be displaced more easily when forced in a lateral direction than when forced into the socket. Continual or frequent displacement of a tooth may lead to it repositioning itself in the socket.

Display Any behavior pattern especially adapted in physical form or frequency to function as a social signal in communication.

Disruptive coloration A color pattern that helps break the outline of an organism.

Disruptive selection Natural selection that occurs when individuals of the most common phenotypes are at a disadvantage; produces contrasting subpopulations.

Dissimilatory nitrate reduction The process in which some bacteria use nitrate as the electron acceptor at the end of their electron transport chain to produce ATP. The nitrate is reduced to nitrite or nitrogen gas.

Dissimilatory reduction The use of a substance as an electron acceptor in energy generation. The acceptor (e.g., sulfate or nitrate) is reduced but not incorporated into organic matter during biosynthetic processes.

Dissociation The breaking up of a molecule into ions when placed in water or other solvents.

Dissolved organic matter Dissolved molecules derived from degradation of dead organisms or excretion of molecules synthesized by organisms

Dissolved oxygen (DO) contentAmount of oxygen dissolved in a given volume of water at a given temperature and atmospheric pressure; usually expressed in parts per million (ppm).

Distad In a direction away from the body.

Distal Away from the point of attachment of a structure on the body (e.g., the toes are distal to the knee).

Distal cytoplasm Distal cytoplasmic layer in the tegument of Monogenea, Digenea, and Cestoidea.

Distal tubule The section of the renal tubule where tubular secretion occurs.

Distillation Boiling of a liquid to evaporation and subsequent condensation of the vapors foD the purposes of purification and concentration.

Distilled spirit Alcoholic beverage with an alcoholic content between 80-100 proof, obtained by distillation of a beer or wine.

Distome Fluke with two suckers, oral and ventral.

Distribution area The geographic range of a taxon.

Allopatric Literally, "other country"; refers to distribution areas of different taxa that do not overlap.

Circumboreal Found throughout the high latitude forests of the northern hemisphere; that is, in North America and Eurasia.

Circumpolar Refers to a distribution area that circles either the north pole or south pole.

Cosmopolitan Worldwide, or nearly so, in distribution.

Disjunct Refers to a fragmented distribution area with two or more geographically separated ranges.

Pantropical Refers to a distribution area that extends through the tropics (that is generally between 23° 30' S and 23° 30' N latitude).

Relict A distribution area that is a mere remnant of a formerly wider range.

Sympatric Literally, "same country"; refers to distribution areas of different species that overlap.

Disturbance A rapid change in an environment that greatly alters a previously persistent biological community

Disulfide bond A linkage between the sulfur atoms of two different amino acids in a protein.

Diurnal Relating to the daytime; occurring during daylight hours. A mammal that is active mostly during daylight hours is said to be "diurnal". Voles tend to be diurnal rather than "nocturnal" which refers to the night.

Diurnal oxygen shifts The changes in oxygen levels that occur in waters when algae produce and use oxygen on a cyclic basis during day and night.

Diurnal tide A tidal pattern with a high and a low tide each day.

Dive reflex the suite of internal responses, including bradycardia and peripheral circulation shutdown, that occurs during dives by an air-breathing vertebrate

Divergent evolution The evolution of one species to a number of different forms. Compare *convergent evolution.*

Divergent plate boundary The boundary between two tectonic plates that are moving apart.

Diversity a) The total number of species present;

b) some index which incorporates both the number of species and the relative abundance of each.

Diversity gradient A regular change in diversity correlated with a geographic space or gradient of some environmental factor

Diving reflex The reflex certain animals have to stay underwater for prolonged periods of time.

Division A taxonomic grouping of related, similar classes; a high- level category below kingdom and above class. Division is generally used in the classification of prokaryotes, algae, fungi, and plants, whereas an equivalent category, phylum, is used in the classification of protozoa and animals.

Dizygotic Twins that arise from two different zygotes, hence, individuals no more closely related than two different-aged siblings.

DNA gun Any one of several methods of inserting a transformed plasmid into a cell.

DNA hybridization The formation of hybrid DNA molecules that contain a strand of DNA from two different species. The number of complementary sequences in common in the two strands is an indication of the degree of relatedness of the species.

DNA ligase An enzyme (usually from the T4 bacteriophage) which catalyzes formation of a phosphodiester bond between two adjacent bases from double-stranded DNA fragments. RNA ligases also exist, but are rarely used in molecular biology.

DNA methylation The addition of methyl groups ($-CH_3$) to bases of DNA after DNA synthesis; may serve as a long-term control of gene expression.

DNA microarrays Solid supports that have DNA attached in highly organized arrays and are normally used to evaluate gene expression.

DNA polymerase An enzyme that catalyses the synthesis of double stranded DNA using a primer and DNA as a template - used in PCR.

DNA probe A chemically synthesized, radioactively labeled segment of nucleic acid used to find a gene of interest by hydrogen-bonding to a complementary sequence.

DNA probe synthesis The production of a single stranded DNA molecule with base sequence designed to be complimentary to DNA it is intended to bind to - may contain radioactively labelled bases which help with location of the probe later.

DNA replication A normal process in cells which are about to divide - DNA double strands dissociate and the enzyme DNA polymerase catalyses the addition of complimentary bases to each strand. Two identical DNA molecules are formed.

DNA sequencing The process of determining the sequence of the bases in a length of DNA - this uses chemical methods and can be done with a machine.

DNA structure The three-dimensional double helix as determined by Watson and Crick. Complimentary base pairing (adenine with thymine and cytosine with guanine) enables many hydrogen bonds to form between the two strands, holding the strands together.>

DNA vaccine A vaccine that contains DNA which encodes antigenic proteins. It is injected directly into the muscle; the DNA is taken up by the muscle cells and encoded protein antigens are synthesized. This produces both humoral and cell-mediated responses.

Dnase Deoxyribonuclease, A class of enzymes which digest DNA. The most common is DNase I, an endonuclease which digests both single and double-stranded DNA.

DNTP deoxyribonucleoside triphosphate.

Doctrine of signatures Early concept that the Creator placed certain items on earth for humans and identified their intended use by their shape.

Dolomite A carbonate sedimentary rock composed of more than 50 percent of the mineral calcium-magnesium carbonate (CaMg(CO3)2).

Domains 1. Compact, self-folding, structurally independent regions of proteins (usually around 100-300 amino acids in length); large proteins may have two or more domains connected by less structured stretches of polypeptide. In the antibody molecule, they are the loops, along with about 25 amino acids on each side, that form compact, globular sections.

2. The primary taxonomic groups above the kingdom level; all living organisms may be placed in one of three domains.

Domestic sewage Wastewater from homes and non-industrial buildings. Compare industrial sewage.

Domesticated plant A plant that has been genetically changed from the wild type due to artificial selection.

Dominance hierarchy A social ranking, formed through agonistic behavior, in which individuals are associated with each other so that some have greater access to resources than do others.

Dominant (1) The most prevalent species in a plant community.

(2) The allele that has its trait expressed.

(3) A gene that has its phenotypic expression appear in the offspring, regardless of the nature of its allelic partner.

Dominant allele In a heterozygote, the allele that is fully expressed in the phenotype.

Dominant plants Those plant species in a community that provide a food base for most of the community; they usually take up the most space and have the largest biomass.

Dominant species A species that exerts an overriding influence in determining the characteristics of a community.

Dominant trait A trait whose phenotype is determined by a single allele at a particular locus.

Dormancy A period of suspended growth and metabolic activity. Many plants, seeds, spores, and some invertebrates become dormant during unfavorable conditions.

Dormant Having reduced metabolic and respiratory activity.

Dorsal Pertaining to or situated near the back; opposite of ventral.

Dorsal nectary organ In the larvae of many species of Lycaenidae (Blue Butterflies) a gland located in the dorsal region of the 7th abdominal segment, it secretes a sweet substance which is attractive to ants.

Dorsal ocellus The simple eye in adult insects and in nymphs and naiads.

Dorsal plate Dorsal plate on the body of a mesostigmatid mite.

Dorsal shield The scutum or sclerotized plate covering all or most of the dorsal surface in males and the anterior portion in females, nymphs, and larvae of hard-backed ticks.

Dorso-central Bristles. The 2 rows of bristles running along the thorax of a fly on the outer side of the acrostichal bristles.

Dorsolateral hinge points During neurulation of bird embryos, regions involved in bending of the lateral neural plate; wedge-shaped neural cells are anchored to the surface ectoderm to form additional furrows near the connection of the neural plate to the remainder of the ectoderm.

Dorsolateral Pertaining to the side and the back (Behler 1988).

Dorso-lateral Towards the sides of the dorsal (upper) surface.

Dorso-ventral Running from the dorsal (upper) to the ventral (lower) surface.

Dorsum The upper surface or back of an animal.

Dot blot A technique for measuring the amount of one specific DNA or RNA in a complex mixture. The samples are spotted onto a hybridization membrane (such as nitrocellulose or activated nylon, etc.), fixed and hybridized with a radioactive probe. The extent of labeling (as determined by autoradiography and densitometry) is proportional to the concentration of the target molecule in the sample. Standards provide a means of calibrating the results.

Double bond A covalent bond sharing two pairs of electrons.

Double circulation A circulation scheme with separate pulmonary and systemic circuits, which ensures vigorous blood flow to all organs.

Double diffusion agar assay (Ouchterlony technique) An immunodiffusion reaction in which both antibody and antigen diffuse through agar to form stable immune complexes, which can be observed visually.

Double fertilization A characteristic of angiosperms in which a pollen tube carries two sperm cells to the female gametophyte in the ovule. One sperm cell fuses with the egg cell and gives rise to a diploid embryo The other sperm cell fuses with the two polar cells to form a triploid cell that develops into the endosperm.

Double helix The shape that two linear strands of DNA assume when bonded together.

Double membrane In mitochondria and plastids, there is a two-layered membrane which surrounds the organelle. This is believed to be the result of endosymbiosis, with the outer membrane coming from the eukaryotic cell, and the inner membrane belonging to the original prokaryote which was "swallowed".

Dourine Disease of horses and other equids caused by Trypanosoma equiperdum.

Down feathers Feathers that provide insulation for adult and immature birds.

Down syndrome A human genetic disease resulting from having an extra chromosome 21, characterized by mental retardation and heart and respiratory defects.

Down stream Identifies sequences proceeding farther in the direction of *expression*; for example, the coding region is downstream from the *initiation codon*, toward the 3' end of an mRNA molecule. Sometimes used to refer to a position within a protein sequence, in which case downstream is toward the *carboxyl* end which is synthesized after the *amino* end during translation.

DPT vaccine A vaccine containing three antigens that is used to immunize people against diphtheria, pertussis or whooping cough, and tetanus.

Drag Resistance to movement through water or any other medium.

Drift net A very long fishing net that is allowed to drift for a long time before it is pulled on board.

Drill core A column of material (e.g., mud, ice, rock) removed from the earth by drilling. Often used as a tool for exploration of natural resources.

Drip irrigation Uses pipe or tubing perforated with very small holes to deliver water one drop at a time directly to the soil around each plant. This conserves water and reduces soil waterlogging and salinization.

Drives Manifestations of specific motivations corresponding to specific needs.

Drone The male honey bee.

Droplet nuclei Small particles (1 to 4 mm in diameter) that represent what is left from the evaporation of larger particles (10 mm or more in diameter) called droplets.

Dropsy An old term for congestive heart failure which produces edema and resulting decrease in manual dexterity, causing one to drop things.

Drought An environmental condition in which precipitation is not sufficient to maximize biological productivity.

Drowned river valley (or coastal plain) estuaryAn estuary that is formed when sea level rose at the end of the last glacial age.

Drumlin Elongated mound of glacial sediment deposited parallel to ice flow.

Drupe A fleshy fruit with a one-carpeled ovary and only one seed. The endocarp is hard and stony, tightly enclosing the seed; the mesocarp is fleshy, and the exocarp is soft and thin; example: cherry.

Dry alkali injection Spraying dry sodium bicarbonate into flue gas to absorb and neutralize acidic sulfur compounds.

Dry fruit Fruit in which the cells of the pericarp are dry (dead) at maturity.

DS "double-stranded"

Dual-gland adhesive organ Organs in the epidermis of most turbellarians, with three cell types; viscid and releasing gland cells and anchor cells.

Duct A tube which carries a secretion onto the surface of skin or mucosa.

Duff layer Organic matter in various stages of decomposition on the floor of the forest.

Dulosis The act of slave making in ants, a species which makes a slave of another is often referred to as Dulotic.

Duo-adhesive gland Platyhelminth tegument gland with two types of cells, one producing an adhesive substance and the other a releasing substance.

Duodenum The first and shortest portion of the small intestine lying between the pyloric end of the stomach and the jejunum.

Duplex A nucleic acid molecule in which two strands are *base paired* with each other.

Duplication An aberration in chromosome structure resulting from an error in meiosis or mutagens; duplication of a portion of a chromosome resulting from fusion with a fragment from a homologous chromosome.

Dyad One of the groups of two chromosomes formed by the division of a tetrad during the first meiotic division.

Dynein A microtubule-dependent motor protein; cytoplasmic dyneins work in organelle transport and mitosis, whereas the closely related ciliary dyneins are attatched to outer doublet microtubules in a cilium or flagellum, providing sliding/bending force used in (for example) sperm propulsion.

Dyspnea Difficult or labored breathing.

Dystrophin Protein making up only 0.002% of all protein in skeletal muscle but which appears vital for proper functioning of the muscle. Sufferers of muscular dystrophy appear to lack dystrophin.

E

E. coli A common Gram-negative bacterium useful for cloning experiments. Present in human intestinal tract. Hundreds of strains of *E. coli* exist. One strain, K-12, has been completely sequenced.

Ear stone (otolith) A calcareous body in the inner ear of fishes and other vertebrates that is used in the maintenance of equilibrium.

Early mRNA Messenger RNA produced early in a virus infection that codes for proteins needed to take over the host cell and manufacture viral nucleic acids.

Earlywood The large, thin-walled xylem cells produced early in the growing season, which appear less dense than latewood.

Ebola virus hemorrhagic fever An acute infection caused by a virus that produces varying degrees of hemorrhage, shock, and sometimes death.

Eccrine Applies to a type of mammalian sweat gland that produces a watery secretion.

Eccrine glands Sweat glands that are linked to the sympathetic nervous system and are widely distributed over the body surface.

Ecdysiotropin Hormone secreted in brain of insects that stimulates prothoracic gland to secrete molting hormone. Prothoracicotropic hormone; brain hormone.

Ecdysis (1) The shedding of the arthropod exoskeleton to accommodate increased body size or a change in morphology (as may occur in molting from immature to adult); to molt; may also refer to the shedding of the outer epidermis of the skin of reptiles.

(2) The shedding of the cuticle in aschelminths in order to grow. Also called molting.

Ecdysone Molting hormone of arthropods, stimulates growth and ecdysis, produced by prothoracic glands in insects and Y organs in crustaceans.

Echinodermata The phylum of coelomate animals whose members are pentaradially symmetrical as adults and possess an endoskeleton covered by epithelium and a water-vascular system. Sea stars, sea urchins, sea cucumbers, sea lilies.

Echinoderms Invertebrates with radial symmetry and a water vascular system.

Echinoidea The class of echinoderms whose members are globular or disk shaped, possess moveable spines, and a skeleton of closely fitting plates. Sea urchins and sand dollars.

Echiura A phylum of protostomate, marine animals whose members burrow in mud or sand or live in rock crevices. They possess a spatula-shaped proboscis and are 15 to 50 cm in length. Spoon worms.

Echolocation A method of locating objects by determining the time it takes for an echo to return and the direction from which it returns. As in bat echolocation.

Eclipse period The initial part of the latent period in which infected host bacteria do not contain any complete virions.

Eclipsed antigen Antigen borne by the parasite that is common to both the host and the parasite but that genetically is of parasite origin.

Eclosion The process whereby the adult form of an insect emerges from the pupa.

Ecocentric A philosophy that claims moral values and rights for both organisms and ecological systems and processes.

Ecocline The gradient between adjacent biomes; a gradient of environmental conditions.

Ecofeminism A pluralistic, nonhierarchical, relationship-oriented philosophy that suggests how humans could reconceive themselves and their relationships to nature in nondominating ways as an alternative to patriarchal systems of domination.

Ecogeographic rule One of the statements from 19th century naturalists recognizing correlations between the morphology of warm blooded animals (mammals and birds) and climatic and/or latitudinal gradients.

Bergmann's rule The colder the climate (thus, usually, the higher the latitude) the larger the body size of a warm blooded animal when compared to close relatives in warmer regions.

Allen's rule The warmer the climate the longer the appendages (ears, legs, wings) of warm blooded animals in comparison with closely related taxa from colder climes.

Golger's rule The drier the climate, the lighter the color of animals relative to closely related taxa of more humid regions. (Yellows and light browns predominate in arid regions; dark browns and blacks in humid regions.)

Ecological adaptation Short-term changing responses expressed by an individual organism to its environment.

Ecological balance Astable balance in the numbers of each species in an ecosystem .In the ecosystem of the mouth this balance is brought about by competition and cooperation between the different organism and the hosts defences which tend to control population size.

Ecological development A gradual process of environmental modification by organisms.

Ecological efficiency The ratio of net productivity at one trophic level to net productivity at the next lower level.

Ecological equivalents Different species that occupy similar ecological niches in similar ecosystems in different parts of the world.

Ecological niche The role a plant or animal plays in its community. The niche of an organism is defined by what it eats, its predators, salt tolerances, light requirements etc. Two species cannot live stably in the same habitat if they occupy identical niches.

Ecological pyramid A graphic representation of the quantitative relationships of numbers of organisms, biomass, or energy flow between the trophic levels of an ecosystem. Because large amounts of energy and biomass are dissipated at every trophic level, these diagrams nearly always take the form of pyramids.

Ecological pyramid of biomass Total biomass of all organisms at each trophic level in a food chain; typically biomass declines with successively higher trophic levels.

Ecological pyramid of energy Total energy content of all organisms at each trophic level in a food chain; since only 5%_20% of energy is passed between trophic levels; energy content declines at successively higher trophic levels.

Ecological pyramid of numbers Number of organisms supported at each trophic level in a food chain; typically, fewer organisms are supported at successively higher trophic levels.

Ecological species concept The idea that ecological roles (niches) define species.

Ecological succession (according to the original theory) The development of an ecosystem through a predictable series of communities until a final, stable community (the climax community) in balance with the regional climate is attained. In its original form, the theory implied that each community altered the habitat and prepared it for invasion by the next, succeeding community.

Ecological time A timescale that focuses on community events that occur on the order of tens to hundreds of years.

Ecology The scientific study of relationships between organisms and their environment. It is concerned with the life histories, distribution, and behavior of individual species as well as the structure and function of natural systems at the level of populations, communities, and ecosystems.

Economic defendability Defense of a resource that yields benefits that outweigh the costs of defending it.

Economic development A rise in real income per person; usually associated with new technology that increases productivity or resources.

Economic growth An increase in the total wealth of a nation; if population grows faster than the economy, there may be real economic growth, but the share per person may decline.

Economic sanctions Using the threat or action of withholding food, medical care, or other critical societal needs to force compliance with one's policies.

Ecosystem A community of plants, animals and other organisms that are linked by energy and nutrient flows and that interact with each other and with the physical environment. Rain forests, deserts, coral reefs, and grasslands are examples of ecosystems.

Ecosystem diversity The diversity of biological communities and their physical environment. Diversity is determined by the species composition, physical structure and processes within an ecosystem. This is the highest level of biodiversity.

Ecosystem management An integration of ecological, economic, and social goals in a unified systems approach to resource management.

Ecosystem restoration To reinstate an entire community of organisms to as near its natural condition as possible.

Ecotage Direct action (guerrilla warfare) or sabotage in defense of nature; also called monkey wrenching.

Ecotone A transition zone between two distinct habitats that contains species from each area, as well as organisms unique to it.

Ecotourism A combination of adventure travel, cultural exploration, and nature appreciation in wild settings.

Ecotype A subdivision of a species; a stage in the formation of a species such that reproductive isolation has occurred.

Ectocommensal Commensal symbiont that lives on the outer surface of its host.

Ectoderm The outer basic layer of tissue in those animals with true tissues. In vertebrates, for instance, the embryonic ectoderm differentiates into the skin and also the nervous system.The outer of the three cell layers which form, as the clump of early embryonic cells begins to differentiate. The ectoderm will form the epidermis of the skin and the nervous system. The other two layers are the mesoderm and the endoderm.

Ectognathous Derived character of most insects; mandibles and maxillae not in pouches.

Ectohormone A substance secreted by an animal to the outside of its body causing a specific reaction, such as determination of physiological development, in a receiving individual of the same species.

Ectolecitha l Yolk for nutrition of the embryo contributed by cells that are separate from the egg cell and are combined with the zygote by envelopment within the eggshell.

Ectomycorrhizae Association of a fungus with a vascular plant root or rhizome system in which the fungus surrounds, but does not penetrate the living cells (protoplast) of the system.

Ectomycorrhizal Referring to a mutualistic association between fungi and plant roots in which the fungus surrounds the root tip with a sheath.

Ectoparasites Types of parasites that live externally on their host; fleas and lice are obligate ectoparasites of man and animals.

Ectopic Infection in a location other than normal or expected.

Ectopic pregnancy Implantation of the embryo in the oviduct or other site outside the uterus typically caused by failure of zona pellucida to keep blastocyst from adhering to oviduct walls; it is a dangerous condition that can cause life-threatening hemorrhage in the mother of the developing embryo. [ectopic: misplaced, not in the normal location.]

Ectoplasm The cortex of a cell or that part of cytoplasm just under the cell surface; contrasts with endoplasm.

Ectoprocta A phylum of animals whose members are colonial and fresh water or marine. Anus ends outside a ring of tentacles. Lophophore used in feeding. Moss animals or bryozoans.

Ectosymbiosis A type of symbiosis in which one organism remains outside of the other organism.

Ectotherm An animal such as a reptile, fish, or amphibian, that must use environmental energy and behavioral adaptations to regulate its body temperature.

Ectothermic Capable of maintaining body temperature by gaining outside heat outside itself.

Ectotympanic Ecto = outer; tympani = a drum.

Edaphic factor A permanent or nearly permanent condition of the substrate that influences the types of plants that grow in an area. For example, substrates may be permanently or seasonally waterlogged, droughty, deficient in essential nutrients, extremely thin, and so forth.

Edema Escape of fluid from blood into interstitial space, causing swelling.

Edge effects A change in species composition, physical conditions, or other ecological factors at the boundary between two ecosystems. Some organisms flourish at this boundary and benefit from processes such as habitat fragmentation that increase edge area. Other organisms are harmed by increasing edge effects.

Edge That part of an ecosystem near the perimeter that is influenced by the environment of the adjacent ecosystem so that it differs in some characteristics from the center of the ecosystem. *Edge effect* refers to changes in species composition, distribution and/or abundance found in the edge relative to the interior.

Edman reaction A chemical reaction used to remove and analyse amino acids one by one from a protein, used in determining the primary structure of a protein, this can be done with a machine.

Eegestion vacuole A membrane-bound vesicle within the cytoplasm of a protist that functions in expelling wastes.

Eegg apparatus Egg cell and adjacent synergids in the embryo sac (female gametophyte) of angiosperms.

Eelectrode stimulation The stimulation of a neuron by artificially depolarizing the membrane with an electrode.

Eelectron A small, negatively charged particle that revolves around the nucleus of an atom. It has a very low mass.

Eemergent plant community Marine plant community in which most leafy growth occurs above sea level.

Eend bulb A tiny swelling on the terminal end of telodendria at the distal end of an axon; also called the synaptic bouton.

Eendemic Peculiar to a certain region or country; native to a restricted area; not introduced.

Eendocrinology The study of the endocrine system and its role in the physiology of an animal.

Eendogenous factors Internal factors regulating an organism's growth and development.

Eenvironmental ethics A search for moral values and ethical principles in human relations with the natural world.

Eenvironmental science The systematic, scientific study of our environment as well as our role in it.

Eenzymes Molecules, usually proteins or nucleic acids, that act as catalysts in biochemical reactions.

Eenzyme-substrate complex The binding of a substrate molecule to the active site of an enzyme.

Eeosinophilia Elevated eosinophil count in the circulating blood; commonly associated with chronic parasite infections.

Eepicaridium First larval stage of the isopod suborder Epicaridea; attaches to a free-living copepod.

Eepidemiology The study of the distribution and causes of disease and injuries in human populations.

Eepigenetics Study of the relationship between genotype and phenotype as mediated by developmental processes.

Eepizootic Massive infection rate among animals other than humans; identical to an epidemic in humans.

Eergosterol The most common sterol; can be converted to vitamin D in the skin of animals exposed to sunlight.

Eesophagus The passage extending from the pharynx to the stomach.

Eethylene Gaseous plant hormone involved in fruit ripening and other aspects of plant growth and development.

Eeuchromatin Part of the chromatin that takes up stain less than heterochromatin, contains active genes.

Eevolutionary duration The length of time that a species or higher taxon exists in geological time.

Eexocrine gland A gland that secretes its product to an epithelial surface, directly or through ducts.

Effacing lesion The type of lesion caused by enteropathogenic strains of E. coli (EPEC) when the bacteria attach to and destroy the brush border of intestinal epithelial cells. The term AE (attaching-effacing) E. coli is now used to designate true EPEC strains that are an important cause of diarrhea in children from developing countries and in traveler's diarrhea.

Effector In a closed system, the element that initiates an action in response to a signal from a sensor. In human systems, a muscle or gland often serves as an effector.

Efferent Leading or conveying away from some organ, for example, nerve impulses conducted away from the brain, or blood conveyed away from an organ; contrasts with afferent.

Effluent Industrial or urban waste discharged into the environment.

Effluent sewerage A low-cost alternative sewage treatment for cities in poor countries that combines some features of septic systems and centralized municipal treatment systems. A septic tank near each residence collects and processes solid wastes_and must be pumped periodically just like a septic tank_while liquid effluent is collected and piped to a central treatment plant. This avoids drainfields in urban areas and yet allows sewer pipes and treatment plants to be downsized, saving money.

Egestion Act of casting out indigestible or waste matter from the body by any normal route.

Egg (1) A large gamete without flagellae that is

fertilized by a sperm cell. An egg cell is also called an ovum.

(2) A complex multicellular structure in which an animal embryo develops.

Egg pod A capsule which encloses the egg mass of grasshoppers and which is formed through the cementing of soil particles together by secretions of the ovipositing female.

Ehrlichiosis A tick-borne (Dermacentor andersoni, Amblyomma americanum) rickettsial disease caused by Ehrlichia chaffeensis. Once inside leukocytes, a nonspecific illness develops that resembles Rocky Mountain spotted fever.

Eicosanoids Are a class of hormones which are all made from phospholipids. They include prostaglandins, thromboxanes, and leucotriens.

Ejaculatory duct In males, a short duct that connects the vas deferens from each testis to the urethra.

Ekman circulation Movement of surface water at an angle from the wind, as a result of the Coriolis effect

Ekman spiral The spiral change in the movement of water in the water column when the water is pushed by wind. The Ekman layer is that part of the water column affected by wind; Ekman transport is the net water movement 90from the wind direction.

El Nino A complex set of changes in the water temperature in the Eastern Pacific equatorial region, producing a warm current; it occurs annually to some degree between October and February, but in some years intensifies and causes unusual storms and destruction of marine life and land ecosystems. (From Spanish for "the child;" meaning the Christ child; it typically begins at Christmas time).

El niño-southern oscillation (ENSO) Condition in which warm surface water moves into the eastern Pacific, collapsing upwelling and increasing surface-water temperatures and precipitation along the west coast of North and South America

Elastic fibres Are long, thin, ribbons-like fibres, sometimes even sheet-like. They are composed of a central core of elastin, a rubbery protein, surrounded by glycoprotein microfilaments. Elastin is found all over the body but particularly in the walls of blood vessels and in our vocal chords.

Elater A cell or part of a cell which assists in dispersing spores. The elaters change shape as they lose or acquire water, and they will then push against surrounding spores.

Elateriform larva A larva with the form of a wireworm; i.e. long and slender, heavily sclerotized, with short thoracic legs, and with few body hairs.

Elbowed antenna Antenna, particularly of ants, in which there is a distinct angle between two of the segments - usually between the 1st and 2nd segments, in which case the 1st segment is usually much longer than the others.

Electric potential The difference in the amount of electric charge between a region of positive charge and a region of negative charge. The establishment of electric potentials across the plasma membrane and across organelle membranes makes possible a number of phenomena, including the chemiosmotic synthesis of ATP, the conduction of nerve impulses, and muscle contraction.

Electrical synapse A synapse at which local currents resulting from electrical activity flow between the two neurons through gap junctions joining them.

Electrochemical gradient The diffusion gradient of an ion, representing a type of potential energy that accounts for both the concentration difference of the ion across a membrane and its tendency to move relative to the membrane potential.

Electrogenic pump An ion transport protein generating voltage across the membrane.

Electrolyte A substance that dissociates into ions when fused or in solution and thus becomes capable of conducting electricity; an ionic solute.

Electromagnetic spectrum The entire radiation spectrum from the high-energy levels of cosmic rays to the low-energy levels of radiowaves; a small portion in the center provides wavelengths visible to humans.

Electron A subatomic particle with a negative charge. Electrons circle the atom's nucleus in regions of space known as orbitals.

Electron acceptor A molecule that forms part of the electron transport system that transfers electrons ejected by chlorophyll during photosynthesis. Part of the energy carried by the electrons is transferred to ATP, part is transferred to NADPH, and part is lost in the transfer system.

Electron carrier A molecule that conveys electrons; one of several membrane proteins in electron transport chains in cells. Electron carriers shuttle electrons during the redox reactions that release energy used to make ATP.

Electron donor Substance that donates or gives up electrons in an oxidation-reduction reaction, becoming oxidized in the process.

Electron microscope (EM) A microscope that focuses an electron beam through a specimen, resulting in resolving power a thousandfold greater than that of a light microscope. A transmission electron microscope (TEM) is used to study the internal structure of thin sections of cells. A scanning electron microscope (SEM) is used to study the fine details of cell surfaces.

Electron shell An energy level at which an electron orbits the nucleus of an atom.

Electron transport 1) A series of coupled oxidation/reduction reactions where electrons are passed like hot potatoes from one membrane-bound protein/enzyme to another before being finally attached to a terminal electron acceptor (usually oxygen or NADPH). ATP is formed by this process.

2) coupled series of oxidation/reduction reactions during which ATP is generated by energy transfer as electrons move from high reducing state to lower reducing state.

Electron transport chain In a cell membrane, electron carriers and enzymes positioned in an organized array that enhance oxidation-reduction reactions; such systems function in the release of energy that is used in ATP formation and other reactions.

Electron transport system Hydrogen ions produced during the 3 preparatory steps of aerobic respiration are carried by nicotinamide adenine dinucleotide (NAD). The hydrogen ion plus one electron form NADH, which is taken to the electron transport system. This transport system is run by a series of 5 molecules. The first, removes the two electrons from NADH,(one comes from the hydrogen atom, leaving behind a hydrogen ion). These two electrons, bounce from the first molecule in the transport system to the second, third, forth and then last one, cytochrome oxidase, which finally places the electrons onto oxygen gas O_2. The electron rich oxygen atoms are attractive to the hydrogen ions and they combine to form water. (Oxygen in the process of aerobic respiration acts therefore as an electron acceptor). In the process of bouncing "downhill" the electrons have released sufficient energy to power up a small battery. This battery has been made by pumping hydrogen ions out of the inner membrane of the mitochondria. The collection of hydrogen ions outside, piles up and their electrical pressure mounts. They want to get back across the membrane, and are allowed, one at a time to pass back through the enzyme ATP synthase. This enzyme sits like a water wheel in the cell membrane, turned by the passage of hydrogen ions. Its turning wheel builds an ATP molecule in every turn. The wheel may be going at about 200 revolutions per second, powering the synthesis of an ATP molecule with each turn. ATP formed in this way takes a while but can be sustained to fuel the body during aerobic exercise. When the demand for power exceeds this rate, the cells have to rely on anaerobic respiration.

Electronegativity The tendency for an atom to pull electrons toward itself.

Electrophoresis A method of separating large molecules (such as DNA fragments or *proteins*) from a mixture of similar molecules. An electric current is passed through a medium containing the mixture, and each kind of molecule travels through the medium at a different rate, depending on its electrical charge and size. Separation is based on these differences. Agarose and acrylamide gels are the media commonly used for electrophoresis of proteins and nucleic acids. A method of separation of molecules in a mixture; an electric current is passed through the mixture which has been loaded on a support medium, usually a gel or sometimes paper, the molecules move through the support - the extent of each molecule's movement is related to its size and its charge.

Electroporation A method for introducing foreign nucleic acid into bacterial or eukaryotic cells that uses a brief, high voltage DC charge which renders the cells permeable to the nucleic acid. Also useful for introducing synthetic peptides into eucaryotic cells.

Electroreceptor A receptor that senses changes in an electrical current usually in the surrounding water; also called an ampullary organ.

Electrostatic A force generated by differences in electric charge of two particles.

Electrostatic attraction The attraction between atoms of opposite charge that holds the atoms together in ionic bonds.

Electrostatic precipitators The most common particulate controls in power plants; fly ash particles pick up an electrostatic surface charge as they pass between large electrodes in the effluent stream, causing particles to migrate to the oppositely charged plate.

Element Building blocks of matter that cannot be broken down into a simpler substance. Each element is composed of just one type of atom.

Elementary body A small, dormant body that serves as the agent of transmission between host cells in the chlamydial life cycle.

Eleocyte Fat-containing cells in annelids that originate from the chlorogogen tissue.

Elephantiasis Disfiguring condition caused by chronic infection with filarial worms Wuchereria bancrofti and Brugia malayi.

Ellipsoidal Elliptical shape.

Elongation cycle The cycle in protein synthesis that results in the addition of an amino acid to the growing end of a peptide chain.

Elongation During protein synthesis, the growth of the polypeptide chain through the addition of amino acids; the second step in translation.

Elver a juvenile eel

Elytra The thickened or leathery front or forewings of insects such as beetles.

Elytron (plural elytra) The tough, horny forewing of a beetle or an earwig.

Emarginate With a distinct notch or indentation in the margin.

Embden-meyerhof pathway A pathway that degrades glucose to pyruvate; the six-carbon stage converts glucose to fructose 1,6-bisphosphate, and the three-carbon stage produces ATP while changing glyceraldehyde 3-phosphate to pyruvate.

Embolium A narrow region along the front margin of the forewing in certain heteropteran bugs: separated from the rest of the corium by a groove or suture.

Embryo 1. the stage of multicellular organism that develops form a zygote before it becomes free-living.

2. specifically, in vertebrates, the period from after the long axis appears until all major structures are represented. In humans, this from about two weeks after fertilization to the end of the seventh or eighth week.

Embryo sac Alternate term applied to the angiosperm female gametophyte contained within a megaspore.

Embryogenesis The origin and development of the embryo; embryogeny.

Embryology The study of development from the egg to the point that all major organ systems have formed.

Embryonic axis The main root/shoot body of a seedling.

Embryonic diapause The arresting of early development to allow young to hatch, or be born, when environmental conditions favor survival.

Embryophore In reference to the eggshell of many cestodes, that portion contributed by the inner envelope, derived from embryonic blastomeres.

Embryophyte Synonym for the Plantae, as here defined. It includes all green photosynthetic organisms which begin the development of the sporophyte generation within the archegonium.

Emergence The appearance of properties in a biological system (at the molecular, cellular, organismal, or species levels) that cannot be deduced from knowledge of the component parts taken separately or in partial combinations; such properties are termed emergent properties.

Emery's rule The rule resulting from the observation that species of social parasite are very closely related to their host.

Emigrate To move from one area to another to take up residence.

Emigration The departure of individuals from a given area

Emission standards Regulations for restricting the amounts of air pollutants that can be released from specific point sources.

Emphysema An irreversible, obstructive lung disease in which airways become permanently constricted and alveoli are damaged or destroyed.

Empodium An outgrowth between the claws of a fly's foot: it may be bristle-like.

Emulsify To produce small droplets of a liquid.

Emulsion A colloidal system in which both phases are liquids.

Enamel prisms Rod-like bundles of hydroxyapatie crystals which are orientated at right angles to the tooth surface. Each prism can be traced from the outside of the enamel all the way to the dentine junction.

Enamel The outer layer hard layer which covers the dentine around the crown of a tooth. Enamel consists of closely packed crystals of hydroxyapatite with very little organic material. A recognisable unit of structure in enamel is the enamel prism.

Enameloid A type of enamel found in fish and reptiles in which the enamel prisms are haphazardly arranged; in contrast enamel prism are parallel to each other and orientated at right angle to the tooth surface.

Enantiomer One of a pair of molecules that are mirror-image isomers of each other.

Enations Flaps of tissue such as those found on psilophytes.

Encapsulation The surrounding of an invading body, such as the egg of a parasite, by insect hemocytes (blood cells) and the formation of a protective capsule.

Encephalitis Infection of the brain, especially by viruses or amebas.

Encrusting Describing an organism that grows as a crust over rocks and other hard surfaces.

Encystation The formation of a cyst.

Encystment The secretion of a protective outer covering that permits some small invertebrates to withstand exposure to extreme environmental stresses, such as desiccation and overcrowding.

End labeling The technique of adding a radioactively labeled group to one end (5' or 3' end) of a DNA strand.

Endangered A species or ecosystem that is so reduced or delicate that it is threatened with or on the verge of extinction. Compare *extinct, extirpated, threatened, vulnerable.*

Endangered species A living species that is in danger of becoming extinct because of small population sizes, poor reproduction, reduced available habitat, or a combination of these factors.Regulation that protects species of animal or plant that has been identified as in danger of becoming extinct because of harmful human activity or environmental factors.

Endemic Native to a particular country, nation, or region.

Endemic disease A disease that is commonly or constantly present in a population, usually at a relatively steady low frequency.

Endemic species Species that are confined to a specific, relatively small geographic area.

Endemic typhus A form of typhus fever caused by the rickettsia Rickettsia typhi that occurs sporadically in individuals who come into contact with rats and their fleas.

Endemism A state in which species are restricted to a single region.

Endemnicity Amount or severity of a disease in a particular geographic area.

Endergonic Chemical reactions that require energy input to begin.

Endergonic reaction A reaction that does not spontaneously go to completion as written; the standard free energy change is positive, and the equilibrium constant is less than one.

Endite Medial process on an arthropod limb.

End-labelling Transfer of 32P to the 5' end(s) of a DNA or RNA molecule, using polynucleotide kinase.

Endocardium The inner layer of the heart rudiment that arises from splanchnic mesoderm and fuses with the epimyocardium to form the heart wall.

Endocarp Innermost layer of the pericarp (fruit wall).

Endochondral Occurring with the substance of cartilage, especially bone formation.

Endochondral ossification The process by which human bones form from cartilage.

Endocommensal Commensal symbiont that lives inside its host.

Endocrine A system of ductless glands secreting hormones into the bloodstream and targeting another organ to produce a response.

Endocrine glands A series of ductless glands in both invertebrates and vertebrates that release hormones into the body through the blood or lymph.

Endocrine system One of eleven major body organ systems in animals; a system of glands that works with the nervous system in controlling the activity of internal organs, especially the kidneys, and in coordinating the long-range response to external stimuli.

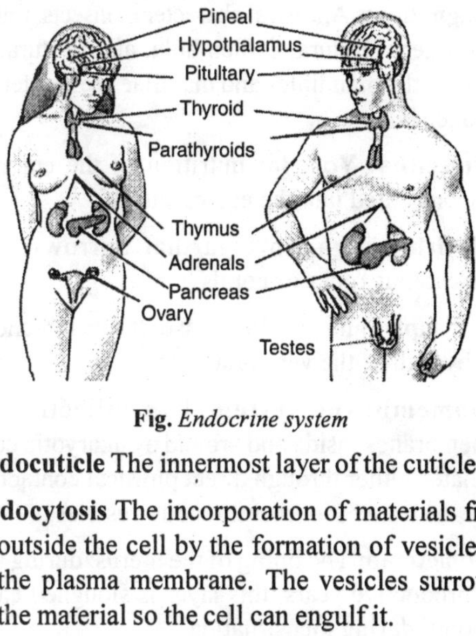

Fig. *Endocrine system*

Endocuticle The innermost layer of the cuticle.

Endocytosis The incorporation of materials from outside the cell by the formation of vesicles in the plasma membrane. The vesicles surround the material so the cell can engulf it.

Endoderm One of the three primary germ layers formed in the embryo, moved into interior by cell movements during gastrulation. In vertebrates, this innermost layer of cells goes on to form the linings of the gut (esophagus, stomach, intestines, rectum, colon), pharyngeal pouch derivatives (tonsils,thyroid, thymus, parathyroid glands), lungs, liver, gall bladder, pancreas. In amniotes, extraembryonic endoderm participates in the formation of the allantois and yolk sac.

Endodyogeny Same as endopolyogeny except that only two daughter cells are formed.

Endogenote The recipient bacterial cell's own genetic material into which the donor DNA can integrate.

Endogenous clock mechanism Any internal processes that are genetically based and that play a role in setting or regulating biological rhythms.

Endogenous infection An infection by a member of an individual's own normal body microbiota.

Endogenous pyrogen A substance such as the lymphokine interleukin-1, which is produced by host cells and induces a fever response in the host. It also is called simply a pyrogen.

Endognathous Ancestral character in insects, found in orders Diplura, Collembola, and Protura, in which the mandibles and maxillae are located in pouches.

Endolecithal Yolk for nutrition of the embryo incorporated into the egg cell itself.

Endolithic alga An alga that burrows into calcareous rocks or corals.

Endolymph Fluid that fills most of the membranous labyrinth of the vertebrate ear.

Endomembrane system The collection of membranes inside and around a eukaryotic cell, related either through direct physical contact or by the transfer of membranous vesicles.

Endometrium The lining of the uterus; during the reproductive years, this layer is sloughed each month during menstruation.

Endomycorrhizae Association of a fungus with a vascular plant root or rhizome system in which the fungus penetrates the living cells of the root.Referring to a mutualistic association of fungi and plant roots in which the fungus penetrates into the root cells and arbuscules and vesicles are formed.

Endonuclease An enzyme which digests nucleic acids starting in the middle of the strand (as opposed to an exonuclease, which must start at an end). Examples include the restriction enzymes, DNase I and RNase A.

Endoparasite A parasite which lives inside its host's body. Most of the ichneumons, are endoparasites during their larval stages.

Endophyte An organism that lives inside another.

Endoplasm The portion of cytoplasm that immediately surrounds the nucleus.

Endoplasmic reticulum A system of membranous tubules and flattened sacs (cisternae) in the cytoplasmic matrix of eucaryotic cells. Rough or granular endoplasmic reticulum (RER or GER) bears ribosomes on its surface; smooth or agranular endoplasmic reticulum (SER or AER) lacks them.Cytoplasmic organelle composed of a system of interconnected membranous tubules and vesicles; ER; rough ER has ribosomes attached to the side of the membrane facing the cytoplasm and smooth ER does not. Rough ER functions in protein synthesis while smooth ER functions in lipid synthesis.

Endopod, endopodite Medial branch of a biramous crustacean appendage(phylum Arthropoda).

Endopolyogeny Formation of daughter cells, each surrounded by its own membrane, while still in the mother cell.

Endopterygote Condition of internal wing bud development in an insect. Also, an insect in which the wing buds develop externally or any insect secondarily wingless but derived from such an ancestor; associated with holometabolous insects.

Endorphins A neuropeptide which has specific binding sites on nerve cells called opiate receptors. When the receptor is activated by endorphins or morphine it reduces the excitability of the post synaptic cell. Peptide receptors are also found on lymphocytes which suggest an association between neuropeptides and the regulation of the immune response.

Endoskeleton A hard skeleton buried within the soft tissues of an animal, such as the spicules of sponges, the plates of echinoderms, and the bony skeletons of vertebrates.A skeleton that lies beneath the surface of the body (e.g., the bony skeleton of vertebrates and the calcium carbonate skeleton of echinoderms); contrasts with exoskeleton.

Endosperm A food storage tissue that provides nutrients to the developing embryo in angiosperms; formed from the triploid cell produced when a sperm cell fertilizes the central cell. Some endosperm is solid (as in corn), some is liquid (as in coconut).

Endospore A thick-coated, resistant cell produced within a bacterial cell exposed to harsh conditions.

Endosteum A layer of bone forming cells, osteoblasts which covers the entire surface of the internal aspect of cortical and spongy bone, separating it from the surrounding connective tissue.

Endostyle A ciliated, mucus-secreting groove in the ventral surface of the pharnyx of non-vertebrate chordates (e.g, tunicates and lancelets); it aids in transporting food to the esophagus. Recent molecular evidence supports the traditional view that the endostyle is homologous to the vertebrate thyroid gland.

Endosymbiont An organism that lives within the body of another organism in a symbiotic association.

Endosymbiont hypothesis The idea whereby the evolution of the eukaryotic cell might have occurred when a large anaerobic amoeboid prokaryote ingested small aerobic bacteria and stabilized them instead of digesting them.

Endosymbiosis When one organism takes up permanent residence within another, such that the two become a single functional organism. Mitochondria and plastids are believed to have resulted from endosymbiosis.

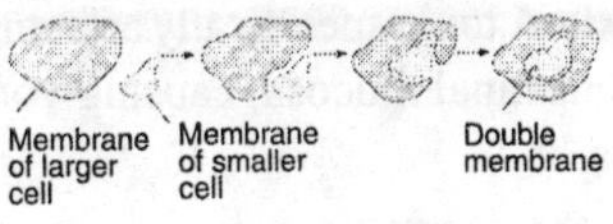

Fig. *Endosymbiosis*

Endothelial cells The epithelial cells of the endothelium which lines blood vessels. The cells are flattened into a pavement stone shape and are usually two or three layers thick.

Endothelium The innermost, simple squamous layer of cells lining the blood vessels; the only constituent structure of capillaries.A layer of epithelial cells forming a tissue that lines the circulatory system and body cavities; derived from the mesoderm of the embryo.

Endotherm (or homeotherm)An organism that can regulate its internal temperature. Compare ectotherm.

Endothermic A reaction that gives off energy. The product is in a lower energy state than the reactants.

Endotherms Animals that have the ability to maintain a constant body temperature over a wide range of environmental conditions.

Endothermy The internal control of body temperature; the ability to generate and maintain internal body heat.

Endotoxin The heat-stable lipopolysaccharide in the outer membrane of the cell wall of gram-negative bacteria that is released when the bacterium lyses, or sometimes during growth, and is toxic to the host.

Energy budget An accounting of the way in which energy coming into an ecosystem from the sun is lost or processed by organisms of the ecosystem.

Energy efficiency A measure of energy produced compared to energy consumed.

Energy flow The movement of energy through a community via feeding relationships.

Energy of activation (E_A) The amount of energy that reactants must absorb before a chemical reaction will start.

Energy pyramid A representation of the loss of useful energy at each step in a food chain.

Energy recovery Incineration of solid waste to produce useful energy.

Energy The capacity to do work (that is, to change the physical state or motion of an object).

Energy-level shell The distribution of electrons around the nucleus of an atom.

Engram (memory trace)The physical manifestation of learning; neural mechanism for storage of information that involves structural synaptic changes and/or chemical changes.

Enhancer DNA sequence in a gene that influences the gene's expression by increasing or decreasing its rate of RNA synthesis (via effects on binding and transcription of the basal transcription apparatus including RNA Polymerase); can be located upstream (5'), downstream (3') or within the protein coding sequence of the gene. A eukaryotic *transcriptional* control element which is a DNA sequence which acts at some distance to enhance the activity of a specific *promoter* sequence. Unlike promoter sequences, the position and orientation of the enhancer sequence is generally not important to its activity. An enhancer is a nucleotide sequence to which transcription factor(s) bind, and which increases the transcription of a gene. It is NOT part of a promoter; the basic difference being that an enhancer can be moved around anywhere in the general vicinity of the gene (within several thousand nucleotides on either side or even within an intron), and it will still function. It can even be clipped out and spliced back in backwards, and will still operate. A promoter, on the other hand, is position- and orientation-dependent. Some enhancers are "conditional" - in other words, they enhance transcription only under certain conditions, for example in the presence of a hormone.

Enkephalin Group of small brain neuropeptides with opiate-like qualities.

Enology The science of wine making.

Enrichment experiments Experimental manipulations that involve providing organisms with particular types of stimuli (e.g., social, sensory, motor) to ascertain the effects later in development.

Enteric bacteria Members of the family Enterobacteriaceae (gram-negative, peritrichous or nonmotile, facultatively anaerobic, straight rods with simple nutritional requirements); also used for bacteria that live in the intestinal tract.

Enterocoel A type of coelom formed by the outpouching of a mesodermal sac from the endoderm of the primitive gut.

Enterocoelic mesoderm formation Embryonic formation of mesoderm by a pouchlike outfolding from the archenteron, which then expands and obliterates the blastocoel, thus forming a large cavity, the coelom, lined with mesoderm.

Enterocoelic pouches Outpocketings of the archenteron (primitive gut) of the gastrula which form the mesodermal somites later in development. The cavities within the pouches will form the body cavity. Found in animals such as *Amphioxus* and many echinoderms. Synonym: coelomic sacs.

Enterocoelomate An animal having an enterocoel, such as an echinoderm or a vertebrate.

Enterocoely Formation of a coelom through outpocketing of the inner portion of the archenteron in some animals (deuterostomes).

Enterohemorrhagic E. coli EHEC strains of E. coli (0157:H7) produce several cytotoxins that provoke fluid secretion in traveler's diarrhea; however, their mode of action is unknown.

Enteron The digestive cavity.

Enteropneusta A class of hemichordates whose members live in burrows in shallow marine water. Their bodies are divided into three regions: proboscis, collar, and trunk. Acorn worms.

Enterotoxin A toxin specifically affecting the cells of the intestinal mucosa, causing vomiting and diarrhea.

Entine The inner layer of a pollen grain shell.

Entire A smooth leaf margin without teeth, lobes, or undulations.

Entner-doudoroff pathway A pathway that converts glucose to pyruvate and glyceraldehyde 3-phosphate by producing 6-phosphogluconate and then dehydrating it.

Entomogenous Growing in or on an insect, for example certain fungi.

Entomology The study of the biology of insects.

Entomopathogenic Insect-attacking organism.

Entomophily Seed plants which are pollinated by insects are said to be entomophilous.

Entoprocta A phylum of aschelminths commonly called entoprocts.

Entozoic Living within another animal; internally parasitic (chiefly parasitic worms).

Entrainment The process by which a biological clock is set or reset by synchronizing with the period of some external, environmental stimulus.

Entropy A measure of the degree of disorganization of a system; how much energy in a system has become so dispersed (usually as heat) so that it is no longer available to do work. The higher the entropy, the more the disorder.

Envelope 1. All the structures outside the plasma membrane in bacterial cells.

2. In virology it is an outer membranous layer that surrounds the nucleocapsid in some viruses.The carton or wax outermost later of the nest of a social insect, particularly those of wasps.

Environment Describes the surroundings in which organisms live. Some physical features of an environment are fairly stable, like trees, rivers, mountains, houses, soil, teeth. Some physical features are changeable, like wind, water, light, pH, food supply. Others features are less predictable, such as the balance in the community of collaborators, competitors and parasites. All forms of life including bacteria in the mouth, have an environment, which has an important influence on their survival. Successful organisms manage to exploit their environment to the best advantage or to adapt to it, perhaps only after several generations, if it becomes a serious challenge to the species.The circumstances or conditions that surround an organism or group of organisms as well as the complex of social or cultural conditions that affect an individual or community.

Environmental grain An ecological term for the effect of spatial variation, or patchiness, relative to the size and behavior of an organism.

Environmental hormones Chemical pollutants that substitute for, or interfere with, naturally-occurring hormones in our bodies; these chemicals may trigger reproductive failure, developmental abnormalities, or tumor promotion.

Environmental impact quotient (EIQ) A relative value that estimates the environmental impact of a pesticide, by taking into account toxicity to natural enemies, wildlife, and humans, degree of exposure, aquatic and terrestrial effects, soil chemistry, etc.

Environmental insults Any factor in the physical environment that inhibits the growth and/or development of an organism.

Environmental justice Combines civil rights with environmental protection to demand a safe, healthy, life-giving environment for everyone.

Environmental literacy Fluency in the principles of ecology that gives us a working knowledge of the basic grammar and underlying syntax of environmental wisdom.

Environmental resistance All the limiting factors that tend to reduce population growth rates and set the maximum allowable population size or carrying capacity of an ecosystem.

Environmental resources Anything an organism needs that can be taken from the environment.

Environmental stress Variously defined as (a) an environmental change to which an organism cannot acclimate and

(b) an environmental change that increases the probability of death

Environmental terrorism Conducting acts of terrorism through the destruction or pollution of the environment.

Environmentalism Active participation in attempts to solve environmental pollution and resource problems.

Enzootic The moderate prevalence of a disease in a given animal population.

Enzyme A protein that controls and helps a chemical reaction to take place, but is not used up in the process. Usually each enzyme is specific for a particular step in a reaction. Enzymes are sensitive to their environment, especially to excessive temperature or pH. A substance, produced by living cells, that is capable of speeding up specific chemical transformations, such as hydrolysis, oxidation, or reduction, but is unaltered itself in the process; a biological catalyst.An organic catalyst, normally a protein formed and secreted by a living cell. complex protein which helps to speed biochemical reactions. Enzymes are important in the construction and degradation of other molecules.

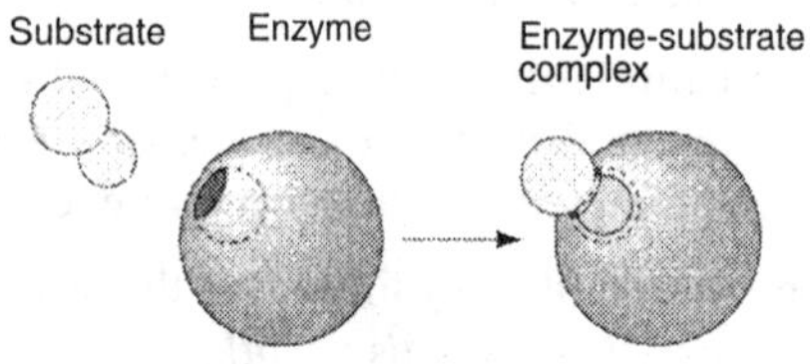

Fig. *Enzyme*

Enzyme-linked immunosorbent assay (ELISA) A technique used for detecting and quantifying specific antibodies and antigens.

Eocytes A group of prokaryotes currently classified among the Archaebacteria but possibly a sister group of eukaryotes.

Eosinophil A polymorphonuclear leukocyte that has a two-lobed nucleus and cytoplasmic granules that stain yellow-red. A mobile phagocyte that is highly antiparasitic.

Ephemeral An annual plant with an extremely short lifespan measured in a few weeks or very few months; characteristic growthform of desert forbs.

Ephyra Refers to castlelike appearance.Miniature medusae produced by asexual budding of a scyphistoma (class Scyphoza, phylum Cnidaria). Ephyrae mature into sexually mature medusae.

Epi G. A prefix meaning on, upon, or over.

Epibenthic Living on the surface of the bottom

Epiblast An outer layer of cells in the embryo of an amniote that forms from the proliferation and movement of cells of the blastoderm.

Epiboly Literally, "over the ball," usually the growth of epidermal ectoderm to cover the surface of the embryo during gastrulation.

Epicenter Point on the Earth's surface directly above the focus of an earthquake.

Epicondyle epi = above; condyl a knuckle.

Epicotyl Portion of the shoot of an angiosperm embryo or seedling above the cotyledons.

Epidemic A disease that suddenly increases in occurrence above the normal level in a given population.

Epidemic hemorrhagic fever Virus disease transmitted by mosquitoes. Also called dengue.

Epidemic spawning Simultaneous shedding of gametes by a large number of individuals

Epidemiology The study of the factors determining and influencing the frequency and distribution of disease, injury, and other health-related events and their causes in defined human populations.Study concerned with all ecological aspects of a disease to explain its transmission, distribution, prevalence, and incidence.

Epidermal Epi = above; derma = skin.

Epidermal growth factor A cytokine that stimulates epithelial cell proliferation.

Epidermis 1. The outermost layer of skin consisting of several layers of epithelial cells&emdash;notably, keratinocytes & emdash;and, in the inner layer of the epidermis, basal cells and melanocytes.

2. The outer layer of cells in the plant body, often covered by a waxy cuticle.The cellular layer of the integument that secretes or deposits a comparatively thick cuticle on its outer surface. The outer, nonvascular layer of skin of ectodermal origin; in invertebrates, a single layer of ectodermal epithelium. A coiled structure found along the posterior border of the testis where mature spermatazoa are received, stored and eventually dispensed into the vas deferens for transport out of the body.

Epididymis Part of the sperm duct that is coiled and lying near the testis.

Epifauna Animals that live on the surface of the substrate. Compare infauna.

Epigaeic Living or foraging primarily above ground, compared to Hypogaeic the opposite.

Epigenesis Theory holding that development is a gradual process of increasing complexity. (This contrasts with preformationism, which holds that the organism is already present in the gamete(s), merely growing and unfolding during development.) For example, organs are formed de novo in the embryo rather than increasing in size from pre-existing structures.

Epiglottis A flap of tissue that closes off the trachea during swallowing.

Epigynous Floral parts (sepals, petals, and stamens) appear to arise from the top of an ovary; the ovary is said to be inferior.

Epilithic A term for organisms that live attached to rocks.

Latin translation:

epi = upon

lith = rock

This term is general to terrestrial and marine habitats, ie. some lichens are epilithic.

Epimastigote Trypanosomatid flagellate similar to a promastigote but with a short undulating membrane, such as in Blastocrithidia.

Epimeron The posterior part of the side wall of any of the three thoracic segments.

Epimyocardium The outer, thicker layer of the heart rudiment which arises from splanchnic mesoderm and fuses with the endocardium to form the heart wall.

Epinasty The unequal growth of petioles causing the leaf blade to curve downward.

Epinephrine A neurotransmitter substance found at all adrenergic synapses (nor epinephrine or epinephrine). It is the most common neurotransmitter in the nervous system, in particular at ganglion cells of the sympathetic nervous system.

Epinotum The first abdominal segment when it is fused with the last thoracic one, relating to the higher thin waisted hymenoptera. Also called a propodeum.

Epipelagic zone The pelagic environment from the surface to a depth of 100 to 200 m (350 to 650 ft).

Epipharynx A component of many insect mouth-parts which is attached to the posterior surface of the labrum or clypeus. In chewing insects it is usually only a small lobe, but in the fleas it is greatly enlarged and used for sucking blood.

Epiphysealdisk A cartilaginous growth line at the ends of long bones.

Epiphyte A plant which grows upon another plant. The epiphyte does not "eat" the plant on which it grows, but merely uses the plant for structural support, or as a way to get off the ground and into the canopy environment.Microalgal organism living on a surface (e.g., on a seaweed frond)

Epipod, epipodite A lateral process on the protopod of a crustacean appendage, often modified as a gill.

Epiproct An appendage arising from the mid-line of the last abdominal segment, just above the anus. In the bristletails and some mayflies it is very long and forms the central 'tail'.

Episome A plasmid that can exist either

independently of the host cell's chromosome or be integrated into it. A plasmid capable of integrating into the bacterial chromosome.

Epistasis The masking of the effects of one gene by the action of another, example: widow's peak masked by the baldness gene.

Episternum The anterior part of the side wall of any of the three thoracic segments.

Epistome Flap over the mouth in some lophophorates bearing the protocoel.

Epitaxy The initiation of crystalformation in a saturated solution by providing a template against which crystal can form. There are specific sites on collagen molecules which appear to function as templates against which hydroxyapatite crystals form.

Epitheca the larger portion of a diatom frustule

Epithelial attachment The cuff of junctional epithelium which joins the gingival sulcus epithelium to the enamel of the tooth. Apical migration of the epithelium down onto the cementum may occur due to ageing or periodontal disease. Loss of attachment produces a periodontal pocket and a new habitat for anaerobic oral bacteria.

Epithelial Belonging to a sheet of tightly joined, polarized cells.

Epithelial tissue The cellular covering of internal and external surfaces of the body; consists of cells joined by small amounts of cementing substances. Epithelium is classified into types based on the number of layers deep and the shape of the superficial cells.

Epithelium A tissue consisting of one or more layers of tightly associated cells forming a covering or lining; may be derived from endoderm, mesoderm, or ectoderm. Layer of cells which lines a body cavity; cells may be ciliated or unciliated, and may be squamous (flat, scale-shaped), cuboidal (cube-shaped), or columnar (column-shaped). Your stomach and cheeks are lined with epithelium.The layer of cells that covers a surface or lines a cavity.

Epitoke Posterior part of a marine polychaete when swollen with developing gonads during the breeding season; contrast with atoke.

Epitope As related to protein antigens, B-cell epitopes consist of the amino acid residues of a protein molecule which interact directly through noncovalent bonds with the amino acid residues of a particular antibody molecule (complementarity determining region). The average epitope probably involves about 15-20 contact amino acid residues, but one or two of these may be critical to the epitope's specificity and the avidity of the antibody-antigen reaction. B-cell epitopes may be either *linear* or *conformational* in nature. T-cell epitopes represent the small, processed peptides which bind to MHC class I and II molecules on the surface of T cells.

Epizootic A disease outbreak within an insect population.

Epizootiology The field of science that deals with factors determining the frequency and distribution of a disease within an animal population.

Epsilon-proteobacteria One of the five subgroups of proteobacteria, each with distinctive 16S rRNA sequences. Slender gram-negative rods, some of which are medically important (Campylobacter and Helicobacter).

Equation With respect to comparative learning, the attempts to match situations and procedures for examining learning behavior in different species.

Equatorial currents Major ocean currents that move parallel to the Equator.

Equilibrium The state of a system in which no further net change is occurring; result of counterbalancing forward and backward processes.

Equilibrium community Also called a disclimax community; a community subject to periodic disruptions, usually by fire, that prevent it from reaching a climax stage.

Equilibrium species Species characterized by low reproduction rates, long development times, large body size, and long adult life with repeated reproductions.

Eras One of the major divisions of the geologic time scale.

ERE estrogen response element. A binding site in a promoter to which the activated estrogen receptor can bind. The estrogen receptor is essentially a transcription factor which is activated only in the presence of estrogens. The activated receptor will bind to an ERE, and transcription of the adjacent gene will be altered.

Ergatogyne Any female member of a eusocial group whose morphological development is somewhere between that of a worker and a queen.

Ergot The spore-producing reproductive body of a fungus that infects grain crops; ergot contains lysergic acid and ergonovine.

Ergotism The disease or toxic condition caused by eating grain infected with ergot; it is often accompanied by gangrene, psychotic delusions, nervous spasms, abortion, and convulsions in humans and in animals.

Erosion The processes by which materials of the Earth's crust are worn away, loosened, or dissolved while being transported from their place of origin.

Erratic A large, isolated boulder left behind by a glacier.

Eruciform (applied to a larva) Caterpillar like; more or less cylindrical with a well developed head and stumpy legs at the rear as well as the true thoracic legs. The caterpillars of butterflies and moths are typical examples.

Erysipelas An acute inflammation of the dermal layer of the skin, occurring primarily in infants and persons over 30 years of age with a history of streptococcal sore throat.

Erythema infectiosum A disease in children caused by the parvovirus B19. This disease is common in children between 4 and 11 years of age and is sometimes called fifth disease, since it was the fifth of six erythematous rash diseases in children in an older classification.

Erythroblastosis fetalis A disease of newborn infants caused when Rh-negative mothers develop antibodies against the Rh-positive blood of the fetus.

Erythrocyte Red blood cell; has hemoglobin to carry oxygen from lungs or gills to tissues; during formation in mammals, erythrocytes lose their nuclei, those of other vertebrates retain the nuclei.

Erythromycin An intermediate spectrum macrolide antibiotic produced by Streptomyces erythreus.

Erythropoiesis Red blood cell (erythrocyte) development, including pluripotential stem cell division, restriction of potential and differentiation into a mature red blood cell containing hemoglobin.

Escarpment A steep or vertical cliff, either above or below sea level.

Eschar A slough produced on the skin by a thermal burn, gangrene, or the anthrax bacillus.

Esker A ridge of glacial sediment deposited by a stream flowing in and under a melting glacier.

Esophagus The muscular tube extending between and connecting the pharynx to the stomach.

Espundia Disease caused by Leishmania braziliensis. Also called chiclero ulcer, uta, pian bois, or mucocutaneous leishmaniasis.

Essential amino acid Amino acid that cannot be synthesized by an organism but must be obtained ready-made in the diet for proper health.

Essential nutrient Nutrient that cannot be synthesized by an organism but must be obtained ready-made in the diet for proper health.

Essential oils Highly volatile and aromatic oils formed in glands or special cells by some plants; probably involved in pollinator attraction or repulsion of herbivores; used in perfumes, soaps, medicine, and food.

Esthete Light sensory receptor on a shell of a chiton (phylum Mollusca).

Estivation A physiological state characterized by slow metabolism and inactivity, which permits survival during long periods of elevated temperature and diminished water supplies.

Estradiol A female sex hormone secreted by the ovary which, in conjunction with progesterone, functions to maintian the uterine mucosa in case fertilization occurs.

Estrogens The primary female steroid sex hormones, which are produced in the ovary by the developing follicle during the first half of the cycle and in smaller quantities by the corpus luteum during the second half. Estrogens stimulate the development and maintenance of the female reproductive system and secondary sex characteristics.

Estrone An animal steroid found in some plants in low concentration.

Estrous cycle The female reproductive cycle of most mammals (not primates). It consists of four stages: proestrus, where the uterine lining begins to thicken; estrus or the "heat" phase of sexual receptivity when ovulation occurs; metestrus, where the corpus luteum forms and secretes hormones to maintain the uterine lining and prevent further ovulation; and diestrus, where if fertilization does not occur, the uterine lining is reabsorbed.The period of highest sexual receptivity (or "heat") in some female mammals that coincides with the time of egg release by the ovarymarked by intense sexual urges.

Estuarine flow Seaward flow of low-salinity surface water over a deeper and higher -alinity layer

Estuarine realms Large coastal water regions that have geographic continuity,are bounded landward by a stretch of coastline with fresh-water input, and are bounded seaward by a salinity front

Estuary A bay or drowned valley where a river empties into the sea. Fresh water mingling with salt water brings in sediment and nutrients and creates a gradient of salinity that makes estuaries among the most diverse and biologically productive ecosystems on earth. A semienclosed body of water that has a free connection with the open sea and within which seawater is diluted measurably with freshwater that is derived from land drainage. An area where fresh water comes into contact with seawater, usually in a partly enclosed coastal body of water; a mix of fresh and salt water where the current of a stream meets the tides; estuarine– adj.

Ethanol Ethyl alcohol is the end product of anaerobic respiration (fermentation) in plants.

Ethidium bromide Intercalates within the structure of nucleic acids in such a way that they fluoresce under UV light. Ethidium bromide staining is commonly used to visualize RNA or DNA in agarose gels placed on UV light boxes. Proper precautions are required, because the ethidium bromide is highly mutagenic and the UV light damaging to the eyes. Ethidium bromide is also included in cesium chloride gradients during ultracentrifugation, to separate *supercoiled* circular DNA from linear and *relaxed* circular DNA.

Ethmoid Like a sieve.

Ethogram An inventory of all of the behavior patterns of a species.

Ethologist A person who studies the whole patterns of animal behavior in natural environments, stressing the analysis of adaptation and the evolution of the patterns.

Ethology The study of whole patterns of animal behavior in natural environments, stressing the analysis of adaptation and the evolution of the patterns.

Ethylene A gaseous plant hormone that stimulates fruit ripening and the dropping of leaves.

Etiolation The abnormal elongation of stems caused by insufficient light or unbalanced hormonal relationships. Etiolated stems often lack chlorophyll.

Eubacteria The subunit of the Monera that includes the true bacteria such as *E. coli.* One of the three major groups of prokaryotes in the Kingdom Monera. The eubacteria have cell walls containing peptidoglycan.

Eucarya The domain that contains organisms composed of eucaryotic cells with primarily glycerol fatty acyl diesters in their membranes and eucaryotic rRNA.

Eucaryote A cell in which the genetic material is confined to the nucleus, in distinction to a procaryote in which the genetic material is dispersed throughout the cell. Other distinctions of eucarytoic cells are the presence of organelles such as the golgi apparatus, endoplasmic reticulum, lysosomes and mitochondria

Euchromatic regions Less densely staining regions of chromosomes that contain active genes.

Euchromatin The more open, unraveled form of eukaryotic chromatin, which is available for transcription.

Euglenoids A group of algae (the division Euglenophyta) or protozoa (order Euglenida) that normally have chloroplasts containing chlorophyll a and b. They usually have a stigma and one or two flagella emerging from an anterior gullet.

Eukaryote A type of cell found in many organisms including single-celled protists and multicellular fungi, plants, and animals; characterized by a membrane-bounded nucleus and other membraneous organelles; an organism composed of such cells. The first eukaryotes are encountered in rocks approximately 1.2-1.5 billion years old. Cell or organism with membrane- bound, structurally discrete *nucleus* and other well- developed subcellular compartments. Eukaryotes include all organisms except viruses, bacteria, and blue- green algae. Compare *prokaryote.* Having a true nucleus; a cell that has membranous organelles, most notably the nucleus.

Eumycota A division of fungi in some classification systems. These are the true fungi consisting of the Zygomycetes, Ascomycetes, Basidiomycetes, and Chytridiomycetes.

Euphausiid Member of an order of holoplanktonic crustacea.

Euphotic zone The upper part of the marine biome where light penetrates and photosynthesis occurs; usually extends to about 200 meters below the water surface.

Euploidy Change in chromosome number from one generation to the next in which there is an addition or deletion of a complete set of chromosomes in the progeny; the most common type is polyploidy.

Euramerica A supercontinent that existed in the Late Silurian through Devonian, formed by the collision of Baltica, Laurentia, and Avalonia; included modern North America, Greenland, Scandinavia, and Europe; also called the "Old Red Continent" for the red color of its oxidized deposits.

Euryhaline Able to tolerate wide ranges of saltwater concentrations.

Euryphagous Eating a large variety of foods.

Eurytopic Refers to an organism with a wide environmental range.

Eusocial Applied to animal societies, such as those of certain insects, in which sterile individuals work on behalf of reproductive individuals.A species which lives in a society such that individuals of the species cooperate in caring for the young, which not all of them have produced; there is a reproductive division of labor, with more or less sterile individuals working on behalf of fecund individuals; and there is an overlap of at least two generations in life stages capable of contributing to colony labor, so that offspring assist parents during some period of their life.

Eustele When a plant's vascular tissue develops in discrete bundles, it is said to have a eustele.

Eutectic chemicals Phase-changing chemicals used

in heat storage systems to store a large amount of energy in a small volume.

Eutely Condition where the body is composed of a constant number of cells or nuclei in all adult members of a species (e.g., rotifers, some nematodes, and acanthocephalans).

Eutherian mammals Placental mammals; those whose young complete their embryonic development within the uterus, joined to the mother by the placenta.

Eutrophic A nutrient-enriched environment. Water bodies or habitats having high concentrations of nutrients.

Eutrophic lake A highly productive lake, having a high rate of biological productivity supported by a high rate of nutrient cycling.

Eutrophication "Runaway" growth of aquatic plants that occurs when agricultural fertilizers containing phosphorus and nitrogen run off into lakes and ponds; also ultimately increases the plant death rate with the result that the bacterial decomposition of the dead plants uses up oxygen, causing fish and other organisms to suffocate.A process in which an aquatic environment accumulates high nutrient levels due to factors such as industrial or urban pollution or run-off of fertilizers from nearby agricultural lands. The nutrients lead to dense blooms of algae and aquatic plants that cloud lake water, deplete specific minerals and dissolved gases, and can cause natural plant and animal populations to decline.Enrichment of a water body with nutrients, resulting in excessive growth of phytoplankton, seaweeds, or vascular plants, and often depletion of oxygen.Succession in a lake. The gradual accumulation of nutrients and plant and animal growth that eventually transforms a lake into a marsh and then a forest. The death of a lake.

Evagination An outpocketing from a hollow structure.

Evaporative cooling The lower speed and hence the lower temperature of molecules remaining in the liquid phase after evaporation of the fastest molecules. The property of a liquid whereby the surface becomes cooler during evaporation, owing to a loss of highly kinetic molecules to the gaseous state.

Evaporite a deposit of salt minerals (e.g., halite, gypsum, anhydrite) left behind by the evaporation of seawater, usually forms within a restricted basin.

Evapotranspiration The combined water loss from both leaf surfaces and from the soil surface; the sum of transpiration and evaporation.

Evenness The component of diversity accounting for the degree to which all species are equal in abundance, as opposed to strong dominance by one or a few species

Evergreens Coniferous trees and broad-leaved plants that retain their leaves year-round.

Evisceration The expulsion of internal organs in sea cucumbers.

Evolution 1) The change in life over time by adaptation, variation, over-reproduction, and differential survival/reproduction, a process referred to by Charles Darwin and Alfred Wallace as natural selection.

2) Descent with modification. All mammals and other living organisms change through time. Evolution specifically refers to change in organisms that is genetic; that is, frequencies of specific genes in the organism's genetic makeup change. This results in changes in genetic characteristics.An explanation of the way present-day and past organisms have been produced. Current evolutionary theory explains this process on the basis of mutations being the starting point for Natural Selection to choose the best varients in a population. Over time, the mutations which are passed on to succeeding generations, accumulate so that speciation occurs. This is the evolution of new species. Change over time. Organic or biological evolution is a series of changes in the genetic composition of a population over time.Darwin's

definition: descent with modification. The term has been variously used and abused since Darwin to include everything from the origin of man to the origin of life.The process by which a species' structural and behavioural characteristics change over many generations, sometimes in response to changes in environmental conditions. "New" species develop in this way. For example, scientists think that whales gradually evolved from land animals.

Evolutionary adaptation The changes occurring in a population of individuals over many generations by processes of natural selection.

Evolutionary clock Defined by the rate at which mutations accumulate within a given gene.

Evolutionary distance A measure of the time since two distinct, or similar species shared a common ancestor. May also be used as a measure of time since reproductively isolated populations of a species were one breeding population. A quantitative indication of the number of positions that differ between two aligned macromolecules, and presumably a measure of evolutionary similarity between molecules and organisms.

Evolutionary footprinting One can infer which portions of a gene are important by comparing the sequence of that gene with its cognates from other species. A plot showing the regions of high conservation will presumably reflect the regions that are functional in all the test species. In theory, the more species involved in the comparison, the more stringent the result can be (i.e. the more the conserved regions will reflect truly important sequences). Care must be taken, however, to use species in which the function of the gene has not diverged excessively, or the outcome will be uninformative.

Evolutionary species concept A single lineage of ancestral-descendant populations that maintains its identity from other such lineages and has its own evolutionary tendencies and historical fate; differs from the biological species concept by explicitly including a time dimension and including asexual lineages.The idea that evolutionary lineages and ecological roles can form the basis of species identification.

Evolutionary systematics The study of the classification of, and evolutionary relationships among, animals; evolutionary systematists attempt to reconstruct evolutionary pathways based on resemblances between animals that result from common ancestry.

Evolutionary taxonomy A system of classification, formalized by George Gaylord Simpson, that groups species into Linnean higher taxa representing a hierarchy of distinct adaptive zones; such taxa may be monophyletic or paraphyletic but not polyphyletic.

Evolutionary tree A diagram showing the evolutionary history of organisms based on differences in amino acid sequences. Organisms with fewer differences are placed closer together while those with more differences are further apart.A diagram which depicts the hypothetical phylogeny of the taxa under consideration. The points at which lineages split represent ancestor taxa to the descendant taxa appearing at the terminal points of the cladogram.

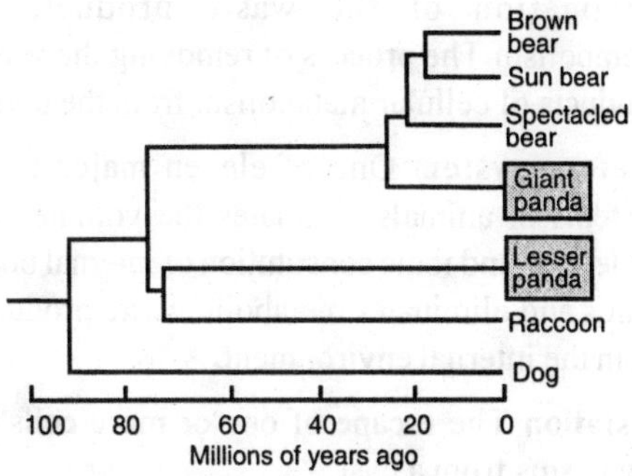

Fig. *Evolutionary tree*

Ewe An adult female sheep.

Exaptation A structure that evolves and functions in one environmental context but that can perform additional functions when placed in some new environment.

Exarate pupa A pupa in which all the appendages, legs etc., are free and capable of movement.

Excavate Hollowed out: applied to the coxae of many beetles, which are hollowed out to receive the femora when the legs are folded.

Excision repair Means by which cells are able to repair certain kinds of damage (dimerized pyrimidines) in their DNA.

Excitation Electron movement caused by light striking the chlorophyll molecule.

Excitatory postsynaptic potential An electrical change (depolarization) in the membrane of a postsynaptic neuron caused by the binding of an excitatory neurotransmitter from a presynaptic cell to a postsynaptic receptor; makes it more likely for a postsynaptic neuron to generate an action potential.

Exclusive economic zone (EEZ) A zone 200 nautical miles (370 km) wide along the coast where nations have exclusive rights to any resource. It was initiated by the United Nations Convention on the Law of the Sea (UNCLOS).

Excretion The act, process, or function of excreting. The elimination of metabolic waste products from an animal's body. The disposal of nitrogen-containing waste products of metabolism.The elimination of the waste products of metabolism.The process of removing the waste products of cellular metabolism from the body.

Excretory system One of eleven major body systems in animals; regulates the volume and molecular and ionic constitution of internal body fluids and eliminates metabolic waste products from the internal environment.

Excystation The escape of one or more cells or organisms from a cyst.

Exergonic Characterized or accompanied by the release of energy; said of chemical reactions that release energy, so that the products have a lower free energy than the reactants.Pertaining to a reaction that gives off energy and occurs spontaneously.

Exergonic reaction A reaction that spontaneously goes to completion as written; the standard free energy change is negative, and the equilibrium constant is greater than one.A spontaneous chemical reaction in which there is a net release of free energy.Chemical reactions that release energy.

Exflagellation Rapid formation of microgametes from a microgametocyte of Plasmodium and related genera.

Exfoliative toxin or exfoliatin An exotoxin produced by Staphylococcus aureus that causes the separation of epidermal layers and the loss of skin surface layers. It produces the symptoms of the scaled skin syndrome.

Exhaustible resources Generally considered the earth's geologic endowment: minerals, nonmineral resources, fossil fuels, and other materials present in fixed amounts in the environment.

Exine Outer covering of pollen grains, often containing sporopollenin, an acid-resistant polysaccharide that allows pollen grains to become fossils. The hard outer coat of a pollen grain.

Existence value The importance we place on just knowing that a particular species or a specific organism exists.

Exit site The location on a ribosome to which an empty tRNA moves from the P site before it finally leaves the ribosome during protein synthesis.Lateral process or joint from the protopod, sometimes referred to as an epipod.

Exobiology The study of evidence relative to the possibility of life on other planets.

Exocarp Outermost layer in the pericarp (fruit wall).

Exocrine A type of gland that releases its secretion through a duct; contrasts with endocrine.

Exocrine glands The secretion passes intoa duct like sweat, saliva and mucous. Glands, such as sweat glands and digestive glands, that secrete their products into ducts that empty onto surfaces, such as the skin, or into cavities, such as the interior of the stomach.

Exocuticle The hard and usually darkened layer of the cuticle lying between the endocuticle and epicuticle.

Exocytosis The cellular secretion of macromolecules by the fusion of vesicles with the plasma membrane.The process by which substances are moved out of a cell; the substances are transported in the cytoplasmic vesicles, the surrounding membrane of which merges with the plasma membrane in such a way that the substances are dumped outside.The process in which a membrane-enclosed vesicle first fuses with the plasma membrane and then opens and releases its contents to the outside.

Exoenzymes Enzymes that are secreted by cells.

Exogenote The piece of donor DNA that enters a bacterial cell during gene exchange and recombination.

Exogenous Arising from a source outside the organism or cell.

Exogenous DNA DNA originating outside an organism.

Exogenous factors External factors regulating an organism's growth and development.

Exon Part of the mRNA as transcribed from the DNA that contains a portion of the information necessary for final gene product. Segment of DNA sequence in a gene that will be transcribed in the nucleus, spliced to other exons, and transported to the cytoplasm as a part of the mature mRNA. The coding region of a eukaryotic gene that is expressed. Exons are separated from each other by introns.The DNA bases that code for an amino acid sequence. Exons are separated by introns that code for no amino acid sequences. Those portions of a genomic DNA sequence which will be represented in the final, mature mRNA. The term "exon" can also be used for the equivalent segments in the final RNA. Exons may include coding sequences, the 5' untranslated region or the 3' untranslated region.The coding sections of eukarotic genes, separated by introns.The *protein*- coding DNA sequences of a *gene*. Compare *introns*.

Exonuclease An enzyme which digests nucleic acids starting at one end. An example is Exonuclease III, which digests only double-stranded DNA starting from the 3' end.

Exopod exopodite. Lateral branch of a biramous crustacean appendage.

Exopterygote Any insect in which the wings develop gradually on the outside of the body, in which there is only a partial metamorphosis and no pupal stage. Condition of external wing bud development in an insect. Also, any insect in which the wing buds develop externally; associated with hemimetabolous insects.Insect in which the wing buds develop externally during nymphal instars; has hemimetabolous metamorphosis.

Exoskeleton A hard encasement on the surface of an animal, such as the shells of mollusks or the cuticles of arthropods, that provides protection and points of attachment for muscles.A hard, jointed, external covering that encloses the muscles and organs of an organism; typical of many arthropods including insects.A skeleton or supportive structure on the outside of an insect body.A supporting structure secreted by ectoderm or epidermis; external, not enveloped by living tissue, as opposed to endoskeleton.An external covering or integument, esp. when hard, as the shell of crustaceans.

Exothermic A reaction where the product is at a higher energy level than the reactants.

Exotoxin A heat-labile, toxic protein produced by a bacterium as a result of its normal metabolism or because of the acquisition of a plasmid or prophage that redirects its metabolism. It is usually released into the bacterium's surroundings.A toxic protein secreted by a bacterial cell that produces specific symptoms even in the absence of the bacterium.

Experience All of the interactions between an organism and its environment, beginning at conception and including both external and internal influences.

Experiment An artificially created situation that is used to test a hypothesis. In a controlled experiment, additional variables that might affect the experiment are prevented from doing so.

Experimental design The prescription of the treatment groups, dependent and independent variables, and sample sizes for the testing of a particular hypothesis.

Experimental hypotheses The biological ideas, both broad and specific, developed from knowledge about previous investigations and our own imagination and creativity, that lead to tests of statistical null hypotheses using experimental designs.

Exploitation A type of competition in which organisms passively use up limited resources; also called scramble competition. Contrast with interference or contest competition, in which organisms defend or otherwise control resources.

Exploratory behavior The spontaneous search for and active investigation of objects, situations, or other organisms in the absence of any homeostatic need.

Exponential growth Growth at a constant rate of increase per unit of time; can be expressed as a constant fraction or exponent. In populations, the increasingly accelerated rate of growth due to the increasing number of individuals being added to the reproductive base. Exponential growth is very seldom approached or sustained in natural populations.

Exponential phase The phase of the growth curve during which the microbial population is growing at a constant and maximum rate, dividing and doubling at regular intervals.

Exponential rate An extremely rapid increase, e.g., in the rate of population growth.

Expressed sequence tag A partial gene sequence unique to a gene that can be used to identify and position the gene during genomic analysis.

Expression To "express" a gene is to cause it to function. A gene which encodes a protein will, when expressed, be transcribed and translated to produce that protein. A gene which encodes an RNA rather than a protein (for example, a rRNA gene) will produce that RNA when expressed.Usually used to refer to the entire process of producing a protein from a gene, which includes *transcription, translation, post-translational modification* and possibly transport reactions.

Expression clone This is a clone (plasmid in a bacteria, or maybe a ? phage in bacteria) which is designed to produce a protein from the DNA insert. Mammalian genes do not function in bacteria, so to get bacterial expression from your mammalian cDNA, you would place its coding region (i.e. no introns) immediately adjacent to bacterial transcription/translation control sequences. That artificial construct (the "expression clone") will produce a pseudo-mammalian protein if put back into bacteria. Often, that protein can be recognized by antibodies raised against the authentic mammalian protein, and vice versa.

Expression vector A plasmid or phage designed for production of a polypeptide from inserted foreign DNA under specific controls. Often an *inducer* is used. The vector always provides a promoter and often the *transcriptional start site, ribosomal binding sequence*, and *initiation codon.* In some cases the product is a *fusion protein.* A special cloning vector used to express recombinant genes in host cells; the recombinant gene is transcribed and its protein synthesized.A vector that allows a DNA sequence cloned into it to be transcribed when the vector is introduced into a cell.A vector which is designed to allow expression (transcription and translation) of the inserted section of DNA. The vector carries a promoter (normally inducible) on one side of the cloning site, and a transcription terminator on the other side. These must be recognisable by the intended host cell. The vector may also carry a ribosome binding sequence (for bacterial expression) and a start codon, depending on the nature of the inserted DNA. Some expression vectors produce fusion proteins.

Expressivity In genetics, the degree to which a particular genotype is expressed in the phenotype of individuals with that genotype.

Extant In existence; still existing; not destroyed or lost.

Exteins Polypeptide sequences of precursor self-splicing proteins that are joined together during formation of the final, functional protein. They are separated from one another by intein sequences, which they flank.

Extermination The extinction of a species that is caused by humans.

External auditory canal The sound channel connecting the external and middle ears.

External cell movement Both flagella and cilia have a 9 + 2 arrangement of microtubules. This arrangement refers to the 9 fused pairs of microtubules on the outside of a cylinder, and the 2 unfused microtubules in the center. Dynein "arms" attached to the microtubules serve as the molecular motors. Defective dynein arms cause male infertility and also lead to respiratory tract and sinus problems.

External costs Expenses, monetary or otherwise, borne by someone other than the individuals or groups who use a resource.

Exteroception Perception of information provided by the exteroceptors.

Exteroceptor A sense organ excited by stimuli from the external world.

Extinct A plant or animal species may entirely cease to exist. Recent examples are the dodo, a large flightless bird which used to live as recently as two hundred years ago, on the island of Mauritius. There are today many species of birds, flowers, fish, insects, large mammals, including certain types of whale, which are threatened with extinction, most as a result of human activity. Happily, the smallpox virus is about to become extinct.A species which no longer exists. The Stellar sea cow is an example of a species which once lived on the Pacific's East Coast and is now extinct. Compare *extirpated*.

Extinction 1. The irrevocable elimination of species; can be a normal process of the natural world as species out-compete or kill off others or as environmental conditions change.

2. In learning, the decrease of response rate or magnitude of response with lack of reinforcement in a learning situation.The elimination of all individuals in a group, both by natural (dinosaurs, trilobites) and human-induced (dodo, passenger pigeon) means.When all the members of a clade or taxon die, the group is said to be extinct.

Extirpated A species no longer existing in Canada, but occurring elsewhere on Earth. Compare *extinct*. Removing, destroying, pulling up extermination or otherwise making extinct.

Extra cellular matrix The supporting surrounding material of a cell including ground substance and fibres.

Extracellular digestion A form of digestion found in annelids, crustaceans, and chordates including vertebrates; takes place within the lumen of the digestive system, and the resulting nutrient molecules are transferred into the blood or body fluid.

Extracellular matrix (ECM) Region outside of metazoan cells which includes compounds attached to the plasma membrane, as well as dissolved substances attracted to the surface charge of the cells. The ECM functions both to keep animal cells adhered together, and well as buffering them from their environment.The component of connective tissue surrounding cells and made up of the amorphous ground substance and fibrous proteins (mainly collagen fibres). The amorphous ground substance consists primarily of glycoproteins and proteoglycans.

Extracellular route Path taken by water through the root in which water moves through the spaces between cell walls of the cortex parenchyma.

Extracutaneous sporotrichosis An infection by the fungus Sporothrix schenckii that spreads throughout the body.

Extraembryonic membranes Four membranes (yolk sac, amnion, chorion, allantois) that support the developing embryo in reptiles, birds, and mammals.

Extreme barophilic bacteria Bacteria that require a high-pressure environment to function.

Extreme environment An environment in which physical factors such as temperature, pH, salinity, and pressure are outside of the normal range for growth of most microorganisms; these conditions allow unique organisms to survive and function.

Extremophiles Microorganisms that grow under harsh or extreme environmental conditions such as very high temperatures or low pHs.

Extrinsic factor An environmental factor such as temperature that influences microbial growth in food.

Extrinsic fibres Refers to those fibres of cementum which are continuous with periodontal ligament fibres. Extrinsic fibres have been trapped in cementum during its formation in order to anchor them.

Extrinsic On the outside; related to some region other than the organ with which it is associated.

Extrusive Igneous. Antonym of intrusive.

Exudate The fluid plasma which leaks out of blood vessels due to an increase in capillary permeability. The increased permeability, is caused by histamine, and bradykinin, which are released in response to tissue damage. The formation of an exudate is the first step in the process of inflammation.

Exuvia The cast-off outer skin of an insect or other arthropod.

Eye-cap Hood formed by the base of the antenna and partly covering the eye in certain small moths.

Eyespot 1. A pigmented photoreceptor in euglenoids. The eyespot senses light and orients the cell for maximum rates of photosynthesis. Term applied to a photosenstive area in starfish.

F

F factor A fertility factor in bacteria, a DNA segment that confers the ability to form pili for conjugation and associated functions required for the transfer of DNA from donor to recipient. May exist as a plasmid or integrated into the bacterial chromosome.

F value The time in minutes at a specific temperature (usually 250°F) needed to kill a population of cells or spores.

F plasmid An F plasmid that carries some bacterial genes and transmits them to recipient cells when the F_ cell carries out conjugation; the transfer of bacterial genes in this way is often called sexduction.

F_1 (first filial generation) The first filial or hybrid offspring in a genetic cross-fertilization.

F1 particle Particle on the inner mitochondrial membrane, which is the site of ATP synthesis by oxidative phosphorylation.

F_2 (second filial generation) Offspring resulting from interbreeding of the hybrid F_1 generation.

Facet A small, facelike articular surface.The surface of an ommatidium - one of the units making up the compound eye. Funnel-shaped opening through the inner membrane complex of the egg of a pentastomid. It receives the product of the dorsal organ.

Facial Faci = the face.

Facilitated diffusion Mediated transport in which a permease makes possible diffusion of a molecule across a cell membrane in the direction of a concentration gradient; contrast with active transport.The spontaneous passage of molecules and ions, bound to specific carrier proteins, across a biological membrane down their concentration gradients.

Facultative anaerobe An organism that makes ATP by aerobic respiration if oxygen is present but that switches to fermentation under anaerobic conditions.Microorganisms capable of switching pathways of respiration, depending on the presence or absence

Facultative symbiont Opportunistic symbiont, establishing a relationship with a host only if the opportunity presents itself but not physiologically dependent on doing so.

FAD Abbreviation for flavine adenine dinucleotide, an electron acceptor in the respiratory chain.

Families 1. In taxonomy, term applied to subcategories within orders.

2. Term applied to a group of similar things, such as languages, chromosomes, etc.A taxonomic grouping of related, similar genera; the category below order and above genus.

Fascicle. A small bundle; the bundle of piercing stylets of insects with piercing sucking mouthparts.

Family planning Controlling reproduction; planning the timing of birth and having as many babies as are wanted and can be supported.

Famines Acute food shortages characterized by large-scale loss of life, social disruption, and economic chaos.

Fas gene The gene that is active in target cells which are susceptible to killing by cells expressing the Fas ligand, a member of the TNF family of cytokines and cell surface molecules.

Fascicle 1. Stylet bundle or combination of mouthparts used to pierce the skin in a blood-feeding arthropod. Composition of a fascicle varies according to group.

2. A small bundle, usually referring to a collection of muscle fibers or nerve axons.

Fascicular cambium The layer of cambium that develops between the xylem and phloem within a vascular bundle.

Fat Triglyceride that is solid at room temperature; usually of animal origin. Organic molecules containing high levels of carbon and hydrogen, but little oxygen. Oils are merely fats in liquid state. 1. Triglycerides that are solid at room temperature.

2. A legendary pool player from Minnesota.

Fate map Diagram that takes the larval or adult structure of an organism and "maps" it onto the region of the embryo from which it arises

Fatigue Loss of efficiency in the performance of a motor act when that act is repeated in rapid succession.

Fat-soluble vitamin Vitamins that can be stored in fatty tissues of the body; vitamins A, D, E, and K.

Fatty acid A long carbon chain carboxylic acid. Fatty acids vary in length and in the number and location of double bonds; three fatty acids linked to a glycerol molecule form fat.Any of a series of saturated organic acids having the general formula CnH2nO2, occurs in natural fats of animals and plants.

Fatty acid synthetase The multienzyme complex that makes fatty acids; the product usually is palmitic acid.

Fault A crack in the earth's crust usually formed when two pieces of crust are moving past each other. A fracture, or large crack, in the Earth's crust where one side moves up/down/sideways relative to the other.

Fault block pieces of crust that have slipped into or alongside a fault.

Fault zone an area with multiple faults.

Fauna Term referring collectively to all animals in an area. The zoological counterpart of flora. The animal life of a given area. A list of all species of animals found in a given area

Fecal coliform bacteria Technically, all the facultative anaerobic gram negative non-spore forming rod shaped bacteria that fermet lactose in EC medium with gas production within 24h at 44.5 degrees C. A measure of bacteria mostly originating from guts that enter waters. Believed to be correlated with disease-causing (pathogenic) bacteria.

Fecal enterococci Enterococci found in the intestine of humans and other warm-blooded animals. They are used as indicators of the fecal pollution of water.

Feces Semisolid material containing undigested foods, bacteria, bilirubin, and water that is produced in the large intestine and eliminated from the body. Frequently noted as "hitting the fan".Waste materials, including undigested food and sloughed-off intestinal cells, that are expelled from the intestinal tract through the anus.

Fecundity The number of eggs produced per female per unit time (often: per spawning season)The physical ability to reproduce.

Feedback A system of control, where work being done is modified by the product. For example the blood pressure is maintained by the strength of the heart beat and the muscle tone of the arterioles. In the walls of the large arteries are receptors sensitive to the degree of stretch in the muscle wall. As the blood pressure increases, the wall are stretched, and the receptor sends signals via the brain to the sympathetic nervous system back to the heart and blood vessels, causing decreased pumping effort and more relaxed muscle tone in the arteries. In chemical reactions the accumulated product slows down the rate of production. For example if the oxygen level of the body falls, the rate of respiration.increases to restore the levels to normal. These control system are thus circular; what is produced returns to control the further production. They are examples of negative feedback, and are common in maintaining stability or homeostasis. Positive feedback is less common as it tends to be unstable. An example is the release by platelets of thrompotaxin. When the levels of thrombotaxin are high, they do not inhibit further production as occurs in a negative feedback system but actually stimulate more platelets to produce more thrombotaxin and so on until there is an explosive increase in the number of sticky platelets. This is useful in an emergency to stop bleeding, but very dangerous when a clot forms inside a blood vessel.

Feedback inhibition A negative feedback mechanism in which a pathway end product inhibits the activity of an enzyme in the sequence leading to its formation; when the end product accumulates in excess, it inhibits its own synthesis.

Feedback loops A sequence of events in which the level of a hormone or related endocrine product circulating in the blood leads to alterations in the rate of production and release of other hormones from one or more endocrine glands. Often, feedback loops involve the hypothalamic portion of the brain where specialized sensory cells monitor circulating blood levels of numerous compounds, including both hormones and products from hormone actions at various body locations.

Feedback mechanisms control mechanisms in organisms and communities in which a change in a given factor either inhibits or stimulates processes controlling the production, release, or use of that factor

Feedback systems Control mechanisms whereby an increase or decrease in the level of a particular factor inhibits or stimulates the production, utilization, or release of that factor; important in the regulation of enzyme and hormone levels, ion concentrations, temperature, and many other factors.

Felsic Term used to describe the amount of light-colored feldspar and silica minerals in an igneous rock. Complement of mafic.

Female defense polygyny Males controlling access to females directly by competing with other males.

Female In organisms with separate sexes, the one which produces eggs.

Femur Podomere of an insect or acarine leg fixed to the trochanter proximally and articulating with the tibia distally in insects and with the patella in acarines.The 3rd (counting out from the body) and often the largest segment of the insect leg.The upper leg bone.

Fen An area of waterlogged soil that tends to be peaty; fed mainly by upwelling water; low productivity.

Feral Describes a species that was once tamed or domesticated and has since reverted to a free-roaming life in the wild.

Fermentation A type of anaerobic respiration that yields carbon dioxide and alcohol; used in commercial fermentation processes, including production of raised bakery dough products and alcoholic beverages. A catabolic process that makes a limited amount of ATP from glucose without an electron transport chain and that produces a characteristic end-product, such as ethyl alcohol or lactic acid.Degradative pathway that begins with glycolysis and ends with the electrons being transferred back to one of the breakdown products or intermediates such as alcohols, acids, and carbon dioxide; does not require molecular oxygen.

Fertile crescent Area between the Tigris and Euphrates Rivers in the Near East; some of the earliest documented sites of agriculture.

Fertility Measurement of actual number of offspring produced through sexual reproduction; usually described in terms of number of offspring of females, since paternity can be difficult to determine.

Fertilization 1. The union of a male and female gamete to form a zygote.

2. the act of insemination, impregnation, or pollination.

3. a process marking the beginning of conception, during which the nuclei of the sperm cell and the egg cell come together and their chromosomes combine. The fusion of two gametes (sperm and ovum) to produce a zygote that develops into a new individual with a genetic heritage derived from both parents. Strictly speaking, fertilization can be divided into the fusion of the cells (plasmogamy) and the fusion of nuclei (karyogamy). The joining or fusion of the male gamete (sperm) and the female gamete (egg) to form a zygote during sexual reproduction.

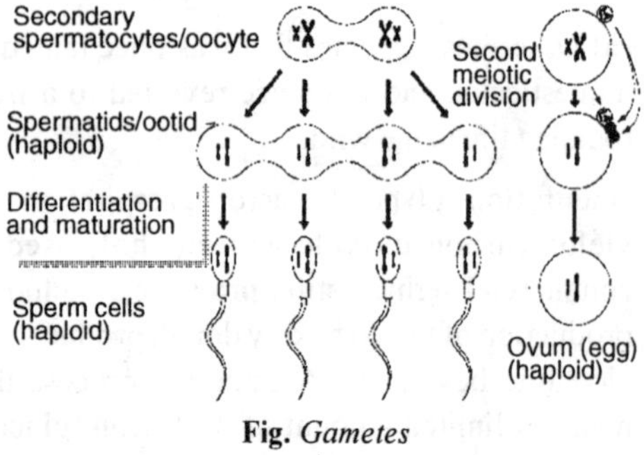

Fig. *Gametes*

Fertilization cone A prominence extending from the surface of some eggs at the moment of, or in some cases allegedly shortly before contact with a sperm.

Fertilization membrane A membrane formed in some aquatic species from the vitelline membrane and cortical granules immediately after sperm penetration into the ovum. It is thought to prevent polyspermy by making the egg impenetrable to other sperm.

Fertilizer burn When the concentration of solutes in the soil is too high due to over fertilization, plasmolysis at the roots occurs, damaging the plant.

Festoons Sclerites on the posterior margin of the opisthosoma of certain hard ticks.

Fetch The span of the sea surface over which the wind blows to form wind-driven waves.

Fetus An unborn or unhatched vertebrate that has passed through the earliest developmental stages; a developing human from about the second month of gestation until birth.

Fever A complex physiological response to disease mediated by pyrogenic cytokines and characterized by a rise in core body temperature and activation of the immune system.

Fiber A lignified cell type that reinforces the xylem of angiosperms and functions in mechanical support; a slender, tapered sclerenchyma cell that usually occurs in bundles.Elongated and thickened cell found in xylem tissue. It strengthens and supports the surrounding cells.Long and narrow sclerenchymous cell; functions in support; an important dietary component that provides bulk; component of fabrics, ropes, and paper.A long thin string-like structure constructed of smaller fibrils and even smaller microfibrils. Examples are collagen, elastic and keratin fibres. Collagen fibres are arranged parallel to each other in a tendon, to give it great resistance to tension (pulling).

Fibril A strand of protoplasm produced by a cell and lying within the cell.

Fibrillar Composed of or pertaining to fibrils or fibers.

Fibrillar flight muscle Insect flight muscle responsible for indirect flight. A single nerve impulse results in many cycles of flight muscle contraction and relaxation.

Fibrin The activated form of the blood-clotting protein fibrinogen, which aggregates into threads that form the fabric of the clot.Protein that forms a meshwork, trapping erythrocytes, to become blood clot. Precursor is fibrinogen.

Fibrinogen A large soluble protein found in blood which is converted into fibrin during blood clotting.

Fibroblast A type of cell in loose connective tissue that secretes the protein ingredients of the extracellular fibers.A term applied to a cell of connective tissue that is separated from similar cells by some degree of matrix material; fibroblasts secrete elastin and collagen protein fibers. Cell of connective tissue which form both the intercellular matrix and fibres.

Fibrocartilage The type of cartilage made up of parallel, thick, compact bundles, separated by narrow clefts containing typical cartilage cells (chondrocytes).

Fibronectin A glycoprotein which is found in the extracellular matrix and is important for the attachments, and therefore the movement of cells.

Fibrosis Deposition of fibrous connective tissue in a localized site, during process of tissue repair or to wall off a source of antigen. The general name for accumulation of scar tissue in the lung.

Fibrous connective tissue The tissue that is made up of fibers that are very densely packed (e.g., tendons and ligaments).

Fibrous protein Insoluble structural protein in which the polypeptide chain is coiled along one dimension. Fibrous proteins constitute the main structural elements of many animal tissues.

Fibrous root A root system found in monocots in which branches develop from the adventitious roots, forming a system in which all roots are about the same size and length.

Fibrous root system A root system in which roots are finely divided and lacking a main axis, common in monocotyledons.

Fibula Fibul = buckle. A leg bone.

Fiddlehead The curled fern frond prior to unrolling and elongation; also known as a crozier.

Fidelity A principle that forbids misleading or deceiving any creature capable of being mislead or deceived. We are to be truthful in our dealings with others.

Field capacity The soil-water storage capacity; the saturated soil profile after gravitational percolation ceases to flow.

Filament A thread-like structure, especially one at the end of an antenna. Long chain of proteins, such as found in hair, muscle, or in flagella.Part of the stamen in a flower that supports the anther.

Filamentous Long, thin, hair-like. A slender or threadlike structure or organism, in the latter consisting of a chain of interconnected cells as in filamentous algae.

Filial An offspring generation, for example, F1, the first filial generation.

Filial imprinting The process by which young animals form a social attachment for a particular stimulus, often the mother.

Filiform Thread-like or hair-like, applied especially to antennae.

Filipodium A type of pseudopodium that is very slender and may branch but does not rejoin to form a mesh.

Filling-in *In vitro* synthesis of DNA (by DNA polymerase) to convert an overhanging end of a double-stranded DNA fragment to a blunt end.

Filoplume feather A small thin feather that probably has sensory functions in birds (pinfeather).

Filter feeder An organism that filters food particles from the surrounding fluid.Organisms such as sponges that feed by removing food from water that filters through their body.Any feeding process by which particulate food is filtered from water in which it is suspended.A porous mesh of cotton cloth, spun glass fibers, or asbestos-cellulose that allows air or liquid to pass through but holds back solid particles.

Filtrate Fluid extracted by the excretory system from the blood or body cavity. The excretory system produces urine from the filtrate after extracting valuable solutes from it and concentrating it.

Filtration Movement of material through a membrane as a result of hydrostatic pressure. The first stage of kidney function; blood plasma is forced, under pressure, out of the glomerular capillaries into Bowman's capsule, through which it enters the renal tubule.The removal of water and solutes from the blood; occurs in the glomerulus of the nephron.

Fimbria A fine, hairlike protein appendage on some gram-negative bacteria that helps attach them to surfaces.

Fin ray Each of the bony spines in the fins of bony fishes.

Final host The host on/in which a parasitic organism either attains sexual maturity or reproduces.

Finite resources Resources that have a limit to their availability; not boundless.

Finlet small median fin on the dorsal and ventral sides of the rear parts of tuna and similar fish

Fire ecology The study of the environmental effects of fire.

Fire-climax community An equilibrium community maintained by periodic fires; examples include grasslands, chaparral shrubland, and some pine forests.

First law of thermodynamics The principle of conservation of energy. Energy can be transferred and transformed, but it cannot be created or destroyedThe total amount of energy in the universe remains constant; more energy cannot be created and existing energy cannot be destroyed; energy can only undergo conversion from one form to another.

First world The industrialized capitalist or market-economy countries of Western Europe, North America, Japan, Australia, and New Zealand.

FISH A *physical mapping* approach that uses fluorescein tags to detect *hybridization* of *probes* with *metaphase chromosomes* and with the less- condensed *somatic interphase* chromatin.

Fish meal A fish-protein supplement used in animal feeds.

Fish protein concentrate (FPC or fish flour)A fish protein supplement for human consumption.

Fission Asexual reproduction in which the cell divides into two (binary fission) or more (multiple fission) daughter parts, each of which becomes an individual organism.Division of single-celled organisms, especially prokaryotes, in which mitosis does not occur. Also used to refer to mitosis in certain unicellular fungi.

Fitness A measure of an individual's ability to survive and reproduce; the chance that an individual will leave more offspring in the next generation than other individuals.Degree of adjustment and suitability for a particular environment. Genetic fitness is relative contribution of one genetically distinct organism to the next generation; organisms with high genetic fitness are naturally selected and become prevalent in a population.The genetic contribution of an individual to succeeding generations relative to the contributions of other individuals in the population.

Fixation The process in which the internal and external structures of cells and organisms are preserved and fixed in position.

Fixed action pattern An innate behavior pattern that is stereotyped, spontaneous, and independent of immediate control, genetically encoded, and independent of individual learning.

Fjord a deep coastal embayment caused by glacial erosion

Flabellate With projecting flaps on one side, applied especially to antennae.

Flabellum Recurved process often found on the first two thoracic exopods of branchiuran crustaceans.

Flaccid Pertaining to a cell or tissue with less than full.

Flagella Flagella are whip-like appendages that undulate to move cells. They are longer than cilia, but have similar internal structures made of microtubules. Prokaryotic and eukaryotic flagella differ greatly.

Flagellar pocket Depression, sometimes long and deep, from which a flagellum arises.

Flagellin Protein which is the primary component of prokaryotic flagella.

Flagellum Hair-like structure attached to a cell, used for locomotion in many protists and prokaryotes. The prokaryotic flagellum differs from the eukaryotic flagellum in that the prokaryotic flagellum is a solid unit composed primarily of the protein flagellin, while the eukaryotic flagellum is composed of several protein strands bound by a membrane, and does not contain flagellin. The eukaryotic flagellum is sometimes referred to as an undulipodium.

Flagging behavior Alarm signaling, as with the use of the tail.

Flame cell Specialized hollow excretory or osmoregulatory structure of one or several small cells containing a tuft of flagella (the "flame") and situated at the end of a minute tubule; connected tubules ultimately open to the outside.

Flank The fleshy part of an animal between the ribs and hip.

Flat or plane warts Small, smooth, slightly raised warts.

Flatworms (phylum Platyhelminthes)Invertebrates that are dorsoventrally flattened and have an incomplete digestive tract, true organs, and organ systems.

Flavin adenine dinucleotide An electron carrying cofactor often involved in energy production (for example, in the tricarboxylic acid cycle and the b-oxidation pathway).

Flavin mononucleotide An electron acceptor in the electron transport scheme or aerobic respiration.

Flavonoids A group of secondary compounds produced by plants and important in chemical identification of those plants; believed to be contained in petals that reflect ultraviolet patterns and thus an element in pollinator attraction.

Fledge 1. Of a young bird, to acquire the feathers needed for flight.

2. to raise a young bird to maturity.

Fleshy fruit Fruit in which the cells of the pericarp are alive at maturity.

Flocculate A process by which dissolved substances come out of solution and aggregate together to form macroscopic particles.

Flock A group of animals, especially birds, that remain together, as for defense from predators or efficiency in locating food.

Flood An overflow of water onto land that normally is dry.

Floodplains Low lands along riverbanks, lakes, and coastlines subjected to periodic inundation.

Flora Term collectively applied to all of the plants in an area. The botanical counterpart of fauna. The plant life of a given area. A list of all species of plants found in a given area., often listing diagnostic features.

Floret One of the small flowers that make up the inflorescences in the composite and grass families.

Florigen The hypothetical flowering hormone. This compound (or compounds) has never been identified chemically.

Flow cytometry Analysis of biological material by detection of the light- absorbing or fluorescing properties of cells or subcellular fractions (i.e., *chromosomes*) passing in a narrow stream through a laser beam. An absorbance or fluorescence profile of the sample is produced. Automated sorting devices, used to fractionate samples, sort successive droplets of the analyzed stream into different fractions depending on the fluorescence emitted by each droplet.

Flow karyotyping Use of flow cytometry to analyze and/or separate *chromosomes* on the basis of their DNA content.

Flower The reproductive structure of angiosperms; a complete flower includes sepals, petals, stamens (male structures), and carpels (female structures).The reproductive structure of the anthophyta or angiosperms.

Flowering plants Plants that have flowers, seeds, and true leaves, stems, and roots.

Flue-gas scrubbing Treating combustion exhaust gases with chemical agents to remove pollutants. Spraying crushed limestone and water into the exhaust gas stream to remove sulfur is a common scrubbing technique.

Fluid feeders Animals such as aphids, ticks, and mosquitoes that pierce the body of a host plant or animal and obtain food from ingesting its fluids.

Fluid mosaic Refers to the model for the structure of membranes, consisting of a bilayer of lipids in which globular proteins are embedded and can move laterally.

Fluid mosaic model The currently accepted model of cell membrane structure, which envisions the membrane as a mosaic of individually inserted protein molecules drifting laterally in a fluid bilayer of phospholipids.

Fluid-feeder An animal that lives by sucking nutrient-rich fluids from another living organism.

Fluidized bed combustion High pressure air is forced through a mixture of crushed coal and limestone particles, lifting the burning fuel and causing it to move like a boiling fluid. Fresh coal and limestone are added continuously to the top of the combustion bed while ash and slag are drawn off below.

Fluke A member of class Trematoda or class Monogenea. Also, certain of the flatfishes (order Pleuronectiformes).

Fluorapatite An apatite crystal in which fluoride has replaced hydroxyl ions.

Fluorescence Longer wavelength emissions which can result form excited electron energy, results in a glowing light.

Fluorescence microscope A microscope that exposes a specimen to light of a specific wavelength and then forms an image from the fluorescent light produced. Usually the specimen is stained with a fluorescent dye or fluorochrome.

Fluorosis Mottling of the teeth caused by an excess of fluoride in the drinking water. A fluorosis index recognises 4 stages of severity.

Fluorescent light The light emitted by a substance when it is irradiated with light of a shorter wavelength.

Flushing time The time required for all of the water of an estuary to be completely exchanged.

FMN Abbreviation for flavin mononucleotide, the prosthetic group of a protein (flavoprotein) and a carrier in the electron transport chain in respiration.

Focal-animal sampling A technique for recording the behavior of individual animals in which the observer concentrates on the activities of one individual for a set time period and then switches to watch another animal for a set time period.

Foci Groups of cells in an embryo that differentiate into specific structures and organs.

Focus The initial point within the Earth that ruptures in an earthquake, directly below the epicenter.

Fold Bent rock strata.

Foliose coral A coral whose skeletal form approximates that of a broad, flattened plate

Follicle Single, dry, dehiscent fruit that splits along one seam; example: milkweed.

Follicles Spherical structures that make up the thyroid gland; contain a gel-like colloid surrounded by a single layer of cells, which secrete thyroglobulin into the colloid.

Follicle-stimulating hormone A protein hormone secreted by the anterior pituitary that stimulates the production of eggs by the ovaries and sperm by the testes.

Fomite An object that is not in itself harmful but is able to harbor and transmit pathogenic organisms. Also called fomes.

Fontanels Membranous areas in the human cranial bones that do not form bony structures until the child is 14 to 18 months old; know as "soft spots."

Food aid Financial assistance intended to boost less-developed countries' standards of living.

Food chain A linear sequence of organisms that exist on successive trophic levels within a natural community, through which energy is transferred by feeding. Primary producers capture energy from the environment (through photo- or chemo-synthesis) and form the base of the food chain. Energy is then passed to primary consumers (herbivores) and on to secondary and tertiary consumers (carnivores and top carnivores) (e.g. phytoplankton -> zooplankton -> herring -> salmon -> killer whales). Once they die, these organisms are in turn consumed and their energy transferred to detrivores and decomposers. Compare *food web*.

Food chain efficiency Amount of energy of some other quantity extracted from a trophic level, divided by the amount of energy produced by the next-lower trophic level

Food chain/food web All the interactions of predator and prey, included along with the exchange of nutrients into and out of the soil. These interactions connect the various members of an ecosystem, and describe how energy passes from one organism to another.

Food intoxication Food poisoning caused by microbial toxins produced in a food prior to consumption. The presence of living bacteria is not required.

Food poisoning A general term usually referring to a gastrointestinal disease caused by the ingestion of food contaminated by pathogens or their toxins.

Food pyramid A way of depicting energy flow in an ecosystem; shows producers (mostly plants or other phototrophs) on the first level and consumers on the higher levels.

Food security The ability of individuals to obtain sufficient food on a day-to-day basis.

Food surpluses Excess food supplies.

Food vacuole An organelle in the cell that functions in intracellular digestion.

Food web A complex network of feeding interrelations among species in a natural ecosystem; more accurate and more complex depiction of energy flow than a food chain.A non-linear network of feeding between organisms that includes many food chains, and hence multiple organisms on each trophic level. For example, both sharks and tuna eat herring, and sharks also eat tuna.

Food-borne infection Gastrointestinal illness caused by ingestion of microorganisms, followed by their growth within the host. Symptoms arise from tissue invasion and/or toxin production.

Foot 1. The basal portion of a moss sporophyte, embedded in the gametophyte.

2.The muscular locomotory structure of molluscs.

Footprinting A technique by which one identifies a protein binding site on cellular DNA. The presence of a bound protein prevents DNase from "nicking" that region, which can be detected by an appropriately designed gel.

Forage crops Crops that are grown as food for domesticated herbivores.

Forager Hunter-gatherer group.

Foraging behavior The process animals use in locating food resources.

Foramen Foram = an opening.

Foramina The plural of foramen, which is a hole, for example Foramen Ovale.

Foraminifera Single-celled protists that secrete a shell or test. Accumulations of the shells of dead foraminifera and other microscopic sea creatures form chalk deposits.

Foraminiferan A member of the class Granuloreticulosea (phylum Sarcomastigophora) bearing a test with many openings.

Foraminiferan ooze A biogenous sediment that consists mostly of the calcareous shells of foraminiferans.

Forb A small, upright soft-stemmed or non-woody plant with broadleaves; the growthform of many common wildflowers.Term for any plant that is not a grass.

Fore reef The outer part of a barrier reefor atoll.

Forebrain The forebrain consists of the diencephalon and telencephalon.

Foregut The anterior part of the alimentary canal from the mouth to the midgut. Fossorial Adapted for digging.

Forest management Scientific planning and administration of forest resources for sustainable harvest, multiple use, regeneration, and maintenance of a healthy biological community.

Forewing The first or anterior pair of insect wings.

Form drag Hydrodynamic drag on an organism caused by its cross-sectional area.

Formation A group of communities in a single region or continent with similar physiognomy (structure) and related climatic and environmental conditions. One of several regional or continental expressions of a given biome.

Formed-element fraction The cellular component of vertebrate blood.

Fossil 1. The remains or traces of prehistoric life preserved in rocks of the Earth's crust.

2. Any evidence of past life. Any evidence of past life, including remains, traces, imprints as well as life history artifacts. Examples of artifacts include fossilized bird's nests, bee hives, etc.Any evidence of pre-existing life which is preserved in the earth's crust. In addition to the remains of entire, or parts of, organisms, fossils also include indirect evidence such as animal tracks and footprints, as well as the impression of leaves.

Fossil fuels Fuels that are formed in the Earth from plant or animal remains; e.g., coal, petroleum, and natural gas.Organic molecules derived from partially decayed plant and animal matter, produced primarily during the Carboniferous period; includes oil, gas, and coal.

Fossil record 1. The observed remains of once-living organisms taken as a whole.

2. the album Meet the Beatles.All of the fossils that have existed throughout life's history, whether they have been found or not.

Fossorial Adapted for digging or burrowing. It often refers to animals that live predominantly underground. Pocket gophers are good examples of Fossorial mammals.

Fouling organisms Organisms that live attached to submerged surfaces such as boats and pilings.

Fouling. Contamination of feeding or respiratory areas of an organism by excrement, sediment, or other matter. Also, accumulation of sessile marine organisms on the hull of a boat or ship so as to impede its progress through the water.

Founder effect The difference in gene pools between an original population and a new population founded by one or a few individuals randomly separated from the original population, as when an island population is founded by one or a few individuals; often accentuates genetic drift.

Founder principle Afew individuals starting a new population may represent an atypical sample of the parent species' gene pool. This "sampling error" leads to the *founder effect*: rapid changes in allele frequencies in the colonizing population and divergence from the parent population.

Fourth world A political/economic category describing very poor nations that have neither market economies nor central planning and are either not developing or are developing very slowly. Also used to describe indigenous communities within wealthier nations.

Fovea A small pit or depression; especially the fovea centralis, a small rodless pit in the retina of some vertebrates, a point of acute vision.

Foveola One of the paired depressions on each side of the vertex in grasshoppers.

Fractal dimension A dimension which is some fraction in between a line(1) and a plane(2), or a plane and a solid (3). These fractal dimensions are useful in describing the quality of natural lines and surfaces, such as coastlines, trees, vascular branching and the patterns of trabecula bone

Fraction I protein Equivalent to RUBP carboxylase; the primary leaf protein in many green plants.

Fragile X syndrome A hereditary mental disorder, partially explained by genomic imprinting and the addition of nucleotides to a triplet repeat near the end of an X chromosome.

Fragmentation Division into smaller units. A type of asexual reproduction whereby a body part is lost and then regenerates into a new organism.

Frameshift mutation A mutation (deletion or insertion, never a simple substitution) of one or more *nucleotides* but never a multiple of 3 nucleotides, which shortens or lengthens a trinucleotide sequence representing a *codon*; the result is a shift from one *reading frame* to another reading frame. The amino acid sequence of the protein downstream of the mutation is completely altered, and may even be much shorter or longer due to a change in the location of the first *termination (stop) codon*. Mutations arising from the loss or gain of a base or DNA segment, leading to a change in the codon reading frame and thus a change in the amino acids incorporated into protein.

Free energy A quantity of energy that interrelates entropy (S) and the system's total energy (H); symbolized by G. The change in free energy of a system is calculated by the equation G = ÄH – T ÄS, where T is absolute temperature.

Free energy change The total energy change in a system that is available to do useful work as the system goes from its initial state to its final state at constant temperature and pressure.

Free energy of activation The initial investment of energy necessary to start a chemical reaction; also called activation energy.

Free-running rhythm The activity cycle that an animal exhibits when placed in a constant environment; its period is different from any known cyclic environmental variable.

Free-threshing grain Grain that separates easily from enclosing bracts.

Freeway space The space between the teeth when the jaw is in a rest position .

Freezing condensation A process that occurs in the clouds when ice crystals trap water vapor. As the ice crystals become larger and heavier, they begin to fall as rain or snow.

Freezing To preserve an organism without any significant alteration to its chemical composition by subjecting it to freezing temperatures.

Frenulum The wing-coupling mechanism found in many moths.

Frequency-dependent selection A decline in the reproductive success of a morph resulting from the morph's phenotype becoming too common in a population; a cause of balanced polymorphism in populations.

Fresh water Water other than seawater; covers only about 2 percent of earth's surface, including streams, rivers, lakes, ponds, and water associated with several kinds of wetlands.

Freshet An increase of water flow into an estuary during the late winter or spring, owing to increased precipitation and snow melt in the watershed

Freshwater biome The aquatic biome consisting of water containing fewer salts than the waters in the marine biome; divided into two zones: running waters (rivers, streams) and standing waters (lakes, ponds).

Freshwater ecosystems Ecosystems in which the fresh (nonsalty) water of streams, rivers, ponds, or lakes plays a defining role.

Frictional drag the resistance created by an animal's body surface when it moves through a fluid medium

Fringing reef a large coral-reef formation that closely borders the shoreline

Frons Upper part of the insect face, between and below the antennae and usually carrying the median ocellus or simple eye. In true flies (Diptera) it occupies almost all of the front surface of the head apart from the eyes.

Front A major discontinuity separating ocean currents and water masses in any combination

Frontal bristles The two vertical rows of bristles running down the face of a fly from the ocelli to the antennae

Frontal lobe The lobe of the cerebral cortex that is responsible for motor activity, speech, and thought processes.

Frontal plane. A plane parallel to the main axis of the body and at right angles to the sagittal plane.

Frontier An unexploited natural area at the leading edge of human settlement.

Frontier mentality The idea that the world has an unlimited supply of resources for human use regardless of the consequences to natural ecosystems and the biosphere.

Fronto-orbital bristles The short row of bristles on each side of a fly's head between the eye and the frontal bristles.

Fructose Six-carbon monosaccharide; often referred to as fruit sugar.

Frugivore Animal which primarily eats fruit. Many bats and birds are frugivores.

Fruit In flowering plants, the structure which encloses the seeds. True fruits develop from the ovary wall, such as bananas and tomatoes, though not all fruits are edible, such as the dry pods of milkweed or the winged fruits of the maple.

Fruiting body A specialized structure that holds sexually or asexually produced spores; found in fungi and in some bacteria.In Dictyostelium, the final structure derived from a grex/slug/ pseudoplasmodium, an aggregation of thousands of myxamoebae forming a single structure composed of spore cells and a stalk.

Frustule The mineral "skeleton" of a diatom or other unicellular organism.The siliceous, box-like cell wall of diatoms.

Fucoxanthin Brown accessory pigment found in and characteristic of the brown algae.Yellowish-brown pigment found in some members of the Chromista, including kelps and diatoms.

Fuel assembly A bundle of hollow metal rods containing uranium oxide pellets; used to fuel a nuclear reactor.

Fuel-switching A change from one fuel to another.

Fuelwood Branches, twigs, logs, wood chips, and other wood products harvested for use as fuel.

Fugitive emissions Substances that enter the air without going through a smokestack, such as dust from soil erosion, strip mining, rock crushing, construction, and building demolition.

Fugitive species A species adapted to colonize newly disturbed habitats

Fumigant A substance which produces a gas, vapor, fume, or smoke intended to kill a pest.

Function Refers to action, how something works. In the case of ecosystem functions we look at photosynthesis, nutrient cycling, population control, dispersal mechanisms, temporal patterns of flowering , breeding, dormancy, and so forth.

Functional group A specific configuration of atoms commonly attached to the carbon skeletons of organic molecules and usually involved in chemical reactions.

Functional neuroanatomy The study of the size, structure, and arrangement of cells within the nervous system, particularly the brain.

Functionalists People who conduct studies involving attempts to discern how the mind works, as opposed to studies of its structure.

Fundamental niche The multidimensional space that an animal would occupy under optimal conditions in the absence of competitors; also called preferred niche. Contrast with realized niche.

Fungi One of the five kingdom classifications; consists of nonphotosynthetic, eukaryotic organisms with cell walls, filamentous bodies, and absorptive nutrition.

Fungicidal A compound that is able to kill a fungus and stop an infection.

Fungicide Any substance that kills or inhibits the growth of a fungus.

Fungus, fungi Any of numerous plants lacking chlorophyll, ranging in form from a single cell to a body of branched filaments. Includes the yeasts, molds, smuts, and mushrooms.

Funicle The portion of the flagellum of the antenna closest to the club.

Funiculus The stalklike structure connecting an ovule to its placental surface within an ovary.

Funnel The siphon of cephalopods.

Furcula The forked spring of a springtail.

Fuscous Smokey grey-brown in colour, normally applied to wings.

Fusiform Spindle shaped; tapering toward each end.

Fusion protein A product of recombinant DNA in which the foreign gene product is juxtaposed ("fused") to either the *carboxyl-terminal* or *amino-terminal* portion of a polypeptide encoded by the vector itself. Use of fusion proteins often facilitates expression of otherwise lethal products and the purification of recombinant proteins.

G

G protein A GTP-binding protein that relays signals from a plasma-membrane signal receptor, known as a G-protein linked receptor, to other signal-transduction proteins inside the cell. When such a receptor is activated, it in turn activates the G protein, causing it to bind a molecule of GTP in place of GDP. Hydrolysis of the bound GTP to GDP inactivates the G protein.

G_1 phase The first growth phase of the cell cycle, consisting of the portion of interphase before DNA synthesis begins.

G_2 phase The second growth phase of the cell cycle, consisting of the portion of interphase after DNA synthesis occurs.

Gabbro Highly mafic igneous plutonic rock, typically dark in color; rough plutonic equivalent of basalt.

Gaia A hypothetical superorganism composed of the Earth's four spheres: the biosphere, hydrosphere, lithosphere, and atmosphere.

Gaia hypothesis A theory that the living organisms of the biosphere form a single, complex interacting system that creates and maintains a habitable Earth; named after Gaia, the Greek Earth mother goddess.

Galactose A 6-carbon sugar.

Galapagos Islands An archipelago on the equator in the Pacific Ocean about 1,000 km west of Ecuador. Charles Darwin's observations of the plant and animal life of these islands were important in the formulation of the theory of evolution by natural selection.

Galea The outer branch of the maxillae, the inner one being the lacinia.

Gall An abnormal growth of a plant caused by the presence in its tissues of a young insect or some other organism. Aphids, gall wasps, and gall midges are among the major gall-causing insects.

Gallbladder The pear-shaped reservoir for bile.

Gametangia Any cell or organ in which gametes are formed.

Gametangium A structure that contains gametes or in which gametes are formed.The reproductive organ of bryophytes, consisting of the male antheridium and female archegonium; a multichambered jacket of sterile cells in which gametes are formed.

Gamete A haploid reproductive cell that develops into a new individual after its union with another gamete.A mature reproductive cell that is capable of fusing with another gamete of the opposite sex to form a zygote. Male gametes are typically known as sperm and female gametes a typically known as eggs.

Gametic meiosis. Meiosis that occurs during formation of the gametes, as in humans and other metazoa.

Gametocyst Cyst produced by some apicomplexan parasites. Sexual reproduction and spore formation occurs within this cyst.

Gametogenesis The formation of gametes by way of meiosis.

Gametogony Multiple fission that forms gametes that fuse to form a zygote. Also called gamogony. Occurs in the class Sporozoea.

Gametophyte The multicellular haploid form in organisms undergoing alternation of generations, which mitotically produces haploid gametes that unite and grow into the sporophyte generation.The haploid stage in the life cycle of an organism undergoing alternations of generations. The gametophyte is multicellular and mitotically produces gametes. In plants, the gametophyte nourishes the zygote and young sporophyte.

Gamma rays That portion of the sun's total range of radiation in which rays are shorter than x rays; below 0.1 nm in length.

Gamma-proteobacteria One of the five subgroups of proteobacteria, each with distinctive 16S rRNA sequences. This is the largest subgroup and is very diverse physiologically; many important genera are facultatively anaerobic chemoorganotrophs.

Gamont Apicomplexan life cycle stage that is committed to undergoing gametogenesis.

Ganglia Clusters of neurons that receive and process signals; found in flatworms and earthworms.

Ganglion A collection of nerve cells usually found outside the central nervous system, from which axons arrive from the periphery and proceed to the spinal cord or brain.

Gangrene The death of tissue on a large scale. May be caused by certain bacteria which spread rapidly through tissues, or by an inadequate blood supply.

Ganoid scales Thick, bony, rhombic scales of some primitive bony fishes; not overlapping.

Gap analysis A biogeographical technique of mapping biological diversity and endemic species to find gaps between protected areas that leave endangered habitats vulnerable to disruption.

Gap junctions Junctions between the plasma membranes of animal cells that allow communication between the cytoplasm of adjacent cells.

Gap phases In the cell cycle, the phases that precede (G_1) and follow (G_2) the synthesis (S) phase in which DNA is replicated; in the G_1 phase, the cell doubles in size, and its enzymes, ribosomes, and other cytoplasmic molecules and structures increase in number; in the G_2 phase, the replicated chromosomes begin to condense and the structures required for mitosis or meiosis are assembled.

Garden city A new town with special emphasis on landscaping and rural ambience.

Gas exchange The movement of oxygen and other gases between the atmosphere and the ocean or between the water or atmosphere and living organisms.

Gas gangrene A type of gangrene that arises from dirty, lacerated wounds infected by anaerobic bacteria, especially species of Clostridium. As the bacteria grow, they release toxins and ferment carbohydrates to produce carbon dioxide and hydrogen gas.

Gas vacuole A gas-filled vacuole found in cyanobacteria and some other aquatic bacteria that provides flotation. It is composed of gas vesicles, which are made of protein.

Gasohol A mixture of gasoline and ethanol.

Gastric caeca The sac-like diverticula at the anterior end of the midgut.

Gaster The hymenopteran abdomen - apart from the 1st segment (the propodeum) which is fused to the thorax. The front part of the gaster often forms a narrow waist.

Gastric pits The folds and grooves into which the stomach lining is arranged.

Gastric shield A chitinized plate in the stomach of a bivalve (phylum Mollusca) on which the crystalline style is rotated.

Gastrin A digestive hormone, secreted by the stomach, that stimulates the secretion of gastric juice.A hormone produced by the pyloric gland area of the stomach that stimulates the secretion of gastric acids.

Gastrodermis In cnidarians, the endodermis which lines the gut cavity. The term is often used instead of endodermis since cnidarians only have two tissue layers instead of three.

Gastroenteritis An acute inflammation of the lining of the stomach and intestines, characterized by anorexia, nausea, diarrhea, abdominal pain, and weakness. It has various causes including food poisoning due to such organisms as E. coli, S. aureus, Campylobacter (campylobacteriosis), and Salmonella species; consumption of irritating food or drink; or psychological factors such as anger, stress, and fear. Also called enterogastritis.

Gastroesophageal sphincter A ring of muscle at the junction of the esophagus and the stomach that remains closed except during swallowing to prevent the stomach contents from entering the esophagus.

Gastrolith Calcareous body in the wall of the cardiac stomach of crayfish and other Malacostraca, preceding the molt.

Gastropoda The class of molluscs characterized by torsion. A shell, when present, is usually coiled. Snails.

Gastrotricha A small phylum of marine and freshwater species of gastrotrichs that inhabit the spaces between bottom sediments.306

Gastrovascular canals Fluid-filled canals opening at the mouth of cnidarians and ctenophores that function in gas exchange and in the distribution of nutrients.

Gastrovascular cavity The central digestive compartment, usually with a single opening that functions as both mouth and anus.Body cavity in certain lower invertebrates that functions in both digestion and circulation and has a single opening serving as both mouth and anus.

Gastrozooid The feeding polyp of a hydroid, a hydranth.

Gastrula The early embryonic stage following the blastula stage during which cell migrations form distinct germ layers. The primitive gut (archenteron) is formed and opens to the exterior through the blastopore. Embryonic stage, usually cap or sac shaped, with walls of two layers of cells surrounding a cavity (archenteron) with one opening (blastopore).

Gastrulation The process in embryonic development whereby the three primary germ layers are laid down and the gut formed through complex cell migrations. Stage in animal development following cleavage characterized by extensive cell movement and rearrangement to form a "three-layered" embryo of ectoderm, mesoderm and endoderm.The embryological process that results in the formation of the gastrula; results in the formation of the embryonic gut, ectoderm, and endoderm.

Gated ion channel A specific ion channel that opens and closes to allow the cell to alter its membrane potential.

Gel electrophoresis A method to analyze the size of DNA (or RNA) fragments. In the presence of an electric field, larger fragments of DNA move through a gel slower than smaller ones. If a sample contains fragments at four different discrete sizes, those four size classes will, when subjected to electrophoresis, all migrate in groups, producing four migrating "bands". Usually, these are visualized by soaking the gel in a dye (ethidium bromide) which makes the DNA fluoresce under UV light.

Gel shift A method by which the interaction of a nucleic acid (DNA or RNA) with a protein is detected. The mobility of the nucleic acid is monitored in an agarose gel in the presence and absence of the protein: if the protein binds to the nucleic acid, the complex migrates more slowly in the gel (hence "gel shift"). A "supershift" allows determination of the specific protein, by virtue of a second shift in mobility that accompanies binding of a specific antibody to the nucleic acid-protein complex.

Gel shift assay (Aka gel mobility shift assay (GMSA), band shift assay (BSA), electrophoretic mobility shift assay (EMSA)) A method by which one can determine whether a particular protein preparation contains factors which bind to a particular DNA fragment. When a radiolabeled DNA fragment is run on a gel, it shows a characteristic mobility. If it is first incubated with a cellular extract of proteins (or with purified protein), any protein-DNA complexes will migrate slower than the naked DNA - a shifted band.

Gel That state of a colloidal system in which the solid particles form the continuous phase and the fluid medium the discontinuous phase.

Gemmae cup A specialized structure on the gametophyte of certain liverworts that produces asexual plantlets (gemmae) capable of starting new gametophytes.

Gemmule Resistant, overwintering capsule formed by freshwater, and some marine, sponges that contains masses of mesenchyme cells; amoeboid mesenchyme cells are released and organize themselves into a sponge. 268

Gena Anterioventral portion of an insect head. For example, genal ctenidium is a row of heavy spines on the gena of a flea.

Genal comb A row of stout spines on the lower border of the cheek of certain fleas.

Gene A discrete unit of hereditary information consisting of a specific nucleotide sequence in DNA (or RNA, in some viruses).A short length of a chromosome, made of DNA, which carries the information for a particular characteristic.A unit of DNA which performs one function. Usually, this is equated with the production of one RNA or one protein. A gene contains coding regions, introns, untranslated regions and control regions.

Gene activators Hormones that can turn a given gene on or off.

Gene amplification The selective synthesis of DNA, which results in multiple copies of a single gene, thereby enhancing expression.

Gene banks Storage for seed varieties for future breeding experiments.

Gene cloning A technique which uses recombinant DNA , inserted into a host cell as a plasmid which reproduces copies of itself, and hence the inserted gene, many times

Gene conversion The alteration of all or part of a gene by a homologous donor DNA that is itself not altered in the process.

Gene data bank A file of information concerning the chromosome location and the base sequences of the genes of a particular species, used to compare newly isolated genes with those reported by other workers. This information is usually stored as computer data, much of which is available on the Internet.

Gene expression The process by which a *genes* coded information is converted into the structures present and operating in the cell. Expressed genes include those that are transcribed into *mRNA* and then translated into *protein* and those that are transcribed into *RNA* but not translated into protein (e.g., *transfer* and *ribosomal RNAs*).

Gene families Groups of closely related *genes* that make similar products.

Gene flow An exchange of genes between two populations of a species, or in extreme cases, between populations of two species (hybridization).

Gene gun A device that uses high-pressure gas or another propellant to shoot a spray of DNA-coated microprojectiles into cells and transform them. Sometimes it is called a biolistic device.

Gene insertion The process by which one or more genes from one organism are incorporated into the genetic makeup of a second organism.

Gene library A collection of organisms, usually viruses, bacteria or yeast, which have been transformed by the addition of extra genes from another species. The library can be screened to find a gene of particular interest to researchers.

Gene locus The position of a gene along a chromosome.

Gene mapping Determination of the relative positions of *genes* on a DNA molecule (*chromosome* or *plasmid*) and of the distance, in *linkage* units or physical units, between them.

Gene pool The sum of all the genetic information carried by members of a population.

Gene product The biochemical material, either *RNA* or *protein*, resulting from expression of a gene. The amount of gene product is used to measure how active a gene is; abnormal amounts can be correlated with disease- causing alleles.

Gene regulation The process by which genes are turned on and off to regulate growth and development of an organism.

Gene synthesizer A machine that can produce a functional gene segment in a short period of time.

Gene therapy The insertion of normal or genetically altered genes into cells through the use of recombinant DNA technology; usually done to replace defective genes as part of the treatment of genetic disorders.

Gene transfer techniques Any one of a number of methods of transferring genes from one organism (species) to another (usually different species). May involve a vector such as a virus or a plasmid, or may use chemical carriers such as liposomes. The method used depends on the nature of the cell which is to receive the genetic information.

Genera Taxonomic subcategories within families (sing.: genus), composed of one or more species.

General fertility rate Representation of population age structure and fecundity; crude birth rate multiplied by the percentage of fecund women (between approximately fifteen and forty-four years of age) by 1000.

General recombination Recombination involving a reciprocal exchange of a pair of homologous DNA sequences; it can occur any place on the chromosome.

Generalist A species that can live in many different habitats and can feed on a number of different organisms. For example, shore crabs on the Pacific coast live in a wide variety of habitats, such as mud, sand and rock, and feed on everything from the algae growing on rocks to invertebrates to detritus. Compare *specialist.*

Generalization In learning, a principle that states if an animal has been conditioned to respond to a certain stimulus, the response will usually also occur to stimuli used during the original acquisition trials.

Generalized niche Niche that has broad requirements; tolerates a range of conditions.

Generalized transduction The transfer of any part of a bacterial genome when the DNA fragment is packaged within a phage capsid by mistake.

Generation Period from any given stage in the life cycle to the same life stage in the offspring. Typically from egg to egg.The group of individuals of a given species that have been reproduced at approximately the same time; the group of individuals of the same genealogical rank.

Generation time The length of time between when an organism is born and when it first reproduces.

Generative nucleus Produced by the haploid microspore nucleus of a pollen grain, the generative nucleus divides mitotically to form two sperm nuclei.

Generator potential A graded potential that travels

only a short distance along the plasma membrane of a sensory cell.

Generic Belonging to the same main group. For example generic medicines are identified by the main group they fall into rather than by their trade names.

Genetic assimilation The disappearance of a species as its genes are diluted through crossbreeding with a closely related species.

Genetic code The arrangement of nucleotides, arranged in triplets known as "codons," which determines the identity and sequence of the amino acids which are used during protein synthesis. The "code" used to translate genetic information contained in the DNA molecule into a gene product.

Second letter

First letter	U	C	A	G	
U	UUU UUC Phenyl-alanine UUA UUG Leucine	UCU UCC UCA UCG Serine	UAU UAC Tyrosine UAA Stop codon UAG Stop codon	UGU CGC Cystelne UGA Stop codon UGG Tryptophan	U C A G
C	CUU CUC CUA CUG Leucine	CCU CCC CCA CCG Proline	CAU CAC HIstidine CAA CAG Glutamine	CGU CGC CGA CGG Arginine	U C A G
A	AUU AUC AUA Isoleucine AUG Methionine initiation codon	ACU ACC ACA ACG Threonine	AAU AAC Asparagine AAA AAG Lysine	AGU AGC Serine AGA AGG Arginine	U C A G
G	CUU CUC CUA CUG Valine	GCU GCC GCA GCG Alanine	CAU CAC Aspartic acid CAA CAG Glutarnic acid	GGU GGC GGA GGG Glycine	U C A G

Fig. *Genetic code*

Genetic divergence The separation of a population's gene pool from the gene pools of other populations due to mutation, genetic drift, and selection. Continued divergence can lead to speciation.

Genetic drift Alele frequency changes in populations caused by random events rather than by natural selection, especially the effects of sampling error on the gene pool of small populations. Random change in allelic frequencies in a population occurring by chance. In small populations, genetic variation at a locus may be lost by chance fixation or a single allelic variant.

Genetic engineering Modifying the genetic structure of one organism by splicing in selected genetic information from another organism.

Genetic erosion Irreversible loss of genetic diversity due to extinction of traditional varieties and wild ancestors of crop plants.

Genetic isolation The absence of genetic exchange between populations or species as a result of geographic separation or of premating or postmating mechanisms (behavioral, anatomical, or physiological) that prevent reproduction.

Genetic locus A location on a chromosome (possibly of a diploid organism with variants that segregate according to the rules of Mendelian heredity)

Genetic map An ordered list of genetic loci (genes or other genetic markers) along a chromosome.

Genetic polymorphism Presence of several genetically controlled variants in a population

Genetic recombination The general term for the production of offspring that combine traits of the two parents.

Genetic screening The use of a number of mutant forms of a species to test for a specific type of change or deficit, as with respect to biological clocks.

Genetic variability Produced by sexual reproduction, all organisms are genetically different from each other.

Genetically engineered microorganisms (GEMs) Bacteria that have been genetically engineered by the insertion or deletion of DNA segments.

Genital atrium Cavity in the body wall of a flatworm into which male and female genital ducts open.

Genital claspers Organs of the male genitalia which serve to hold the female during copulation.

Genital herpes A sexually transmitted disease caused by the herpes virus; results in sores on the mucus membranes of the mouth or genitals.

Genital slit The genital opening of cetaceans.

Genitalia The copulatory organs of insects and other animals. The shape and arrangement of the genitalia are often used to distinguish closely related and otherwise very similar species.

Genitointestinal canal Duct connecting the oviduct and intestine of some polyopisthocotylean Monogenea.

Genome All the DNA in a haploid set of chromosomes (nuclear genome), organelle (mitochondrial genome, chloroplast genome) or virus (viral genome, which in some viruses consists of RNA rather than DNA).Complete haploid complement of DNA (including all genes) from the chromosomes of the nucleus of an organism. The total DNA contained in each cell of an organism. Mammalian genomic DNA (including that of humans) contains 6x109 base pairs of DNA per diploid cell. There are somewhere in the order of a hundred thousand genes, includıng coding regions, 5' and 3' untranslated regions, introns, 5' and 3' flanking DNA. Also present in the genome are structural segments such as telomeric and centromeric DNAs and replication origins, and intergenic DNA.

Genome projects Research and technology development efforts aimed at *mapping* and *sequencing* some or all of the *genome* of human beings and other organisms.

Genomic blot A type of Southern blot specifically used to analyze a mixture of DNA fragments derived from total genomic DNA. Because genomic DNA is very complicated, when it has been digested with restriction enzymes, it produces a complex set of fragments ranging from tens of bp to tens of thousands of bp. However, any specific gene will be reproducibly found on only one or a few specific fragments. A million identical cells will produce a million identical restriction fragments for any given gene, so probing a genomic Southern with a gene-specific probe will produce a pattern of perhaps one or just a few bands.

Genomic clone A piece of DNA taken from the genome of a cell or animal, and spliced into a bacteriophage or other cloning vector. A genomic clone may contain coding regions, exons, introns, 5' flanking regions, 5' untranslated regions, 3' flanking regions, 3' untranslated regions, or it may contain none of these...it may only contain intergenic DNA (usually not a desired outcome of a cloning experiment!).

Genomic DNA The DNA contained in the chromosomes of a cell.

Genomic equivalence Concept that each cell in the body has the same genetic material and therefore all the information necessary to create a complete organism. Animal cloning from a somatic nucleus 'proves' this idea.

Genomic imprinting The parental effect on gene expression. Identical alleles may have different effects on offspring, depending on whether they arrive in the zygote via the ovum or via the sperm.

Genomic library A DNA library which contains DNA fragments hopefully representing each region of the genome of an organism, virus, etc, cloned into individual vector molecules for subsequent selection and amplification. The DNA fragments are usually very small in size compared with the genome. Such libraries are *cDNA libraries* when prepared from RNA viruses.

Genomics The study of the molecular organization of genomes, their information content, and the gene products they encode.

Genotype The genetic makeup of an organism. The actual appearance of an individual (the phenotype) depends on the interaction between different forms (or alleles) of genes and between the genotype and the environment. Compare *phenotype*.Two uses: one is a verb, the other a noun. To 'genotype' (verb) is to example polymorphisms (e.g. RFLPs, microsatellites, SNPs) present in a sample of DNA.

Genotypic ratio The relative numbers of progeny in each genotypic category produced by a genetic cross.

Genus The level of the taxonomic hierarchy above the species but below the family level.A group of closely related species (plural: genera). The name of the genus is incorporated into the scientific names of all the member species: *Pieris napi* and *Pieris rapae*, for example, both belong to the genus *Pieris*.

GEO Genetically engineered organism.

Geographic information system (GIS) A data management system that organizes and displays digital map data from remote sensing and aids in the analysis of relationships between mapped features.

Geographic isolation The separation of a population from the rest of its species due to some physical barrier, such as a mountain range, an ocean, or great distance.

Geographic range The total area occupied by a population.

Geologic maps Maps that show the types and ages of rock of an area. These maps are used by paleontologists to find areas that are likely to contain fossils they are interested in.

Geological time scale A time scale established by geologists that reflects a consistent sequence of historical periods, grouped into four eras: Precambrian, Paleozoic, Mesozoic, and Cenozoic.

Geological time The span of time that has passed since the formation of the Earth and its physical structures; also, a timescale that focuses on events on the order of thousands of years or more.

Geometric growth Growth that follows a geometric pattern of increase, such as 2, 4, 8, 16, etc.

Georeceptor A specialized nerve ending that responds to the force of gravity.

Geostrophic flow Movement of water in the oceans as a combined response to the Coriolis effect and gravitational forces created by an uneven sea surface

Geothermal energy Energy drawn from the internal heat of the earth, either through geysers, fumaroles, hot springs, or other natural geothermal features, or through deep wells that pump heated groundwater.

Geotropism Plants' response to gravity: roots grow downward, showing positive geotropism, while shoots grow upward in a negative response.

Germ cells Collective term for cells in the reproductive organs of multicellular organisms that divide by meiosis to produce gametes.

Germ layer. In the animal embryo, one of three basic layers (ectoderm, endoderm, mesoderm) from which the various organs and tissues arise in the multicellular animal.

Germ plasm Genetic material that may be preserved for future agricultural, commercial, and ecological values (plant seeds or parts or animal eggs, sperm, and embryos).

Germ ring The advancing ring of animal pole micromeres migrating over the vegetal hemisphere in amphibian embryos.

Germicide An agent that kills pathogens and many nonpathogens but not necessarily bacterial endospores.

Germinal disc In megalecithal eggs, the small area on the fertilized egg to which cleavage is restricted. The blastodisc results from meroblastic cleavage and develops into the embryo.

Germinal epithelium The epithelium that covers the surface of a mammalian ovary.

Germinal vesicle The greatly enlarged nucleus of an oocyte.

Germinate To resume growth and increase metabolic activity,

Germination The process by which a seedling emerges and develops from a seed, or by which a sporeling emerges and develops from a spore.

Germplasm Entire genetic makeup of an organism.

Gerontology The scientific problems of aging in all their aspects, including clinical, biological, and sociological.

Gestation Period of development of the young in viviparous animals, from the time of fertilization of the ovum until birth. Refers to the length of time between when a fertilized egg implants in the wall of a female's uterus and birth of the young. In many small mammals, such as various mice, gestation is only about a month, while in larger mammals such as deer, it can be about 7 months.

Gestation period The portion of the reproductive cycle in a female mammal extending from fertilization to birth of its offspring.

Gestation ggap junction A type of intercellular junction in animal cells that allows the passage of material or current between cells.

Ggerm Embryo of a cereal grain.

Ggluon Subatomic particle that binds quarks together in the nucleus of an atom.16

Gguyot A flat-topped seamount.

Giabrous Without hairs.

Giardiasis A common infection of the lumen of the small intestine with the flagellated protozoan Giardia lamblia, and spread via contaminated food and water and by direct person-to-person contact.

Gibberellins A group of hormones that stimulate cell division and elongation in plants. Gibberellic acid (GA), the first of this class to be discovered, causes bolting (extreme elongation) of stems. GA is also applied to certain plants to promote larger fruits.

Gid (GID)Disorientation caused by cysticercia in the brain; usually manifested by staggering or whirling.

Gill A localized extension of the body surface of many aquatic animals, specialized for gas exchange. Breathing organ possessed by many aquatic creatures, including numerous young insects. Insect gills are usually very fine outgrowths from the body and they contain numerous air-tubes, or tracheae. Oxygen passes into the tubes from the water by diffusion.

Gill arches Stiffenings which support the flesh between the gill slits of chordates. In most vertebrates, the first gill arches have been modified to form the jaw, and in tetrapods, the inner ear bones.

Gill cover (or operculum)The flap of bony plates that covers the gills of bony fishes.

Gill filaments A thin-walled, fleshy extension of a gill arch that contains vessels carrying blood to and from gas exchange surfaces.

Gill lamellae Thin plates of tissue on gill filaments that contain the capillary beds across which gases are exchanged.

Gill rakers Finger-like processes on the gill arches in fishes and larval amphibians.

Gill slit A slitlike or porelike opening connecting the pharynx of a chordate with the outside of the body. Gill slits may contain the gills and be used for gas exchange, as in most fish, but may also be used for filter-feeding, or may be highly modified in land-dwelling vertebrates.One of several openings in the pharyngeal region of chordates. Gill slits allow water to pass from the pharynx to the outside of the body. In the process, water passes over gills or suspended food is removed in a filter-feeding mechanism. Opening or clefts between the gill arches in fish. Water taken in by the mouth passes through the gill slits and bathes the gills. Also, rudimentary grooves in the neck region of embryos of air-breathing vertebrates such as humans; a characteristic of chordates.

Gillnet A net set upright in the water to catch fish by entangling their gills in its mesh.

Gingival crevice fluid A secretion found in the gingival sulcus, formed by the cells attaching the gingival epithelium to the tooth.

Gingival sulcus A potential space between the gingival margin and the tooth, lined by non-keratinised epithelium. The depth of the sulcus is normally between 1 and 2 mm in health.

Gingivitis An inflammation of the gingival mucosa,

due to the increase in the virulence or mass of bacteria in the gingival sulcus, or to reduced resistance of the host.

Ginkgo A decidious gymnosperm tree (Ginkgo biloba) with distinctive fan-shaped leaves, native to eastern China.Group of seed plants today restricted to a single genus (*Ginkgo biloba*); ginkgos were more diverse during the Mesozoic Era.

Girdle A silken thread supporting the midsection of a pupa.

Girth Circumference, as of a tree trunk.

GIS geographic information system A system that allows automatic location of information suitable for mapping. Usually involves a software system that takes geographic position data and other data (e.g., type of bottom sediment) in order to create a map. Data on processes (e.g., current speed) can be incorporated to make a geographic model of flow.

Gland A structure composed of modified epithelial cells specialized to produce one or more secretions that are discharged to the outside of the gland.A collection of cells secreting a specific product such as insulin or sweat.

Glass A non-crystaline rock that results from very rapid cooling of magma.

Glass sponges Deep-water sponges with a skeleton of fused silica spicules.

Glial cells Nonconducting cells that serve as support cells in the nervous system and help to protect neurons.

Gliding motility A type of motility in which a microbial cell glides along when in contact with a solid surface.

Global environmental citizenship A shift in our attention from pollution in a specific place to a concern about the life-support systems of the whole planet.

Global regulatory systems Regulatory systems that simultaneously affect many genes and pathways.

Global warming Predicted increase in the earth's oceanic and atmospheric temperature, owing to additions of carbon dioxide to the atmosphere, as a result of human activities

Globular protein A polypeptide chain folded into a roughly spherical shape.

Globulins A large group of compact proteins with high molecular weight; includes immunoglobulins (antibodies).

Glochidium A larval stage of freshwater bivalves in the family Unionidae; it lives as a parasite on the gills or fins of fishes.

Glomerulus A ball of capillaries surrounded by Bowman's capsule in the nephron and serving as the site of filtration in the vertebrate kidney.A tuft of capillaries projecting into a renal corpuscle in a kidney. Also, a small spongy mass of tissue in the proboscis of hemichordates, presumed to have an excretory function. Also, a concentration of nerve fibers situated in the olfactory bulb.

Glossa (plural glossae) One of a pair of lobes at the tip of the labium or lower lip: usually very small, but long in honey bees and bumble bees, in which the two glossae are used to suck up nectar.

Glucagons A peptide hormone secreted by pancreatic endocrine cells that raises blood glucose levels; an antagonistic hormone to insulin.A hormone released by the pancreas that stimulates the breakdown of glycogen and the release of glucose, thereby increasing blood levels of glucose. Glucagon and insulin work together to maintain blood sugar levels.

Glucans Polysaccharides composed of glucose units held together by glycosidic linkages. Some types of glucans have a(1_3) and a(1_6) linkages and bind bacterial cells together on teeth forming a plaque ecosystem.

Glucocorticoids One of the two major hormones secreted by the adrenal medulla. The most common glucocorticoid is cortisol (hydrocortisone) but they all share the common effect of increasing blood glucose concentration. They may achieve this at the cost of body protein stores, by converting amino acids into glucose. Cortisol also converts fatty acids into glucose. Any type of stress, including trauma, infection, fear, anxiety or malnutrition causes an increase in cortisol secretion. Cortisol stabilises the membrane of lysosomes, which are then unlikely to rupture, a process which stimulates inflammation. Cortisol therfore inhibits inflammation. Cells like neutrophils, are less able to protect the body from foreign proteins. Stress therefore reduces the bodies ability to cope with infection. Malnutrition not only stunts mental and physical development but also allows viral, bacterial and parasitic infections to flourish.

Gluconeogenesis Synthesis of glucose from protein or lipid precursors.

Glucosamine A glucose or galactose molecule with an amine group attached.

Glucose A molecule of great importance to life as it provides a ready source of energy for both plant and animal cells. Glucose can only be formed in plants with the aid of sunlight. This process of photosynthesis sustains all animal life on earth. The glucose molecule is formed by a ring of a six carbon atoms. It is progressively broken down in a process called glycolysis during both aerobic and anaerobic respiration into ATP

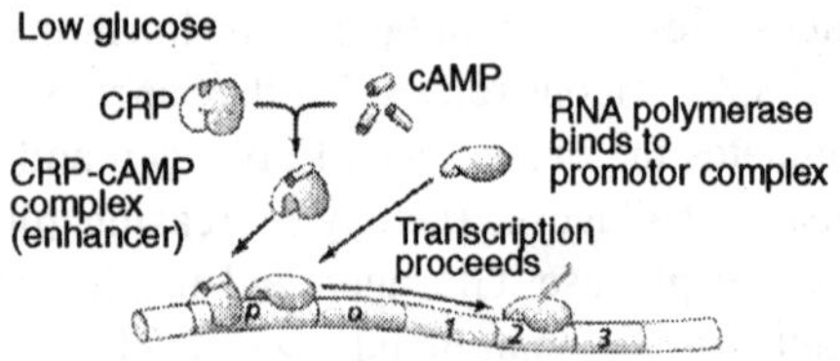

Fig. *Glucose*

Glucuronic acid A glucose molecule with an acid carboxyl group. One of the two molecules which makes up the repeating disaccharide unit of glycosaminoglycans other molecule is a glucosamine.

Glumes Pair of bracts at the base of a spikelet in a grass flower.

Gluten Protein complex in endosperm of wheat and some other cereals that is essential in making a leavened bread.

Glycerol An organic molecule to which fatty acids are attached to form a fat.

Glycine One of 20 amino acids commonly found in proteins.

Glycocalyx A network of polysaccharides extending from the surface of bacteria and other cells.Finely filamentous layer containing carbohydrate, found on the outer surface of many cells, from 7.5 to 200.0 nm thick.

Glycogen An extensively branched glucose storage polysaccharide found in the liver and muscle of animals; the animal equivalent of starch.A polysaccharide constituting the principal form in which carbohydrate is stored in animals; animal starch.A polysaccharide made up of repeated glucose units. Animals make glycogen and store it in liver and muscles.

Glycolipids Organic molecules similar in structure to fats, but in which a short carbohydrate chain rather than a fatty acid is attached to the third carbon of the glycerol molecule; as a result, the molecule has a hydrophilic "head" and a hydrophobic "tail." Glycolipids are important constituents of the plasma membrane and of organelle membranes.

Glycolysis The anaerobic breakdown (that is, in the absence of oxygen) of carbohydrates, in which a molecule of glucose is converted by a series of steps to two molecules of lactic acid, yielding energy in the form of ATP. The compound *adenosine triphosphate*, or ATP, is used by cells as a versatile source of chemical energy.

The sequence of reactions from glucose to pyruvic acid that is common to both aerobic and anaerobic carbohydrate catabolism. The breakdown of glucose in series of metabolic steps. Energy in the form of ATP is released even if there is no oxygen available as in anaerobic respiration . In the presence of oxygen as in aerobic respiration the breakdown is more complete and yields more energy.

Glycoprotein A membrane-bound protein which has attached branching carbohydrates. These may function in cell-cell recognition, such as in human blood groups and immune system response, as well as in resisting compression of cells.Are proteins which have many sugar molecules attached to them. They are an important component of saliva where they provide lubrication for the teeth. They also have a wide range of other functions in connective tissues. Examples are fibronectin, osteonectin, osteopontin and interferon. Glycoproteins are also found in cell membranes where they define part of the cells identity. The four major blood groups are defined by glycoproteins on the cell membranes of red blood cells.Polysaccharides formed of sugars linked to proteins. On the outer surface of a membrane, they act as receptors for molecular signals originating outside the cell.

Glycosaminoglycans (GAGs) Large to huge molecules of the connective tissue matrix, made up of repeating disaccharide units linked to a protein core. The disaccharide units are made of glucosamine and glucuronic acid. The position of a sulphate molecule on the glucosamine determines the type of GAG.

Glycosylation The covalent addition of sugar moities to N or O atoms present in the side chains of certain amino acids of certain proteins, generally occuring within the Golgi apparatus during secretion of a protein. Glycosylation sites are only partially predictable by current computer searches for relevant motifs in protein sequence. Glycosylation may have profound but very unpredictable effects on the folding, stability, and antigenicity of secreted proteins. Glycosylation is a property of eukaryotic cells, and differs among different cell types (i.e., it may be very different in yeast or insect cells used for protein expression, when compared with Chinese hamster ovary (CHO) cells).

Glyoxylate cycle A modified tricarboxylic acid cycle in which the decarboxylation reactions are bypassed by the enzymes isocitrate lyase and malate synthase; it is used to convert acetyl-CoA to succinate and other metabolites.Metabolic pathway that functions to convert fatty acids or acetate to carbohydrate.

Glyoxysomes A subcellular microbody present in the cytoplasm of many oil seeds. Enzymes packaged in the glyoxysome convert lipids to carbohydrates during the germination process.

Gnathobase The expanded and hardened base of the appendage of many arthropods, notably trilobites, crustaceans, and marine cheliceramorphs. Used to macerate food items before ingestion.

Gnathopod Prehensile appendages of some Crustacea, such as the second and third thoracic legs of Amphipoda and the first thoracic legs of some Isopoda.

Gnathosoma The anterior part of the body of mites and ticks which bears the mouth and mouthparts.

Gnathostomata A superclass of vertebrates whose members possess hinged jaws and paired appendages. Notochord may be replaced by the vertebral column.

Gnetales Group of seed plants restricted to three genera today (*Gnetum*, *Ephedra*, and *Welwitschia*); the possible outgroup for flowering plants.

Gnetophytes Taxa of the division Gnetophyta, a group of gymnosperms.

Gnotobiotic Animals that are germfree (microorganism free) or live in association with one or more known microorganisms.

Goblets Markings on the stigmatal plates of certain hard ticks.

Golden brown algae Common name applied to the protistan division Chrysophyta.

Golgi apparatus A cell organelle which is part of the inner cell membrane. It collects and stores the products from the endoplasmic reticulum. It is prominent in actively secreting cells.A membranous, vesicular structure which is in continuity with the *endoplasmic reticulum* of eukaryotic cells and generally in close proximity to the nucleus, the Golgi plays an important role in the posttranslational *processing* and transport of secreted proteins.Stack of intracellular membrane compartments (organelle) that function to modify and package secreted and integral membrane proteins; during spermatogenesis, the golgi apparatus causes the formation of an acrosomal vesicle in the sperm head.

Golgi complex Organelles in animal cells composed of a series of flattened sacs that sort, chemically modify, and package proteins produced on the rough endoplasmic reticulum.

Gomphosis A form of tooth attachment in which the root is help in a bony socket by a fibrous ligament.

Gonad An organ that produces gametes (ovary in the female and testis in the male).

Gonadotropin A hormone that stimulates activity in the gonads.

Gonadotropin-releasing hormone A hormone produced by the hypothalamus that controls the secretion of luteinizing hormone.

Gonadotropins Hormones that stimulate the activities of the testes and ovaries; a collective term for follicle-stimulating and luteinizing hormones.Hormones produced by the anterior pituitary that affect the testis and ovary; include follicle-stimulating hormone and luteinizing hormone.

Gonads Glands responsible for the production of gametes and where certain gonadal hormones are produced. These consist of the ovaries in females and the testes in males.The male and female sex organs.

Gonangium Reproductive zooid of hydroid colony (Cnidaria).

Gondwana Name applied to the ancient (Paleozoic-early Mesozoic) southern hemisphere supercontinent that rifted apart to form present-day Antarctica, India, Africa, Australia, and South America. The southern part of Pangaea.

Gonial angle The angle made by the posterior part of the ramus and the lower border of the mandible.

Gonidia Reproductive cells of colonial eukaryotes such as Volvox.

Gonochoristic Having separate sexes; that is, an individual is either male or female, but never both. "Dioecious" means the same thing.

Gonococci Bacteria of the species Neisseria gonorrhoeae-the organism causing gonorrhea.

Gonoduct Duct leading from a gonad to the exterior.

Gonopore A genital pore found in many invertebrates.

Gonorrhea An acute infectious sexually transmitted disease of the mucous membranes of the genitourinary tract, eye, rectum, and throat. It is caused by Neisseria gonorrhoeae.A sexually transmitted disease that is caused by a bacterium that inflames and damages epithelial cells of the reproductive system.

Gonotyl Muscular sucker or other perigenital specialization surrounding or associated with the genital atrium of a digenetic trematode.

Gonozooid A polyp of a hydrozoan cnidarian that produces medusae.

Goosecoid A homeodomain transcription factor expressed in the dorsal lip of the blastopore; may be a key factor is specifying these cells as the organizer of the embryo. (Its name is derived from the two Drosophila homedomain proteins to which its homoeodomain is similar: gooseberry and bicoid).

Gorget A patch on the throat of a bird or other animal, distinguished by color, texture, etc.

Gorgonians Colonial anthozoans that secrete a skeleton made of protein.

G-protein linked receptor A signal receptor protein in the plasma membrane that responds to the binding of a signal molecule by activating a G protein.

GPS global positioning system An electronic device that uses positioning signals from satellites in order to locate precisely latitutude and longitude. Now used nearly exclusively for locating ship sampling stations at sea, but also useful for locations near and on shore.

Graaflan follicle A follicle found in mammalian ovaries that is mature and ready to release an ovum. After ovulation, it becomes a corpus luteum.

Grab Benthic sampling device with two or more curved metal plates designed to converge when the sampler hits bottom and grab a specified volume of bottom sediment

Grade A level of organismal complexity or adaptive zone characteristic of a group of evolutionarily related organisms.

Graded potential A local voltage change in a neuron membrane induced by stimulation of a neuron, with strength proportional to the strength of the stimulus and lasting about a millisecond.

Graded signal A form of communication that is analog, or varying in intensity or frequency, providing quantitative information about the strength of the stimulus.

Gradual metamorphosis A type of insect development in which there is no prolonged resting stage (pupa). The three stages are: egg, nymph, and adult.

Gradualism A component of Darwin's evolutionary theory postulating that evolution occurs by the temporal accumulation of small, incremental changes, usually across very long periods of geological time; it opposes claims that evolution can occur by large, discontinuous or macromutational changes.

Grafting A method of asexual propagation characterized by placing a shoot (the scion) onto the rootstock (the stock) of another plant.

Graft-versus-host disease A disease that results when mature post-thymic T cells in donor grafts (e.g., bone marrow) recognize the host as foreign and attack it.

Grain (1) The texture of wood, produced by the kinds of xylem cells present.

(2) The fruit of a member of the grasses.

Gram stain A differential staining procedure that divides bacteria into gram-positive and gram-negative groups based on their ability to retain crystal violet when decolorized with an organic solvent such as ethanol.

Graminoid An herb with grass-like morphology; a growthform typified by true grasses (Graminae) and by sedges (Cyperacaeae).

Gram-negative Denoting bacterial cells that do not stain with crystal violet and iodine.

Gram-positive Denoting bacterial cells that are readily stained with crystal violet and iodine.

Grana flattened saclike structures inside chloroplasts containing chlorophyll and other photosynthetic enzymes

Granite Highly felsic igneous plutonic rock, typically light in color; rough plutonic equivalent of rhyolite. Granite is actually quite rare in the U.S.; often the term is applied to any quartz-bearing plutonic rock.

Granodiorite Lgneous plutonic rock, less felsic than granite, typically light in color; rough plutonic equivalent of dacite.

Granular/granule cell Most abundant neuron type in the cerebellar cortex of the vertebrate brain.

Granulocytes White blood cells (neutrophils, eosinophils, and basophils) bearing "granules" (vacuoles) in their cytoplasm that stain deeply.

Granuloma, granulomatous tissue Repaired area of a body marked by fibrous connective tissue (fibrosis). Also, fibrous connective tissue surrounding an antigen source.

Granum Structures within chloroplasts, seen as green granules with the light microscope and as a series of stacked disk-shaped membranes with an electron microscope; the grana contain the chlorophylls and carotenoids and are the sites of the light reactions of photosynthesis.

Grass A member of the Graminae family of flowering plants.

Grassland Region in which the climate is dry for long periods of the summer, and freezes in the winter. Grasslands are characterized by grasses and other erect herbs, usually without trees or shrubs. Grasslands occur in the dry temperate interiors of continents, and first appeared in the Miocene.

Grasslands biome Occurs in temperate and tropical regions with reduced rainfall or prolonged dry seasons; characterized by deep, rich soil, an absence of trees, and large herds of grazing animals.

Gravel Inorganic soil particles larger than 2.0 mm in diameter.

Gravid Describing a female that is carrying developing offspring within its body.

Gravitropism A response of a plant or animal in relation to gravity.

Gray crescent A dark arching band that forms on the surface of the amphibian zygote opposite the point of sperm penetration; forms in the region where gastrulation will occur.Region of the egg seen in some amphibian zygotes formed by the 30 degree shift of cortical cytoplasm toward the site of sperm entry; it marks the future site of gastrulation initiation (dorsal lip of the blastopore).

Grazer Refers to a mammal that predominantly feeds on non-woody vegetation such as grasses and forbs (non-woody leafy plants). Elk are predominantly grazers except for a short period of time during the winter when they are browsers. Typically grazers do eat some browse, and browsers do eat some grass.

GRE glucocorticoid response element A binding site in a promoter to which the activated glucocorticoid receptor can bind. The glucocorticoid receptor is essentially a transcription factor which is activated only in the presence of glucocorticoids. The activated receptor will bind to a GRE, and transcription of the adjacent gene will be altered.

Great ocean conveyor A global circulation pattern in which water cycles throughout the ocean basins.

Great whales Large whales: the sperm and the baleen whales.

Green algae Seaweeds in which chlorophyll is not green beard effectA mechanism for recognizing kin that does not require previous experience. Recognition genes enable an organism to identify and behave altruistically towards other organisms bearing that same gene.

Green gland Excretory gland of certain Crustacea; the antennal gland.

Green plans Integrated national environmental plans for reducing pollution and resource consumption while achieving sustainable development and environmental restoration.

Green political parties Political organizations based on environmental protection, participatory democracy, grassroots organization, and sustainable development.

Green revolution Dramatically increased agricultural production brought about by "miracle" strains of grain; usually requires high inputs of water, plant nutrients, and pesticides.

Greenhouse effect The heating that occurs when gases such as carbon dioxide trap heat escaping from the Earth and radiate it back to the surface; so-called because the gases are transparent to sunlight but not to heat and thus act like the glass in a greenhouse.

Greenhouse gases Gases released from the earth's surface through chemical and biological processes that interact with the chemicals in the stratosphere to decrease the release of radiation from the earth. It is believed that this leads to global warming.

Gregarious settling Settlement of larvae that have been attracted to members of their own species

Greywacke Sandstone composed of poorly sorted angular clasts.

Griseofulvin An antibiotic from Penicillium griseofulvum given orally to treat chronic dermatophytic infections of skin and nails.

Gross national product (GNP) The sum total of all goods and services produced in a national economy. Gross domestic product (GDP) is used to distinguish economic activity within a country from that of off-shore corporations.

Gross primary production The total amount of photosynthesis accomplished in a given period of time.

Gross productivity A measure of the rate at which energy is assimilated by the organisms in a trophic level, a community, or an ecosystem.

Ground meristem The basic or fundamental tissue of the apical meristem; dermal tissues surround the ground meristem, and the provascular strands are embedded in it.

Ground parenchyma The basic ground tissue consisting of living parenchymal cells.

Ground substance A jelly-like substance which surrounds cells and provides, with fibre, a supportive matrix around each cell. It consists of water and huge molecules which helps transport nutrients to cells and carries away cell products.

Ground system Plant tissue system, composed mainly of parenchyma cells with some collenchyma and sclerenchyma cells, that occupies the space between the epidermis and the vascular system; is involved in photosynthesis, water and food storage, and support; one of the four main tissue systems in plants.

Ground tissue Includes primary tissues of parenchyma, sclerenchyma, and collenchyma that make up much of the bulk of the primary plant body; function in support, photosynthesis, and storage; also known as fundamental tissue.

Ground tissue system A tissue of mostly parenchyma cells that makes up the bulk of a young plant and fills the space between the dermal and vascular tissue systems.

Groundwater Water found underground as a result of rainfall, ice and snow melt, submerged rivers, lakes, and springs. This water often carries minerals. These minerals can accumulate in the remains of buried organisms and eventually cause fossilization.

Group selection Selection that operates on two or more genetic lineages (groups) as units; broadly defined, this includes kin and interdemic selection.

Group translocation A transport process in which a molecule is moved across a membrane by carrier proteins while being chemically altered at the same time.

Growth An irreversible increase in size or volume of a cell, tissue, organ, or organism.

Growth factor A protein that must be present in the extracellular environment (culture medium or animal body) for the growth and normal development of certain types of cells.

Growth hormone (GH) A peptide hormone produced by the anterior pituitary that is essential for growth.

Growth ring A concentric layer of wood (secondary xylem) formed in one growing season through cell divisions of the vascular cambium. Also termed annual ring.Features of woody stems produced by plants growing in areas with seasonal (as opposed to year-long) growth. The growth ring marks the position of the vascular cambium at the cessation of the previous year's growth.

Growthform The overall morphology of a plant species, including its stature, leaf type, and habit. The most basic growtforms are trees, shrubs, forbs, and graminoids; but there are many more specialized growthforms such as epiphytes, lianas, and stem succulents.

Guanine (G) A nitrogenous base, one member of the *base pair* G- C (guanine and *cytosine*). A purine derivative, 2-amino-6-oxypurine, found in nucleosides, nucleotides, and nucleic acids.A nitrogen-containing base found in DNA molecules. Pairs with cytosine.A white crystalline purine base, C5H5N5O, occurring in various animal tissues and in guano and other animal excrements.

Guard cell A specialized epidermal plant cell that forms the boundaries of the stomata.

Guild A group of species, possibly unrelated taxonomically, that exploit overlapping resources

Guillain-barr syndrome A relatively rare disease affecting the peripheral nervous system, especially the spinal nerves, but also the cranial nerves. The cause is unknown, but it most often occurs after an influenza infection or flu vaccination. Also called French Polio.

Gular flutter The type of breathing experienced by some birds. Rapid movement of the throat region promotes evaporative water loss.

Gular fold A transverse fold of skin across the throat.

Gully erosion Removal of layers of soil, creating channels or ravines too large to be removed by normal tillage operations.

Gustation The act of tasting or the sense of taste.

Gustatory Having to do with the sensation of taste, the chemoreception of molecules in solution by specialized epithelial receptor cells.

Gut Body cavity formed between the mouth and anus in which food is digested and nutrients absorbed; it consists of the mouth, pharynx, esophagus, stomach, intestine, and anus, though some animals do not have all these regions.

Gut-associated lymphoid tissue (GALT) The defensive lymphoid tissue present in the intestines.

Guttation Root pressure forcing water out of the leaves, usually overnight when stomates are closed.

Gymnosperm A vascular plant that bears naked seeds not enclosed in any specialized chambers.

Gynandromorph An individual creature with a mixture of male and female characteristics. One half of the body may be male and the other half female. This is particularly noticeable when it occurs among the blue butterflies and others in which the sexes are differently coloured.

Gynandry In a hermaphroditic organism, maturation first of the female gonads and then of the male organs. Also called protogyny.

Gynocophoral canal Longitudinal groove in the ventral surface of a male schistosome fluke.

Gynoecium The collective term referring to the carpels or ovule-bearing structures, within a flower.

Gynostemium The central reproductive stalk of an orchid, which consists of a stamen and pistil fused together.

Gyre A large, nearly circular system of wind-driven surface currents that center around latitude 30in both hemispheres.

H

H-2 complex Term for the MHC in the mouse.

Habit The general growth pattern of a plant. A plant's habit may be described as creeping, trees, shrubs, vines, etc.

Habitat The place in which individuals of a particular species can usually be found.A location which has a suitable environment for an organism to live in. Caves are natural habitats for bats, trees for birds; oral surfaces and crevices, for some bacteria. The immediate space where an animal or plant lives and has food, water and protection. Habitat loss, which includes the destruction, degradation, or fragmentation of habitats, is the primary cause of decreasing biodiversity.

Habitat disruption A disturbance of the physical environment in which a population lives.

Habitat imprinting The tendency of an animal to prefer a place to live based on early experience in that habitat.

Habitat manipulation Manipulation of agricultural areas and surrounding environment with the aim of conserving or augmenting populations of natural enemies (e.g., the planting of a refuge for natural enemies).

Habitat restoration The recovery of stressed or destroyed habitats.

Habitat selection The choice of an animal's place to live. Habitat selection involves the interaction of physiological and psychological factors.

Habituation The relatively persistent waning of a response that results from repeated presentations not followed by any form of reinforcement.

Habitus Body-build, general appearance.

Hadal zone The bottom below 6,000 m (20,000 ft).

Hadley cells Circulation patterns of atmospheric convection currents as they sink and rise in several intermediate bands.

Haemolymph The blood plasma or liquid part of the blood, though generally synonymous for blood of insects.

Haemostasis The prevention of blood loss through a damaged vessel wall. There are three main mechanisms, vasoconstriction, formation of a platelet plug and blood clotting.

Hair A long slender filament. Applied especially to such filamentous appendages of the skin.

Hair bulb The base of a hair; contains cells that divide mitotically to produce columns of hair cells.

Hair cell An epithelial cell that bears modified cilia on one end; when the cilia are deflected, they change the receptor cell's potential.

Hair root The portion of a hair that extends from the skin's surface to the hair bulb.

Hairpin A helical (duplex) region formed by base pairing between adjacent (inverted) complementary sequences within a single strand of RNA or DNA.

Half-life The average time required for the disappearance or decay of one-half of any amount of a given substance.

Haller's organ Depression on the first tarsi of ticks; functions as an olfactory and humidity receptor.

Hallucinogen Psychoactive drug capable of altering moods and perceptions of time and space.

Hallux L. Digit number one of the foot.

Halobacteria or extreme halophiles A group of archaea that have an absolute dependence on high NaCl concentrations for growth and will not survive at a concentration below about 1.5 M NaCl.

Halocline Depth zone within which salinity changes maximally

Halophile Organism which lives in areas of high salt concentration. These organisms must have special adaptations to permit them to survive under these conditions.A group of archaebacteria that are able to tolerate high salt concentrations.

Halophilic Pertaining to microorganisms that tolerate high concentrations of saline media.

Halophyte A salt-tolerant terrestrial plant.

Halter In Diptera, small club-shaped structure on each side of the metathorax representing the hindwings; believed to be sense organs for balancing; also called balancer.

Haltere One of the club-shaped 'balancers' found on each side of the metathorax among the true flies (Diptera). They are the much-modified hind wings. Vestigial wing on the metathorax of a fly of the order Diptera; necessary for balance during flight.

Halzoun Disease resulting from blockage of the nasopharynx by a parasite. Also called marrara.

Hamuli Large hooks on the opisthaptor of a monogenetic trematode; referred to as anchors by American authors.The minute hooks on the front edge of the hind wing of bees and other hymenopterans, used to link the front and hind wings together. The hook which holds the springtail's spring in place is also called the hamula.

Hantavirus pulmonary syndrome The disease in humans caused by the pulmonary syndrome hantavirus. Deer mice shed the virus in their feces, humans inhale the virus and first develop ordinary flulike aches and pains. Within a few days the hantavirus causes lung damage and capillary leakage. After about a week the infected person enters a crisis phase and may die.

Haplodiploidy The mode of sex determination by which males develop from haploid (unfertilized) eggs and females from diploid (fertilized) eggs; characteristic of the insect order Hymenoptera.

Haploid Cells that contain only one member of each homologous pair of chromosomes (haploid number = n). At fertilization, two haploid gametes fuse to form a single cell with a diploid number of chromosomes. The reduced, or N, number of chromosomes, typical of gametes, as opposed to the diploid, or 2N, number found in somatic cells. In certain groups, mature organisms may have a haploid number of chromosomes

Haploid cell a cell containing only one of each type of chromosome characteristic of its species

Haploid life cycle Occurs when the only multicellular stage in an organism's life cycle is haploid.

Haploid-diploid life cycle Occurs when a multicellular diploid phase, or sporophyte, alternates with a multicellular haploid phase, or gametophyte. Only plants and certain algae possess this kind of life cycle, which is also called "alternation of generations".

Haptens Molecules of small molecular weight

(usually) that are immunogenic only when attached to carrier molecules, usually proteins.

Haptera Short, sturdy rootlike structures that form the holdfast of seaweeds.

Haptonema Peg-like structure unique to the Prymnesiophyta; its function is not known.

Harborage transmission The mode of transmission in which an infectious organism does not undergo morphological or physiological changes within the vector.

Hardwood A term often used in reference to all woody dicots; more accurately, wood having a high specific gravity.

Hardy-weinberg equilibrium. Mathematical demonstration that the Mendelian hereditary process does not change the populational frequencies of alleles or genotypes across generations, and that changes in allelic or genotypic frequencies requires factors such as natural selection, genetic drift in finite populations, recurring mutation, migration of individuals among populations, and nonrandom mating.

Hardy-weinberg formula The application of the binomial expansion to determine the frequency of two alleles in the absence of any evolutionary forces.

Hardy-weinberg law Law that states that the frequencies of genotypes in a population at a locus are determined by random mating and allele frequency

Harem A large group of females herded by a large male pinniped for the purpose of mating; harems also occur in some fishes.

Harmful algal bloom A bloom of (usually) planktonic microalgae belonging to a strain of a species that has a toxic harmful to marine organisms or humans consuming marine organisms.

Haustellate Adapted for sucking liquids rather than biting solid food.

Haustorium In parasitic fungi, a nutrient-absorbing hyphal tip that penetrates the tissues of the host but remains outside the host cell membranes.

Haversian system The basic structural arrangement of compact bone consisting of: a Haversian canal which contains nerves and blood supply, lamellae which are concentric rings of matrix (produced by osteocytes within lacunae), and canaliculi which provide a link between osteocytes.

Hay fever Allergic rhinitis; a type of atopic allergy involving the upper respiratory tract.

Hazardous waste Any discarded material containing substances known to be toxic, mutagenic, carcinogenic, or teratogenic to humans or other life-forms; ignitable, corrosive, explosive, or highly reactive alone or with other materials.

HDL (high density lipoprotein) Transport molecule that removes excess cholesterol from the body's tissues to the liver for degradation and elimination.

Head Many individual flowers tightly compressed into the shape of a single large flower. The anterior body region of insects which bears the mouthparts, eyes, and antennae.

Head mesenchyme Mesoderm that will give rise, along with cranial neural crest cells, to connective tissue, bone and musculature in the head.

Head-foot The body region of a mollusc that contains the head and is responsible for locomotion as well as retracting the visceral mass into the shell.325

Health A state of physical and emotional well-being; the absence of disease or ailment.

Healthy carrier An individual who harbors a pathogen, but is not ill.

Heap-leach extraction A technique for separating gold from extremely low-grade ores. Crushed ore is piled in huge heaps and sprayed with a dilute alkaline-cyanide solution, which percolates through the pile to extract the gold, which is separated from the effluent in a processing plant. This process has a high potential for water pollution.

Heart The multicellular, chambered, muscular structure that pumps blood through the circulatory system by alternately contracting and relaxing.

Heart urchins Burrowing sea urchins with a flattened test and short spines.

Heartwood The wood found in the center of a tree trunk; often a darker color due to the accumulation of resins, oils, gums, and other metabolic by-products, which prevent water movement through this tissue.

Heartworm disease A parasitic infection in dogs caused by the nematode Dirofilaria immitis.

Heat A form of energy transferred from one body to another because of a difference in temperatures.

Heat capacity The amount of heat energy that must be added or subtracted to change the temperature of a body; water has a high heat capacity.

Heat of vaporization The amount of heat required to change a given amount of a liquid into a gas; 540 calories are required to change 1 gram of liquid water into vapor.

Heat-shock proteins Proteins produced when cells are exposed to high temperatures or other stressful conditions. They protect the cells from damage and often aid in the proper folding of proteins.

Heath A member of the Ericaceae or heather family. Usually leathery-leaved shrubs preferring acidic or low-nutrient substrates and often tolerant of cold.

Heavy metals A group of toxic metals: mercury, lead, and others.

Hectocotylus Specialized, and sometimes autonomous, arm that serves as a male copulatory organ in cephalopods.

Helical In virology this refers to a virus with a helical capsid surrounding its nucleic acid.

Helicases Enzymes that use ATP energy to unwind DNA ahead of the replication fork.

Helix-loop-helix A protein structural motif characteristic of certain DNA-binding proteins.

Helper T cell (T_H) A type of T cell that is required by some B cells to help them make antibodies or that helps other T cells respond to antigens or secrete lymphokines or interleukins.

Helpers-at-the-nest Individuals other than the parents that forego reproduction to assist rearing the young of others.

Hemadsorption The adherence of red blood cells to the surface of something, such as another cell or a virus.

Hemagglutination The agglutination of red blood cells by antibodies.

Hemal system Strands of tissue found in echinoderms. The hemal system is of uncertain function. It may aid in the transport of large molecules or coelomocytes, which engulf and transport waste particles within the body.

Hematophagous Feeding or subsisting on blood.

Hematopoiesis hema = blood; poie = produce. The formation and development of blood cells.

Heme The iron-containing group of heme proteins such as hemoglobin and the cytochromes.

Hemelytra The front wing of an insect of the order Hemiptera.

Hemerythrin A red, iron-containing respiratory pigment found in the blood of some polychaetes, sipunculids, priapulids, and brachiopods.

Hemicellulose Polysaccharide in plant cell walls that cross-link cellulose fibrils.

Hemichordata The phylum of marine, wormlike animals whose members have an epidermal nervous system and pharyngeal gill slits. Acorn worms and pterobranchs. Invertebrates with a dorsal nerve cord and gill slits. Includes the acorn worms, or enteropneusts.

Hemi-elytron The forewing of a heteropteran bug, differing from the beetle elytron in having the distal portion membranous.

Hemimetabola Insects with simple metamorphosis, with no pupal stage.

Hemimetabolous metamorphosis A type of insect metamorphosis in which immature insects are different in form and habitats from the adult. It is different from holometabolous metamorphosis in that there is a gradual series of changes in form during the transition from immature to adult.

Hemipenes The paired copulatory organs lying laterally in a cavity in the base of the tail in snakes and lizards.

Hemiptera The true bugs, a large order of insects including bedbugs, cicadas, and aphids, with mouth parts adapted for piercing and sucking and with mandibles in the form of long stylets lying in a trough-like labium.

Hemizygous Having one or more genes that have no allele counterparts. Usually applied to genes on the male's X chromosome (in humans).

Hemocoel Main body cavity of arthropods, the embryonic development of which differs from that of a true coelom but that includes a vestige of a true coelom.

Hemocyanin A nonheme, blue respiratory pigment that is found in the plasma of many molluscs and arthropods and is composed of monomers, each of which contains two atoms of copper and can bind one molecule of oxygen.

Hemocyte Any blood corpuscle or formed element of the blood.

Hemoflagellate A flagellated protozoan parasite that is found in the bloodstream.

Hemoglobin An iron-containing respiratory pigment occurring in vertebrate red blood cells and in blood plasma of many invertebrates; a compound of an iron porphyrin heme and globin proteins.The oxygen-carrying pigment of the erythrocytes, formed by the developing erythrocyte in the bone marrow. It is a complex protein composed of four heme groups and four globin polypeptide chains.

Hemolymph In invertebrates with an open circulatory system, the body fluid that bathes tissues.

Hemolysin A substance that causes hemolysis (the lysis of red blood cells). At least some hemolysins are enzymes that destroy the phospholipids in erythrocyte plasma membranes. The disruption of red blood cells and release of their hemoglobin. There are several types of hemolysis when bacteria such as streptococci and staphylococci grow on blood agar. In a-hemolysis, a narrow greenish zone of incomplete hemolysis forms around the colony. A clear zone of complete hemolysis without any obvious color change is formed during b-hemolysis.

Hemolytic uremic syndrome A kidney disease characterized by blood in the urine and often by kidney failure. It is caused by enterohemorrhagic strains of Escherichia coli O157:H7 that produce a Shiga-like toxin, which attacks the kidneys.

Hemophilia A human sex-linked recessive genetic disorder that results in the absence of certain blood-clotting factors, usually Factor VII. Hemophiliacs suffer from an inability to clot their blood.

Hemorrhagic fever A fever usually caused by a specific virus that may lead to hemorrhage, shock, and sometimes death.

Hemozoin Insoluble digestion product of malaria parasites produced from hemoglobin.

Hensen's node The thickened area at the anterior end of the primitive streak through which cells invaginate to form the notochord. It is homologous to the dorsal lip of the blastopore in amphibian embryos. Regional thickening of cells at the top (anterior) of the primitive groove through which gastrulating cells migrate anteriorally to form tissues in the future head and neck. Found in birds, reptiles and mammals, it is the functional equivalent of the dorsal lip of the blastopore in amphibians. Also known as the primitive knot.

Heparin sulphate A glycosaminoglycan which is unusual in that it is stored inside the cell (mast cells) surrounding the liver. Heparin prevents blood clotting.

Hepatic portal vessel A large circulatory channel that conveys nutrient-laden blood from the small intestine to the liver, which regulates the blood's nutrient content.

Hepatitis A A type of hepatitis that is transmitted by fecal-oral contamination; it primarily affects children and young adults, especially in environments where there is poor sanitation and overcrowding. It is caused by the hepatitis A virus, a single-stranded RNA virus.

Hepatitis B A potentially serious viral disease that affects the liver; can be transmitted through sexual contact or through contact with infected blood.

Hepatitis C About 90% of all cases of viral hepatitis can be traced to either HAV or HBV. The remaining 10% is believed to be caused by one and possibly several other types of viruses. At least one of these is hepatitis C (formerly non-A, non-B).

Hepatitis D The liver diseases caused by the hepatitis D virus in those individuals already infected with the hepatitis B virus.

Hepatitis E The liver disease caused by the hepatitis E virus. Usually, a subclinical, acute infection results; however, there is a high mortality in women in their last trimester of pregnancy.

Hepatosplenomegaly Swollen liver and spleen.

Herb Any non-woody vascular plant; a category of plants including both forbs and graminoids. Hence the term "herbaceous" layer .

Herbaceous 1. of or relating to an herb. 2. soft and green, rather than woody.Without woody tissues; typical of most annuals and biennials; herbaceous perennials die back to the soil level each year.

Herbal A text that describes plants that are useful medicinally and in other ways.

Herbalism The use of plants for treating ailments.

Herbarium A permanent collection of dried and pressed plants that provides information on the location and identification of the local flora.

Herbicide A plant growth regulator that inhibits growth or kills a plant when applied at relatively low concentrations.

Herbivore A plant-eating animal. Sea urchins re a good example of a marine herbivore as they feed on kelp. Literally, an organism that eats plants or other autotrophic organisms. The term is used primarily to describe animals.Term pertaining to a heterotroph, usually an animal, that eats plants or algae. Herbivores function in food chains and food webs as primary consumers.

Herd immunity The resistance of a population to infection and spread of an infectious organism due to the immunity of a high percentage of the population.

Heritability (in the broad sense)The amount of the variance in a trait in a population that is due to genetic factors; defined as the ratio of the variance due to genetic factors to the sum of variance due to both genetic and environmental factors; also called the degree of genetic determination.A measure of the extent to which genes inherited from parents determine phenotypes; defined as the ratio of the additive genetic variance to total phenotypic variance; also called realized heritability.

Heritable Capable of being inherited, such as a physical trait.

Heritable character A morphological character whose given state can be explained partially in terms of the genotype of the individual

Hermaphrodite An organism with both male and female functional reproductive organs.

Hermaphroditism may refer to an aberration in unisexual animals; monoecy implies that this is the normal condition for the species.

Hermaphroditic Having both male and female sex organs in one individual.

Hermaphroditism A state characterized by the presence of both male and female reproductive organs in the same animal.

Hermatypic coral A reef-building coral.

Hermatypic Relating to reef-forming corals.

Herpetic keratitis An inflammation of the cornea and conjunctiva of the eye resulting from a herpes simplex virus infection.

Hesperidium A berry with a thick leathery peel (exocarp and mesocarp) and a fleshy endocarp arranged in sections.

Heterocercal In some fishes, a tail with the upper lobe larger than the lower, and the end of the vertebral column somewhat upturned in the upper lobe, as in sharks.

Heterochromatic regions Having inactive genes. Inactive regions of chromosomes are said to be heterochromatic.

Heterochromatin Chromatin that stains intensely and appears to represent inactive genetic areas.

Heterochrony Evolutionary change in the relative time of appearance or rate of development of characteristics from ancestor to descendant.

Heterochrony Evolutionary changes in the timing or rate of development.

Heterocyst A specialized cell of certain cyanobacteria; usually larger and thick-walled, this cell excludes oxygen and allows for anaerobic nitrogenase activity.

Heterodont Having teeth differentiated into incisors, canines, and molars for different purposes.

Heteroduplex DNA Generated by base pairing between complementary single strands derived from different parental *duplex* molecules; heteroduplex DNA molecules occur during genetic *recombination* in vivo and during *hydridization* of different but related DNA strands in vitro. Since the sequences of the two strands in a heteroduplex differ, the molecule is not perfectly base-paired; the *melting* temperature of a heteroduplex DNA is dependent upon the number of mismatched base pairs..

Heterogametic sex The sex with two different chromosomes, such as males in humans and *Drosophila*.

Heterogeneity The fact or state of being dissimilar, as in composition, source, quantity, dimensions, and so on.

Heterogeneous nuclear RNA (hnRNA) The RNA transcript of DNA made by RNA polymerase II; it is then processed to form mRNA.

Heterogonic life cycle Life cycle involving alternation of parasitic and free-living generations.

Heterolactic fermenters Microorganisms that ferment sugars to form lactate, and also other products such as ethanol and CO2.

Heteromerous (of beetles) Having unequal numbers of tarsal segments on the three pairs of legs.

Heteromorphic A condition in the life cycle of all modern plants in which the sporophyte and gametophyte generations differ in morphology.

Heterophile reaction Antigen-antibody reaction in which the antibody was not specifically elicited by the antigen to which it binds.

Heterosporangiate Producing two different kinds of sporangia, specifically microsporangia and megasporangia. Compare with heterosporous.

Heterosporous Producing two different sizes or kinds of spores. These may come from the same or different sporangia, and may produce similar or different gametophytes. Contrast with homosporous, and compare with heterosporangiate.

Heterotherm An animal whose body temperature fluctuates markedly; "cold-blooded."

Heterotroph An organism that is unable to synthesize organic compounds (and thus its energy) from the environment and therefore fulfils its energy requirements by feeding on other organisms or organic matter. Compare *autotroph.*

Heterotrophic algae Algae that take up organic molecules as a primary source of nutrition

Heterotrophic nitrification Nitrification carried out by chemoheterotrophic microorganisms.

Heterotrophic Refers to organisms, such as animals, that depend on preformed organic molecules from the environment (or another organism) as a source of nutrients/energy.

Heterotrophic theory A theory of the origin of life which proposes that the first organisms obtained nutrition from the spontaneous formation of organic molecules derived from primordial gases.

Heteroxenous Describes a parasite that lives within more than one host during its life cycle.

Heterozygosity The presence of different *alleles* at one or more *loci* on *homologous chromosomes*.

Heterozygote An organism in which homologous chromosomes contain different allelic forms (often dominant and recessive) of a locus; derived from a zygote formed by union of gametes of dissimilar allelic constitution.

Heterozygote advantage A mechanism that preserves variation in eukaryotic gene pools by conferring greater reproductive success on heterozygotes over individuals homozygous for any one of the associated alleles.

Hexacanth Oncosphere; a six-hooked larva hatching from the egg of a eucestode.

Hexactinellida The class of sponges whose members are characterized by triaxon siliceous spicules, which are sometimes formed into an intricate lattice. Cup or vase shaped. Scyconoid body form. Glass sponges.

Hexamerous Six parts, specifically, symmetry based on six or multiples thereof.

Hexapoda The class of mandibulate arthropods whose members are characterized by having three pairs of legs. Commonly called insects. Hexapods often have wings and a body divided into head, thorax, and abdomen. Insecta has been used as an alternate class name.

Hexon or hexamer A capsomer composed of six protomers.

Hexose A 6-carbon sugar.

HFR strain A bacterial strain that donates its genes with high frequency to a recipient cell during conjugation because the F factor is integrated into the bacterial chromosome.

Handicap hypothesis The hypothesis that apparently deleterious sexual "ornaments" possessed by males in many species are attractive to females because they indicate that the males have such vigor (good genes) that they can survive even with the handicap.

Hemoglobinuria Bloody urine.

Hepatic Pertaining to the liver.

Hypophysis Pituitary body.

Hibernacula The places in which an animal hibernates or overwinters; winter quarters.

Hibernation A dormant, sleeplike state, with a lower body temperature and slower heart and breathing rate, that is characteristic of various animals during the winter months in cold climates, such as bears, bats, certain birds, snakes, frogs, and turtles; this state tends to protect against cold weather and to reduce the need for food. Ground squirrels are good examples of hibernating mammals as some hibernate for about 9 months. Hibernators do arouse from hibernation periodically, but usually stay in their hibernation chamber or nest.

Hierarchical system. A scheme arranging organisms into a series of taxa of increasing inclusiveness, as illustrated by Linnean classification.

High oxygen diffusion environment A microbial environment in close contact with air and through which oxygen can move at a rapid rate (in comparison with the slow diffusion rate of oxygen through water).

High tide The highest level reached by the rising tide.

High-energy molecule A molecule whose hydrolysis under standard conditions makes available a large amount of free energy (the standard free energy change is more negative than about 27 kcal/mole); a high-energy molecule readily decomposes and transfers groups such as phosphate to acceptors.

High-level waste repository A place where intensely radioactive wastes can be buried deep in the ground and remain unexposed to groundwater and earthquakes for tens of thousands of years. The first such site chosen in the United States is Yucca Mountain, Nevada, although it is not certain that a waste repository will ever be built there.

Highly stratified estuary An estuary having a distinct surface layer of fresh or very-low-salinity water, capping a deeper layer of higher salinity, more oceanic water

High-quality energy Intense, concentrated, and high-temperature energy that is considered high-quality because of its usefulness in carrying out work.

Hill reaction The splitting of a molecule of water during the light reactions of photosynthesis; the photolysis of water.

Hilum Scar on a seed indicating where it was attached to the ovary.

Hindbrain Includes the medulla oblongata, cerebellum, and pons.

Hindgut The posterior part of the alimentary canal between the midgut and anus.

Hindwings The second pair of wings of an insect.

Hinge point During neurulation, region of initial bending of the neural plate to fold into the neural tube; wedge-shaped hinge point cells are anchored to underlying notochord (medial hinge point) or surface ectoderm (dorsolateral hinge point).

Hippocampus Part of the limbic system, it seems to provide a spatial map, useful in the event of a sudden need to escape from an unpleasant sensory experience.

Hirudinea The class of annelids whose members are characterized by bodies with 34 segments, each of which is subdivided into annuli. Anterior and posterior suckers present. Leeches.

Histamine Is a product of the amino acid, histidene and is released by damaged cells. Histamine causes an increase in capillary permeability and vasodilation, two vascular events which are the first stages of inflammation. Histamine is also a neurotransmitters substance, released at nerve synapses mostly in the hypothalamus.

Histocompatibility antigens Cell surface glycoproteins that differ from individual to individual; their recognition as foreign by a host organism is responsible for rejection of grafted foreign tissue.

Histogenesis Formation and development of tissue.

Histogram A multiple-bar diagram representing the frequency distribution of a group as a function of some variable. The frequency of each class is proportional to the length of its associated bar

Histology The microscopic study of cells and tissues and their organization into organs.

Histone Any of several simple proteins found in cell nuclei and complexed at one time or another with DNA. Histones yield a high proportion of basic amino acids on hydrolysis; characteristic of eukaryotes.

Histone proteins Proteins associated with DNA in eukaryote chromosomes.

Histoplasmosis A systemic fungal infection caused by Histoplasma capsulatum var capsulatum.

Histosiphon Same as stylostome.The tube formed by the host as a result of the feeding of a chigger secreting salivary fluids, the chigger partially digests skin tissues, which induces the host to form a proteinaceous tube walling off the injury.

Histozoic Dwelling within the tissues of a host.

HIV Abbreviation of human immunodeficiency virus, the infectious agent that causes AIDS; HIV is an RNA retrovirus.

Hives An eruption of the skin.

HnRNA heterogeneous nuclear RNA Refers collectively to the variety of RNAs found in the nucleus, including primary transcripts, partially processed RNAs and snRNA. The term hnRNA is often used just for the unprocessed primary transcripts, however.

Holdfast A structure produced by some bacteria and algae that attaches the cell to a solid object.

Holistic medicine The treatment of the whole entity, the human body, spirit, and mind especially accepted in Eastern and primitive cultures.

Holoblastic cleavage Complete cleavage - major pattern of embryonic cell divison in which cytokinesis completely separates cells during division; it is typically seen in smaller eggs containing moderate (mesolecithal) to sparse (isolecithal) yolk. Examples of eggs that divide holoblastically include those of amphibians, mammals, non-vertebrate chordates, echinoderms, most molluscs, annelids, flatworms, nematodes.

Holoenzyme A complete enzyme consisting of the apoenzyme plus a cofactor.

Holometabola The higher insects which have complex metamorphosis.

Holometabolous metamorphosis A type of insect metamorphosis in which immatures, called larvae, are different in form and habitats from the adult; the last larval molt results in the formation of a pupa; radical cellular changes in the pupal stage end in adult emergence.

Holophytic nutrition Occurs in green plants and certain protozoa and involves synthesis of carbohydrates from carbon dioxide and water in the presence of light, chlorophyll, and certain enzymes.

Holoplankton Organisms spending all their life in the water column and not on or in the sea bed

Holoptic With the eyes touching or almost touching on the top of the head: used mainly when describing flies (Diptera).

Holothuroidea The class of echinoderms whose members are elongate along the oral-aboral axis, have microscopic ossicles embedded in a muscular body wall, and have circumoral tentacles. Lack rays. Sea cucumbers.

Holotype The type specimen of a species is the actual insect from which the original description of that species was produced. If several specimens were used for this purpose, one of them should have been designated as the type. Because the type can be of only one sex, it is usual to designate a certain individual of the opposite sex as the allotype. The original type specimen is then called the holotype. These 'type specimens' are very important in taxonomy and classification.

Holozoic nutrition In this type of nutrition, nutrients (such as bacteria) are acquired by phagocytosis and the subsequent formation of a food vacuole or phagosome.

Home range. The area over which an animal ranges in its activities. Unlike territories, home ranges are not defended.

Homeobox A short stretch of *nucleotides* whose *base sequence* is virtually identical in all the *genes* that contain it. It has been found in many organisms from fruit flies to human beings. In the fruit fly, a homeobox appears to determine when particular groups of genes are expressed during development.

Homeosis Evolutionary alteration in the placement of different body parts.

Homeostasis Control of an organism's internal environment. Water content, temperature, acid-base balance, level of oxygen and carbon dioxide, adequate supply of energy are some of the many factors in the organism which require monitoring and control. A common form of control is feedback.

Homeotherm An organism that regulates its body temperature despite changes in the external environmental temperature

Homeotic genes Genes, identified through mutations, that give developmental identity to specific body segments.

Homing The ability of an animal to return to its home site after being displaced.

Hominid Humans and closely related primates; includes modern and fossil forms, such as the australopithecines, but not the apes.

Hominoid Relating to the Hominoidea, a superfamily of primates to which the great apes and humans are assigned.

Homocercal A tail with the upper and lower lobes symmetrical and the vertebral column ending near the middle of the base, as in most teleost fishes.

Homodont A dentition in which all the teeth are the same shape (from "home" = same).

Homogeneous Similar or uniform structure or composition throughout.

Homogonic life cycle Life cycle in which all generations are parasitic or all are free living. There is no (or little) alternation of the two.

Homolactic fermenters Organisms that ferment sugars almost completely to lactic acid.

Homologies Similarities in DNA or *protein* sequences between individuals of the same species or among different species.

Homologous Organs or parts which exhibit similarity in structure, in position with reference to other parts, and in mode of development, but not necessarily similarity of function, are said to be homologous.

Homologous chromosomes Chromosome pairs of the same length, centromere position, and staining pattern that possess genes for the same characters at corresponding loci. One homologous chromosome is inherited from the organism's father, the other from the mother.

Homologous recombination The exchange of sequence between two related but different DNA (or RNA) molecules, with the result that a new "chimeric" molecule is created. Several mechanisms may result in recombination, but an essential requirement is the existence of a region of *homology* in the recombination partners. In DNA recombination, breakage of single strands of DNA in the two recombination partners is followed by joining of strands present in opposing molecules, and may involve specific enzymes. Recombination of RNA molecules may occur by other mechanisms.

Homologous structures Body parts in different organisms that have similar bones and similar arrangements of muscles, blood vessels, and nerves and undergo similar embryological development, but do not necessarily serve the same function; e.g., the flipper of a whale and the forelimb of a horse.

Homology Similarity of parts or organs of different organisms caused by evolutionary derivation from a corresponding part or organ in a remote ancestor, and usually having a similar embryonic origin. May also refer to a matching pair of chromosomes. Serial homology is the correspondence in the same individual of repeated structures having the same origin and development, such as the appendages of arthropods. Adj., homologous.

Homonym A scientific name which has been given to two different species. When such an instance comes to light one of the species must be given another name.

Homoplasy Phenotypic similarity among characteristics of different species or populations (including molecular, morphological, behavioral or other features) that does not accurately represent patterns of common evolutionary descent (nonhomologous similarity); it is produced by evolutionary parallelism, convergence and/or reversal, and is revealed by incongruence among different characters on a cladogram or phylogenetic tree.

Homopolymer tailing Extension of the 3' ends of a piece of double-stranded DNA using the enzyme terminal transferase and a single deoxynucleoside triphosphate as substrate, resulting in a 3' overhang composed of a single base repeat. If DNA which is to be annealed to this is tailed with the complementary base, mutually cohesive ends are produced. The tailed molecules are unable to circularise because their ends are not cohesive. Sometimes used for insertion of DNA into vectors.

Homosporous Producing only one size or kind of spore. Contrast with heterosporous.

Homothetogenic fission Mitotic fission across the rows of ciiia of a protozoan.

Homozygote An organism in which the pair of alleles for a trait is composed of the same genes (either dominant or recessive but not both). Adj., homozygous.

Homozygous Having two identical alleles for a given trait.

Honeydew The sugary liquid discharge from the anus of certain insects (Homoptera) such as aphids and scales.

Hood Dorsal wall of the camerostome that extends over the capitulum.

Horizons Soil zones from the surface downward that reveal visible horizontal layers of soil.

Horizontal gene transfer The process in which genes are transferred from one mature, independent organism to another.

Hormogonia Small motile fragments produced by fragmentation of filamentous cyanobacteria; used for asexual reproduction and dispersal.

Hormone 1. An organic molecule synthesized by a plant that exerts, even in low concentrations, profound regulation of growth and/or development.

2. Chemical products of ductless glands that are carried by the circulatory system and that influence various physiological processes in the body.

Horns Consist of an inner, boney core covered by an outer sheath that is much like our fingernails. Horns are not shed annually as are antlers. They are found in the Bovidae family (bighorn sheep and mountain goats), and in pronghorns.

Horsehair worms Pseudocoelomate animals that belong to the phylum Nematomorpha. Also known as Gordian worms or hairworms (Gordius is the name for an ancient king who tied an intricate knot).

Horseshoe crabs Arthropods with a large horseshoe-shaped carapace.

Horticultural therapy The activities of propagating, growing, and planting plants is therapeutic for a variety of people including the elderly and mentally challenged.

Host The body of an organism that harbors another organism. It can be viewed as a microenvironment that shelters and supports the growth and multiplication of a parasitic organism.

Host organism An organism that provides lodging for a parasite.

Host plant resistance The relative amount of heritable qualities possessed by a plant that reduces the degree of damage to the plant by a pest or pests.

Host restriction The degradation of foreign genetic material by nucleases after the genetic material enters a host cell.

Host specificity Degree to which a parasite is able to mature in more than one host species.

Host strain (bacterial) The bacterium used to harbor a plasmid. Typical host strains include HB101 (general purpose *E. coli* strain), DH5? (ditto), JM101 and JM109 (suitable for growing M13 phages), XL1-Blue (general-purpose, good for blue/white lacZ screening). Note that the host strain is available in a form with no plasmids, or it may have plasmids present. Hundreds, perhaps thousands, of host strains are available.

Hot desert Deserts of the American Southwest and Mexico; characterized by extreme summer heat and cacti.

Human chorionic gonadotropin (hCG) A peptide hormone secreted by the chorion that prolongs the life of the corpus luteum and prevents the breakdown of the uterine lining.

Human ecology The study of the interactions of humans with the environment.

Human gene therapy Insertion of normal DNA directly into cells to correct a genetic defect.

Human genome initiative Collective name for several projects begun in 1986 by to (1) create an ordered set of DNA segments from known chromosomal locations, (2) develop new computational methods for analyzing genetic map and DNA sequence data, and (3) develop new techniques and instruments for detecting and analyzing DNA. This initiative is now known as the Human Genome Program.

Human genome project Project to determine the DNA base sequence of every gene in the human genome.

Human herpesvirus 6 (HHV-6, type A and B) HHV-6 was discovered in 1986 and was initially called the human B-lymphotropic virus. The virus was later shown to have a marked tropism for CD41 T cells and was renamed HHV-6. HHV-6 is genetically similar to cytomegalovirus. HHV-6 causes exanthem subitum (roseola infantum or sixth disease) in infants and has been suspected of involvement in many conditions, including opportunistic infections in immunocompromised patients, hepatitis, lymphoproliferative diseases, synergistic interactions with HIV, lymphadenitis, and chronic fatigue syndrome.

Human immunodeficiency virus (HIV) The retrovirus that attacks T-cells in the human immune system, destroying the body's defenses and allowing the development of AIDS.

Human leukocyte antigen complex (HLA) An antigen on the surface of cells of human tissues and organs that is recognized by the immune system cells and therefore is important in graft rejection and regulation of the immune response. This is the same as MHC class II.

Human resources Human wisdom, experience, skill, labor, and enterprise.

Humeral angle The front basal part of the wing, close to its attachment to the body.

Humeral vein A small cross-vein running from the costa to the sub-costa in the humeral (basal) region of the wing.

Humidity Water vapor in the atmosphere.

Humoral The type of immunity that results from the presence of soluble antibodies in blood and lymph; also known as antibody-mediated immunity.

Humoral immune response Binding of antigen with soluble antibody in blood serum. Also, the entire process by which the body responds to an antigen by producing antibody to that antigen.

Humoral immunity The type of immunity that fights bacteria and viruses in body fluids with antibodies that circulate in blood plasma and lymph, fluids formerly called humors.

Humus Sticky, brown, insoluble residue from the bodies of dead plants and animals; gives soil its structure, coating mineral particles and holding them together; serves as a major source of plant nutrients.

Hunter-gatherers Human social group secures food sources from wild resources such as hunting animal prey or collecting edible plants from the wild.

Huntington disease A progressive and fatal disorder of the nervous system that develops between the ages of 30 and 50 years; caused by an expansion of a trinucleotide repeat and inherited as a dominant trait.

Hyaline Protein released by cortical granules forming a coating around the sea urchin egg;

the hyaline layer provides support for the blastomeres during cleavage.

Hyaluronic acid The largest glycosaminoglycan known, it plays an important role of restricting the flow of water in tissues, particularly in synovial fluid where it acts as a lubricant..

Hybrid (1) Offspring of two parents that differ in one or more inheritable characteristics.

(2) Offspring of two different varieties or of two different species.

Hybrid zone A region where two related populations that diverged after becoming geographically isolated make secondary contact and interbreed where their geographical ranges overlap.

Hybridization The reaction by which the pairing of complementary strands of nucleic acid occurs. DNA is usually double-stranded, and when the strands are separated they will re-hybridize under the appropriate conditions. Hybrids can form between DNA-DNA, DNA-RNA or RNA-RNA. They can form between a short strand and a long strand containing a region complementary to the short one. Imperfect hybrids can also form, but the more imperfect they are, the less stable they will be (and the less likely to form). To "anneal" two strands is the same as to "hybridize" them.

Hybridoma A *clone* of plasmacytoma cells which secrete a monoclonal antibody; usually produced by fusion of peripheral or splenic plasma cells taken from an immunized mouse with an immortalized murine plasmacytoma cell line (fusion partner), followed by *cloning* and *selection* of appropriate antibody-producing cells.

Hydatid cyst A type of cyst formed by juveniles of certain tapeworms (Echinococcus) in their vertebrate hosts.

Hydatid sand Sediment in a hydatid cyst formed by free protoscolices.

Hydranth Nutritive zooid of hydroid colony.

Hydraulic skeleton The use of body fluids in open circulatory systems to give support and facilitate movement; muscles contracting in one part of the body force body fluids into some distant tissue space, thus causing a part of the body to extend or become turgid.

Hydrocarbon An organic molecule that contains only carbon and hydrogen and has its carbons bonded in a linear fashion.

Hydrocaulus Stalks or "stems" of a hydroid colony, the parts between the hydrorhiza and the hydranths.

Hydrocoel Second or middle coelomic compartment in echinoderms; left hydrocoel gives rise to water vascular system.

Hydrogen An element that is one of the constituents of water, organic matter, and many other chemicals. Hydrogen gas (H_2) is composed of two hydrogen atoms.

Hydrogen bonds A weak force holding two molecules containing hydrogen together, each of which has a covalent bond with another atom. For example, water is a molecule made up of two hydrogen atoms attached covalently to an oxygen atom. The hydrogen proton is however not completely balanced and is still attracted to the oxygen atom of a neighbouring molecule. Molecules of water are thus held together by hydrogen bonds, which accounts for the unusually high boiling point of water considering its low molecular weight. Hydrogen bonds hold protein molecules in shape by linking up various sections. When proteins are heated, these bonds collapse and the protein is physically altered, even though its chemical composition remains unchanged. When an egg is heated the white part rapidly gels, indicating a change in the shape of the protein. The process is not reversible.. so an egg cannot be uncooked.

Hydrogen ion A single proton with a charge of +1. The dissociation of a water molecule (H_2O) leads to the generation of a hydroxide ion (OH^-) and a hydrogen ion (H^+).

Hydrogen potential Same as pH; refers to the number of charged hydrogen atoms or protons present, acidity of the soil.

Hydrogen sulfide (H_2S) The gas that is produced in anoxic sediments.

Hydrogenate Add hydrogen atoms to a molecule.

Hydrogenation Addition of one or more hydrogens to monunsaturated and polyunsaturated fatty acid chains.

Hydrogenosome A microbodylike organelle that contains a unique electron transfer pathway in which hydrogenase transfers electrons to protons (which act as the terminal electron acceptors) and molecular hydrogen is formed.

Hydrographic Referring to the arrangement and movement of bodies of water, such as currents and water masses.

Hydroid The polyp form of a cnidarian as distinguished from the medusa form. Any cnidarian of the class Hydrozoa, order Hydroida.

Hydrological cycle The cycling of water between reservoirs in oceans, lakes, and groundwater, and the atmosphere.

Hydrolysis The decomposition of a chemical compound by the addition of water; the splitting of a molecule into its groupings so that the split products acquire hydrogen and hydroxyl groups.

Hydrolyze Break a bond in a molecule by adding water.

Hydrophilic Water-loving. Term applied to polar molecules that can form a hydrogen bond with water.

Hydrophilicity plot A computer plot which examines the relative summed hyrophobicity/ hydrophilicity of adjacent *amino acid* sidechains (usually within a moving window of about 6 amino acid residues) along the primary sequence of a polypeptide chain. Values for the contribution of sidechains of each the 20 common amino acids to hydrophobicity/ hydrophilicity have been developed by Hopp & Woods, and Kyte & Doolittle, and these plots are often named after these workers. Generally, hydrophobic regions of proteins are considered likely to be in the interior of the native protein, while hydrophilic domains are likely to be exposed on thc surface and thus possibly antigenic sites (epitopes). At best, these are crude predictions.

Hydrophobic Hydrophobic compounds do not dissolve easily in water, and are usually non-polar. Oils and other long hydrocarbons are hydrophobic.

Hydrophytic leaves The leaves of plants that grow in water or under conditions of abundant moisture.

Hydroponics A system of growing plants in a liquid nutrient solution without soil.

Hydrorhiza Rootlike stolon that attaches a hydroid to its substrate.

Hydrostatic pressure The pressure exerted by a fluid (gas or liquid), defined as force per unit area. For example, the hydrostatic pressure of one atmosphere (1 atm) is 14.7 lb/in2.

Hydrostatic skeleton . A mass of fluid or plastic parenchyma enclosed within a muscular wall to provide the support necessary for antagonistic muscle action; for example, parenchyma in acoelomates and perivisceral fluids in pseudocoelomates serve as hydrostatic skeletons.

Hydrostatic skeleton Fluid-filled closed chambers that give support and shape to the body in organisms such as jellyfish and earthworms. No to be confused with the water-vascular system of echinoderms.

Hydrothermal vent A place on the seafloor, generally associated with spreading centers, where warm to super-hot, mineral-rich water is released; may support a diverse community of organisms.

Hydrothermal vents Deep, oceanic regions where

the tectonic plates of the earth's crust are moving apart. They are characterized by occasional lava flows and hot water springs.

Hydroxyapatite One of the apatites which is the main salt of bone and teeth.

Hydroxyl group A functional group consisting of a hydrogen atom joined to an oxygen atom by a polar covalent bond. Molecules possessing this group are soluble in water and are called alcohols.

Hydrozoa The class of cnidarians whose members have epidermally derived gametes, mesoglea without wandering amoeboid cells, and gastrodermis without nematocysts. Medusae, when present, with a velum. Hydra, Obelia, Physalia.

Hygrophilus Moisture loving.

Hygroreceptor A receptor found in insects that detects the water content of air.

Hymenium The layer of fertile cells that produces spores in a fungal fruiting body.

Hyomandibular Bone derived from the hyoid gill arch, forming part of articulation of the lower jaw of fishes, and forming the stapes of the ear of amniotic vertebrates.

Hyperapolysis Detachment of a tapeworm proglottid while still immature, before eggs are formed.

Hyperendemic disease A disease that has a gradual increase in occurrence beyond the endemic level, but not at the epidemic level, in a given population; also may refer to a disease that is equally endemic in all age groups.

Hyperinfection Condition in Strongyloides infections in which filariform juveniles repenetrate mucosa of the small intestine and proceed with migration.

Hypermetamorphosis Type of metamorphic development in which different larval instars have markedly dissimilar body forms.

Hypermutation A rapid production of multiple mutations in a gene or genes through the activation of special mutator genes. The process may be deliberately used to maximize the possibility of creating desirable mutants.

Hyperosmotic Refers to a solution whose osmotic pressure is greater than that of another solution to which it is compared; contains a greater concentration of dissolved particles and gains water through a selectively permeable membrane from a solution containing fewer particles; contrasts with hypoosmotic.

Hyperparasitism Parasitism of a parasite by another parasite.

Hyperphagia A condition in which an animal does not stop eating when it normally would.

Hyperplasia An increase in the size of an organ due to an increase in the numbers of cells. The developing embryo increases in size due to cell division. The cells of some tissues retain the ability to divide throughout life, like the epithelium and connective tissues but muscle and nerve cells lose their ability to divide soon after birth. When hyperplasia is uncontrolled it produces a tumour which may be benign if it is well contained, not destructive and does not spread. But it may be malignant, destroys normal tissue, and spread all over the body.

Hyperpolarization An electrical state whereby the inside of the cell is made more negative relative to the outside than at the resting membrane potential. A neuron membrane is hyperpolarized if a stimulus increases its voltage from the resting potential of –70 mV, reducing the chance that the neuron will transmit a nerve impulse.

Hypersaline Extremely salty, having much more salt than normal seawater.

Hypersensitivity A condition of increased immune sensitivity in which the body reacts to an antigen with an exaggerated immune response that usually harms the individual. Also termed an allergy.

Hypertension High blood pressure; blood pressure consistently above 140/90.

Hyperthermophile A bacterium that has its growth optimum between 80 degrees C and about 113 degrees C. Hyperthermophiles usually do not grow well below 55 degrees C.

Hypertonic A solution having a greater number of solute particles than another solution to which it is compared.

Hypertrophy An increase in the size of an organ due to an increase in the size of each cell. Muscles increase in size due to hypertrophy.

Hyphae The multinucleate or multicellular filaments that make up the mycelium (body) of a fungus.

Hypnozoite Dormant exoerythrocytic form found in certain Plasmodium species.

Hypoblast An inner layer of cells that results from the proliferation and movement of cells in the blastoderm of an avian or reptilian embryo.

Hypocotyl hook The hooked portion of a hypocotyl formed at germination that assists a dicot shoot in pulling itself through the soil crust.

Hypodermis The cellular layer lying beneath and secreting the cuticle of annelids, arthropods, and certain other invertebrates.

Hypognathous Having a vertical head and face with the mouth-parts at the bottom.

Hypogynous Floral whorls (sepals, petals, stamens) inserted below the ovary of a flower.

Hypoosmotic Refers to a solution whose osmotic pressure is less than that of another solution with which it is compared or taken as a standard; contains a lesser concentration of dissolved particles and loses water during osmosis; contrasts with hyperosmotic.

Hypophagia A condition in which an animal does not eat as much as it normally would.

Hypopharynx A component of the insect mouth-parts arising behind the mouth and just in front of the labium or lower lip. Usually short and tongue-like in species with biting jaws, but often drawn out to form a tube for the salivary duct in those species with sucking mouths.

Hypophysis This term may be used in two different manners. It may be used to describe an embryonic portion of a developing gland, as in, 1) The invagination of the stomodeum which fuses with the infundibulum to form the pituitary gland. Synonym: Rathke's pocket. It may also describe a complete adult gland, as in,

2) An endocrine gland formed from an ingrowth from the stomodeum (Rathke's pocket) and the infundibulum. Synonym: pituitary gland.

Hypoplasia Reduced formation of a tissue during development. Enamel hypoplasia may be recognised as pits and depressions in the enamel and may be cause by fluorosis.

Hypopleural bristles A curved row of bristles on the side of the thorax of certain true flies just below and in front of the haltere and just above the base of the hind leg.

Hypostome Name applied to structure in various invertebrates (such as mites and ticks), located at posterior or ventral area of mouth.

Hypothalamus This small body of nerve cells controls the activity of the pituitary gland, the source of several hormones which control the activity of other hormones, including ACTH which in turn controls the level of glucocorticoid secretion. The hypothalamus also has powerful connections with the other members of the limbic system, from which nerve pathways descend to control nerves in the spinal cord. The influence of the limbic system on the hypothalamus explains the raised levels of glucocorticoids in response to emotional stress. This bridge with the peripheral nervous system, provides a link between the emotional state of a person, as influenced by the activity of the limbic system, and the excitability of neurones in the spinal cord, to in-coming pain impulses.

Hypotheca the smaller portion of a diatom frustule

Hypothesis A temporary working explanation or supposition based on accumulated facts and suggesting some general principle or relation of cause and effect; a postulated solution to a scientific problem that must be tested and if not validated, discarded.

Hypothetico-deductive Scientific process of making a conjecture and then seeking empirical tests that potentially lead to its rejection.

Hypothyroidism Listlessness and other metabolic symptoms caused by low thyroid hormone levels.

Hypotonic A comparison between two solutions, indicating that one has a lower osmotic pressure, or is less salty than the other.

Hypoxic Having a low oxygen level.

Hysterosoma In mites, the posterior part of the body when there is a demarcation of the body between the second and third pair of legs.

I

Iapetus ocean A relatively small ocean that existed between the continents of Laurentia, Baltica, and Avalonia from the Late Precambrian to the Devonian.

Ice age Interval of geologic time between 2 million and 10,000 years ago during which the northern hemisphere experienced several episodes of continental glacial advance and retreat along with a climatic cooling. The icing over of Antarctica was also completed during this time.

Ichnology The study of trace fossils.

Ichthyologist Biologists who specialize in the study of fish behavior, anatomy, physiology, and evolution.

Ick Serious disease of freshwater fishes, caused by the ciliate protozoan Ichthyophthirius multifiliis.

Icosahedral In virology this term refers to a virus with an icosahedral capsid, which has the shape of a regular polyhedron having 20 equilateral triangular faces and 12 corners.

Icterus Yellowing of the skin and other organs because of bile pigments in the blood.

Ideal free distribution The distribution of individuals among resource patches of different quality that equalizes the net rate of gain of each individual. Assumes that organisms are free to move and have ideal knowledge about patch quality.

Identification The process of determining that a particular isolate or organism belongs to a recognized taxon.

Idiosoma Posterior of the two basic parts of the body of a mite or tick, bearing the legs and most internal organs.

Idiotype A set of one or more unique epitopes in the variable region of an immunoglobulin that distinguishes it from immunoglobulins produced by different plasma cells.

IgA Immunoglobulin A; the class of immunoglobulins that is present in dimeric form in many body secretions (e.g., saliva, tears, and bronchial and intestinal secretions) and protects body surfaces. IgA also is present in serum.

IgD Immunoglobulin D; The class of immunoglobulins found on the surface of many B lymphocytes; thought to serve as an antigen receptor in the stimulation of antibody synthesis.

IgE Immunoglobulin E; the immunoglobulin class that binds to mast cells and basophils, and is responsible for type I or anaphylactic hypersensitivity reactions such as hay fever and asthma. IgE is also involved in resistance to helminth parasites.

IgG Immunoglobulin G; the predominant immunoglobulin class in serum. Has functions such as neutralizing toxins, opsonizing bacteria, activating complement, and crossing the placenta to protect the fetus and neonate.

IgM Immunoglobulin M; the class of serum antibody first produced during an infection. It is a large, pentameric molecule that is active in agglutinating pathogens and activating complement. The monomeric form is present on the surface of some B lymphocytes.

Igneous rocks Crystalline minerals solidified from molten magma from deep in the earth's interior; basalt, rhyolite, andesite, lava, and granite are examples.

Ileum The third and last section of the small intestine.

Imaginal disk An island of undifferentiated cells in an insect larva, which are committed (determined) to form a particular organ during metamorphosis to the adult.

Imago The adult and sexually mature insect.

Imagochrysalis Quiescent stage between the nymph and adult in the life cycle of a chigger mite.

Imbibition The capillary movement of water into germinating seeds and into substances such as wood and gelatin, which swell as a result.

Imitation An act that occurs when an animal immediately copies the actions of another animal while they are in each other's presence.

Immediate hypersensitivity Biological manifestation of an antigen-antibody reaction in which the maximal response is reached in a few minutes or hours. Intradermal injection of the antigen produces local swelling and redness with heavy infiltration of polymorphonuclear leukocytes. Intravenous injection may produce anaphylactic shock and death.

Immobilization The incorporation of a simple, soluble substance into the body of an organism, making it unavailable for use by other organisms.

Immovable joint A joint in which the bones interlock and are held together by fibers or bony processes that prevent the joint from moving; e.g., the bones of the cranium.

Immune complex The product of an antigen-antibody reaction, which may also contain components of the complement system.

Immune cross reaction Binding of an antibody or cell receptor site with an antigen other than the one that would provide an exact "fit"; that is, an antigen-antibody reaction in which the antigen is not the same one that stimulated the production of that antibody.

Immune response A highly specific defensive reaction of the body to invasion by a foreign substance or organism; consists of a primary response in which the invader is recognized as foreign, or "not-self," and eliminated and a secondary response to subsequent attacks by the same invader. Mediated by two types of lymphocytes B cells, which mature in the bone marrow and are responsible for antibody production, and T cells, which mature in the thymus and are responsible for cell-mediated immunity.

Immune surveillance The still somewhat hypothetical process in which lymphocytes such as natural killer (NK) cells recognize and destroy tumor cells; other cells with abnormal surface antigens (e.g., virus-infected cells) also may be destroyed.

Immune system The defensive system in a host consisting of the nonspecific and specific immune responses. It is composed of widely distributed cells, tissues, and organs that recognize foreign substances and microorganisms and acts to neutralize or destroy them.

Immunity The body's response to a foreign antigen, either ingested as food, or as part of a foreign organism. There are two major ways the body defends itself; one is by antibody, production, the so called humeral response, as the antibodies circulate in the blood and the fluid between cells. The other is the cellular response, as it involves the cells of the immune system, the family of leucocytes. The particular leucocyte responsible for immune specificity is the lymphocytc . In total cell mass there are as many lymphocytes as there are liver or brain cells. During development there are millions of B (from the Bone marrow) lymphocytes made, each with a different cell membrane ligand, specific for any one of millions of antigens. The lymphocytes are circulating all the time so that they can have the chance to meet up with a foreign antigen. As soon as an antigen has been recognised by one of these cells, and bound to the cell ligand, it stimulates the cell to reproduce millions of copies of itself. All the daughter cells are clones of the original cell. These B lymphocyte daughters, migrate to the site where the antibody is needed. Instead of making an antigen for the membrane these cells make large amounts of soluble antibody. They are now recognisable as plasma cells. T lymphocytes (having spent time in the Thymus) comprises the cell mediated response to an antigen. They are of two types, Killer T cells and Helper T Cells. Most T lymphocytes are helpers and they regulate the response of the B lymphocytes . The killer T cells are however capable of recognising the foreign antigen on the surface of a cell, and then killing the entire cell. The immune response is part of a less specific defense and healing response of the body known as inflammation.

Immunization The process of making one immune. The induction of protective immunity by administration of either (1) a vaccine or toxoid (active immunization) or

(2) preformed antibodies (passive immunization).

Immunoblotting The electrophoretic transfer of proteins from polyacrylamide gels to nitrocellulose sheets to demonstrate the presence of specific proteins through reaction with labeled antibodies.

Immunodeficiency The inability to produce a normal complement of antibodies or immunologically sensitized T cells in response to specific antigens.

Immunodiffusion A technique involving the diffusion of antigen and/or antibody within a semisolid gel to produce a precipitin reaction where they meet in proper proportions. Often both the antibody and antigen diffuse through the gel; sometimes an antigen diffuses through a gel containing antibody.

Immunoelectrophoreses The electrophoretic separation of protein antigens followed by diffusion and precipitation in gels using antibodies against the separated proteins.

Immunofluorescence A technique used to identify particular antigens microscopically in cells or tissues by the binding of a fluorescent antibody conjugate.

Immunogenic Refers to any substance that is antigenic; that is, stimulates production of antibody or cell-mediated immunity.

Immunoglobulin Any of a group of plasma proteins, produced by plasma cells, that participates in the immune response by combining with the antigen that stimulated its production.

Immunological Pertaining to the immune response, in which a protein antibody is synthesized by an organism to counteract some pathogenic factor.

Immunology The branch of science that deals with the immune system and attempts to understand the many phenomena that are responsible for both acquired and innate immunity. It also includes the use of antibody-antigen reactions in other laboratory work (serology and immunochemistry).

Immunopathology The study of diseases or conditions resulting from immune reactions.

Immunoprecipitation A process whereby a particular protein of interest is isolated by the addition of a specific antibody, followed by centrifugation to pellet the resulting immune complexes. Often, staphylococcal proteins A or G, bound to sepharose or some other type of macroscopic particle, is added to the reaction mix to increase the size and ease collection of the complexes. Usually, the precipitated protein is subsequently examined by *SDS-PAGE.*

Immunotoxin A monoclonal antibody that has been attached to a specific toxin or toxic agent (antibody 1 toxin 5 immunotoxin) and can kill specific target cells.

Imperfect flower Unisexual flower; either staminate or pistillate.

Imperfect stage The asexual phase in a fungus life cycle characterized by the production of asexual spores.

Impermeable Having the property of restricting the passage of substances.

Impetigo This superficial cutaneous disease, most commonly seen in children, is characterized by crusty lesions, usually located on the face; the lesions typically have vesicles surrounded by a red border. It is the most frequently diagnosed skin infection caused by S. pyogenes (impetigo can also be caused by S. aureus).

Implantation The process in which the blastocyst embeds in the endometrium.

Impressions Prints or marks made when an organism's body has been compressed (flattened). Impressions are different from compressions because no thin organic material is left behind.

Imprinting A process that occurs when an animal learns to make a particular response to only one type of animal or object. The sensory modes used for establishing such a connection can be visual, auditory, olfactory, or some combination of these, depending upon the animal.

In situ hybridization Use of a DNA or RNA probe to detect the presence of the *complementary DNA* sequence in cloned bacterial or cultured *eukaryotic* cells.

In vitro Experiments which are carried out in a laboratory as distinct from in vivo experiments

in vitro fertilization Fertilization of ova in laboratory containers followed by artificial implantation of the early embryo in the mother's uterus.

In vivo Experiments which are carried out in live animals as distinct from in vitro experiments

Indirect pulp cap A dressing, usually calcium hydroxide, placed against the pulpal wall of a deep cavity, in order to encourage affected dentine to remineralise. The cavity is closed with a temporary filling material and re-opened after 6 weeks to assess the state of the pulp.

Induction Cell differentiation which is brought about by the influence of cytokines released by cells of another type.

Infected dentine- dentine which has been damaged beyond repair by caries and is infected by large numbers of caries bacteria.

Inflammation A whole complex of events which occur in sequence, in response to injury. Tissue damaged by bacteria, chemicals, heat, trauma etc, release histamine and bradykinin and serotonin which cause an increase in capillary permeability and vasodilation. Both these factors contribute to the formation of a fluid exudate in the damaged tissue, which includes fibrinogen and therefore soon clots into a firm gel. This process has the effect of walling off the bacteria or toxic substances causing the damage, or at least it slows down their spread into surrounding tissues. Local macrophages, begin their phagocytic activity but their numbers are small. Damaged tissues also release interleukin, messengers which are transported all the way to the bone marrow, where millions of leucocytes are stored.. These stores now release leucocytes, mostly neutrophils into the blood. The neutrophils gather at the site of damage because the endothelial cells of the local capillary walls have become sticky to leucocytes. This stickiness is specific for leucocytes and is the work of selectins expressed on the cell membrane of the endothelial cell. The leucocytes begin to catch and roll along the endothelium until they are brought to a standstill. The increased permeability of the endothelial cells allows leucocytes to wriggle out of the capillary and migrate into the damaged area. This migration is also dependent on a process know as chemotaxis, in which cytokine messages from the damaged cells attract the leucocytes to come to their aid. After several days the battle zone is filled with dead bacteria, dead tissue cells, dead neutrophils and macrophages. This dead mass of tissue is called pus. The end of the event may be the gradual resorption of pus by fresh macrophages, or the pus, now under some pressure, may force its way somewhere else. Pus from the apex of a tooth may escape laterally through the alveolar bone and mucosa, where it is recognisable as a "gum boil".Ten days after a foreign protein is detected for the first time, the bodies immune system has mounted a more specific defense. Antibodies are produced by B lymphocytes and T lymphocytes have been alerted to the invasion.

Insulin An endochrine hormone produced in the spleen which controls the amount of sugar in the blood by a) transporting it into cells and promoting glycolysis b) converting it into glycogen for storage in the liver and muscles, and c) converting it into fats . Without sufficient insulin, glucose accumulates in the blood and urine and the cells of the body are starved, a condition known as diabetes. The control of insulin production is another example of a feedback system.

Insulin-like growth factor (IGF) A cytokine that influences growth hormone activity. **Integrins** The cell surface receptor molecules which match up with parts of the matrix protein ligands to allow adhesion of the cell to the matrix. The attachment of cells to matrix proteins also influences the cells behaviour, by the expression of genes. The integrins are a family of proteins, found doing the same job on all animal cells. Their importance in maintaining the structural integrity of cells led to the name integrins.

Intercalated ducts Ducts which carry saliva between the tubules and the striated ducts .

Interferon A glycoprotein produced by cell s which mobilise the T lymphocytes to inhibit viruses and the growth of cancer cells.

Interleukins A variety of compounds(about 20) that are produced by lymphocytes, macrophages, and monocytes. They regulate the cell mediated response of the immune system. Interleukin-1 is involved in the triggering of the immune response, starting acute inflammation and maintaining chronic inflammation. Interleukin-2 is produced by helper T cells and induces proliferation of immune cells, both T and B. Interleukin-3 promotes the differentiation and proliferation of stem cells of the leucocyte family.Interleukin -6 produced by various cells including tumour cells and acts as a stimulant of plasma proteins and B and T cells. Interleukin - 12 is produced by a range of cells. It activates T cells and natural killer cells. It promotes the response to a range of pathogens including HIV of Interleukin-2. It appears to be one of the most promising Interleukins for the control of viral, bacterial and protozoal infections.

Intermediate filaments Unlike microfilaments and microtubules, they are verystable. Instead of being stacked proteins, as in actin, intermediate filaments are built of interlocking proteins. A dense sheet of intermediate filaments strengthens the nucleus. Skin cells are filled with keratin, which at the last moment, just before they die. they cross link, to provide a really insoluble barrier layer of the skin. The cross linkage is between the sulphur atoms of cysteine, one of keratin's amino acids .

Interproximal wear Loss of enamel on the adjacent surfaces of teeth which is due to continual friction between the two surfaces as teeth move against each other.

Intratubular dentine Dentine formed inside the tubule by the odontoblast process in response to tooth wear, ageing or arrested caries.

Intrinsic fibres - refers to those fibres of cementum which were laid down by cementoblasts.

Ionised The loss or gain of an electron from an atom which makes it no longer neutral but an electrically charged ion. If the electron leaves the atom it becomes a positively charged ion, such when calcium or sodium becomes ionised (Ca^+,Na^+). If the electron is gained, the atom becomes relatively negatively charged such as when chlorine or a phosphate group of atoms lose an electron (Cl^-, PO_4^-). Ionized atoms or groups of atoms are more reactive than when they are neutral.

Ions An atom or molecules which has a net electrical charge This may be caused by the temporary loss (positive ion) or gain (negative ion) of an electron. A calcium ion is written Ca^+.

Ipsilateral The same side as distinct from contralateral. Often used to refer to the teeth, joint or muscles on the same side as chewing is occurring.

Inbreeding depression In a small population, an accumulation of harmful genetic traits (through random mutations and natural selection) that lowers viability and reproductive success of enough individuals to affect the whole population.

Inbreeding Mating between closely related individuals. This can lead to a loss of vability in the population due to accumulation of undesirable alleles passed on from a common ancestor.

Incidence In epidemiology, the number of new cases of a disease per unit time; that is, a rate measurement. Contrast with prevalence.

Incidental parasite Accidental parasite.

Inclusion bodies Granules of organic or inorganic material lying in the cytoplasmic matrix of bacteria.

Inclusion conjunctivitis An acute infectious disease that occurs throughout the world. It is caused by Chlamydia trachomatis that infects the eye and causes inflammation and the occurrence of large inclusion bodies.

Inclusive fitness The relative number of an individual's alleles that are passed on from generation to generation, either as a result of his or her own reproductive success, or that of related individuals.

Incomplete dominance An interaction between alleles in which both alleles are expressed more or less equally, and the phenotype of the heterozygote is different from either homozygote. Referring to the phenotypic expression for a given trait demonstrating a blending of the genetic messages from the allele partners controlling that trait; no dominant allelic partner.

Incomplete flower Lacking one or more of the four floral parts (sepals, petals, stamens, pistil).

Incomplete linkage Two genes that are carried on the same chromosome but are far enough apart that crossing-over between them occurs.109

Incomplete metamorphosis or Simple metamorphosis Metamorphosis in which the wings (when present) develop externally during the immature stage and there is no prolonged resting stage (i.e. pupa) preceding the last molt; stages included are the egg, nymphal, and adult. Also called gradual or partial metamorphosis, and paurometabolous development.

Incomplete protein Protein that lacks the full complement of essential amino acids in the correct proportions.

Incubate Of a hen, to sit on and hatch eggs.

Incubation period The period after pathogen entry into a host and before signs and symptoms appear.

Incubatory carrier An individual who is incubating a pathogen in large numbers but is not yet ill.

Incus One of the three bones comprising the middle ear of mammals.

Indeciduate placenta A small round flat cake Type of placenta where the uterine lining is not shed with the placenta due to the "looser" connection of the chorionic villi to the endometrium. There are three types of indeciduate placenta which reflect the arrangement of the chorionic villi: diffuse, intermediate and cotyledonary.

Independent data points A reference to the fact that each of the data points collected to use in a statistical test of a hypothesis must be independent of every other data point.

Independent variables The variables that define the treatment conditions in an experiment; literally, the variables that we manipulate to test their effects on behavior.

Indeterminate cleavage A type of embryonic development in deuterostomes, in which each cell produced by early cleavage divisions retains the capacity to develop into a complete embryo.

Indeterminate growth A type of growth characteristic of plants, in which the organism continues to grow as long as it lives.

Indeterminate Pertaining to a stem that produces unrestricted vegetative growth; the stem does not terminate in flowering.

Index case The first disease case in an epidemic within a given population.

Indicator organism An organism whose presence indicates the condition of a substance or environment, for example, the potential presence of pathogens. Coliforms are used as indicators of fecal pollution.

Indigenous Pertains to organisms that are native to a particular region; not introduced.

Indirect development In arthropods, refers to development in which larva or nymph hatches from an egg and is distinctly different in body form from the adult; that is, development with metamorphosis.

Indirect fitness A measure of the potential for an individual to contribute copies of the genes it possesses to future generations through its influence on the reproduction of nondescendant relatives. Only the portion of the nondescendant kins' reproduction due to helping by the relative is included in the measure of its inclusive fitness.

Indirect flight Insect flight accomplished by flight muscles acting on the body wall. Changes in shape of the thorax cause wing movements. A single nerve impulse results in many cycles of the wings; also called asynchronous flight.

Individual distance The fixed minimal distance an animal keeps between itself and other members of its species.

Individual variation Phenotypic diversity within a population.

Indoleacetic acid (IAA) The naturally occurring auxin.

Indo-west Pacific region The tropical Indian and west and central Pacific oceans.

Induced fit The change in shape of the active site of an enzyme so that it binds more snugly to the substrate, induced by entry of the substrate.

Induced ovulators Species in which copulation is necessary to effect ovulation.

Inducer A small molecule, such as IPTG, that triggers gene transcription by binding to a regulator protein, such as LacZ.

Inducer gene Gene encoding the repressor protein of the lac operon; when lactose binds the repressor protein, the lac operon is induced.

Inducible enzyme An enzyme present only when induced by some particular substrate.

Induction The developmental process in which the fate of a group of cells is determined by interaction with another group of cells. One group (the inducing cells) controls the differentiation of the other group (the induced cells).

Inductor In embryology, a tissue or organ that causes the differentiation of another tissue or organ.

Indusium A thin membrane that covers the sori of many ferns; the membrane often breaks at the time of spore maturation.

Industrial ecology The study of the ecology of industrial societies with a major focus on material cycling, energy flow, and the ecological impacts of such societies.

Industrial fishery A fishery in which the catch is used for purposes other than direct human consumption.

Industrial revolution Advances in science and technology that have given us power to understand and change our world.

Industrial sewage Wastewater from industries. Compare domestic sewage.

Industrial timber Trees used for lumber, plywood, veneer, particleboard, chipboard, and paper; also called roundwood.

Inertial confinement A nuclear fusion process in which a small pellet of nuclear fuel is bombarded with extremely high-intensity laser light.

Infauna Animals that burrow in the substrate. Compare epifauna.

Infection The invasion of a host by a microorganism with subsequent establishment and multiplication of the agent. An infection may or may not lead to overt disease.

Infection thread A tubular structure formed during the infection of a root by nitrogen-fixing bacteria. The bacteria enter the root by way of the infection thread and stimulate the formation of the root nodule.

Infectious disease Any change from a state of health in which part or all of the host's body cannot carry on its normal functions because of the presence of an infectious agent or its products.

Infectious disease cycle The chain or cycle of events that describes how an infectious organism grows, reproduces, and is disseminated.

Infectious dose 50 (ID50) Refers to the dose or number of organisms that will infect 50% of an experimental group of hosts within a specified time period.

Infectious mononucleosis An acute, self-limited infectious disease of the lymphatic system caused by the Epstein-Barr virus and characterized by fever, sore throat, lymph node and spleen swelling, and the proliferation of monocytes and abnormal lymphocytes.

Infectivity Infectiousness; the state or quality of being infectious or communicable.

Inferior Below a point of reference (e.g., the mouth is inferior to the nose in humans).

Inferior ovary Ovary lies below the attachment of the sepals, petals, and stamens; an epigynous flower.

Infiltration The process of water percolation into the soil and pores and hollows of permeable rocks.

Inflammation A reaction to the invasion of microorganisms through the skin or through the epithelial layers of the respiratory, digestive, or urinary system; characterized by four signs: redness, swelling, heat, and pain.

Inflammatory response The body's reaction to invading infectious microorganisms; includes an increase in blood flow to the affected area, the release of chemicals that draw white blood cells, an increased flow of plasma, and the arrival of monocytes to clean up the debris.

Inflorescence A flower cluster with a definite arrangement

Influenza or flu An acute viral infection of the respiratory tract, occurring in isolated cases, epidemics, and pandemics. Influenza is caused by three strains of influenza virus, labeled types A, B, and C, based on the antigens of their protein coats.

Informatics The study of the application of computer and statistical techniques to the management of information. In *genome* projects, informatics includes the development of methods to search databases quickly, to analyze DNA sequence information, and to predict *protein* sequence and structure from DNA sequence data.

Information In communication, a measure of the uncertainty of the behavioral response of an individual. In learning, refers to the fact that for successful classical conditioning to occur there must be some intuitive knowledge of the relationship between the CS and US.

Infraciliature All cilia, basal bodies, and their associated fibrils in a ciliate protozoan.

Infrapopulation All individuals of a single parasite species in one host.

Infrared That portion of the sun's total range of radiation having wavelengths immediately longer than the longest of the visible spectrum (red); between approximately 750 nm and1 m in length.

Infundibulum 1) Anatomically, any funnel-shaped passage.

2) Embryologically, a ventral evagination of the floor of the diencephalon that forms the posterior lobe of the pituitary gland.

Infusoriform larva Ciliated larva produced by an infusorigen within a dicyemid mesozoan.

Infusorigen Mass of reproductive cells within a rhombogen.

Ingestion The intake of water or food particles by "swallowing" them, taking them into the body cavity or into a vacuole.

Ingestive feeders Animals that ingest food through a mouth.

Ingoldian fungi Aquatic hyphomycetes that often have a characteristic tetraradiate hyphal development form and which sporulate under water. Discovered by the British mycologist, C. T. Ingold, in the 1940s.

Ingression The migration of individual cells from surface layers into the interior of the embryo.

Ingroup In a cladistic analysis, the set of taxa which are hypothesized to be more closely related to each other than any are to the outgroup.

Inguinal amplexus Sexual embrace of anurans; the grasping of the female's body by the male's forelimbs from a dorsal position just above the hind legs. The grip is usually quite strong, and is not released until oviposition is completed.

Inheritance of acquired characteristics Lamarck's view that features acquired during an organism's lifetime would be passed on to succeeding generations, leading to inheritable change in species over time.

Inhibition In the context of modern classical conditioning, the learning of negative relationships between the CS and US.

Inhibitory postsynaptic potential (IPSP) An electrical charge (hyperpolarization) in the membrane of a postsynaptic neuron caused by the binding of an inhibitory neurotransmitter from a presynaptic cell to a postsynaptic receptor; makes it more difficult for a postsynaptic neuron to generate an action potential.

Initiation The first step in translation; occurs when a messenger RNA molecule, a ribosomal subunit, and a transfer RNA molecule carrying the first amino acid bind together to form a complex; begins at the start codon on mRNA.

Initiation codon The *codon* at which translation of a polypeptide chain is initiated. This is usually the first AUG triplet in the *mRNA* molecule from the 5' end, where the ribosome binds to the cap and begins to scan in a 3' direction. However, the surrounding sequence context is important and may lead to the first AUG being bypassed by the scanning ribosome in favor of an alternative, downstream AUG. Also called a "start codon". Occasionally other codons may serve as initiation codons, e.g. UUG.

Ink sac A gland found in some cephalopods that secretes a dark fluid Used to discourage predators.

Innate A characteristic based partly or wholly on genetic or epigenetic constitution.

Innate releasing mechanism In ethology, a circuit within an animal's brain that is hypothesized to respond to a specific stimulus, setting in motion, or "releasing," the sequence of movements that constitute a fixed action pattern.

Inner cell mass A cluster of cells in a mammalian

blastocyst that protrudes into one end of the cavity and subsequently develops into the embryo proper and some of the extraembryonic membranes.

Inner ear A paired, sound-sensitive organ in vertebrates.

Inoculative release The release of relatively small numbers of natural enemies that are expected to colonize, reproduce, and spread naturally throughout an area.

Inorganic Part of or derived from non-biological material. A chemical compound that does not include a carbon chain. Compare *organic*.

Inorganic molecules (compounds) Pertaining to compounds that are not hydrocarbons or their derivatives. Compounds other than organic compounds.

Inositol trisphosphate (IP_3) (in-nos-i-tahl) The second messenger, which functions as an intermediate between certain nonsteroid hormones and the third messenger, a rise in cytoplasmic Ca^{2+} concentration.

Inquiline. A creature that shares the home of another species without having any obvious effect on that species.

Insect growth regulator (IGR) A substance, natural or synthetic, that controls or modifies insect growth processes.

Insect resistant (plants) Tolerant of, or resistant to, insect attack (as in plants). individuals from strains whose ancestors had not been exposed to the pesticide.

Insecta A 'class' of the 'phylum' Arthropoda, distinguished by adults having three body regions: head, thorax, and abdomen; and by having the thorax three-segmented with each segment bearing a pair of legs.

Insecticide A chemical that kills insects.

Insecticide resistance Genetically inherited ability to withstand doses of pesticide which would kill individuals from strains whose ancestors had not been exposed to the pesticide.

Insectivorous plants Plants which actively or passively capture insects as an adaptation to acquire nitrogen. Examples include the venus flytrap, sundew, and pitcher plant.

Insectivorous Refers to an animal that feeds primarily on insects, such as bats.

Insects Arthropods with three pairs of legs and one pair of antennae. Few are marine, an exception being the water strider.

Insert In a complete plasmid clone, there are two types of DNA - the "vector" sequences and the "insert". The vector sequences are those regions necessary for propagation, antibiotic resistance, and all those mundane functions necessary for useful cloning.

Insertion A type of mutation in which a new DNA base is inserted into an existing sequence of DNA bases. This shifts the reference frame in protein synthesis, resulting (sometimes) in altered amino acid sequences.

Insertion sequence A small bacterial transposon carrying only the genetic functions involved in *transposition*. There are usually *inverted repeats* at the ends of the insertion sequence.

Insight learning The ability of an animal to perform a correct or appropriate behavior on the first attempt in a situation with which it has had no prior experience.

Insolation Incoming solar radiation.

Instar The stage of an arthropod's life cycle between molts (shedding of the exoskeleton). As an example, the third instar is the stage of the life cycle between the second and third molting cycles. Some arthropods molt throughout their entire life and may have as many as 30 instars. Insects tend to have a set number of instars, the number varying by species. The instars succeed one another until the final mature instar, when the organism stops molting.

Instinct Stereotyped, predictable, genetically programmed behavior. Learning may or may not be involved.

Instrumental conditioning Trial-and-error learning. The reinforcement of certain behaviors in animals leads to an animal repeating the behavior.

Instrumental value Value or worth of objects that satisfy the needs and wants of moral agents. Objects that can be used as a means to some desirable end.

Insulin A vertebrate hormone that lowers blood glucose levels by promoting the uptake of glucose by most body cells and the synthesis and storage of glycogen in the liver; also stimulates protein and fat synthesis; secreted by endocrine cells of the pancreas called islets of Langerhans.

Intangible resources Abstract commodities, such as open space, beauty, serenity, genius, information, diversity, and satisfaction.

Integral protein Protein that spans or penetrates the lipid bilayer of cell membranes.

Integrated pest management (IPM) IPM is an ecologically based pest-control strategy that relies on natural mortality factors, such as natural enemies, weather, cultural control methods, and carefully applied doses of pesticides.

Integration The incorporation of one DNA segment into a second DNA molecule to form a new hybrid DNA. Integration occurs during such processes as genetic recombination, episome incorporation into host DNA, and prophage insertion into the bacterial chromosome.

Integrin adhesive protein of the extracellular matrix in animals.

Integrins A large and broadly distributed family of a/b heterodimers. Integrins are cellular adhesion receptors that mediate cell-cell and cell-substratum interactions. Integrins usually recognize linear amino acid sequences on protein ligands.

Integument The outer covering of the insect body that includes the cuticle and the epidermis.

Integumentary exchange The exchange of gases through the integument. Also cutaneous exchange.

Integumentary system The skin and its derivatives (hair, nails, feathers, horns, antlers, and glands), which in multicellular animals protect against invading foreign microorganisms and prevent the loss or exchange of internal fluids.

Inteins Internal intervening sequences of precursor self-splicing proteins that separate exteins and are removed during formation of the final protein.

Intelligence A collection of mental capacities including imagination, problem-solving ability, memory, the ability to use information gained from past experiences, perceptiveness, and behavioral flexibility. The processes by which animals obtain information about their environment, retain it and use the information to make decisions during the course of their behavioral activities.

Intensity Number of parasites in an infected host (infrapopulation). Mean intensity is the average number of parasites per infected host.

Intention movement The preparatory movements that an animal makes prior to a complete behavior pattern, as with wing movements before taking flight.

Interbedded Describes beds (layers) of rock lying between or alternating with beds of a different kind of rock.

Intercalary cartilage A phalanx-like cartilaginous element inserted between the ultimate and penultimate phalanges in the digits of the frog families.

Intercalary meristem A type of meristem present at the base of the blade and/or sheath of many monocots.

Intercalary vein An additional longitudinal vein, arising at the wing margin and running inwards but not directly connected to any of the major veins.

Intercalate To insert or interpose

Intercalated disk Membranous boundary between adjacent cardiac muscle cells.

Intercalating agents Molecules that can be inserted between the stacked bases of a DNA double helix, thereby distorting the DNA and inducing insertion and deletion mutations.

Intercellular Occurring between body cells.

Interdemic selection The selection of entire breeding populations as the basic units.

Interdigitating dendritic cell Special dendritic cells in the lymph nodes that function as potent antigen-presenting cells and develop from Langerhans cells.

Interfascicular cambium The layer of cambium that develops between vascular bundles and connects with the fascicular cambium to form the vascular cambium of woody tissues.

Interference A form of competition in which organisms defend or otherwise control limited resources; also called contest competition. Contrast with exploitative competition.

Interferon A glycoprotein that has nonspecific antiviral activity by stimulating cells to produce antiviral proteins, which inhibit the synthesis of viral RNA and proteins. Interferons also regulate the growth, differentiation, and/or function of a variety of immune system cells. Their production may be stimulated by virus infections, intracellular pathogens (chlamydiae and rickettsias), protozoan parasites, endotoxins, and other agents.

Intergenic Between two genes; e.g. intergenic DNA is the DNA found between two genes. The term is often used to mean non-functional DNA (or at least DNA with no known importance to the two genes flanking it). Alternatively, one might speak of the "intergenic distance" between two genes as the number of base pairs from the polyA site of the first gene to the cap site of the second. This usage might therefore include the promoter region of the second gene.

Interkinesis The activities occurring between meiosis I and meiosis II; similar to interphase but without chromosome replication.

Interleukin A glycoprotein produced by macrophages and T cells that regulates growth and differentiation, particularly of lymphocytes. Interleukins promote cellular and humoral immune responses.

Interleukin-1 A cytokine produced by macrophages that stimulates T helper lymphocytes.

Interleukin-2 A lymphokine produced by T helper lymphocytes that leads to proliferation of T helper cells and other T lymphocytes.

Intermediary meiosis Meiosis that occurs neither during gamete formation nor immediately after zygote formation, resulting in both haploid and diploid generations, such as in foraminiferan protozoa.

Intermediate filament The chemically heterogeneous group of protein fibers, the specific proteins of which can vary with cell type. One of the three most prominent types of cytoskeletal filaments.

Intermediate filaments A class of cytoplasmic filaments of animal cells so named originally because their diameter (nominally 10nm) in muscle cells was intermediate between thick (myosin) and thin (actin) filaments. Unlike microfilaments and microtubules, the protein subunits of intermediate filaments show considerable diversity and tissue specificity. They provide tensile strength for the cell.

In neural cells, intermediate filaments are called neurofilaments.

Intermediate host The host which harbors the immature stages or the asexual stages of a parasite, a separate organism to that which harbours the sexual stage.

Intermediate inheritance. Neither of alternate alleles of a gene are completely dominant, and heterozygote shows a condition intermediate between or different from homozygotes for each allele.

Intermediate layer Of the three main layers of the ocean, the one below the surface, or mixed, layer. It includes the main thermocline.

Intermediate lobe The area in the pituitary gland between the anterior and posterior lobes.

Intermittent parasite Temporary parasite.

Intermittent reinforcement Schedules of reinforcement that involve providing rewards, not on every trial, but on usually one of four other types of schedules: fixed ratio, fixed interval, variable ratio, or variable interval.

Internal costs The expenses, monetary or otherwise, borne by those who use a resource.

Internal environment In multicellular organisms, the aqueous environment that is outside the cells but inside the body.

Internal fertilization 1. broadly, the reproductive condition in animals, such as birds and mammals, when the egg is fertilized within the female's body. 2. specifically, the union of the nuclei of the egg cell and the sperm cell, resulting in the disappearance of their nuclear membranes and the combination of their chromosomes.

Internalizing costs Planning so that those who reap the benefits of resource use also bear all the external costs.

International whaling commission (IWC) An agency that regulates whaling around the world.

Interneuron An association neuron; a nerve cell within the central nervous system that forms synapses with sensory and motor neurons and integrates sensory input and motor output.

Internode The region of a stem between two nodes, when there is no branching of the vascular tissue.

Internuncial processes Cytoplasmic channels that connect one part of a cell to another, such as those linking the distal cytoplasm to the tegumental cytons in many flatworms.

Interobserver reliability A method for checking on the agreement between two or more people watching the same animals behave. This becomes important both as a check on the definitions developed for particular behavior patterns and when we have two or more individuals gathering the data in a particular investigation and we wish to know that they are taking down their records of the data in as nearly similar a manner as possible.

Interphase The period in the cell cycle when the cell is not dividing. During interphase, cellular metabolic activity is high, chromosomes and organelles are duplicated, and cell size may increase. Interphase accounts for 90% of the time of each cell cycle.

Interplanting The system of planting two or more crops, either mixed together or in alternating rows, in the same field; protects the soil and makes more efficient use of the land.

Intersexual selection Selection on characteristics of one sex (usually males) based on mate choices made by members of the other sex (usually females).

Interspecies hydrogen transfer The linkage of hydrogen production from organic matter by anaerobic heterotrophic microorganisms to the use of hydrogen by other anaerobes in the reduction of carbon dioxide to methane. This avoids possible hydrogen toxicity.

Interspecific competition One species subtly depresses another species' survival or breeding success through reduction of critical resources.

Interstitial Situated in the interstices or spaces between structures such as cells, organs, or grains of sand.

Interstitial animal An animal that occupies the spaces (interstices) between sediment particles

Interstitial cell A cell that is located in the spaces of a tissue, in particular, between follicles of the ovary and between seminiferous tubules of the testis. In the testes these cells produce testosterone: the male sex steroid.

Interstitial fauna Animals living between sediment particles.

Interstitial fluid Fluid surrounding the cells in body tissues; provides a path through which nutrients, gases, and wastes can travel between the capillaries and the cells.

Interstitial Interstitial water The water contained between sediment particles.

Intertidal The coastal zone measuring from the lowest to the highest tide mark. The intertidal zone is subject to alternating periods of flooding and drying.

Intertidal zone The vertical extent of the shoreline between the high and low tide lines.

Intertriginous candidiasis A skin infection caused by Candida species. Involves those areas of the body, usually opposed skin surfaces, that are warm and moist (axillae, groin, skin folds).

Intestine The portion of the digestive tract between the stomach and anus; it is the region where most of the nutrients and absorbed.

Intine The inner layer of the pollen wall.

Intoxication A disease that results from the entrance of a specific toxin into the body of a host. The toxin can induce the disease in the absence of the toxin-producing organism.

Intracellular Occurring within a body cell or within body cells.

Intracellular digestion A form of digestion in which food is taken into cells by phagocytosis; found in sponges and most protozoa and coelenterates.

Intracellular parasites Viruses that enter a host cell and take over the host's cellular machinery to produce new viral particles.

Intracellular route Path taken by water through the cells of the root between the epidermis and the xylem, moving through plasmodesmata.

Intraepidermal lymphocytes T cells found in the epidermis of the skin that express the gd T-cell receptor.

Intralecithal cleavage Cleavage in which the nuclei undergo several divisions within the yolk mass without concurrent cytokinesis; common in arthropods.

Intramembranous ossification The process of bone formation from mesenchyme. This process typically occurs in the flat bones of mammalian crania.

Intranuclear inclusion body A structure found within cells infected with the cytomegalovirus.

Intrasexual competition Competition between the sexes.

Intrasexual selection Selection on characteristics of the sexes based on competition among them (usually males) for access to members of the other sex (usually females).

Intraspecific competition In a community, competition for resources among members of the same species.

Intraspecific nest parasitism The addition of eggs to another female's nest.

Intrinsic factors Food-related factors such as moisture, pH, and available nutrients that influence microbial growth.

Intrinsic growth rate Exponential growth rate of a population, that is, the difference between the density-independent components of the birth and death rates of a natural population with stable age distribution.

Intrinsic rate of increase The difference between the number of births and the number of deaths, symbolized as r_{max}; the maximum population growth rate.

Introduced species An organism that has been brought into an area, usually by humans, where it does not normally occur. Introduced species often compete with and cause problems for native species. Introduced species are also called exotic, nonnative, and alien species.

Introduction (classical biological control) The importation of a natural enemy from a foreign country or continent, usually to control a pest also of foreign origin.

Introgression The transplantation of genes between species resulting from fertile hybrids mating successfully with one of the parent species.

Intron Introns are portions of genomic DNA which ARE transcribed (and thus present in the primary transcript) but which are later spliced out. They thus are not present in the mature mRNA.

Note that although the 3' flanking region is often transcribed, it is removed by endonucleolytic cleavage and not by splicing. It is not an intron.

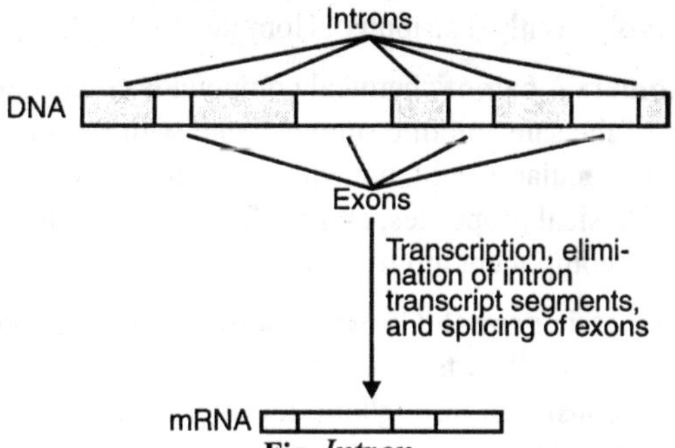

Fig. *Intron*

Introvert A tubular, eversible extension of the head, bearing the mouth at its tip.

Intrusion Magma (and the rock it forms) that has pushed into pre-existing rock.

Inundative release The release of relatively large numbers of natural enemies to suppress pest populations, without the expectation that the natural enemies will colonize and spread throughout the area.

Invagination The infolding of a sheet of cells, much like the indenting of a hollow rubber ball when poked.

Invasive species An introduced species that outcompetes native species for space and resources. Scotch Broom is an invasive species that outcompetes local vegetation and results in a monoculture, and hence a decrease in local diversity.

Invasiveness The ability of a microorganism to enter a host, grow and reproduce within the host, and spread throughout its body.

Inversion An aberration in chromosome structure resulting from an error in meiosis or from mutagens; reattachment in a reverse orientation of a chromosomal fragment to the chromosome from which the fragment originated.

Invertebrate An animal without a backbone; invertebrates make up 95% of animal species.

Inverted repeats Two copies of the same or related sequence of DNA repeated in opposite orientation on the same molecule. Adjacent inverted repeats constitute a *palindrome*.

Involucre Whorl of bracts that subtend a flower or an inflorescence.

Involuntary Nervous system Stimulates smooth and cardiac muscle and glands of the body.

Involution A process of cellular migration involving movement of a sheet of cells over an edge and then spreading inwards in the opposite direction. This occurs, for example, during amphibian gastrulation.

Iodinophilous vacuole Vacuole within a protozoan that stains readily with iodine.

Ion An atom that has lost or gained electrons from its outer shell and therefore has a positive or negative charge, respectively; symbolized by a superscript plus or minus sign and sometimes a number, e.g., H^+, Na^{+1}, Cl^{-2}.

Ionic bond A chemical bond in which atoms of opposite charge are held together by electrostatic attraction.

Ionize To split a molecule into two or more parts, each part becoming an electrically charged particle.

Ionizing radiation High-energy electromagnetic radiation or energetic subatomic particles released by nuclear decay.

Ionophore Lipid soluble substance that forms a channel or acts as a carrier in a lipid bilayer membrane to allow specific ions can move across the membrane. A23187 is an example of a calcium ionophore.

Ionosphere The lower part of the thermosphere.

Iridescent Displaying a shining, rainbowlike range of colors.

Iridocyte A fish skin cell that contains reflecting crystals made of guanine.

Iridophore A chromatophore with light-reflecting crystals.

Irregular symmetry Bilateral symmetry; having only one plane through which the structure (flower) could be cut to result in mirror image halves; zygomorphy.

Irrigation The application of water to agricultural crops by transporting it to the crop through canals, pipes, and sprinklers.

Irritability The ability of a cell to undergo a change in membrane potential.

Irritants Corrosives (strong acids), caustics (alkaline reagents), and other substances that damage biological tissues on contact.

Island arc A curved chain of islands that rise from the sea floor, usually near a continent. The convex side usually faces the open ocean, while the concave side usually faces the continent, e.g., the Aleutian Islands in Alaska; volcanic arc- syn.

Island biogeography The study of rates of colonization and extinction of species on islands or other isolated areas based on size, shape, and distance from other inhabited regions.

Island effects Reductions in species diversity caused by reduction in ecosystem area.

Island of fertility The region in a desert ecosystem directly beneath a tree or shrub. Leaf fall and accumulation of litter result in nutrient cycling directly around the plant, even though the bare soil between plants may be depleted of nutrients.

Isogametes Outwardly similar male and female gametes.

Isogamous Having haploid gametes that are similar in size, structure and motility.

Isohaline Having the same salinity.

Isolating mechanisms Mechanisms that prevent genetic exchange between individuals of different populations or species; they prevent mating or successful reproduction even when mating occurs; may be behavioral, anatomical, or physiological.

Isolecithal Pertaining to a zygote (or ovum) with yolk evenly distributed. Homolecithal.

Isomers A pair of chemical compounds composed of the same atomic subunits, and with the same molecular weight, but with different chemical or physical properties, and a different arrangement of component atoms in space.

Isometric growth Growth that occurs at the same rate for all parts of an organism so that its shape is consistent throughout development. Contrast with allometric growth.

Isomorphic generations Alternating generations in which the sporophytes and gametophytes look alike, although they differ in chromosome number.

Isopods Small, dorsoventrally flattened crustaceans such as the sea louse.

Isoprenes Five-carbon compounds that are the basic unit of terpenes.

Isotonic Having the same overall concentration of dissolved substances as a given reference solution

Isotopes Atoms with the same atomic number but different numbers of neutrons; indicated by adding the mass number to the element's name, e.g., carbon 12 or ^{12}C.

Isotype A variant form of an immunoglobulin (e.g., an immunoglobulin class, subclass, or type) that occurs in every normal individual of a particular species. Usually the characteristic antigenic determinant is in the constant region of H and L chains.

Isozyme Different chemical forms of the same enzyme; thought to be important in adaptation to environmental extremes.

Iteroparity The production of offspring by an organism in successive bouts. Contrast with semelparity.

J

J chain A polypeptide present in polymeric IgM and IgA that links the subunits together.

J curve A growth curve that depicts exponential growth; called a J curve because of its shape.

Jaccard coefficient An association coefficient used in numerical taxonomy; it is the proportion of characters that match, excluding those that both organisms lack.

Jacket cells Ciliated somatoderm of an orthonectid mesozoan.

Jacobson's (vomeronasal) organ Olfactory receptor present in most reptiles; blind-ending sacs that open through the secondary palate into the mouth cavity; they are used to sample airborne chemicals.

Jaw Often loosely applied to any movable, toothed structures at or near the mouth of an animal, such as the scolecodonts of annelids. In vertebrates, the jaw is derived from the first gill arch.

Jawless fishes (class Agnatha)Fishes that lack jaws and paired fins: hagfishes (or slime eels) and lampreys.

Jejunum Portion of the small intestine located between the duodenum and the ileum.

Jerboas A small, nocturnal, social rodent of the family Dipodidae found in Old World deserts, having enlarged hind limbs that are modified for leaping.

Jet streams Powerful winds or currents of air that circulate in shifting flows; similar to oceanic currents in extent and effect on climate.

Johnston's organ A sense organ located in the second antennal segment of many insects and particularly well developed in male mosquitoes and certain other Diptera.

Johnston's organ mechanoreceptor (auditory receptor) Found at the base of the antennae of male mosquitoes and midges.

Joint Strictly speaking, an articulation between neighbouring parts, such as the femur and tibia of the leg, but the word is commonly used as a synonym of segment - meaning any of the divisions of the body or its appendages.

Jointed When stiff body parts are connected by a soft flexible region, the body is said to be jointed.

Joule A unit of energy. One joule is the energy expended in 1 second by a current of 1 amp flowing through a resistance of 1 ohm.

Jugum A narrow lobe projecting from the base of the forewing in certain moths and overlapping the hind wing, thereby coupling the two wings together.

Junctional epithelium The epithelium which seals the base of the gingival sulcus against the tooth.

Jurassic period Middle period of the Mesozoic Era, between 185-135 million years ago. Characterized by the (possible) origin of angiosperms and the continued split of the worldwide supercontinent of Pangaea.

Juvenile hormone Hormone produced by the corpora allata of insects; among its effects are maintenance of larval or nymphal characteristics during development.

Juxtaglomerular apparatus (JGA) Specialized tissue located near the afferent arteriole that supplies blood to the kidney glomerulus; the JGA raises blood pressure by producing renin, which activates angiotensin.

K

Kala-azar Disease caused by Leishmania donovani. Also called Dumdum fever or visceral leishmaniasis.

Kame A short, steep-sided knoll of glacial sediment.

Karst A type of topography formed by dissolution of rocks like limestone and gypsum that is characterized by sinkholes, caves, and subterranean passages.

Karyogamy A process of fusion of the nuclei of two cells; the second step in syngamy.

Karyokinesis The division of a cell nucleus into two daughter nuclei each having an identical complement of chromosomes.

Karyokinesis Division of the nucleus during the cell cycle.

Karyotype A photomicrograph of an individuals *chromosomes* arranged in a standard format showing the number, size, and shape of each chromosome type; used in low resolution *physical mapping* to correlate gross chromosomal abnormalities with the characteristics of specific diseases.

KB Abbreviation for kilobase, one thousand bases.

Keel A narrow ridge: also called a carina

Kelp A common name for any of the larger members of the order Laminariales of the brown algae.

Kelp forest Marine ecosystem dominated by large kelps. These forests are restricted to cold and temperate waters, and are most common along the western coasts of continents. Kelp forests first appeared in the Miocene.

Kentrogon Larva in the crustacean order Rhizocephala that is attached to its host crab; formed after the cypris larva molts and its appendages and carapace are discarded.

Keratan sulphate A glycosaminoglycan found in cartilage, with<> chondroitin sulphate.

Keratin A fibrous polymer which is notas strong as collagen but less soluble. It forms the strong and water-resistant properties of skin, nails, hair, horn and beaks.

Keratinized scale Scales containing keratin, which is a hard, tough, non-soluble protein produced in the epidermis of both reptiles and amphibians.

Keratinocytes Cells of the epidermis which secrete the protein keratin. They become progressively flattened as they mature and are eventually are off.

Keratitis Inflammation of the cornea of the eye.

Kettle lakes Lakes formed as water fills a hole formerly occupied by a block of stranded ice.

Keystone predator The dominant predator or the top predator that has a major influence on community structure. For example, sea otters are a keystone predator in kelp beds. Sea otters eat urchins that feed on kelp which house a huge diversity of other organisms. If sea otter populations are lowered in an area the kelp beds are generally reduced and urchin barrens appear.

Keystone species A species that is of exceptional importance in maintaining the species diversity of a community; when a keystone species is lost, the diversity of the community decreases and its structure is significantly altered.

Kidney In vertebrates, the organ that regulates the balance of water and solutes in the blood and the excretion of nitrogenous wastes in the form of urine.

Kidney stones Crystallized deposits of excess wastes such as uric acid, calcium, and magnesium that may form in the kidney.

Kilobase Unit of length for DNA fragments equal to 1000 *nucleotides*.

Kilocalorie The energy needed to heat 1000 grams of water from 14.5 to 15.5 degrees C.

Kin selection The selection of genes due to individuals assisting the survival and reproduction of nondescendant relatives who possess the same genes by common descent.

Kinase A kinase is in general an enzyme that catalyzes the transfer of a phosphate group from ATP to something else. In molecular biology, it has acquired the more specific verbal usage for the transfer onto DNA of a radiolabeled phosphate group. This would be done in order to use the resultant "hot" DNA as a probe.

Kinesis Movements by an organism in random directions in response to stimulus.

Kinetic energy The energy of motion, which is directly related to the speed of that motion. Moving matter does work by transferring some of its kinetic energy to other matter.

Kinetid Axoneme of a cilium of flagellum together with its basal fibrils and organelles. Also called a mastigont.

Kinetochore Structure that attaches a eukaryotic chromosome to the spindle apparatus for separation during mitosis; adjacent to the centromere.

Kinetodesma Fibril arising from the kinetosome of a cilium in a ciliate protozoan, and passing along the kinetosomes of cilia in that same row.

Kinetodesmose Compound fiber joining cilia into rows.

Kinetoplast Conspicuous part of a mitochondrion in a trypanosome; usually found near the kinetosome.

Kinetosome Centriole from which an axoneme arises. Also called a basal body or blepharoplast.

Kinety All the kinetosomes and kinetodesmata of a row of cilia.

Kingdom The highest level of classification of life; the most widely accepted classification system includes five kingdoms: Monera, Protista, Fungi, Plantae, and Animalia.

Kinin A type of local hormone that is released near its site of origin; also called parahormone or tissue hormone.

Kinorhyncha The phylum of aschelminths that contains members called kinorhynchs; small elongate worms found exclusively in marine environments where they live in mud and sand.

Kirby-bauer method A disk diffusion test to determine the susceptibility of a microorganism to chemotherapeutic agents.

Klenow fragment The large fragment of E. coli DNA polymerase I which lacks 5' -> 3' exonuclease activity. Very useful for sequencing reactions, which proceed in a 5' -> 3' fashion (addition of nucleotides to templated free 3' ends of primers).

Kleptoparasitised An animal that obtained its food mainly by scavenging the food supplies of another species, especially one with which it habitually lives in close proximity.

Klinefelter syndrome In humans, a genetically determined condition in which the individual has two X and one Y chromosome. Affected individuals are male and typically tall and infertile.

Knock-out The excision or inactivation of a gene within an intact organism or even animal (e.g., "knock-out mice"), usually carried out by a method involving *homologous recombination.*

Knock-out experiment A technique for deleting, mutating or otherwise inactivating a gene in a mouse. This laborious method involves transfecting a crippled gene into cultured embryonic stem cells, searching through the thousands of resulting clones for one in which the crippled gene exactly replaced the normal one (by homologous recombination), and inserting that cell back into a mouse blastocyst. The resulting mouse will be chimaeric but, its germ cells will carry the deleted gene. A few rounds of careful breeding can then produce progeny in which both copies of the gene are inactivated.

Knockout gene A genetically engineered mutant gene that is introduced into an embryo to study the specific effects of that gene.

Knot Commonly seen as an irregular formation in a longitudinal section of wood formed when a branch becomes surrounded (buried) by formation of secondary xylem.

Known resources Minerals or other useful environmental materials or services that are identified and partially mapped; may not be environmentally or socially acceptable or economically feasible to exploit.

Koch's blue bodies Schizonts of Theileria parva in circulating lymphocytes.

Koch's postulates A set of four criteria for determining whether a specific pathogen is the cause of a disease.

Koplik's spots Lesions of the oral cavity caused by the measles (rubeola) virus that are characterized by a bluish white speck in the center of each.

Korean hemorrhagic fever An acute infection caused by a virus that produces varying degrees of hemorrhage, shock, and sometimes death.

Kranz anatomy A specialized leaf anatomy found in C4 plants in which the vascular bundle is surrounded by bundle sheath cells.

Kreb cycle The end stage of aerobic respiration. Kreb's cycle is a circular series of reactions taking place in the matrix of mitochondria in which acetyl CoA is broken down into carbon dioxide and hydrogen atoms. The hydrogen atoms are used to produce ATP via the electron transport system.

Krill Any of the small, pelagic, shrimplike crustaceans. Krill are an important source of energy in antarctic food webs.

K-selection The concept that in certain (K-selected) populations, life history is centered around producing relatively few offspring that have a good chance of survival.

K-strategist Species of organism that uses a survival and reproductive "strategy" characterized by low fecundity, low mortality, and longer life and with populations approaching the carrying capacity of the environment, controlled by density-dependent factors.

Kupffer cells Phagocytic epithelial cells lining the sinusoids of the liver.

Kwashiorkor Malnutrition caused by diet high in carbohydrate and extremely low in protein.

L

Labellum The expanded tip of the labium, used by many flies to mop up surface fluids.

Labia majora The outer folds of skin that cover and protect the genital region in women.

Labia minora Thin membranous folds of skin outside the vaginal opening.

Labial Concerning the labium.

Labial palp 1. Chemosensory appendage found on the labium of insects (Arthropoda).

2. Flaplike lobe surrounding the mouth of bivalve molluscs that directs food toward the mouth.

Labial palpus The labial palps, One of the pair of sensory appendages (feeler-like and 2 to 5 segments long) of the insect labium.

Labial scale Scale that borders the lip in reptiles.

Labium The posterior mouthpart of insects. It is often referred to as the "lower lip," is chemosensory, and was derived evolutionarily from paired head appendages (Hexapoda, Arthropoda).

Labrum The 'upper lip' of the insect mouth-parts: not a true appendage but a movable sclerite on the front of the head.

Labrum-epipharynx A mouthpart composed of the labrum and epipharynx and usually elongate.

Labyrinth Vertebrate internal ear, composed of a series of fluid-filled sacs and tubules (membranous labyrinth) suspended within bone cavities (osseous labyrinth).

Labyrinth organ One of a pair of equilibrium organs in vertebrates that contains three fluid-filled semicircular canals.

Labyrinthodont A group of Paleozoic amphibians containing the temnospondyls and the anthracosaurs.

Lachrymal Secreting or relating to tears.

Lacinia The inner branch of the maxilla, the outer one being the galea.

Lactation The production of milk by the mammary glands.

Lacteal Noun, one of the lymph vessels in the villus of the intestine. Adj., relating to milk.

Lactic acid The end product of anaerobic respiration in animals, causes muscle soreness.

Lactic acid fermentation A fermentation that produces lactic acid as the sole or primary product.

Lactose Milk sugar; a disaccharide composed of glucose and galactose.

Lactose intolerance A genetic trait characterized by the absence of the enzyme lactase, which breaks down lactose, the main sugar in milk and other dairy products.

Lacuna A small space. More specifically: 1) The cavity enveloping and occupied by chondrocytes and osteocytes;

2) The spaces which develop in the placenta and which will later fuse to form the intervillous lake.

Lacunar system System of canals in the body wall of an acanthocephalan, functioning as a circulatory system.

Lag phase The initial growth phase, during which cell number remains relatively constant prior to rapid growth.

Lagena Portion of the primitive ear in which sound is translated into nerve impulses; evolutionary beginning of cochlea.

Lager Pertaining to the process of aging beers to allow flavor development.

Lagging strand A discontinuously synthesized DNA strand that elongates in a direction away from the replication fork.

Lagoon A shallow, sheltered body of water separated from the open sea by coral reefs, sand bars, and/ or barrier islands .

Lamarckism. Hypothesis, as expounded by Jean Baptiste de Lamarck, of evolution by the acquisition during an organism's lifetime of characteristics that are transmitted directly to offspring.

Lamella One of the two plates forming a gill in a bivalve mollusc. One of the thin layers of bone laid concentrically around an osteon (Haversian canal). Any thin, platelike structure.

Lamella-bone The microscopic structure of cortical bone gives it the appearance of concentric or parallel plates (from Latin, lamella, the diminutive of lamina, meaning a plate or leaf).

Lamellate Possessing lamellae: applied especially to antennae.

Lamina Any broad and flattened region of a plant or alga, which allows for increased photosynthetic surface area.

Lamina dura The name given to the radiographic appearance of a dense layer of bone around the tooth root. It represents the dense cortical bone lining the tooth socket.

Lamina propria The layer of loose connective tissue underneath the epithelium of mucosa, which provides physical and nutritional support.

Laminae The scales for the epidermal plates of the shell of turtles.

Laminar flow The movement of a fluid where movement of the entire fluid is regular and with parallel streamlines.

Laminarin A beta-glucan polysaccharide produced by many chromists through photosynthesis.

Laminin An adhesive molecule of connective tissue related to fibronectin andtenascin.

Lancefield system One of the serologically distinguishable groups (as group A, group B) into which streptococci can be divided.

Lancelets Chordates with the three basic chordate characteristics but lacking a backbone.

Land ethic A philosophy or attitude about land that provides for its protection or more careful management.

Land farming The addition of waste material, such as a hydrocarbon waste, to the soil surface so that it will be degraded. The soil may be moistened or mixed to stimulate the desired degradation process.

Land races Traditional varieties of plant crops.

Land reform Democratic redistribution of landownership to recognize the rights of those who actually work the land to a fair share of the products of their labor.

Land rehabilitation A utilitarian program to repair damage and make land useful to humans.

Landfills Land disposal sites for solid waste;

operators compact refuse and cover it with a layer of dirt to minimize rodent and insect infestation, wind-blown debris, and leaching by rain.

Landscape A mosaic of repeated ecosystems in a given geographic area. The land is heterogeneous, but there are structural and functional relationships among the matrix and the various patches and corridors.

Landscape ecology The study of the reciprocal effects of spatial pattern on ecological processes. A study of the ways in which landscape history shapes the features of the land and the organisms that inhabit it as well as our reaction to, and interpretation of, the land.

Landscape epidemiology Approach to epidemiology that employs all ecological aspects of a nidus. By recognizing certain physical conditions, the epidemiologist can anticipate whether a disease can be expected to exist.

Landslide The sudden fall of rock and earth from a hill or cliff. Often triggered by an earthquake or heavy rain.

Langerhans cell Cell found in the skin that internalizes antigen and moves in the lymph to lymph nodes where it differentiates into a dendritic cell. Cells are active in the immune response of the skin and mucous membrane. They act as sentries, detecting the presence of foreign antigens on the surface of the epithelium. They do not contain keratin and are thus sometimes called clear cells.

Langmuir cells Parallel pairs of surface ocean convection cells driven by winds.

Lanugo Thin and closely spaced hairs which are the first hairs in the human embryo. This type of hair is usually shed before birth and is replaced by the short and silky vellus.

Lappets. Lobes around the margin of scyphozoan medusae (phylum Cnidaria).

Large intestine That part of the digestive system between the ileocecal valve of the small intestine and the anus; removes salt and water from undigested food and releases feces through the anus.

Larva A stage in the development of many insects and other organisms including sea urchins and sponges. In sponges, sexual reproduction results in the production of motile ciliated larvae.

Larvacea A group of planktonic tunicates that secrete a gelatinous house, used to strain unsuitable particles (large particles are rejected). An inner filter apparatus of the house, the so-called food trap or particle-collecting apparatus, is used to retain food particles.

Larval instars Any of the different immature feeding stages of an insect that undergoes holometabolous metamorphosis.

Larynx Modified upper portion of respiratory tract of air-breathing vertebrates, bounded by the glottis above and the trachea below; voice box; adj., laryngeal relating to the larynx.

Last glacial maximum (LGM) the time that the last major continental glacial advance in the Northern Hemisphere reached its maximum extent, about 18,000 years ago.

Late mRNA Messenger RNA produced later in a virus infection, which codes for proteins needed in capsid construction and virus release.

Latent heat of evaporation The amount of heat energy that is needed to evaporate a substance, that is, to change it from a liquid to a gas.For water, it is 540 cal/g.

Latent heat of fusion The heat that must be extracted from a liquid to freeze it to a solid at the same temperature. For water, it is 80 cal/g.

Latent heat of melting The amount of heat energy needed to melt a substance, that is, to change it from a solid to a liquid.

Latent heat of vaporization the heat energy required to convert a liquid to a gas at the same temperature.

Latent learning Associations made with neither immediate reinforcement or reward nor with particular behavior evident at the time of learning.

The animal may store and use such information in appropriate situations at a later time.

latent period The initial phase in the one-step growth experiment in which no phages are released.

latent virus infections Virus infections in which the virus quits reproducing and remains dormant for a period before becoming active again.

Lateral bud Bud found in the axil of a leaf, also called an axillary bud.

Lateral line A system of canals and sensory cells on the sides of fishes that helps them detect vibrations in the water.

Lateral line system A mechanoreceptor system consisting of a series of pores and receptor units (neuromasts) along the sides of the body of fishes and aquatic amphibians; detects water movements made by an animal itself and by other moving objects.

Lateral meristem Meristems that give rise to secondary tissues, typically by the formation of radial files of cells and thereby increasing the diameter and circumference of an axis over time; the vascular cambium and cork cambium.

Lateral ocellus The simple eye in holometabolous larvae. Also called stemma (pl., stemmata).

Lateral oviduct In insects, one of the paired lateral ducts of the female genital system connected with the ovary.

Lateral root A root that arises from another, older root. Often termed secondary or branch roots.

Lateral-line system 1. A line of sensory receptors along the side of some fishes and amphibians used to detect water movement (phylum Chordata).

2. The external manifestation of a lateral excretory canal of nematodes (phylum Nematoda).

Laterite Group of hard, red soils from topical areas that show intense weathering and leaching of bases and silica, leaving aluminum hydroxides and iron oxides; adj. lateritic.

Latewood The small, thick-walled xylem cells produced at the end of the growing season that appear as a dense ring of wood adjacent to the thin-walled earlywood.

Latex Milky juice exuded from some plants.

Latitude the angular distance north or south of the equator of a position on the earth's surface; measured in degrees.

Latitudinal diversity gradient The decrease in species richness that occurs as one moves away from the equator.

Latitudinal gradient As latitude increases, a gradient of cooler, drier conditions occurs.

Lattice A crystalline-like structure caused by the precise orientation of molecules in a solid or liquid.

Laurasia A supercontinent that existed from the Jurassic to Early Tertiary after splitting from Pangea; composed of Laurentia, Baltica, Avalonia, (modern North America, Scandinavia, Greenland, Western and Central Europe); eventually fragmented into Eurasia and North America in the Tertiary with the opening of the North Atlantic Ocean.

Laurentia A separate continental plate that existed from the Late Precambrian to Silurian, consisting of the major part of North America, northwest Ireland, Scotland, Greenland, and pieces of Norway and Russia.

Laurer's canal Usually blind canal extending from the base of the seminal receptacle of a digenetic trematode. It probably represents a vestigial vagina.

Lava Any molten material that is extrusive or volcanic, or the rock that forms from a molten extrusive.

Law of independent assortment Mendel's second law, stating that each allele pair segregates independently during gamete formation; applies when genes for two traits are located on different pairs of homologous chromosomes.

Law of segregation Mendel's first law, stating that

allele pairs separate during gamete formation, and then randomly re-form pairs during the fusion of gametes at fertilization.

Law of the minimum Holds that population growth is limited by the resource in shortest supply.

Lay botanist A nonscientist who understands botanical principles and appreciates the value of the plant world.

Layering A method of asexual propagation in which portions of the stem are wounded and covered with a medium, usually soil, to stimulate the production of adventitious roots.

LD50 A chemical dose lethal to 50 percent of a test population.

LDL (low density lipoprotein) Transports fats and cholesterol to the body cells, including the cells lining the bloodstream.

L-dopa A chemical related to dopamine that is used in the treatment of Parkinson's disease.

Leaching The dissolving of minerals and other elements in soil or rocks by the downward movement of water.

Leader sequence A nontranslated sequence at the 5_ end of mRNA that lies between the operator and the initiation codon; it aids in the initiation and regulation of transcription.

Leading strand The new continuous complementary DNA strand synthesized along the template strand in the mandatory 5' to 3' direction.

Leaf An organ found in most vascular plants; it consists of a flat lamina (blade) and a petiole (stalk). Many flowering plants have additionally a pair of small stipules near the base of the petiole.

Leaf margin Refers to the edge of a leaf, the area between the apex and base.

Leaf primordial Young leaves, recently formed by the shoot apical meristem, located at the tip of a shoot.

Leaf scars A scar left on a stem or twig when the leaf abscises.

Leaf sheath The lower part of a blade or petiole that invests the stem more or less completely.

Leaf trace The strand of vascular tissue which connects the leaf veins to the central vascular system of the stem.

Leaf veins Vascular tissue in leaves, arranged in a net-like network (reticulate vennation) in dicots, and running parallel (parallel vennation) to each other in monocots.

Leaf-area index A numerical index of the ratio of leaf area to ground area in a plant community.

Leaflet In a compound leaf, the individual blades are called leaflets.

Leafy greens Lettuces and other leaf vegetables.

Learning The process that leads to modification in individual behavior as the result of experience.

Learning curve A graphical presentation of the response measures in a learning situation. This may be depicted as correct or incorrect responses, or as a proportion of the trials given to the animal on which it gave correct or incorrect responses.

Learning set The acquisition of a learning strategy by the animal; the ability to transfer learning between problems of the same type.

Least toxic Having a minimal toxic effect upon non-target organisms.

Leaves The site of photosynthesis; one of the three major organs in plants.

Lecithotrophic larva A planktonic-dispersing larva that lives off yolk supplied via the egg.

Lectin A protein molecule which bindson to a specific sequence of sugars. Bacteria may use lectin attachments to bind onto each other or oral surfaces.

Lectin complement pathway The lectin pathway for complement activation is triggered by the binding of a serum lectin (mannan-binding lectin; MBL) to mannose-containing proteins or to carbohydrates on viruses or bacteria.

Leeches Segmented worms that are specialized predators and parasites.

Leeward The side of an island opposite from the one facing a persistent wind

Legionnaires' disease (legionellosis) A pulmonary form of legionellosis, resulting from infection with Legionella pneumophila.

Legume Simple, dry, dehiscent fruit that splits along two seams, a pod. Member of the Fabaceae; a type of bean or pea.

Leishman-donovan (L-D) body Amastigote in the Trypanosomatidae.

Leishmanias Zooflagellates, members of the genus Leishmania, that cause the disease leishmaniasis.

Leishmaniasis The disease caused by the protozoa called leishmanias.

Leishmaniasis Infection by a species of Leishmania.

Lek An area where animals assemble for communal courtship display and mating.

Lemniscus One of a pair of internal projections of the epidermis from the neck region of Acanthocephala, which functions in fluid control in the protrusion and invagination of the proboscis.

Lentic Of or relating to standing water such as swamp, pond, or lake.

Lentic ecosystem A freshwater ecosystem that has standing water (e.g., lakes or ponds).

Lenticel Isolated areas of loosely arranged cells in the cork surfaces of stems, roots, and fruit that allow interchange of gases between internal tissues and the atmosphere through the periderm.

Lepidoptera A large order of scaly-winged insects including the butterflies, skippers, and moths, often brightly colored and having a coiled sucking proboscis.

Lepidosaurs A lineage of diapsid reptiles that appeared in the Permian and that includes the modern snakes, lizards, amphisbaenids, and tuataras, and the extinct ichthyosaurs.

Lepospondyls A group of Paleozoic amphibians distinguished by the possession of spool-shaped vertebral centra.

Lepromatous (progressive) leprosy A relentless, progressive form of leprosy in which large numbers of Mycobacterium leprae develop in skin cells, killing the skin cells and resulting in the loss of features. Disfiguring nodules form all over the body.

Leprosy or Hansen's disease A severe disfiguring skin disease caused by Mycobacterium leprae.

Leptocephalous larva The leaf-shaped larva of freshwater eels and other fishes.

Leptocephalus pl. leptocephali. Transparent, ribbonlike migratory larva of the European or American eel.

Leptotene The earliest stage in prophase 1 of of meiosis, when the thread-like chromosomes begin to condense and become visible.

Lesion An area of tissue that has been destroyed by an agent such as electric current or a chemical.

Less-developed countries Nonindustrialized nations characterized by low per capita income, high birth and death rates, high population growth rates, and low levels of technological development.

Lethargic The state of being abnormally drowsy or stupor.

Leucine zipper A motif found in certain proteins in which Leu residues are evenly spaced through an ?-helical region, such that they would end up on the same face of the helix. Dimers can form between two such proteins. The Leu zipper is important in the function of transcription factors such as Fos and Jun and related proteins.

Leucocytes Un pigmented (white) cells of the blood. Those with granular cytoplasm are neutrophils, eosinophils and basophils. The agranulocytes are lymphocytes and monocytes.

Leucon The sponge body form that has an extensively branched canal system; the canals lead to chambers lined by choanocytes.

Leucoplast Colorless plastid typically associated with starch formation and storage.

Leucotriens Concerned with signalling between cells of the immune system and a member of the eicosanoid family of hormones.

Leukemia A progressive, malignant disease of blood-forming organs, marked by distorted proliferation and development of leukocytes and their precursors in the blood and bone marrow. Certain leukemias are caused by viruses (HTLV-1, HTLV-2).

Leukemism Presence of white pelage or plumage in animals with normally pigmented eyes and skin.

Leukocidin A microbial toxin that can damage or kill leukocytes.

Leukocyte Any of several kinds of white blood cells (for example, granulocytes, lymphocytes, monocytes), so-called because they bear no hemoglobin, as do red blood cells.

Leukorrhea White, puslike discharge resulting from infection.

Leukotriene A type of prostaglandin produced by various white blood cells involved in the inflammatory and immune responses and in allergic reactions.

Liana A woody-stemmed vine. Lianas are rooted in the substrate and use trees or shrubs as support; often their leaves and flowers reach the canopy of layer of the vegetation. A type of growthform.

Library A library might be either a genomic library, or a cDNA library. In either case, the library is just a tube carrying a mixture of thousands of different clones - bacteria or phages. Each clone carries an "insert" - the cloned DNA.

A cDNA library is usually just a mixture of bacteria, where each bacteria carries a different plasmid. Inserted into the plasmids (one per plasmid) are thousands of different pieces of cDNA (each typ. 500-5000 bp) copied from some source of mRNA, for example, total liver mRNA. The basic idea is that if you have a large enough number of different liver-derived cDNAs carried in those bacteria, there is a 99% probability that a cDNA copy of any given liver mRNA exists somewhere in the tube. The real trick is to find the one you want out of that mess - a process called screening.

A genomic library is similar in concept to a cDNA library, but differs in three major ways - 1) the library carries pieces of genomic DNA (and so contains introns and flanking regions, as well as coding and untranslated);

2) you need bacteriophage or cosmids, rather than plasmids, because...

3) the inserts are usually 5-15 kb long (in a library) or 20-40 kb (in a cosmid library). Therefore, a genomic library is most commonly a tube containing a mixture of phages. Enough different phages must be present in the library so that any given piece of DNA from the source genome has a 99% probability of being present.

Librigenae The "free cheeks"; separate, detachable portions of the trilobite cephalon.

Lichens Autotrophic organisms composed of a fungus (sac or club fungus) and a photosynthetic unicellular organism (e.g., a cyanobacterium or alga) in a symbiotic relationship; are resistant to extremes of cold and drought and can grow in marginal areas such as Arctic tundra.

Liebig's law of the minimum Living organisms and populations will grow until lack of a resource begins to limit further growth.

Life cycle The sequence of events that occurs during the lifetime of an individual organism.

Life-cycle analysis Evaluation of material and energy inputs and outputs at each stage of manufacture, use, and disposal of a product.

Life expectancy The average number of years lived by a group of individuals after reaching a given age; the probable number of years of survival for an individual of a given age.

Life history The age at sexual maturity, age at death, and age at other events in an individual's lifetime that influence reproductive traits.

Life history pattern A group of traits, such as size

and number of offspring, length of maturation, age at first reproduction, and the number of times reproduction occurs, that affect reproduction, survival, and the rate of population growth.

Life range The larger geographic area that an animal utilizes over the course of its life.

Life span The longest period of life reached by a type of organism.

Life table A table summarizing statistics of a population, such as survival and reproduction, all broken down according to age classes

Life zone Belts of vegetation that are similar in structure and species composition in both latitudinal and elevational expressions.

Ligaments Dense parallel bundles of connective tissue that strengthen joints and hold the bones in place.

Ligand A protein molecule which binds to another specific protein molecule. The forces of the bond are week and thus protein-ligand bonds depend on close fit of one molecule to the other, so as to capture as many bonding sites as possible. Ligands are specific for a particular protein. They are found on cell surfaces of microorganisms where they assist in cell adhesion. They are also sights on cell membranes onto which protein messengers attach such ascytokines.

Ligand-gated ion channel receptor A signal receptor protein in a cell membrane that can act as a channel for the passage of a specific ion across the membrane. When activated by a signal molecule, the receptor either allows or blocks passage of the ion, resulting in a change in ion concentration that usually affects cell functioning.

Ligase An enzyme, T4 DNA ligase, which can link pieces of DNA together. The pieces must have compatible ends (both of them blunt, or else mutually compatible sticky ends), and the ligation reaction requires ATP.

Ligation *In vitro* reaction used to make the covalent phosphodiester linkage between two double stranded DNA ends. The reaction requires the presence of a phosphate group on the free 5' end and a -OH group on the free 3' end of the strands to be joined. The reaction is catalysed by the enzyme DNA ligase.The process of splicing two pieces of DNA together. In practice, a pool of DNA fragments are treated with ligase in the presence of ATP, and all possible splicing products are produced, including circularized forms and end-to-end ligation of 2, 3 or more pieces. Usually, only some of these products are useful, and the investigator must have some way of selecting the desirable ones.

Light harvesting antenna A complex of several hundred chlorophyll and carotenoid molecules that form a part of each photosystem.

Light intensity The strength of light rays; the degree of brightness dependent on the number of photons striking a given area at a point in time.

Light microscope (LM) An optical instrument with lenses that refract (bend) visible light to magnify images of specimens.

Light reaction that part of the photosynthetic process that, in the presence of light, captures energy to form ATP and NADPH2 to be used to synthesize complex organic molecules in the dark reaction.The steps in photosynthesis that occur on the thylakoid membranes of the chloroplast and convert solar energy to the chemical energy of ATP and NADPH, evolving oxygen in the process.

Light The electromagnetic radiation produced by the sun that heats and illuminates the earth.

Light-dependent reactions The reactions of the first stage of photosynthesis, in which light energy is captured by chlorophyll molecules and converted to chemical energy stored in ATP and NADPH molecules.

Light-independent reactions The carbon-fixing reactions of the second stage of photosynthesis; energy stored in ATP and NADPH by the light-dependent reactions is used to reduce carbon from carbon dioxide to simple sugars; light is not required for these reactions.

Lignified Impregnated with lignin, such as the secondary cell walls of woody plants.

Lignin A complicated organic molecule found as an important constituent of many secondary cell walls; imparts strength and rigidity to the cellulose microfibrils.

Lignite Soft, brown to black coal of low BTU value with original plant components still discernible.

Ligulae Name given to the lobes at the tip of the labium: usually divided into glossae and paraglossae.

Limax form Form of pseudopodial movement in which entire organism moves without extending a discrete pseudopodium.

Limbic system A ring of structures around the thalamus which play a major role in pain as well as other types of behaviour. The limbic system includes the hypothalamus, hippocampus, amygdala, septum and cingulum. The limbic system plays an important role in pain at the level of motivation to avoid it. It thus operates at a slightly higher level than the reticular formation with strong connections to the thalamus and cortex.

Limestone A carbonate sedimentary rock composed of more than 50 percent of the mineral calcium carbonate (CaCO3).

Limiting factors Chemical or physical factors that limit the existence, growth, abundance, or distribution of an organism. The principle of limiting factors states that for each physical factor in the environment, there are both minimum and maximum limits (tolerance limits) beyond which a particular species cannot survive. The single factor closest to a tolerance limit for a given species at a given time is the critical factor.

Limiting resource An essential factor whose short supply prevents the growth of a population.

Limnetic zone The open-water zone of a lake or pond; extends to the depth of effective light penetration.

Limnologist A scientist who studies freshwater biology.

Limnology The study of river system ecology and life.

Limy Describes sediments, soils, or rocks that contain a significant amount of lime (calcium oxide, CaO).

Lineage Any continuous line of descent; any series of organisms connected by reproduction by parent of offspring.

Linear epitope An epitope formed by a series of amino acids which are adjacent to each other within the primary structure of the protein. Such epitopes can be successfully modelled by synthetic peptides, but comprise only a small proportion of all epitopes. The minimal epitope size is about 5 amino acid residues. Also called a sequential epitope.

Lingual Pertaining to the tongue.

Linkage group Genes linked to the same chromosome that tend to be inherited together.

Linkage map A genetic map based on the frequencies of recombination between markers during crossing over of homologous chromosomes. The greater the frequency of recombination between two genetic markers, the farther apart they are assumed to be.

Linkage The proximity of two or more *markers* (e.g., *genes, RFLP* markers) on a *chromosome*; the closer together the markers are, the lower the probability that they will be separated during DNA repair or replication processes (binary fission in *prokaryotes, mitosis* or *meiosis* in *eukaryotes*), and hence the greater the probability that they will be inherited together.

Linked gene Gene located on the same chromosome as another gene.

Linker A short *oligodeoxyribonucleotide*, usually representing a specific *restriction endonuclease recognition sequence*, which may be *ligated* onto the termini of a DNA molecule to facilitate cloning. Following the ligation reaction, the product is digested with the endonuclease, generating a DNA fragment with the desired *sticky* or *blunt ends*.

Lip cell Thin-walled cells that interrupt the annulus in fern sporangia.

Lipase An enzyme that accelerates the hydrolysis or synthesis of fats.

Lipid A fat, oil, or fatlike compound that usually has fatty acids in its molecular structure. An organic compound consisting mainly of carbon and hydrogen atoms linked by nonpolar covalent bonds. Examples include fats, waxes, phospholipids, and steroids that are insoluble in water.One of the four classes of organic macromolecules. Lipids function in the long-term storage of biochemical energy, insulation, structure and control. Examples of lipids include the fats, waxes, oils and steroids (e.g. testosterone, cholesterol).

Lipofectin A commercially marketed liposome suspension which is mixed with DNA or RNA to facilitate uptake of the nucleic acid by eukaryotic cells.

Lipopolysaccharide A molecule containing both lipid and polysaccharide, which is important in the outer membrane of the gram-negative cell wall.

Lipoprotein A protein bonded to a lipid; includes the low-density lipoproteins (LDLs) and high-density lipoproteins (HDLs) that transport fats and cholesterol in blood.

Liposome Artificial lipoid particle used to deliver antiparasitic drugs directly to macrophages (which eat the particles).

Liquid metal fast breeder A nuclear power plant that converts uranium 238 to plutonium 239; thus, it creates more nuclear fuel than it consumes. Because of the extreme heat and density of its core, the breeder uses liquid sodium as a coolant.

Listeriosis A sporadic disease of animals and humans, particularly those who are immunocompromised or pregnant, caused by the bacterium Listeria monocytogenes.

Lithogenous sediment A marine sediment that is derived from the breakdown, or weathering, of rocks.

Lithosphere The crust and the top part of the mantle that covers the earth's surface. It is broken into separate lithospheric plates.

Lithotroph An organism that uses reduced inorganic compounds as its electron source.

Litter Dead organic material such as branches, tree trunks, and dry grass that accumulates on the floor of a forest. Litter acts as additional fuel in a forest fire, producing increased destruction.Leaf litter, or forest litter, is the detritus of fallen leaves and bark which accumulate in forests.

Littoral Pertaining to the shore. Noun, that portion of the sea floor between the extent of high and low tides, intertidal; in lakes, the shallow part from the shore to the lakeward limit of aquatic plants.

Littoral zone The shallow-water zone of a lake or pond; light penetrates to the bottom, and the area is occupied by rooted plants such as water lilies, rushes, and sedges.

Liver A large gland of a dark-red color. It carries out many vital functions such as the formation of urea, manufacture of plasma proteins, synthesis of amino acids, synthesis and storage of glycogen, and many others.

Liver diverticulum Endodermal rudiment of the liver and associated glands and ducts.

Liverwort A group of the division Bryophyta charcterized by a small, inconspicuous, liver-shaped thallus and that lives in moist environments.

Llemma Bract in a grass flower.

Llien The spleen.

Llogistic population growth A model describing population growth that levels off as population size approaches carrying capacity.

Lmago The adult insect (Plural imagines)

Loam A mixture of sand-, silt-, and clay-sized soil particles.

Lobe-finned Fish with muscular fins containing large jointed bones that attach to the body; one of the two main types of bony fish.

Lobopodium A blunt, lobelike pseudopodium that is commonly tubular, and is composed of both ectoplasm and endoplasm.

Local chemical messenger A chemical that acts on nearby cells.

Local mate competition Competition for mates that may arise among siblings depending on the population size and operational sex ratio and that may cause the evolution of skewed offspring sex ratios.

Local resource competition model In female philopatric primates, females giving birth to a disproportionate number of males that will disperse and not compete with each other, or high-ranking females giving birth to a disproportionate number of daughters that are able to compete for resources effectively.

Localize Determination of the original position *(locus)* of a *gene* or other *marker* on a chromosome.

Locomotion-fear dichotomy A process during the imprinting process in young birds whereby they initially are more likely to locomote and follow a stimulus object and later develop a fear toward objects; thus, imprinting occurs when the locomotor tendency is high and fear of objects is low.

Loculi Shallow, suckerlike depressions in an adhesive organ of a flatworm.

Locus The position on a *chromosome* of a *gene* or other chromosome *marker*; also, the DNA at that position. The use of *locus* is sometimes restricted to mean regions of DNA that are *expressed.*

Loess An extremely fertile, fine-grained loam composed of quartz, feldspar, hornblende, mica, and clay; deposited by the wind during the Pleistocene Age. It originates in arid regions from glacial outwash.

Logistic equation. A mathematical expression describing an idealized sigmoid curve of population growth.

Logistic growth Growth rates regulated by internal and external factors that establish an equilibrium with environmental resources; species may grow exponentially when resources are unlimited but slowly as the carrying capacity is reached.

Logistic growth model A model of population growth in which the population initially grows at an exponential rate until it is limited by some factor; then, the population enters a slower growth phase and eventually stabilizes.

Logistic population growth Population growth that is modulated by the population size relative to carrying capacity. Population growth declines as population approaches carrying capacity, and is negative when population size is greater than carrying capacity

Long-day plant A plant that flowers, usually in late spring or early summer, only when the light period is longer than a critical length.

Longevity The length or duration of life; compare to survivorship.

Longitude the angular distance of a position on the earth's surface east or west of the Greenwich Prime Meridian; measured in degrees.

Longshore current A current moving parallel to a shoreline

Loop of Henle The long hairpin turn, with a descending and ascending limb, of the renal tubule in the vertebrate kidney; functions in water and salt reabsorption.

Looping movement The type of locomotion exhibited by leeches and some insect larvae whereby they alternate temporary points of attachment to move forward.

Loose connective tissue The type of tissue in which the matrix contains strong, flexible fibers of the protein collagen that are interwoven with fine, elastic, and reticular fibers.

Lophocyte Type of sponge amebocyte that secretes bundles of fibrils.

Lophophore Complex ring of hollow tentacles used as a feeding organ. The tentacles are covered by cilia, which generate a current to bring food particles into the mouth. The structure is only found in the brachiopods, phoronids, and bryozoans.

Lophotrichous A cell with a cluster of flagella at one or both ends.

Lorica A vase-shaped or cup-shaped outer covering. Found in many protists, including some flagellates, ciliates, chrysophytes, and choanoflagellates, as well as in some animal cells.

Loricifera A phylum of aschelminths. The most recent animal phylum to be described; members are commonly called loriciferans.

Lotic ecosystems Flowing water ecosystems. They include brooks, streams, and rivers.

Lotic Of or pertaining to running water, such as a brook or river.

Lottery hypothesis The hypothesis that unpredictability plays a key role in the development of communities.

Low oxygen diffusion environment An aquatic environment in which microorganisms are surrounded by deep water layers that limit oxygen diffusion to the cell surface. In contrast, microorganisms in thin water films have good oxygen transfer from air to the cell surface.

Low tide the lowest level reached by the falling tide

Low-head hydropower Small-scale hydro technology that can extract energy from small headwater dams; causes much less ecological damage.

Lowland Large area of relatively low relief, usually applied to coastal regions that do not rise high above sea level. *upland, highland* – ant.

Low-quality energy Diffuse, dispersed energy at a low temperature that is difficult to gather and use for productive purposes.

LPS-binding protein A special plasma protein that binds bacterial lipopolysaccharides and then attaches to receptors on monocytes, macrophages, and other cells. This triggers the release of IL-1 and other cytokines that stimulate the development of fever and additional endotoxin effects.

Lsd Lysergic acid diethylamide; a synthetic hallucinogenic derivative of lysergic acid.

Lubrication Helping two surfaces to slide over each other.

Luciferase Enzyme which activates luciferin to produce bioluminescence.

Luciferin Compound whose activated form emits light.

Lulus locally unwanted land uses such As toxic waste dumps, incinerators, smelters, airports, freeways, and other sources of environmental, economic, or social degradation.

Lumbar Relating to or near the loins or lower back.

Lumbosacral Lumb = the loin; sacrum = sacrum.

Lumen A term used when referring to the cavity or passageway within a tubular structure (ex. the interior space of a seminiferous tubule).

Lungs Sac-like structures of varying complexity where blood and air exchange oxygen and carbon dioxide; connected to the outside by a series of tubes and a small opening. In humans, the lungs are situated in the thoracic cavity and consist of the internal airways, the alveoli, the pulmonary circulatory vessels, and elastic connective tissues.

Lungfish A type of lobe-finned fish that breathe by a modified swim bladder (or lung) as well as by gills.

Lungworm The common name for any of various parasitic nematodes that infect the lungs of vertebrates, principally domestic animals.

Lunules Small, suckerlike discs on the anterior margin of some copepods in the family Caligidae, functioning as organs of adhesion.

Luteal phase The second half of the ovarian cycle when the corpus luteum is formed; occurs after ovulation.

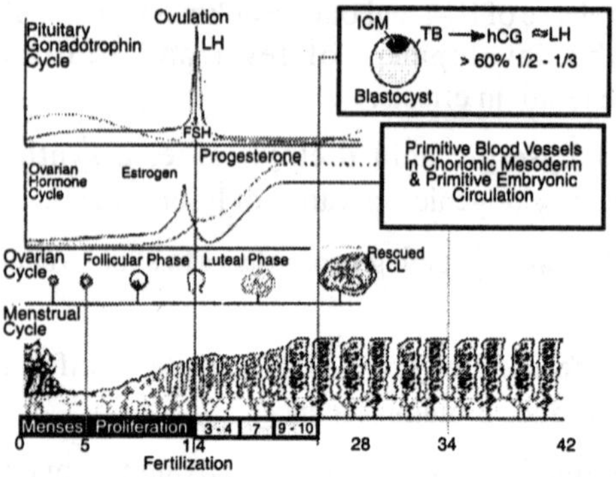

Fig. *Luteal phase*

Luteinizing hormone A hormone secreted by the anterior pituitary gland that stimulates the secretion of testosterone in men and estrogen in women.

Lycine One of 20 aminoacids common in proteins. It is a common amino acid of collagen and like proline must be hydroxylated by ascorbic acid in order to allow the formation of bonds which will hold the triple helix together.

Lycophora Ten-hooked larva that hatches from the egg of a cestodarian tapeworm. Also called a decacanth.

Lycophyta A division of "lower" (i.e., non-seed-producing) vascular plants, commonly referred to as club and spike mosses.

Lyme disease A tick-borne disease caused by the spirochete Borrella burgdorferi.

Lymph The colorless fluid, derived from interstitial fluid, in the lymphatic system of vertebrate animals.

Lymph hearts Contractile enlargements of vessels that pump lymph back into the veins; found in fish, amphibians, and reptiles.

Lymph node A mass of spongy tissues, separated into compartments; located throughout the lymphatic system, lymph nodes remove dead cells, debris, and foreign particles from the circulation; also are sites at which foreign antigens are displayed to immunologically active cells.

Lymphatic circulation A secondary circulatory system that collects fluids from between the cells and returns it to the main circulatory system; the circulation of the lymphatic system, which is part of the immune system.

Lymphatic system A network of glands and vessels that drain interstitial fluid from body tissues and return it to the circulatory system.

Lymphocytes White cells involved in the immune response. B lymphocytes are so called because they mature in bone while T lymphocytes mature in the thymus. Both cells look alike until they recognise a foreign antigen. The B cell starts to make antibodies while the T lymphocytes accumulate vesicles loaded with cytotoxic agents. On contact with a foreign cell, the lymphocytes changes shape so that all it vesicles are pointed at the enemy. The release of cytotoxic agents need to be carefully controlled. One of the methods by which the enemy cell is killed is by agents which make holes in its cell membrane. Enemy cells maybe bacteria, or the bodies own cells which have ingested viruses or they may be cancer cells, or the cells of transplanted organs .

Lymphogranuloma venereum A sexually transmitted disease caused by Chlamydia trachomatis serotypes L1-L3, which affect the lymph organs in the genital area.

Lymphokine A biologically active glycoprotein (e.g., IL-1) secreted by activated lymphocytes, especially sensitized T cells. It acts as an intercellular mediator of the immune response and transmits growth, differentiation, and behavioral signals.

Lyon hypothesis Idea proposed by Mary Lyon that mammalian females inactivate one or the other X-chromosome during early embryogenesis. This deactivated chromosome forms the Barr body.

Lysergic acid diethylamide (LSD) A derivative of lysergic acid alkaloids, first isolated from ergot; strongly hallucinogenic.

Lysis Disintegration of a cell by rupture of its plasma membrane.

Lysogenic bacteria Bacteria carrying a bacteriophage integrated into the bacterial chromosome. The virus may subsequently set up an active cycle of infection, causing lysis of the bacterial cells.

Lysogenic cycle A type of phage replication cycle in which the viral genome becomes incorporated into the bacterial host chromosome as a prophage.

Lysogeny The state in which a phage genome remains within the bacterial host cell after infection and reproduces along with it rather than taking control of the host and destroying it.

Lysosomes Small membrane bound vesicles in the cytoplasm of cells which contain toxic enzymes. When a cell dies, these membranes rupture and the enzymes are released. They break down the cells structure, and the debris is removed. The lysosome also contains cytokines which summon inflammatory cells and stimulate inflammation. The contents of lysosomes can be released by macrophages and neutrophils both to kill bacteria and viruses, and to stimulate inflammation.

Lysozyme An enzyme in perspiration, tears, and saliva that attacks bacterial cell walls.

Lytic cycle A type of viral replication cycle resulting in the release of new phages by death or lysis of the host cell.

M

M cell Specialized cell of the intestinal mucosa and other sites, such as the urogenital tract, that delivers the antigen from the apical face of the cell to lymphocytes clustered within the pocket in its basolateral face.

M phase The mitotic phase of the cell cycle, which includes mitosis and cytokinesis.

M13 A bacteriophage which infects certain strains of *E. coli* . The salient feature of this phage is that it packages only a single strand of DNA into its capsid. If the investigator has inserted some heterologous DNA into the M13 genome, copious quantities of single-stranded DNA can subsequently be isolated from the phage capsids. M13 is often used to generate templates for DNA sequencing.

Macroalgae The multicellular brown, red, and green algae.

Macrobenthos (macrofauna or macroflora) Benthic organisms (animals or plants) whose shortest dimension is greater than or equal to 0.5 mm

Macroelements Essential elements needed for plant growth in relatively large amounts; C, H, O, P, K, N, S, Ca, Fe, Mg.

Macroevolution The combination of events associated with the origin, diversification, extinction, and interactions of organisms which produced the species that currently inhabit the Earth. Large scale evolutionary change such as the evolution of new species (or even higher taxa) and extinction of species.

Macrofauna Animals whose shortest dimension is greater than or equal to 0.5 mm

Macrofibril An aggregation of microfibrils in the cell wall, visible with the light microscope.

Macrogamete The larger of the two gamete types in a heterogametic organism, considered the female gamete.

Macrolide antibiotic An antibiotic containing a macrolide ring, a large lactone ring with multiple keto and hydroxyl groups, linked to one or more sugars.

Macromeres Large blastomeres found in the vegetal hemisphere of eggs formed by unequal cleavage (ex. the large blastomeres of the vegetal pole of the amphibian embryo).

Macromolecules Large molecules made up of many small organic molecules that are often referred to as monomers; e.g., carbohydrates, lipids, proteins, and nucleic acids. Macromolecules are polymers of monomers.

Macromolecule vaccine A vaccine made of specific, purified macromolecules derived from pathogenic microorganisms.

Macronucleus In ciliates, the large nucleus that carries up to several hundred copies of the genome and controls metabolism and asexual reproduction.

Macronutrient An essential nutrient for which an animal has a large minimal daily requirement (greater than 100 mg) (e.g., calcium, phosphorus, magnesium, potassium, sodium, and chloride along with carbohydrates, lipids, and proteins).

Macroparasite Large parasite that does not multiply in the host of interest. Examples are cestodes, trematodes, and most nematodes in their definitive hosts.

Macrophage The name for a large mononuclear phagocytic cell, present in blood, lymph, and other tissues. Macrophages are derived from monocytes. They phagocytose and destroy pathogens; some macrophages also activate B cells and T cells. Cells derived from monocytes which have the ability to phagocytose foreign particles and dead tissue and to move through tissue, or to remain fixed in one place. There are many macrophages in the spleen where they remove dead red blood cells from the circulation.

Macrophage migration inhibitory factor Cytokine released by sensitized lymphocytes that tends to inhibit migration of macrophages in the immediate vicinity, thus contributing to accumulation of larger numbers of macrophages close to the site of MIF release.

Macrophyte An individual alga large enough to be seen easily with the unaided eye

Macroplankton The component of the plankton that consists of large organisms 2 to 20 cm in size.

Macropores Large pore spaces caused by invertebrates and larger animals, including reptiles and mammals, that permeate the soil. They drain water not held by capillarity.

Macrorestriction map Map depicting the order of and distance between sites at which *restriction enzymes* cleave *chromosomes*.

Macroscopic Objects or organisms that are large enough to be seen with the naked eye.

Madreporite Sievelike structure that is the intake for the water-vascular system of echinoderms.

Maduromycosis A subcutaneous fungal infection caused by Madurella mycetoma; also termed an eumycotic mycetoma.

Madurose The sugar derivative 3-O-methyl-D-galactose, which is characteristic of several actinomycete genera that are collectively called maduromycetes.

Mafic Term used to describe the amount of dark-colored iron and magnesium minerals in an igneous rock. Complement of felsic.

Maggot A vermiform larva; a larva without legs and without well-developed head capsule.

Magma Molten rock generated within the Earth; forms intrusive (solidifies below the surface) and extrusive (solidifies above the surface) igneous rocks.

Magnetic anomalies Magnetic bands in the sea floor that run parallel to the mid-ocean ridge .

Magnetic confinement A technique for enclosing a nuclear fusion reaction in a powerful magnetic field inside a vacuum chamber.

Magnetite A magnetic, iron-containing material.

Magnetosomes Magnetite particles in magnetotactic bacteria that are tiny magnets and allow the bacteria to orient themselves in magnetic fields.

Magnoliid Any member of the basal assemblage of flowering plants.

Mainstream flow The flow in a part of the fluid (e.g., in a tidal creek) that is well above the bottom or well away from a surface and essentially not under the influence of the boundary layer.

Maintenance energy The energy a cell requires simply to maintain itself or remain alive and functioning properly. It does not include the energy needed for either growth or reproduction.

Major histocompatibility complex (MHC) A large

set of cell surface molecules in each individual, encoded by a family of genes, that serves as a unique biochemical marker of individual identity. It can trigger T-cell responses that may lead to rejection of transplanted tissues and organs. MHC molecules are also involved in the regulation of the immune response and the interactions between immune cells.

Major mineral Mineral requirement needed in relatively large amounts.

Major salivary glands Are three large glands on each side of the face; the parotid, submandibular and submaxillary.

Malacostraca The class of crustaceans whose members are characterized by having appendages modified for crawling along the substrate, as in lobsters, crayfish, and crabs. Alternatively, the abdomen and body appendages may be used in swimming, as in shrimp.

Malaria A serious infectious illness caused by the parasitic protozoan Plasmodium. Malaria is characterized by bouts of high chills and fever that occur at regular intervals.

Malate Four-carbon organic acid found with aspartate to be the first products of the C4 pathway of photosynthesis.

Male dominance polygyny A mating system in which males compete and acquire dominance ranks that influence their access to females, with higher-ranking males often having greater mating activity.

Male In organisms with separate sexes, the one which produces sperm.

Male parasitism The permanent attachment of a male to a female in some deep-sea fishes.

Malignant tumor A mass of cancerous cells that have left their site of origin, migrated through the body, invaded normal tissues, and are growing out of control.

Malleus One of the three bones of the inner ear. The others are the stapes and the incus.

Malleus The ossicle attached to the tympanum in middle ears of mammals.

Malnourishment A nutritional imbalance caused by lack of specific dietary components or inability to absorb or utilize essential nutrients.

Malpighian tubules Blind tubules opening into the hindgut of nearly all insects and some myriapods and arachnids, and functioning primarily as excretory organs.

Malthusian growth A population explosion followed by a population crash; also called irruptive growth.

Mamelon Ventral, serrated projection on the ventral surface of a male nematode of the family Syphaciidae. Its function is unknown.

Mammalia The class of vertebrates whose members are at least partially covered by hair, have specialized teeth, and are endothermic. Young are nursed from mammary glands. The mammals.

Mammal-like reptiles Group of Permian-Triassic reptiles having some possible mammalian features, notably a more prominent dentary (tooth-bearing) bone and reduction of the incus and malleus (which are part of the reptilian jaw along with the dentary). The mammal-like reptiles are thought to have been the reptile group from which the mammals later evolved.

Mammals Vertebrates that have hair and mammary glands.

Mammary gland The breast. In female mammals, the mammary glands produce and secrete milk to nourish developing young.

Man and biosphere (MAB) program A design for nature preserves that divides protected areas into zones with different purposes. Critical ecosystem functions and endangered wildlife are protected in a central core region where limited scientific study is the only human access allowed. Ecotourism and research facilities are located in a relatively pristine buffer zone around the core, while sustainable resource harvesting and permanent habitation are allowed in multiple-use peripheral regions.

Mandibles Third pair of appendages from the anterior in Crustacea; second pair in Insecta; primarily function in feeding; derived from appendages on primitive fourth (first postoral) somite.

Mandibulate Having mandibles suited for biting and chewing.

Mangal a tropical community of mangrove plants and associated organisms

Manganese nodules Lumps of minerals (including manganese and other valuable minerals) that are found on the sea floor beyond the continental shelf.

Mange Dermatitis caused by species of mites, often designated with the causative organism. For example, Sarcoptes causes sarcoptic mange.

Mangroves Shrubs and trees that live along the seashore in tropical and subtropical regions and tolerate inundation by seawater.

Mannoxylic Wood in which there is a great deal of parenchyma tissue among the xylem is called mannoxylic. Cycads and pteridosperms have mannoxylic wood. Contrast with pycnoxylic.

Mantle (1) The semi-liquid region between the crust and core of the earth.

(2) The outer layer of tissue that secretes the shell of molluscs.

Mantle cavity The space between the mantle and the visceral mass of molluscs.

Mantle In mollusks, a membranous or muscular structure that surrounds the visceral mass and secretes a shell if one is present.

Manubrium The portion projecting from the oral side of a jellyfish medusa, bearing the mouth; oral cone; presternum or anterior part of sternum; handle-like part of malleus of ear.

Marasmus Malnutrition, especially of infants, caused by a diet deficient in both calories and protein.

Marburg viral hemorrhagic fever An acute infection caused by a virus that produces varying degrees of hemorrhage, shock, and sometimes death.

Margin The edge of a flattened structure; in leaves, the lateral edge of the blade.

Marginal bodies Sensory pits or short tentacles between the marginal loculi of the opisthaptor of an aspidogastrean trematode.

Marginal cell One of a number of cells bordering the front margin of the wing in the outer region.

Marginal meristem The meristem along the margin of a leaf primordium responsible with the formation and shape of the leaf.

Marginal value theorem A model that predicts when an organism should cease foraging in one patch and travel to another.

Marginal zone Region near the equator of the amphibian blastula, where the animal and vegetal hemispheres meet; gastrulation begins among these cells.

Mariculture The culture of marine organisms. In open mariculture (or semi-culture), organisms are cultured in natural environments; in closed mariculture (or intensive mariculture), organisms are cultured in a controlled environment.

Marine Living in or pertaining to the sea.

Marine archaeology The discovery, salvage, and interpretation of material remains of humankind's past that have been preserved in the sea.

Marine biome The aquatic biome consisting of waters containing 3.5% salt on average; includes the oceans and covers more than 70% of the Earth's surface; divided into benthic and pelagic zones.

Marine natural products Chemical compounds that are obtained from marine organisms.

Marine protected area A conservation geographic unit designed to protect crucial communities and to provide reproductive reserves for fisheries that hopefully will disperse over wider areas.

Marine snow Fragile organic aggregates, resulting from the collision of dissolved organic molecules or from the degradation of gelatinous substances such as larvacean houses. Usually enriched with microorganisms.

Marine terrace A platform of marine deposits (typically sand, silt, gravel) sloping gently seaward. Such a platform may be exposed along the coast, forming cliffs, due to uplift and/or the lowering of sea level, e.g., Marine terraces of coastal Southern California.

Maritime culture A human culture with a very close relationship with the sea.

Marker An identifiable physical location on a *chromosome* (e.g., *restriction enzyme cutting site, gene*) whose inheritance can be monitored. Markers can be expressed regions of DNA (genes) or some segment of DNA with no known coding function but whose pattern of inheritance can be determined.

Marl A loose, crumbly deposit consisting of clay and calcium carbonate and formed in marine or freshwater conditions.

Marrara Nasopharyngeal blockage by a parasite. Also called halzoun.

Marsh Wetland without trees; in North America, this type of land is characterized by cattails and rushes.

Marsupials Pouched mammals. The young develop internally, but are born while in an embryonic state and remain in a pouch on the mother's abdomen until development is complete; this group includes kangaroos, koalas, and opossums. One of the three reproductive "strategies" of living mammals g-laying and placental being the other two), marsupials finish development in a pouch or under hairy coverings attached to the mother.

Mashing The process in which cereals are mixed with water and incubated in order to degrade their complex carbohydrates (e.g., starch) to more readily usable forms such as simple sugars.

Mass The fundamental unit of measurement equivalent to the weight of a substance when compared with the weight of hydrogen.

Mass burn Incineration of unsorted solid waste.

Mass extinction A time during which extinction rates are generally accelerated so that more than 50% of all species then living become extinct; results in a marked decrease in the diversity of organisms. Mass extinctions are thought to have occurred numerous times in Earth history, often from a variety of reasons: impacts, tectonism, changes in primary productivity of the seas, etc.

Mass number Sum of the number of protons and neutrons in the nucleus of an atom; designated by a superscript to the upper left of the elemental symbol.

Mass reared Produced in large numbers, as in natural enemies produced for release programs.

Mass spectrometry A chemical technique used to investigate aspects of the structure of molecules.

Mast cell A type of noncirculating white blood cell, found in connective tissue, that is the major protagonist in allergic reactions; when an allergen binds to complementary antibodies on the surface of a mast cell, large amounts of histamine are released from the cell.

Mastax The pharyngeal apparatus of rotifers used for grinding ingested food.

Mastication The process of preparing food for swallowing and digestion by chewing it.

Mastigoneme Small hair-like filaments found on the "hairy" flagellum of the Chromista.

Mastigont Axoneme of a cilium or flagellum together with its basal fibrils and organelles.

Mastigophora The protozoan subphylum where members possess one or more flagella that are used for locomotion; autotrophic, heterotrophic, or saprozoic.

Mastitis Infection of the udder of cattle.

Mastoid A bony process of the skull.

Maternal condition model In polygynous mammals, females that are high ranking, well nourished, or otherwise in good condition giving birth to a greater proportion of sons since their sons will outreproduce their daughters and vice versa for females that are in poor condition.

Mating system The species-typical pattern of mate finding, reproduction, and parenting of offspring.

Matric potential The water potential component caused by the attraction of water molecules to a hydrophilic matrix.

Matrix Comes from the Latin word "mater"meaning mother. It is a structure which encloses or holds something within it. Cells are held or enclosed in a matrix of fibres, water and large molecules called the ground substance .

Matrix vesicles Small bubble-like structures containing calcium-binding phospholipids and alkaline phosphatase. Crystals of hydroxyapatite from inside the vesicle which ruptures and releasing crystals into the surrounding osteoid or predentine so as to start mineralising it.

Matter Anything that has mass and occupies space.

Matter cycling The flow of matter through various organisms and the physical environment of an ecosystem.

Maturation To complete the natural development of an animal system. Improvement in the performance of behavior occurs as parts of the nervous system and other structures complete development.

Maurer's clefts Blotches on the surface of an erythrocyte infected with Plasmodium falciparum.

Maxillae (second maxillae)Fifth pair of appendages in Crustacea, primarily feeding in function, derived from appendages on primitive sixth (third postoral) somite; homologous to labium in insects. The maxillae of insects are the third pair of head appendages, homologous to maxillules of Crustacea.

Maxillary Concerning or to do with the maxillae.

Maxillary gland In malacostracan crustaceans, the excretory organs are located near the maxillary segments and are termed maxillary glands.

Maxillary teeth Teeth located on the two principal dermal bones of the upper jaw in reptiles and amphibians.

Maxilliped One of the pairs of head appendages located just posterior to the maxilla in crustaceans, a thoracic appendage that has become incorporated into the feeding mouthparts.

Maxillopodan eye Naupliar eye of crustacean class Maxillopoda; has a tapetum (crystalline reflective layer).

Maxillules (first maxillae)Fourth pair of appendages in Crustacea, primarily feeding in function; derived from appendages on primitive fifth (second postoral) somite; homologous to maxillae in insects.

Maximum sustainable yield (MSY) The maximum number of a food or game population that can be harvested without harming the population's ability to grow back.

Meadow An open clearing in a landscape in which low-growing, herbaceous plants dominate.

Mean growth rate constant (k) The rate of microbial population growth expressed in terms of the number of generations per unit time.

Measles (rubeola) A highly contagious skin disease that is endemic throughout the world. It is caused by a morbilli virus in the family Paramyxoviridae, which enters the body through the respiratory tract or through the conjunctiva.

Mechanical control Control of pests by physical means such as the use of screens or row covers.

Mechanical vector Vector that transmits disease organism by mechanical means only. Contrast with biological vector.

Mechanoreceptor A sensory receptor that detects physical deformations in the body's environment associated with pressure, touch, stretch, motion, and sound.

Meconium A dark green mucilaginous material in the intestine of the full-term fetus, being a mixture of the secretions of the intestinal glands and some amniotic fluid.

Media The longitudinal vein running through the central region of the wing in most insects: often the 4th and abbreviated to M.

Medial hinge point cells The midline neural plate cells in birds derived from the anterior midline of Hensen's node and epiblast region that are involved in the intial bending of the neural plate during neurulation. These cells become wedge-shaped, and are anchored to the notochord.

Medial On or near the plane that divides a bilateral animal into mirror images.

Median (parietal) eye A photoreceptor located middorsally on the head of some vertebrates; it is associated with the vertebrate epithalamus.

Median oviduct In insects, the single duct formed by the merging of the paired lateral oviducts; this duct opens posteriorly into a genital chamber or vagina.

Mediastinum The membranes enclosing the tissues and organs between the two lungs.

Mediated transport. Transport of a substance across a cell membrane mediated by a carrier molecule in the membrane.

Medical mycology The discipline that deals with the fungi that cause human disease.

Mediterranean climate areas Specialized landscapes with warm, dry summers; cool, wet winters; many unique plant and animal adaptations; and many levels of endemism.

Medulla 1. A term referring to the central portion of certain organs; e.g., the medulla oblongata of the brain and the adrenal medulla, which synthesizes epinephrine and norepinephrine.

2. In more common usage, the area in the brain that regulates breathing, heartbeat, blood pressure and similar activities.

Medulla oblongata The lowest part of the vertebrate brain; a swelling of the hindbrain dorsal to the anterior spinal cord that controls autonomic, homeostatic functions, including breathing, heart and blood vessel activity, swallowing, digestion, and vomiting.

Medusa The floating, flattened, mouth-down version of the cnidarian body plan. The alternate form is the polyp.

Megabase (MB) Unit of length for DNA fragments equal to 1 million *nucleotides* and roughly equal to 1 *cM.*

Megacolon Flabby distended colon caused by chronic Chagas' disease.

Megaesophagus Distended esophagus caused by chronic Chagas' disease.

Megagametogenesis The production of megagametes (large gametes) from megaspores in the ovules of angiosperms.

Megagametophyte The gametophyte stage containing eight haploid nuclei within the embryo sac.

Megakarocytes Cells found in the bone marrow that produce platelets.

Megalopolis Also known as a megacity or supercity; megalopolis indicates an urban area with more than 10 million inhabitants.

Megapascal (MPa) A unit of pressure equivalent to 10 atmospheres of pressure.

Megaplankton The component of the plankton that consists of very large organisms over 20 cm in size.

Megasporangium Sporangium that contains megaspores.

Megaspore In plants which are heterosporous, the larger kind of spore is called a megaspore; it usually germinates into a female (egg-producing) gametophyte. Contrast with microspore. Four haploid cells produced by meiosis in the ovule of a flower. Usually, three of these cells degenerate, with the remaining cell becoming the female gametophyte phase of the plant's life cycle. Large (palynologists consider the megaspores to generally be above 200 micrometers in diameter) spores that develop into the megagametophyte, which in turn produces eggs.

Megaspore mother cell Cells that undergo meiosis to produce megaspores.

Megasporocyte Also known as the megaspore mother cell; this diploid cell undergoes meiosis to produce four haploid megaspores in the ovules of angiosperms.

Megasporogenesis The process of megaspores (large spores) being produced via meiosis in the ovules of angiosperms.

Megawatt (MW) Unit of electrical power equal to one thousand kilowatts or one million watts.

Mehlis' glands Unicellular mucous and serous glands surrounding the ootype of a flatworm.

Meiobenthos Benthic organisms (animals or plants) whose shortest dimension is less than 0.5 mm but greater than or equal to 0.1 mm.

Meiofauna Animals whose shortest dimension is less than 0.5 mm but greater than or equal to 0.1 mm.

Meiosis A two-stage type of cell division in sexually reproducing organisms. In meiosis, a diploid cell divides to produce four haploid cells, each with half the original chromosome content. For this reason, meiosis is often called a "reduction division". In organisms with a diploid life cycles, the products of meiosis are usually called gametes. In organisms with an alternation of generations, the products of meiosis are caled spores.

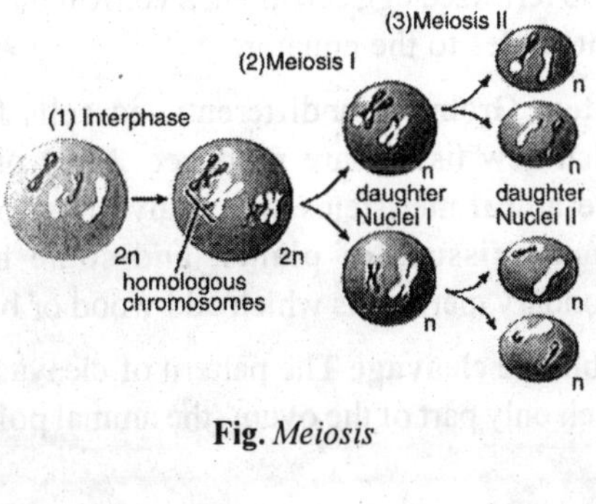

Fig. *Meiosis*

Meiotic Cell division that reduces the number of chromosomes from diploid to haploid.

Meissner's corpuscles Sensory receptors concentrated in the epidermis of the fingers and lips that make these areas very sensitive to touch.

Mélange A body of rocks consisting of large blocks (mappable size) of different rocks jumbled together with little continuity of contacts.

Melanin A pigment that gives the skin color and protects the underlying layers against damage by ultraviolet light; produced by melanocytes in the inner layer of the epidermis.

Melanism Dark coloration of the skin, fur, or feathers because of the presence of the pigment melanin.

Melanocytes The cells in the inner layer of the epidermis that produce melanin.

Melanophore Black or brown chromatophore containing melanin.

Melatonin A hormone secreted by the pineal gland. Functions in regulating photoperiodicity.

Melon A fatty structure on the forehead of some cetaceans that is used to direct sound waves emitted during echolocation.

Melting The dissociation of a duplex nucleic acid molecule into single strands, usually by increasing temperature.

Melting temperature The temperature at which double-stranded DNA separates into individual strands; it is dependent on the G 1 C content of the DNA and is used to compare genetic material in microbial taxonomy.

Membrane Semi-fluid structure which bounds all cells, and partitions the interior of eukaryotic cells. It consists primarily of two lipid layers, with proteins "dissolved" in the lipids.

Membrane attack complex (MAC) The complex complement components (C5b-C9) that create a pore in the plasma membrane of a target cell and leads to cell lysis. C9 probably forms most of the actual pore.

Membrane filter technique The use of a thin porous

filter made from cellulose acetate or some other polymer to collect microorganisms from water, air, and food.

Membrane potential The charge difference between the cytoplasm and extracellular fluid in all cells, due to the differential distribution of ions. Membrane potential affects the activity of excitable cells and the transmembrane movement of all charged substances.

Membrane-attack complex (MAC) A large cylindrical multiprotein complex formed by the complement system; kills invading microorganisms by embedding in their plasma membrane, creating a pore through which fluid flows, ultimately causing the cell to burst.

Membrane-disrupting exotoxin A type of exotoxin that lyses host cells by disrupting the integrity of the plasma membrane.

Membranelle Short, transverse rows of cilia, fused at their bases, serving to move food particles toward the oral groove of a protozoan.

Membranous Thin and transparent (in reference to wings); thin and pliable (in reference to integument).

Memory The capacity of an organism to form lasting connections based on past experiences; the ability to store and use information.

Memory B cell A lymphocyte capable of initiating the antibody-mediated immune response upon detection of a specific antigen molecule for which it is genetically programmed. It circulates freely in the blood and lymph and may live for years.

Memory cell A clone of long-lived lymphocytes, formed during the primary immune response, that remains in a lymph node until activated by exposure to the same antigen that triggered its formation. Activated memory cells mount the secondary immune response.

Meninges Any of three membranes (arachnoid, dura mater, pia mater) that envelop the vertebrate brain and spinal cord. Also, solid connective tissue sheath enclosing the central nervous system of some vertebrates.

Meningitis A condition that refers to inflammation of the brain or spinal cord meninges (membranes). The disease can be divided into bacterial (septic) meningitis (caused by bacteria) and aseptic meningitis syndrome (caused by nonbacterial sources).

Menisci A crescent-shaped fibrocartilage.

Meniscus The curved top surface of a column of liquid.

Menopause In the human female, that time of life when ovulation ceases; cessation of the menstrual cycle.

Menses The period of shedding of the lining (endometrium) of the uterus and associated fluids if an ovum is not fertilized, most notably in primates.

Menstrual cycle The reproductive cycle of female primates. The cycle consists of four stages: menstrual, where the uterine lining is shed; proliferative, where the endometrium begins to thicken; secretory, where the endometrium thickness is maintained and so becomes suitable for implantation; and ischemic, where if fertilization does not occur, the endometrium begins to break down.

Menstruation The process in which the uterine endometrium breaks down and sheds cells, resulting in bleeding; occurs approximately once a month. The first day marks the beginning of the menstrual and ovarian cycles.

Menziesia A type of shrub

Meridional In reference to cleavage, the splitting of the fertilized egg cell along a vertical plane at right angles to the equator.

Meristem Group of undifferentiated cells from which new tissues are produced. Most plants have apical meristems which give rise to the primary tissues of plants, and some have secondary meristems which add wood or bark.

Meroblastic cleavage The pattern of cleavage in which only part of the ovum (the animal pole) is divided into cells. It is also called incomplete cleavage and is usually observed in embryos with large amounts of yolk.

Merogones Egg fragments (in sea urchins) that can divide and develop, even if they have only a haploid nucleus.

Merophytes Group of cells which have all been produced from the same initial cell. Leaves and stems in particular are often built from specific patterns of merophytes.

Meroplankton Organisms that spend part of their time in the plankton but also spend time in the benthos (e.g., planktonic larvae of benthic invertebrates)

Merostomata The class of arthropods whose members are aquatic and possess book gills on the opisthosoma. Eurypterids (extinct) and horseshoe crabs.

Merozoite A very small trophozoite at the stage just after cytokinesis has been completed in multiple fission of a protozoan.

Mesencephalon The mid region of the developing embryonic brain. Synonym: midbrain.

Mesenchyme Embryonic connective tissue derived from mesoderm and neural crest cells. It later forms predominantly connective tissues, but may also differentiate into epithelial, muscular and nervous tissues.

Mesendoderm A tissue layer that gives rise to both mesoderm and endoderm.

Mesenterial filament Any of the long, thin tubes attached to the gut of corals and other cnidarians that are involved in digestion and absorption.

Mesenteries Infoldings of gastroderm and mesoglea extending into the gastrovascular cavity of cnidarians; sheets of peritoneum from which the digestive tract is suspended in coelomates.

Mesial drift A gradual movement of all the posterior teeth in a mesial direction. It occurs only if there has been interproximal wear between the teeth. The drift is not a passive one however, as it has been shown that during chewing, the bite force has a a mesial component.

Mesic Referring to a region that receives adequate precipitation to maintain biological productivity.

Mesocarp The middle layer of the fruit wall, located between the exocarp and endocarp.

Mesocercaria Juvenile stage of the digenetic trematode Alaria. It is an unencysted form between the cercaria and the metacercaria.

Mesocoel Middle body coelomic compartment in some deuterostomes, anterior in lophophorates, corresponds to hydrocoel in echinoderms.

Mesoderm The middle of the three cell layers which form, as the clump of early embryonic cells begins to differentiate. The mesoderm will form the muscles, blood system, connective tissue, including bone and dentine, the kidneys and the dermis of the skin. The other two layers are the ectoderm and theendoderm.

Mesoglea The layer of jellylike or cement material between the epidermis and gastrodermis in cnidarians and ctenophores; also may refer to jellylike matrix between epithelial layers in sponges.

Mesohyl Gelatinous matrix surrounding sponge cells; mesoglea, mesenchyme.

Mesokaryotic Nuclear condition unique to the dinoflagellates in which the chromosomes remain permanently condensed.

Mesolecithal Pertaining to a zygote (or ovum) having a moderate amount of yolk concentrated in the vegetal pole.

Mesomere Blastomere exhibiting a size intermediate between a macromere and a micromere.

Mesonephros The second excretory organ formed in the embryo, caudal to the pronephros. The pronephric ducts become mesonephric or Wolffian ducts. This kidney is fully functional in adult amphibians and fish. The duct portion becomes the epididymis of the testes in amniotes.

Mesonotum The dorsal surface of the 2nd thoracic segment - the mesothorax: usually the largest thoracic sclerite.

Mesopelagic zone Also called the "twilight zone" of the ocean, this area from 200m to 1000m in depth, is bordered by the photic zone above and darkness below. It's in this zone where we start to see bioluminescence on all sorts of animals.

Mesophyll Photosynthetic middle layer in the blade of a leaf; typically composed of palisade and spongy parenchyma.

Mesophyte A plant that grows in soils that contain moderate or intermediate amounts of moisture.

Mesophytic leaves The leaves of plants that grow under moderately humid conditions with abundant soil and water.

Mesoplankton The component of the plankton that consists of organisms 0.2 to 2 mm in size .

Mesopleuron The sclerite or sclerites making up the side wall of the mesothorax.

Mesoscutellum Hindmost of the three major divisions of the mesonotum, often triangular or shield-shaped: usually abbreviated to scutellum.

Mesoscutum The middle and usually the largest division of the mesonotum.

Mesosome The portion of the body in lophophorates and some deuterostomes that contains the mesocoel.

Mesosphere The atmospheric layer above the stratosphere and below the thermosphere; the middle layer; temperatures are usually very low.

Mesosternum The ventral surface or sclerite of the mesothorax.

Mesothorax The middle of the three thoracic segments of an insect; usually contains the second pair of legs and the first pair of wings.

Mesovarium A supporting mesentary (membrane) from the uterus that holds the mammalian ovaries in place.

Mesozoic era The period of geologic time beginning 245 million years ago and ending 65 million years ago; the age of the dinosaurs and cycads, the Mesozoic falls between the Paleozoic and Cenozoic Eras and includes the Triassic, Jurassic, and Cretaceous Periods.

Messenger RNA A single-stranded polyribonucleotide; formed in the nucleus from a DNA template and carries the transcribed genetic code to the ribosome where the genetic code is translated into protein.

Metabolic channeling The localization of metabolites and enzymes in different parts of a cell.

Metabolic control engineering Modification of the controls for biosynthetic pathways without altering the pathways themselves in order to improve process efficiency.

Metabolic energy That energy obtained from ATP produced in metabolism.

Metabolic pathway A series of individual chemical reactions in a living system that combine to perform one or more important functions. The product of one reaction in a pathway serves as the substrate for the following reaction. Examples include glycolysis and Kreb's cycle.

Metabolic pathway engineering (MPE) The use of molecular techniques to improve the efficiency of pathways that synthesize specific products.

Metabolic rate The overall rate of biochemical reactions in an organism. Often estimated by rate of oxygen consumption in aerobes

Metabolise To obtain energy by breaking down a molecule into smaller components, as in respiration or to store energy by building a more complex molecule out of smaller components. It always takes place in a number of stages, controlled by enzymes. Each step in the process follows a predictable "metabolic" pathway for that reaction.

Metabolism A group of processes that includes digestion, production of energy (respiration), and synthesis of molecules and structures by organisms; the sum of the constructive (anabolic) and destructive (catabolic) processes.

Metabolites Chemical substances required in metabolism.

Metacentric Chromosome with centromere at or near the middle.

Metacercaria Stage between the cercaria and adult in the life cycle of most digenetic trematodes; usually encysted and quiescent.

Metacestode Developmental stage of a cestode after metamorphosis of the oncosphere; a juvenile cestode.

Metachromatic granules Granules of polyphosphate in the cytoplasm of some bacteria that appear a different color when stained with a blue basic dye. They are storage reservoirs for phosphate. Sometimes called volutin granules.

Metacoel Posterior coelomic compartment in some deuterostomes and lophophorates; corresponds to somatocoel in echinoderms.

Metacommunication Communication about communication, where one signal changes the meaning of signals that follow.

Metacryptozoite Merozoite developed from a cryptozoite.

Metacyclic Stage in the life cycle of a parasite that is infective to its definitive host.

Metacyst Cystic stage of a parasite that is infective to a host.

Metamerism Division of the body along the anteroposterior axis into a serial succession of segments, each of which contains identical or similar representatives of all the organ systems of the body; primitively in arthropods, including externally a pair of appendages and internally a pair of nerve ganglia, a pair of nephridia, a pair of gonads, paired blood vessels and nerves, and a portion of the digestive and muscular systems.

Metamorphic rock Any rock derived from other rocks by chemical, mineralogical and structural changes resulting from pressure, temperature or shearing stress.

Metamorphism The process of altering the chemical or mineralogical composition of a rock through different amounts of heat and pressure below the surface of the Earth; metamorphose- v; metamorphic - adj.

Metamorphosis Name given to the changes that take place during an insect's life as it turns from a young animal to an adult. These changes may be gradual and not too large, as in the grasshopper, and metamorphosis is then said to be partial or incomplete. On the other hand, the changes may be much greater and they may take place in one big step - as in the butterflies and moths, which change from caterpillars to adults during the pupal stage. Metamorphosis of this kind is said to be complete.

Metanauplius Later naupliar larvae of some crustaceans; that is, occurring after several naupliar stages but before another larval type or preadult in the developmental sequence.

Metanephridium An excretory organ found in many invertebrates; it consists of a tubule that has one end opening at the body wall and the opposite end in the form of a funnel-like structure that opens to the body cavity.

Metanephros Embryonic renal organs of vertebrates arising behind the mesonephros; the functional kidney of reptiles, birds, and mammals. It is drained from a ureter.

Metanotum The dorsal surface of the metathorax. It is often very small and its sub-divisions are usually obscured.

Metaphase Stage in mitosis when chromosomes become aligned in the middle of the cell and firmly attached to the mitotic spindle but have not yet segregated toward opposite poles.

Metaplasia The transformation of one differentiated cell type into another.

Metapleuron The sclerite or sclerites making up the side wall of the metathorax.

Metapodosoma Portion of the podosoma that bears the third and fourth pairs of legs of a tick or mite.

Metapolar cells Posterior tier of cells in the calotte of a dicyemid mesozoan.

Metapopulation A group of interconnected subpopulations, usually of subequal size. The features of individuals now founnd in one subpopulation might have been determined by conditions affecting them when they were located in another subpopulation.

Metasome The portion of the body in lophophorates and some deuterostomes that contains the metacoel.

Metastasis The process in which cancer cells break away from the original tumor mass and establish new tumor sites elsewhere in the body.

Metasternum The ventral surface or sclerite of the metathorax.

Metatarsus The basal segment of the tarsus or foot: usually the largest segment.

Metathorax The posterior of the three segments of an insect thorax; it usually contains the third pair of walking legs and the second pair of wings (Arthropoda).

Metazoa The Subkingdom of the Kingdom Animalia that includes all mulitcellular animals developing from embryos and possessing differentiated cells which form into tissues.

Methanogen A bacterium that obtains energy from CO_2 and H_2 and forms methane.

Methanogens Strictly anaerobic archaeons that derive energy by converting CO_2, H_2, formate, acetate, and other compounds to either methane or methane and CO_2.

Methionine The amino acid coded for by the initiation codon; all polypeptides begin with methionine, although post-translational reactions may remove it.

Methylotroph A bacterium that uses reduced one-carbon compounds such as methane and methanol as its sole source of carbon and energy.

Metraterm Muscular, distended termination of the uterus of a digenetic trematode.

Micelles Structures formed when bile salts surround digested fats in order to enable the water-insoluble fats to be absorbed by the epithelial cells lining the small intestine.

Michaelis constant A kinetic constant for an enzyme reaction that equals the substrate concentration required for the enzyme to operate at half maximal velocity.

Microaerophile A microorganism that requires low levels of oxygen for growth, around 2 to 10%, but is damaged by normal atmospheric oxygen levels.

Microarray technology Profiling of gene expression by measuring binding of RNA from growing cells to an array of function-specific oligonucleotides attached to an inert surface.

Microatoll A small, flat, atoll-shaped coral structure generally found in protected coral lagoons.

Microbe A microscopic organism.

Microbenthos Benthic organisms (animals or plants) whose shortest dimension is less than 0.1 mm

Microbial dietary adjuvant A substance added to the diet to stimulate specific microbial processes and populations.

Microbial ecology The study of microorganisms in their natural environments, with a major emphasis on physical conditions, processes, and interactions that occur on the scale of individual microbial cells.

Microbial insecticide A preparation of microorganisms (e.g., viruses or bacteria) or their products used to suppress insect pest populations.

Microbial loop The mineralization of organic matter synthesized by photosynthetic phytoplankton through the activity of microorganisms such as bacteria and protozoa. This process "loops" minerals and carbon dioxide back for reuse by the primary producers and makes the organic matter unavailable to higher consumers.

Microbial mat A firm structure of layered

microorganisms with complementary physiological activities that can develop on surfaces in aquatic environments.

Microbiology The study of organisms that are usually too small to be seen with the naked eye. Special techniques are required to isolate and grow them.

Microbivory The use of microorganisms as a food source by organisms that can ingest or phagocytose them.

Microbody Membrane-bound organelles that are the site of certain enzymatic conversions; example: peroxisomes and glyoxisomes.

Microenvironment The immediate environment surrounding a microbial cell or other structure, such as a root.

Microevolution A small-scale evolutionary event such as the formation of a species from a preexisting one or the divergence of reproductively isolated populations into new species.

Microfauna Animals whose shortest dimension is less than 0.1 mm

Microfibril An elongated strand of cellulose molecules.

Microfilaments Are the smallest filaments of the cytoskeleton. The filaments are made of hundreds of actin molecules, stacked in a line. They can be quickly broken down or extended. Actin filamanents give the cell its shape and help to change it. When cells move, in embryology and repair, or just during the continual patrol of lymphocytes, they must hold on to something in order to crawl. The filaments serve to anchor one part of the cell, via fibronectin to the cell matrix, so the rest of the cell can pull itself towards the anchor. Lymphocytes move and scavenge by sticking out arms and feet to help them crawl and engulf foreign particles. Muscle cells change their shape by using stacks of actin filaments as a ladder on which myosin climbs.

Microfilaria The prelarval stage of filarial worms. Found in the blood of humans and the tissues of the vector.

Microfossil A very small fossil, best studied with the aid of a microscope, e.g. foraminifera, radiolarians, and small vertebrate fossils such as teeth. *macrofossil*– ant.

Microgamete The smaller of the two gamete types in a heterogametic organism, considered the male gamete.

Microgametocyte Cell that gives rise to microgametes.

Microgametogenesis The production of microgametes (small gametes) from microspores in the anthers of flowering plants.

Microgametophyte Stage of the plant life cycle that develops from or within a microspore. The microgametophyte produces sperm in specialized structures known as antheridia.

Micro-hydro generators Small power generators that can be used in low-level rivers to provide economical power for four to six homes, freeing them from dependence on large utilities and foreign energy supplies.

Micromeres Small blastomeres; in the sea urchin embryo, the four small cells that result from the fourth cleavage of the vegetal tier of cells (half the cells resulting from that tier's cleavage).

Micron (m) One one-thousandth of a millimeter; about 1/25,000 of an inch. Now largely replaced by micrometer (mm).

Microneme One of the types of structures composing the apical complex in the phylum Apicomplexa, slender and elongate, leading to the anterior and thought to function in host cell penetration.

Microniscus Intermediate larval stages of the isopod suborder Epicaridea, parasitic on free-living copepods.

Micronucleus The smaller of the two nuclei in ciliate protozoa. Micronuclei are diploid and involved only in genetic recombination and the regeneration of macronuclei.

Micronutrients Elements that are required by plants in very small quantities, but are toxic in large quantities: iron, manganese, molybdenum, copper, boron, zinc, and chloride.

Microorganisms Single celled animals which may range from the very small viruses, through bacteria and fungi to almost visible protozoa.

Microparasite Small (or very small) parasite that multiplies within the host of interest. Examples are protistan and prokaryotic parasites.

Microphyll A kind of leaf, specifically one which has a single, unbranched vein in it. Microphylls are only found in the lycophytes.

Microplankton The component of the plankton that consists of organisms 20 to 200 mm (0.02 to 0.2 mm) in size.

Micropores Spaces between soil particles that hold water by means of capillary forces.

Micropyle A minute opening or group of openings into the insect egg through which the spermatozoa enter in fertilization.

Microsatellite A microsatellite is a simple sequence repeat (SSR). It might be a homopolymer ('...TTTTTTT...'), a dinucleotide repeat ('....CACACACACACACA.....'), trinucleotide repeat ('....AGTAGTAGTAGTAGT...') etc. Due to polymerase slip (a.k.a. polymerase chatter), during DNA replication there is a slight chance these repeat sequences may become altered; copies of the repeat unit can be created or removed. Consequently, the exact number of repeat units may differ between unrelated individuals. Considering all the known microsatellite markers, no two individuals are identical. This is the basis for forensic DNA identification and for testing of familial relationships (e.g. paternity testing).

Microscopic Objects or organisms that are too small to be seen with the naked eye.

Microscopy Examination with a microscope.

Microspora The protozoan phylum characterized by members having unicellular spores; intracellular parasites in nearly all major animal groups. Examples: microsporeans (Nosema).

Microsporangia Structures of the sporophyte in which microspores are produced by meiosis. In flowering plants the microsporangia are known as anther sacs.

Microspores Four haploid cells produced by the meiotic division in the pollen sacs of flowers or microsporangia of gymnosperms. Microspores undergo mitotic division and become encased in a thick protective wall to form pollen grains. Small, size usually less than 200 micrometers, spores produced by meiosis. Microspores either germinate into the male gametophyte or have the male gametophyte develop inside the microspore wall.

Microspore mother cell Cells in the microsporangium that undergo meiosis to produce microspores. In flowering plants the microspore is known as the pollen grain, and contains a three-celled male.

Microsporidia Single-celled life forms, related to Protozoa.

Microsporocyte Also known as the microspore mother cell, this diploid cell undergoes meiosis to produce four haploid microspores in the anthers of angiosperms.

Microsporogenesis The process of microspores (small spores) being produced via meiosis in the anthers of angiosperms.

Microsporophyll A leaflike structure giving rise to one or more microsporangia.

Microthrix (microtriches) Minute projections of the tegument of a cestode.

Microtopography Topography on a smaller scale

Microtrichia Minute hairs projecting from the integument, they are formed around cellular filaments.

Microtubules Microtubules are cylindrical tubes, 20-25 nm in diameter.

They are composed of subunits of the protein

tubulin — these subunits are termed alpha and beta. Microtubules act as a scaffold to determine cell shape, and provide a set of "tracks" for cell organelles and vesicles to move on.

Microtubules also form the spindle fibers for separating chromosomes during mitosis. When arranged in geometric patterns inside flagella and cilia, they are used for locomotion.

Microvilli Hair-like projections on the surface of the epithelial cells of the villi in the small intestine; increase the surface area of the intestine to improve absorption of digested nutrients.

Mictic eggs Pertaining to the haploid eggs of rotifers. If it isn't fertilized, the egg develops parthenogenetically into a male; if fertilized, mictic eggs secrete a heavy shell and become dormant, hatching in the spring into amictic females.

Midbrain A network of neurons that connects with the forebrain and relays sensory signals to other integrating centers.

Middens A large refuse heap containing such materials discarded materials, food remains, bones, and so on.

Middle lamella A thin layer of adhesive extracellular material, primarily pectins, found between the primary walls of adjacent young plant cells.

Midgut The middle part of the alimentary canal and the main site of digestion and absorption.

Mid-ocean ridge The continuous chain of volcanic submarine mountains that extends around the earth. It includes the Mid-Atlantic Ridge and East Pacific Rise.

Midrib The central large vein of a leaf.

Midwater Pertaining to the mesopelagic zone.

Miescher's tubules Sarcocysts; tissue cysts of Sarcocystis.

Migration 1. In general, any movement of an animal from one location to another.

2. specifically, a predictable, recurring group movement that is characteristic of the members of a given species, and that occurs regularly in response to seasonal changes in temperature, precipitation, and so on.

Milankovitch cycles Periodic variations in tilt, eccentricity, and wobble in the earth's orbit; Milutin Milankovitch suggested that it is responsible for cyclic weather changes.

Miliary tuberculosis An acute form of tuberculosis in which small tubercles are formed in a number of organs of the body because of dissemination of M. tuberculosis throughout the body by the bloodstream. Also known as reactivation tuberculosis.

Milpa agriculture An ancient farming system in which small patches of tropical forests are cleared and perennial polyculture agriculture practiced and is then followed by many years of fallow to restore the soil; also called swidden agriculture.

Mimicry A phenomenon in which one species benefits by a superficial resemblance to an unrelated species. A predator or species of prey may gain a significant advantage through mimicry.

Mine tailings The decomposed outcrop of a bed or vein of valuable material.

Mineral In nutrition, one of many chemical elements, other than carbon, hydrogen, oxygen, and nitrogen, that an organism requires for proper body functioning.

Mineralization The release of inorganic nutrients from organic matter during microbial growth and metabolism.

Mineralocorticoids A group of steroid hormones produced by the adrenal cortex that are important in maintaining electrolyte balance.

Minerals Trace elements required for normal metabolism, as components of cells and tissues, and in nerve conduction and muscle contraction.

Minimal inhibitory concentration (MIC) The lowest concentration of a drug that will prevent the growth of a particular microorganism.

Minimum dynamic area The amount of suitable habitat needed to sustain a viable population.

Minimum viable population size (MVP) The smallest number of individuals needed to perpetuate a population.

Minisatellite An oligonucleotide, used as a gene probe, which is produced to reflect the repeats of nucleotide base sequences seen in many eukaryotic genes. Minisatellites are typically between 13 -19 nucleotides in length.

Minor salivary glands Are microscopic glands under the surface of the oral mucosa. They are found throughout the lining mucosa of the mouth including the tongue.

Minus, or negative, strand The virus nucleic acid strand that is complementary in base sequence to the viral mRNA.

Miracidium The ciliated, free-swimming first stage larva of a digenean trematode that undergoes further development in the body of a snail.

Missense mutation The most common type of mutation involving a base-pair substitution within a gene that changes a codon, but the new codon makes sense in that it still codes for an amino acid.

Mississippian Part of the Carboniferous period, beginning around 360 million years ago.

Mite Any of several minute invertebrates belonging to the phylum Arthropoda, class Arachnida.

Mitigation Repairing or rehabilitating a damaged ecosystem or compensating for damage by providing a substitute or replacement area.

Mitochondria A cell organelle found in eucaryotic cells which produces ATP as a product of the kreb cycle and the electron transport system Cells requiring large amounts of energy, such as secreting odontoblasts, have large numbers of mitochondria. Mitochondria are self replicating and contain their own DNA for this purpose. Self-replicating organelles, bounded by two membranes, that are found in the cytoplasm of all eukaryotic cells and produce cellular energy in the the form of ATP via the oxidative phosphorilation reactions.

Mitochondrial DNA The DNA in the mitochondria of eukaryotes. This DNA codes for many of the organelles functions and is replicated indepently of the nuclear DNA. Mitochondrial DNA (mtDNA) appears to contain no introns.

Mitochondrial matrix The compartment of the mitochondrion enclosed by the inner membrane and containing enzymes and substrates for the Krebs cycle.

Mitochondrion Complex organelle found in most eukaryotes; believed to be descended from free-living bacteria that established a symbiotic relationship with a primitive eukaryote. Mitochondria are the site of most of the energy production in most eukaryotes; they require oxygen to function. See: double membrane.

Mitosis A process of nuclear division in eukaryotic cells conventionally divided into five stages: prophase, prometaphase, metaphase, anaphase, and telophase. Mitosis conserves chromosome number by equally allocating replicated chromosomes to each of the daughter nuclei.A type of nuclear division which leads to two identical daughter nuclei in the resulting daughter cells. Occurs in eukaryotes constantly during the processes of growth and repair. Asexual cell reproduction.

Mitotic activity The rate of mitosis, and hence cell division. The mitotic activity of basal cells in an epithelium must match the rate of desquamation.

Mitotic apparatus Collectively, the asters, spindle, centrioles, and microtubules of a dividing cell.

Mitotic spindle A network of microtubules formed during prophase. Some microtubules attach to the centromeres of the chromosomes and help draw the chromosomes apart during anaphase.

Mitotic spindle apparatus Microtubule-based structure present during mitosis to which chromosomes attach and are separated toward opposite poles of the dividing cell.

Mixed acid fermentation A type of fermentation carried out by members of the family Enterobacteriaceae in which ethanol and a complex mixture of organic acids are produced.

Mixed perennial polyculture Growing a mixture of different perennial crop species (where the same plant persists for more than one year) together in the same plot; imitates the diversity of a natural system and is often more stable and more suitable for sustainable agriculture than monoculture of annual plants.

Mixed semidiurnal tide A tidal pattern with two successive high tides of different heights each day.

Mixing depth The water depth to which wind energy evenly mixes the water column

Mixoplankton Planktonic organisms that can be classified at several trophic levels. For example, some ciliates can be photosynthetic but also can ingest other plankton and are heterotrophic.

Mixotrophic Refers to microorganisms that combine autotrophic and heterotrophic metabolic processes (they use inorganic electron sources and organic carbon sources).

Mmetazoa Multicellular animals.

Modal action pattern (MAP) A spatiotemporal behavior pattern that is common to members of a species; different individuals perform the pattern in a recognizably similar fashion.

Moderately stratified estuary An estuary in which seaward flow of surface low-salinity water and moderate vertical mixing result in a modest vertical salinity gradient

Modern synthesis A comprehensive theory of evolution emphasizing natural selection, gradualism, and populations as the fundamental units of evolutionary change; also called neo-Darwinism.

Modification A change, as in morphology, usually associated with a functional advantage.

Modified atmosphere packaging (MAP) Addition of gases such as nitrogen and carbon dioxide to packaged foods in order to inhibit the growth of spoilage organisms.

Molarity A common measure of solute concentration, referring to the number of moles of solute in 1 L of solution.

Mold Any of a large group of fungi that cause mold or moldiness and that exist as multicellular filamentous colonies; also the deposit or growth caused by such fungi. Molds typically do not produce macroscopic fruiting bodies.Fossils formed when the sediment surrounding a buried organism hardens. When the organism decays, its impression is left in the rock and can be seen if the rock is broken open.

Mole Avogadro's number (6.02×10^{23} atoms) of a substance.One gram molecular weight of any substance; that is, the molecular weight of any substance in grams.

Molecular biology The branch of biology concerned with the structure and development of biological systems in terms of the biochemistry of their molecular components. Molecular biology overlaps a wide variety of other disciplines in use of molecular techniques e.g. PCR, RFLP analysis, etc.

Molecular chaperones Proteins that aid in the proper folding of unfolded polypeptides or partly denatured proteins and often also help transport proteins across membranes.

Molecular chronometers Nucleic acid and protein sequences that gradually change over time in a random fashion and at a steady rate, and which therefore can be used to determine phylogenetic relationships.

Molecular formula A type of molecular notation indicating only the quantity of the constituent atoms.

Molecular genetics The study of the biochemical structure and function of DNA.

Molecular weight size marker A piece of DNA of known size, or a mixture of pieces with known size, used on electrophoresis gels to determine the size of unknown DNA's by comparison.

Genetic marker A known site on the chromosome. It might for example be the site of a locus with some recognizable phenotype, or it may be the site of a polymorphism that can be experimentally discerned.

Molecular weight The sum of the atomic weights of the constituent atoms in a molecule.

Molecules Units of two or more atoms held together by chemical bonds. The combination of atoms by chemical bonds with the component atoms in definite porportions, such as water (two H to one O).

Mollusca The phylum of coelomate animals whose members possess a head-foot, visceral mass, mantle, and mantle cavity. Most molluscs also possess a radula and a shell. The molluscs. Bivalves, snails, octopuses, and related animals.

Molt A process in arthropods, such as crustaceans and insects, as well as in some vertebrates, such as snakes, whereby the organism sheds its outer exoskeleton or skin periodically as it grows. Also refers to changing the pelage or feathers in a mammal or bird.

Monera Prokaryotic kingdom that includes (in the most widely accepted classification system) archaebacteria, eubacteria, and cyanobacteria. Members of this kingdom were among the first forms of life over 3.5 billion years ago.

Monestrous A species in which the female is receptive for only a few days once each year.

Moniliform (of antennae) Composed of bead-like segments, each well separated from the next.

Monitored, retrievable storage Holding wastes in underground mines or secure surface facilities such as dry casks where they can be watched and repackaged, if necessary.

Monkey wrenching Environmental sabotage such as driving large spikes in trees to protect them from loggers, vandalizing construction equipment, pulling up survey stakes for unwanted developments, and destroying billboards.

Monocarpic Denoting a plant that flowers only once.

Monoclonal antibody An antibody with very specific and often unique binding specificity which is secreted by a biologically cloned line of plasmacytoma cells in the absence of other related antibodies with different binding specificities. Differs from *polyclonal antibodies*, which are mixed populations of antibody molecules such as may be present in a serum specimen, within which many different individual antibodies have different binding specificities.

Monocots One of the two major types of flowering plants; characterized by having a single cotyledon, floral organs arranged in threesd or multiples of three, and parallel-veined leaves; include grasses, cattails, lilies, and palm trees. One of the two major groups in the Angiosperms, monocots are characterized by having a single seed leaf (cotyledon), flower parts in 3's or multiples of 3, monoaperturate pollen (although some dicots also have this feature), parallel veins in their leaves, and scattered vascular bundles in their stems.

Monoculture agroforestry Intensive planting of a single species; an efficient wood production approach, but one that encourages pests and disease infestations and conflicts with wildlife habitat or recreation uses.

Monoculture The growth of only one species in a given area; such as a cornfield or other agricultural field.

Monocytes Remain in the blood only a short time before they migrate into the tissues particularly where dead tissue must be removed, where they are calledmacrophages.

Monocyte-macrophage system The collection of fixed phagocytic cells (including macrophages, monocytes, and specialized endothelial cells) located in the liver, spleen, lymph nodes, and bone marrow. This system is an important component of the host's general nonspecific defense against pathogens.

Monoecious Hermaphroditic; an individual that contains reproductive systems of both sexes.

Monoestrus Refers to a mammal having only one estrus cycle in a year. Elk are a good example.

Monogamy A mating pattern in which males and females are paired one to one for at least one reproductive season; usually associated with parental care by both parents. (Monogamous) - Prolonged and exclusive pair bond with a single member of the opposite sex in order to raise young.

Monogenea The class of Platyhelminthes that has members that are called monogenetic flukes; most ectoparasites on vertebrates (usually on fishes, occasionally on turtles, frogs, copepods, squids); one life-cycle form in only one host; bear an opisthaptor. Examples: Disocotyle, Gyrodactylus, Polystoma.

Monogononta A class of rotifers containing members that possess one ovary; mastax not designed for grinding; produce mictic and amictic eggs. Example: Notommata.

Monohybrid cross A mating between two individuals heterozygous for one particular trait.

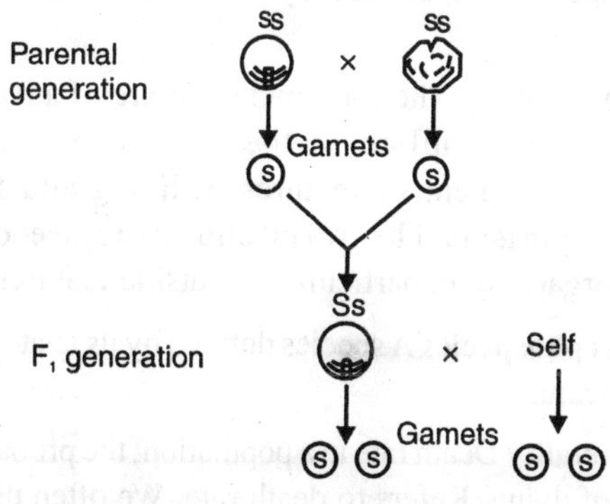

Fig. *Monohybrid cross*

Monokine A generic term for a cytokine produced by mononuclear phagocytes (macrophages or monocytes).

Monomer An organic chemical unit linked to other units (usually by a covalent bond formed by the removal of water) to produce a larger molecule (macromolecule) known as a polymer.

Monophagous Feeding upon only one kind of food, for example one species or one genus of plants.

Monophyletic Term applied to a group of organisms which includes the most recent common ancestor of all of its members and all of the descendants of that most recent common ancestor. A monophyletic group is called a clade.

Monophyletic group A group of organisms descended from a common ancestor. For example: your immediate family may be considered such a group, being descended from a common ancestral group (grandparents, etc.).

Monophyly The condition that a taxon or other group of organisms contains the most recent common ancestor of the group and all of its descendants; contrasts with polyphyly and paraphyly.

Monoplacophora The class of molluscs whose members have a single, arched shell; a broad, flat foot; and certain serially repeated structures. Neopilina.

Monoplacophorans A small group of molluscs that are thought by some to represent a link with invertebrates that show segmentation.

Monosaccharide A simple sugar that cannot be decomposed into smaller sugar molecules; the most common are pentoses (such as ribose) and hexoses (such as glucose). Simple carbohydrates, usually with a five- or six-carbon skeleton; e.g., glucose and fructose. A carbohydrate composed of a single sugar unit, such as glucose, ribose, deoxyribose, etc.

Monostome Fluke that lacks a ventral sucker.

Monotreme An egg-laying mammal, represented by the platypus and echidna. Egg-laying mammals; e.g., the spiny anteater and the duck-billed platypus.

Monotrichous Having a single flagellum.

Monounsaturated fat Composed of fatty acid chains in which there is only a single C-C double bond; examples: canola oil and olive oil.

Monoxenous Living within a single host during a parasite's life cycle.

Monozoic Tapeworms with a single proglottid, do not undergo strobilation to form chain of proglottids.

Monozygotic Twins that arise from a single zygote, hence, two genetically identical individuals.

Monsoon A seasonal reversal of wind patterns caused by the different heating and cooling rates of the oceans and continents.

Monsoonal Describes a climate pattern with a wind system that changes direction with the seasons; this pattern is dominant over the Arabian Sea and Southeast Asia.

Montane coniferous forests Coniferous forests of the mountains consisting of belts of different forest communities along an altitudinal gradient.

Moraine A mound or ridge of sediment deposited by a glacier; lateral **moraine**- deposited to the side of a glacier; terminal moraine- deposited to the front of a glacier; ground moraine- deposited on the land surface.

Moral agents Beings capable of making distinctions between right and wrong and acting accordingly. Those whom we hold responsible for their actions.

Moral subjects Beings that are not capable of distinguishing between right or wrong or that are not able to act on moral principles and yet are susceptible of being wronged by others. This category assumes some rights or inherent values in moral subjects that gives us duties or obligations towards them.

Morbidity Illness or disease.

Morbidity rate Measures the number of individuals who become ill as a result of a particular disease within a susceptible population during a specific time period.

Mordant A substance that helps fix dye on or in a cell.

More-developed countries (MDC) Industrialized nations characterized by high per capita incomes, low birth and death rates, low population growth rates, and high levels of industrialization and urbanization.

Morph A distinct phenotypic variant within a population.

Morphine A pain-relieving and addictive compound derived from the opium poppy (Papaver somniferum).

Morphogen A substance, such as bicoid protein, that provides positional information in the form of a concentration gradient along an embryonic axis.

Morphogenic fieled An environment in which the shape or pattern of a developing organ is determined.

Morphological Pertaining to form and structure, at any level of organization.

Morphological convergence The evolution of basically dissimilar structures to serve a common function. For example: the wings of birds and insects.

Morphological species concept The idea that species are defined by measurable anatomical criteria.

Morphology The form and structure of anything, usually applied to the shapes, parts, and arrangement of features in living and fossil organisms.The form and structure of an organism, in particular its outside features.

Morphospecies A species defined by its anatomical features.

Mortality Death rate in a population; the probability of dying. Refers to death rate. We often use the term "mortality factors", which are factors that cause the death of an animal. Starvation could be a mortality factor.

Mortality rate The ratio of the number of deaths

from a given disease to the total number of cases of the disease.

Morula A stage in the embryonic development of some animals that consists of a solid ball of cells.

Mosaic An organism whose tissues are made up of two or more genetically different types.

Mosaic cleavage. Embryonic development characterized by independent differentiation of each part of the embryo; determinate cleavage.

Mosaic development A pattern of development, such as that of a mollusk, in which the early blastomeres each give rise to a specific part of the embryo. In some animals, the fate of the blastomeres is established in the zygote.

Mosaic evolution A pattern of evolution where all features of an organism do not evolve at the same rate. Some characteristics are retained from the ancestral condition while others are more recently evolved.

Moss A group of terrestrial, nonvascular plants; the dominant plant body is the gametophyte with the sporophyte embedded in it.

Most probable number (MPN) The statistical estimation of the probable population in a liquid by diluting and determining end points for microbial growth.

Motif A recurring pattern of short sequence of DNA, RNA, or protein, that usually serves as a recognition site or active site. The same motif can be found in a variety of types of organisms.

Motile Able to move oneself about, capable of self-locomotion.

Motility Ability to move.

Motivation Internal processes that arouse and direct behavior. Today this refers more to examination of the consequences of behavior and feedback to the animal concerning the consequences of those actions.

Motoneurone Nerve cells with their cell bodies in the brain stem or spinal cord, which transmit impulses along their axons to effector organs, including endocrine, exocrine glands and muscles fibres. The axons of most motoneurones have many branches, each of which ends at a neuromuscular junction. The group of muscle cells innervated by one motoneurone is called a motor unit

Motor pathways The portion of the peripheral nervous system that carries signals from the central nervous system to the muscles and glands.

Motor neurons Neurons that receive signals from interneurons and transfer the signals to effector cells that produce a response. Nerve cells connected to a muscle or gland. Sometimes also known as effector neurons.

Motor output A response to the stimuli received by the nervous system. A signal is transmitted to organs that can convert the signals into action, such as movement or a change in heart rate.

Motor units Consist of a motor neuron with a group of muscle fibers; form the units into which skeletal muscles are organized; enable muscles to contract on a graded basis.

Moult To moult is to shed the outer covering of the body - the exoskeleton.

Mountain tundra That portion of tundra vegetation confined to alpine meadows. Low-growing grasses, sedges, and forbs with a very short growing season; permafrost is typical.

Mouth Front opening of the digestive tract, into which food is taken for digestion. In flatworms, the mouth is the only opening into the digestive cavity, and is located on the "belly" of the worm.

MPF maturation/mitosis promoting factor Heterodimeric protein that initiates prophase of mitosis and alters the activity of diverse proteins that function in mitosis by phosphorylation; active MPF contains both cyclin B (large) and cyclin-dependent kinase 1 (small) subunits. Activity is regulated partly by the synthesis and degradation of cyclin B subunit.

MRNA A single-stranded polynucleotide molecule. The sugar in the nucleotides is ribose, and the bases are adenine, uracil, cytosine and guanine. The order of the bases of mRNA is determined during transcription of the DNA code in the preduction of mRNA. The mRNA code determines the order of amino acid linking during translation at the ribosome.

MTOC (microtubule organizing center) MTOCs are bundles of protein tubes which may be found at the base of a eukaryotic flagellum. In animals, they also function in creating the arrays of microtubules that pull the chromosomes apart during mitosis.

Mucigel A slimy material secreted by and covering the root cap and root hairs.

Mucilaginous Containing a mucilage, usually composed of mucopolysaccharides.

Mucin Any of a group of glycoproteins secreted by certain cells, especially those of salivary glands.

Mucociliary blanket The layer of cilia and mucus that lines certain portions of the respiratory system; it traps microorganisms up to 10 mm in diameter and then transports them by ciliary action away from the lungs.

Mucoperiosteum A type of oral mucosa which has a fibrous lamina propria , no submucosa, and is attached to the underlying periosteum of bone. The attached gingiva is a mucoperiosteum.

Mucosal-associated lymphoid tissue (MALT) The defensive immune lymphoid tissue located in the intestinal mucosa.

Mucous-bag suspension feeder Suspension feeder employing a sheet or bag of mucus to trap particles nonselectively

Mucron Apical anchoring device on an acephaline gregarine protozoan.

Mucus A thick, lubricating fluid produced by the mucous membranes that line the respiratory, digestive, urinary, and reproductive tracts; serves as a barrier against infection and, in the digestive tract, moistens food, making it easier to swallow.Sticky secretion used variously for locomotion, lubication, or protection from foreign particles.Viscid, slippery secretion rich in mucins produced by secretory cells such as those in mucous membranes. Adj., mucous.

Mud flat A muddy bottom that is exposed at low tide.

Muellerian mimicry Evolution by one species to resemble the coloration, body shape, or behavior of an unrelated species that is protected from predators by a venomous stinger, bad taste, or some other adaptation.

Mulch Protective ground cover, including manure, wood chips, straw, seaweed, leaves, and other natural products, or synthetic materials, such as heavy paper or plastic, that protect the soil, save water, and prevent weed growth.

Mullerian mimicry Occurs when two similar species are both distasteful to predators.

Müller's larva. Free-swimming ciliated larva that resembles a modified ctenophore, characteristic of certain marine polyclad turbellarians.

Muller's ratchet The steady accumulation of mutations in a population of asexual organisms over time.

Multicellular Organisms composed of multiple cells and exhibiting some division of labor and specialization of cell structure and function.

Multicistronic message An *mRNA transcript* with more than one *cistron* and thus encoding more than one *polypeptide.* These generally do not occur in eukaryotic organisms, due to differences in the mechanism of translation initiation.

Multicopy plasmids Present in bacteria at amounts greater than one per chromosome. Vectors for cloning DNA are usually multicopy; there are sometimes advantages in using a single copy plasmid.

Multi-drug-resistant strains of tuberculosis (MDR-TB) A multi-drug-resistant strain is defined as Mycobacterium tuberculosis resistant to isoniazid and rifampin, with or without resistance to other drugs.

Multigene family A collection of genes with similar or identical sequences, presumably of common origin.

Multigenic (polygenic) inheritance Inheritance in which the genetic control for a trait results in the phenotypic expression varying continuously.

Multinucleate Cells having more than one nucleus per cell.

Multiple alleles A condition in which more than two alleles exist for a given trait.

Multiple cloning site An artificially constructed region within a *vector* molecule which contains a number of closely spaced *recognition sequences* for *restriction endonucleases*. This serves as a convenient site into which foreign DNA may be inserted.

Multiple fission Asexual reproduction by the splitting of a cell or organism into many cells or organisms.

Multiple fruit A fruit derived from the fusion of the ovaries of several flowers in an inflorescence; example: pineapple.

Multiple use Many uses that occur simultaneously; used in forest management; limited to mutually compatible uses.

Multiplexing A *sequencing* approach that uses several pooled samples simultaneously, greatly increasing sequencing speed.

Multivoltine Having more than one brood or generation per season.

Mumps An acute generalized disease that occurs primarily in school-age children and is caused by a paramyxovirus that is transmitted in saliva and respiratory droplets. The principal manifestation is swelling of the parotid salivary glands.

Muscle Bundle of contractile cells which allow animals to move. Muscles must act against a skeleton to effect movement.

Muscle fibers Long, multinucleated cells found in skeletal muscles; made up of myofibrils. One of the four major groups of vertebrate cell/tissue types. Muscle cells contract/relax, allowing movement of and/or within the animal.

Muscle tissue The type of tissue that allows movement. The three kinds are skeletal, smooth, and cardiac. Tissue made of bundles of long cells called muscle fibers.

Muscular system One of eleven major body organ systems in animals; allows movement and locomotion, powers the circulatory, digestive, and respiratory systems, and plays a role in regulating temperature.

Muscularis mucosae Smooth muscle fibers around the mucosa of the gut wall, surrounding the lamina propria and surrounded by the submucosa.

Mushroom Common name for a group of fungi (Basidiomycetes) that produce an aboveground reproductive structure.

Muskeg 1. A bog characterized by scattered and stunted evergreens.

2. broadly, any area of wetland vegetation.

Must The juices of fruits, including grapes, that can be fermented for the production of alcohol.

Mutagen A chemical or physical agent that interacts with DNA and causes a mutation.

Mutagenesis he creation of mutations.

Mutagens Agents which increase the rate mutations in DNA. Examples are X-rays, UV light and many chemicals.

Mutation A change in the order of nucleotide bases on a gene, which alters the configuration of the protein produced, and thus may alter the behaviour of the cell. A mutation may cause a cell to die, or become cancerous. Mutations in bacteria and viruses help them to evade detection by their hosts.

Mutation pressure The constant resupplying of mutations to a gene pool due to a base mutation rate.

Mutation rate The average occurrence of mutations in a species per a given unit of time.

Mutational analysis An analysis that examines the effects of specific mutants for the effects on particular phenomena with respect to morphology, physiology, or behavior.

Mutualism A form of symbiosis in which both species benefit. A type of symbiosis where both organisms benefit. The classic example is lichens, which is a symbiosis between an alga and a fungus. The alga provides food and the fungus provides water and nutrients.

Mutualistic onferring reciprocal benefit to individuals of two different associated species

Mycelium The mass of interwoven filaments of hyphae in a fungus.

Mycetome Specialized organ in some insects that bears mutualistic bacteria.

Mycobiont The fungal partner in a mutualistic relationship such as mycorrhizae or lichens.

Mycolic acids Complex 60 to 90 carbon fatty acids with a hydroxyl on the b-carbon and an aliphatic chain on the a-carbon; found in the cell walls of mycobacteria.

Mycologist A person specializing in mycology; a student of mycology.A person who studies fungi.

Mycology The science and study of fungi.

Mycoplasma Bacteria that are members of the class Mollicutes and order Mycoplasmatales; they lack cell walls and cannot synthesize peptidoglycan precursors; most require sterols for growth; they are the smallest organisms capable of independent reproduction.

Mycoplasmal pneumonia A type of pneumonia caused by Mycoplasma pneumoniae. Spread involves airborne droplets and close contact.

Mycorrhiza Occurs when a fungus (basidiomycete or zygomycete) weaves around or into a plant's roots and forms a symbiotic relationship. Fungal hyphae absorb minerals from the soil and pass them on to the plant roots while the fungus obtains carbohydrates from the plant (pl.: mycorrhizae).

Mycorrhizae Symbiotic association between a fungus and the roots or rhizoids of a plant.

Mycorrhizal fungi The symbiotic relationship between certain nonpathogenic or weakly pathogenic fungi and the living cells of roots of certain higher plants.

Mycorrhizosphere The region around a mycorrhizal fungus in which nutrients released from the fungus increase the microbial population and its activities.

Mycosis Any disease caused by a fungus.

Mycotoxicology The study of fungal toxins and their effects on various organisms.

Mycotoxin A toxic compound formed by the hyphae of common molds growing under a variety of conditions, especially in contaminated foods.

Myelin sheath Layers of specialized glial cells, called Schwann cells, that coat the axons of many neurons.

Myelin The fatty covering of myelinated nerves which appears white to the naked eye. The parts of the brain and spinal cord, in which myelinated nerves run, has therefore been called the "white" matter as distinct from the "grey" matter composed of nerve cells. Myelin also contains about 20 % of proteins whose prime role is to mediate adhesion between adjacent Schwann cells. These cell membrane glycoproteins are also members of the immunoglobulin family of cell surface proteins. Defects in the these surface proteins may cause them to act as antigens to the immune system. The disease multiple sclerosis is caused by antibodies to the myelin proteins, which results in inflammation and loss of myelin.

Myelinated Nerves axons which are completely wrapped in a sheath of myelin by schwann cells. One cell wraps about a millimetre of nerve axon. Myelinated nerve axons carry impulses faster then unmeyelinated nerves as the impulse jumps across the myelin sheath of each adjacent Schwann cell to that of the next

Myeloma cell A tumor cell that is similar to the cell type found in bone marrow. Also, a malignant, neoplastic plasma cell that produces large quantities of antibodies and can be readily cultivated.

Myiasis Infestation of the body by the larvae of flies.

Myoblast An embryonic muscle fibre forming cell.

Myocyte Contractile cell (pinacocyte) in sponges.

Myofibril A fibril collectively arranged in longitudinal bundles in muscle cells (fibers); composed of thin filaments of actin and a regulatory protein and thick filaments of myosin.Striated contractile microfilaments in skeletal muscle cells.

Myogenesis Differentiation of skeletal muscle.

Myogenic Originating in muscle, such as heartbeat arising in vertebrate cardiac muscle because of inherent rhythmical properties of muscle rather than because of neural stimuli.

Myoglobin The oxygen-transporting pigment of muscle tissue.

Myomere The muscle plate or portion of a somite that develops into voluntary muscle.

Myosin Myosin constitutes the thick filament or myofilaments of the striated muscle cells.

Each individual molecule of myosin (mol wt. 500 000) is shaped like a double-headed golf club with a hinged shaft.

Myositis Inflammation of a striated or voluntary muscle.

Myotome That part of a somite destined to form muscles; the muscle group innervated by a single spinal nerve.

Myriapods Members of the four noninsect classes of the subphylum Uniramia. Includes centipedes, millipedes, pauropods, and symphylans.

Myxamoeba A free-living amoeboid cell that can aggregate with other myxamoeba to form a plasmodium or pseudoplasmodium. Found in cellular slime molds and the myxomycetes.

Myxini The class of vertebrates whose members are fishlike, jawless, without paired appendages, and possess four pairs of tentacles around the mouth. Hagfishes.

Myxobacteria A group of gram-negative, aerobic soil bacteria characterized by gliding motility, a complex life cycle with the production of fruiting bodies, and the formation of myxospores.

Myxospores Special dormant spores formed by the myxobacteria.

Myxozoa The protozoan phylum characterized by members having spores of multicellular origin; the myxozoans.

Myzorhynchus Apical stalked, suckerlike organ on the scolex of some tetraphyllidean cestodes.

N

N- or C-terminal The amino acids which form polypeptides are joined by peptide bonds between the amino group of one amino acid and the carboxy group of the next. Therefore each polypeptide has a free amino group at one end (N-terminal) and a free carboxy group at the other (C-terminal).

Nacre Innermost lustrous layer of mollusc shell, secreted by mantle epithelium. Adj., nacreous.

NAD^+ (nicotinamide adenine dinucleotide) Abbreviation of nicotinamide adenine dinucleotide, a coenzyme present in all cells that helps enzymes transfer electrons during the redox reactions of metabolism. A coenzyme that acts as an electron acceptor; particularly important in respiration.

NADP Abbreviation of nicotinamide adenine dinucleotide phosphate, a coenzyme that functions as an electron acceptor in the lightdependent reactions of photosynthesis.

Nagana Disease of ruminants caused by Trypanosoma brucei brucei or T. congolense.

Naiad The aquatic immature stage of any hemimetabolous insect.

Nail The horny cutaneous plate of the dorsal surface of the distal end of a finger or toe.

Naked grain A grain that separates easily from the surrounding bracts.

Nanoplankton The component of the plankton that consists of very small organisms 2 to 20 mm (0.002 to 0.2 mm) in size; they are too small to catch in a standard plankton net.

Nape The back of the neck.

Napkin (diaper) candidiasis Typically found in infants whose diapers are not changed frequently and are therefore not kept dry. Caused by Candida species of fungi.

Narcotic Any psychoactive compound that is dangerously addictive; a compound that induces central nervous system depression resulting in numbness, lethargy, and/orsleep.

Nares Openings into the nasal cavity, both internally and externally, in the head of a vertebrate.

Naris The openings of the nasal cavity.

Narrow-spectrum drugs Chemotherapeutic agents that are effective only against a limited variety of microorganisms.

Nasal gland Salt-excreting gland in the nostrils of seabirds and reptiles

Nasolabial groove A depression or trough running from the external naris (nostril) to the edge of the upper lip in Plethodontid (lungless) salamanders.

Nastic movement A plant's response to a stimulus in which the direction of the response is independent of the direction of the stimulus. Non-directional plant movements.

Nasutus. A type of soldier caste in certain termites; this form bears a median frontal rostrum through which it ejects a defensive fluid; the jaws are small or vestigial.

Natal Relating to or occurring at birth.

Natality Refers to births. Specifically, it refers to the birth rate of an animal, while mortality specifically refers to the death rate.

Native A species which is a natural member of a biotic community. An indigenous species. (The term implies that humans were not involved in the dispersal or colonization of the species.)

Native gel An electrophoresis gel run under conditions which do not denature proteins (i.e., in the absence of SDS, urea, 2-mercaptoethanol, etc.).

Native species A species that occurs naturally in an area (i.e. is not introduced). Compare *introduced species, invasive species*.

Natural landscaping A landscaping design that uses carefully contrived meadows, forests, and natural-looking ponds and marshes to create an illusion of wild nature.

Nauplius Typically the earliest larval stage(s) of crustaceans; has only three pairs of appendages: antennules, antennae, and mandibles_all primarily of locomotive function.

Natural attenuation The decrease in the level of an enviromental contaminant that results from natural chemical, physical, and biological processes.

Natural classification A classification system that arranges organisms into groups whose members share many characteristics and reflect as much as possible the biological nature of organisms.

Natural control The suppression of pest populations by naturally occurring biological and environmental agents.

Natural enemies Living organisms found in nature that kill, weaken, or reduce the reproductive potential of other organisms.

Natural history The study of where and how organisms carry out their life cycles.

Natural increase Crude death rate subtracted from crude birth rate.

Natural killer (NK) cell A non-T, non-B lymphocyte present in nonimmunized individuals that exhibits MHC-independent cytolytic activity against tumor cells.

Natural law A theory that has been tested many thousands of times and found always to be true, e.g., the law of gravity.

Natural logarithm The logarithm (log) to base *e* where $e \sim 2.718$. The natural logarithm is represented by the symbol ln.

Natural resources Goods and services supplied by the environment.

Natural selection Differential success in the reproduction of different phenotypes resulting from the interaction of organisms with their environment. Evolution occurs when natural selection causes changes in relative frequencies of alleles in the gene pool.The main mechanism of evolutionary change. In a given population of organisms, there are heritable traits that enable some members to contribute a larger number of offspring than others. If these offspring also have a greater reproductive success, then the genetic composition of the population is altered, thus evolution. The process of differential survival and reproduction of fitter genotypes; can be stabilizing, directional, or disruptive. Better adapted individuals are more likely to survive to reproductive age and thus leave more offspring and make a larger contribution to the gene pool than do less fit individuals. The differential survival and reproductive successes of individuals in a variable population that powers the evolutionary process. When all individuals survive and reproduce (except for chance occurrences) natural selection works at a lower rate, if at all.

Naturally acquired active immunity The type of immunity that develops when an individual's immunologic system comes into contact with an appropriate antigenic stimulus during the course of normal activities; it usually arises as the result of recovering from an infection.

Naturally acquired passive immunity The type of temporary immunity that involves the transfer of antibodies from one individual to another.

Navicular navicul = a little ship. A carpal bone shaped like a boat.

Navigation The process by which an animal uses various cues to determine its position in reference to a goal.

Neap tides The tides with a small tidal range. They occur around the times when the moon is in quarter. Compare spring tides.

Neascus Strigeoid metacercaria with a spoon-shaped forebody.

Necrosis Cell or tissue death

Nectar A sweet, syrupy exudate produced by some flowers as an attractant for pollinators.

Nectary A gland that secretes nectar.

Nephrostome Ciliated, funnel-shaped opening of a nephridium.

Nerve cell body The largest part of a neuron that typically contains the nucleus.

New towns Experimental urban environments that seek to combine the best features of the rural village and the modern city.

Necrotizing fasciitis disease that results from a severe invasive group A streptococcus infection. Necrotizing fasciitis is an infection of the subcutaneous soft tissues, particularly of fibrous tissue, and is most common on the extremities. It begins with skin reddening, swelling, pain, and cellulitis, and proceeds to skin breakdown and gangrene after 3 to 5 days.

Nectar sugary fluid that attracts insects to plants.

Nectar guide Color patterns present on petals that direct insects toward the nectar; often not visible to human eye.

Nectarines Nectar-secreting organs in flowering plants that serve as insect feeding stations and thus attract insects, which then assist in the transfer of pollen.

Nectary A floral gland which secretes nectar.

Negative feedback A primary mechanism of homeostasis, whereby a change in a physiological variable that is being monitored triggers a response that counteracts the initial fluctuation.

Negative feedback control Occurs when information produced by the feedback reverses the direction of the response; regulates the secretion of most hormones.

Negative feedback loop A biochemical pathway where the products of the reaction inhibit production of the enzyme that controlled their formation.

Negative staining A staining procedure in which a dye is used to make the background dark while the specimen is unstained.

Negri bodies Masses of viruses or unassembled viral subunits found within the brain neurons of rabies-infected animals.

Nekton Pelagic organisms that are free-swimming and so whose movements are independent of the tides, currents and waves. Such animals include fish, whales, squid, crabs and shrimps. The distribution of nekton is limited by temperature and nurtient supply and decreases with decreasing depth. Compare *benthic*, *plankton*.

Nektonic organisms "Swimmers"; one of the two main types of organisms in the pelagic zone of the marine biome.

Nematocyst An organelle characteristic of the Cnidaria that is used in defense, food gathering, and attachment.The stinging structure of cnidarians.

Nematoda The phylum of aschelminths that contains members commonly called either roundworms or nematodes. Tripoblastic, bilateral, vermiform, unsegmented, and pseudocoelomate.Invertebrates with a cylindrical body, a conspicuous body cavity, and a complete digestive tract.

Nematogen State in the life cycle of a dicyemid mesozoan.

Nematomorpha The phylum of aschelminths commonly called horsehair worms.

Nemertea The phylum that has members commonly called the proboscis worms; elongate, flattened worms found in marine mud and sand; triploblastic; complete digestive tract with anus; closed circulatory system.

Neo-darwinism A modified version of Darwin's evolutionary theory that eliminates elements of the Lamarckian inheritance of acquired characteristics and pangenesis that were present in Darwin's formulation; this theory originated with August Weissmann in the late nineteenth century and, after incorporating Mendelian genetic principles, has become the currently favored version of Darwinian evolutionary theory.

Neolithic The Stone Age period following the advent of agriculture.

Neo-malthusian A belief that the world is characterized by scarcity and competition in which too many people fight for too few resources. Named for Thomas Malthus, who predicted a dismal cycle of misery, vice, and starvation as a result of human overpopulation.

Neonate Newly hatched or born.

Neopterygian Any of a large group of bony fishes that includes most modern species.

Neoteny An evolutionary process by which organismal development is retarded relative to sexual maturation; produces a descendant that reaches sexual maturity while retaining a morphology characteristic of the preadult or larval stage of an ancestor.

Neotony Retention of larval or embryonic characteristics past the time of reproductive maturity.

Nephric nephros = kidney.

Nephridiopore An external excretory opening in invertebrates.

Nephridium One of the segmentally arranged, paired excretory tubules of many invertebrates, notably the annelids. In a broad sense, any tubule specialized for excretion and/or osmoregulation; with an external opening and with or without an internal opening.

Nephron Functional unit of kidney structure of vertebrates, consisting of a Bowman's capsule, an enclosed glomerulus, and the attached uriniferous tubule.

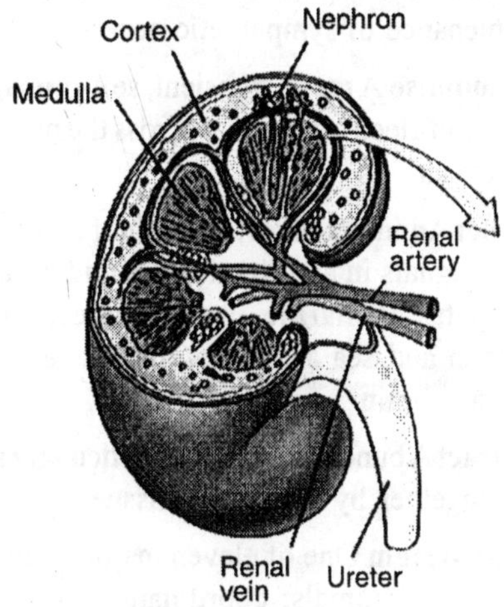

Fig. *Nephron*

Nephrotome A stalk-like connection between somites and lateral plate mesoderm which anteriorly forms the pronephos and posteriorly forms mesenchyme which develops into the mesonephros and metanephros. Synonyms: intermediate mesoderm, mesomere, nephromere.

Neritic ecosystems The marine ecosystem that consists of relatively shallow water extending from the littoral zone to the edge of the continental shelves.

Neritic Portion of the sea overlying the continental

shelf, specifically from the subtidal zone to a depth of 200 m.

Neritic zone The shallow regions of the ocean overlying the continental shelves.

Nerve A ropelike bundle of neuron fibers (axons and dendrites) tightly wrapped in connective tissue.

Nerve cell A cell specialized to originate or transmit nerve impulses.

Nerve cord A dorsal tubular cord of nervous tissue above the notochord of a chordate.

Nerve fiber A filamentous process extending from the cell body of a neuron and conducting the nerve impulse; an axon.

Nerve growth factor A cytokine that promotes the growth and repair of sensory nerves and maintenance of sympathetic nerves.

Nerve impulse A rapid, transient, self-propagating change in electric potential across the membrane of an axon.

Nerve net An interconnected mesh of neurons that sends signals in all directions; found in radially symmetrical marine invertebrates, such as jellyfish and sea anemones, that have no head region or brain.

Nerve tract A bundle of neurons, particularly axons, held together by connective tissue.

Nervous system One of eleven major body organ systems in animals; coordinates and controls actions of internal organs and body systems, receives and processes sensory information from the external environment, and coordinates short-term reactions to these stimuli.

Nervous tissue The type of tissue composed of individual cells called neurons and supporting neuroglial cells.

Nested hierarchy. A pattern in which species are ordered into a series of increasingly more inclusive clades according to the taxonomic distribution of synapomorphies.

Nested PCR Very sensitive method for amplfication of DNA, which takes part of the product of a single *PCR* reaction (after 30-35 cycles), and subjects it to a new round of PCR using a different set of *PCR primers* which are nested within the region flanked by the original primer pair.

Nesting density The amount of bird nesting in the area.

Net energy yield Total useful energy produced during the lifetime of an entire energy system minus the energy used, lost, or wasted in making useful energy available.

Net plankton Plankton that is caught in a standard plankton net.

Net primary production In a community or an ecosystem, the increase in the amount of plant or algal material between the beginning and end of a specified time period, such as a growing season.

Net productivity In a trophic level, a community, or an ecosystem, the amount of energy (in calories) stored in chemical compounds or the increase in biomass (in grams or metric tons) in a particular period of time; it is the difference between gross productivity and the energy used by the organisms in respiration.

Net secondary productivity (NSP) The rate at which consumer and decomposer biomass is produced in a community.

Netted venation A type of vein arangement in leaves in which the profusely branching veins form an interconnecting network; also termed reticulate venation.

Neural crest A band of cells along the border where the neural tube pinches off from the ectoderm; the cells migrate to various parts of the embryo and form the pigment cells in the skin, bones of the skull, the teeth, the adrenal glands, and parts of the peripheral nervous system.Cells derived from the ectoderm layer in the embryo. This layer folds to form a neural tube which will later become the spinal cord of the animal. Cells from the crest of this fold, leave the ectoderm and migrate into the mesoderm layer. These neural crest cells are thus of ectomesenchymal origin. They migrate to form the dental mesenchyme, supportive cells of the nervous system, the adrenal cortex and melanocytes of the skin.

Neural groove Dorsal, longitudinal groove that forms in a vertebrate embryo; bordered by two neural folds; preceded by the neural-plate stage and followed by the neural-tube stage.

Neural plate Thickened strip of ectoderm in early vertebrate embryos that forms along the dorsal side of the body and gives rise to the central nervous system.

Neural tube A tube of tissue formed by a thickening and rolling up of the neural plate during embryonic neurulation. It will later form the brain and spinal cord of the animal. Hollow cylindrical structure of neuroepithelial cells that will give rise to the brain and spinal cord; an ectodermal derivative.

Neurobiology The branch of biology concerned with the structure and function of cells of the nervous system.

Neuroblast Dividing neuronal precursor cell

Neurocoele The cavity or lumen of the neural tube.

Neuroendocrine system The combination of the nervous and endocrine systems.

Neurofibril node Regular gaps in a myelin sheath around a nerve fiber; formerly called node of Ranvier.

Neurogenic Originating in nervous tissue, as does the rhythmical beat of some arthropod hearts.

Neuroglia Tissue supporting and filling the spaces between the nerve cells of the central nervous system.

Neurohormone A chemical transmitter produced by nervous tissue. Uses the bloodstream or other body fluids for distribution to its target site.

Neurolemma Delicate nucleated outer sheath of a nerve cell; sheath of Schwann.

Neurolemmocyte The cell that surrounds a fiber of a peripheral nerve and forms the neurolemmal sheath and myelin; formerly called Schwann cell.

Neuromast Cluster of sense cells on or near the surface of a fish or amphibian that is sensitive to vibratory stimuli and water.

Neuromeres Segments of the brain which likely reflect the segmental origins of the vertebrates.

Neuromodulator A chemical agent that is released by a neuron and diffuses through a local region of the central nervous system, acting on neurons within that region; generally has the effect of modulating the response to neurotransmitters.

Neuromuscular junction The junction between an axon terminal of a motor neuron and a muscle fiber innervated by that motor neuron; the axon terminal of a motor neuron is typically branched, forming neuromuscular junctions with a number of different muscle fibers.The synapse between the axon terminal of a motoneurone and a skeletal muscle fibre. The release at this synapse of the neurotransmitter substance, acetyl choline causes the muscle to contract.

Neuron A nerve cell; the fundamental unit of the nervous system, having structure and properties that allow it to conduct signals by taking advantage of the electrical charge across its cell membrane.A nerve cell that consists of a cell body and its processes.Highly specialized cells that generate and transmit bioelectric impulses from one part of the body to another; the functional unit of the nervous system. A cell of the nerve tissue having a cell body input zone of dendrites and an output zone of an axon (of varying length). The electrochemical nerve impulse/message is transmitted by neurons.

Neuropeptides Compounds which have extremely potent affect to excite or depress nerve cells, in very low concentration . In this regard they are distinctly different from neurotransmitters. They include substances such as prostaglandins, substance P , endorphins and enkephalins. Neuropeptides also have receptors on lymphocytes and thus influence the immune response.

Neuropodium Lobe of parapodium nearer the ventral side in polychaete annelids.

Neuropore The anterior or posterior openings of the neural tube that will close at a later stage of development.

Neurosecretions Hormonelike chemical substances that are produced by specialized neurons that affect various physiological processes in the body.

Neurosecretory cells Hypothalamus cells that receive signals from other nerve cells, but instead of signaling to an adjacent nerve cell or muscle, they release hormones into the bloodstream.

Neurotoxin A toxin that is poisonous to or destroys nerve tissue; especially the toxins secreted by C. tetani, Corynebacterium diphtheriae, and Shigella dysenteriae.Chemical that paralyzes nerves. Neurotoxins are produced by a variety of organisms, most notably some of the heterotrophic dinoflagellates.Poison which interferes with nerve function, usually by affecting the flow of ions through the cell membrane.

Neurotransmitters Chemicals released from the tip of an axon into the synaptic cleft when a nerve impulse arrives; may stimulate or inhibit the next neuron. The chemical that crosses the synaptic cleft and causes the transmission of the nerve message in an adjacent neuron or the stimulation of an effector cell (muscle or gland).Chemicals which are secreted into a synapse in order to transmit an electrical nerve impulse from a nerve axon of one cell to the cell body of another nerve cell. They bind to receptors on nerve cells and are produced rapidly in high concentrations at nerve synapses. In these respects they differ from neuropeptides. There are many different neurotransmitter substances. They include acetyl choline, epinephrine, histamine, serotonin GABA and glutamate.

Neurula Vertebrate embryo during neurulation.

Neurulation Organogenesis of the nervous system in vertebrate embryos during which dorsal neuroectoderm cells of the neural plate (typically) roll up to form the neural tube which gives rise to central the nervous system. The process of the formation of the neural plate and subsequently, the neural tube.

Neuston Planktonic organisms associated with the air-water interface.

Neustonic The microorganisms that live at the atmospheric interface of a water body.

Neutral Soil pH of 7, neither acid nor alkaline.

Neutral variation Genetic diversity that confers no apparent selective advantage.

Neutrophils Members of the family of white blood cells which are involved in non-specific phagocytosis of bacteria and foreign material. Neutrophils are also called polymorphonuclear leucocytes (PMN) because of their multi-lobed nucleus. Neutrophils also release toxic enzymes, stored safely in their cytoplasm as lysosomes and released in the presence of foreign proteins. These toxic enzymes may do more harm to the host tissues than the bacteria they were released to kill.

Niche The portion of the environment which a species occupies, defined in terms of the conditions under which an organism can survive, and may be affected by the presence of other competing organisms.

Niche overlap The extent to which two species require similar resources; specifies the strength of the competition between the two species.

Nick In *duplex* DNA, this refers to the absence of a *phosphodiester bond* between two adjacent *nucleotides* on one strand.

Nick translation A method for incorporating radioactive isotopes (typically 32P) into a piece of DNA. The DNA is randomly nicked by DNase I, and then starting from those nicks DNA polymerase I digests and then replaces a stretch of DNA. Radiolabeled precursor nucleotide triphosphates can thus be incorporated. A method for introducing labeled *nucleotides* into a double-stranded DNA molecule which involves making small *nicks* in one strand with DNase, and then repairing with DNA polymerase I.

Nicotine adenine dinucleotide phosphate ($NADP^+$) A substance to which electrons are transferred from photosystem I during photosynthesis; the addition of the electrons reduces NADP, which acquires a hydrogen ion to form NADPH, which is a storage form of energy that can be transferred to the Calvin Cycle for the production of carbohydrate.

Nictitating membrane Third eyelid, a transparent membrane of birds and many reptiles and mammals, that can be pulled across the eye.

Nidus Specific locality of a given disease; result of a unique combination of ecological factors that favors the maintenance and transmission of the disease organism.

Nihilists Those who believe the world has no meaning or purpose other than a dark, cruel, unceasing struggle for power and existence.

Nit The egg of a louse.

Nitrate-forming bacteria Bacteria that convert nitrites into compounds that can be used by green plants to build proteins.

Nitrification The oxidation of ammonia or ammonium to nitrites and nitrates, as by nitrifying bacteria.

Nitrifying bacteria Chemolithotrophic, gram-negative bacteria that are members of the family Nitrobacteriaceae and convert ammonia to nitrate and nitrite to nitrate.

Nitrogen (N_2)A colorless and tasteless gas that is an essential constituent of proteins.

Nitrogen base The basic component of nucleic acids; composed of a nitrogen-containing molecule of purine or pyrimidine.

Nitrogen cycle Worldwide circulation and reutilization of nitrogen atoms, chiefly due to metabolic processes of living organisms; plants take up inorganic nitrogen and convert it into organic compounds (chiefly proteins), which are assimilated into the bodies of one or more animals; bacterial and fungal action on nitrogenous waste products and dead organisms return nitrogen atoms to the inorganic state.

Nitrogen fixation Reduction of molecular nitrogen to ammonia by some bacteria and cyanobacteria, often followed by nitrification, the oxidation of ammonia to nitrites and nitrates by other bacteria. The metabolic process in which atmospheric molecular nitrogen is reduced to ammonia; carried out by cyanobacteria, Rhizobium, and other nitrogen-fixing bacteria.

Nitrogen fixing bacteria Bacteria that convert nitrogen from the atmosphere or soil solution into ammonia that can then be converted to plant nutrients by nitrite- and nitrate-forming bacteria.

Nitrogen oxides Highly reactive gases formed when nitrogen in fuel or combustion air is heated to over 650°C in the presence of oxygen or when bacteria in soil or water oxidize nitrogen-containing compounds.

Nitrogen oxygen demand (NOD) The demand for oxygen in sewage treatment, caused by nitrifying microorganisms.

Nitrogen saturation point The point at which mineral nitrogen, when added to an ecosystem, can no longer be incorporated into organic matter through biological processes.

Nitrogenase The enzyme that catalyzes biological nitrogen fixation. An enzyme, unique to certain prokaryotes, that reduces N_2 to NH_3.

Nitrogenous base A nitrogen- containing molecule having the chemical properties of a base.

Nitrous oxide NO_2, an atmospheric gas that traps infrared radiation and also can form nitric acid in rain droplets, resulting in acid rain.

Nitrogenase The enzyme involved in the conversion of atmospheric nitrogen gas into ammonia.

Nocturnal Animals whose primary activity occurs during the dark portion of the daily cycle.

Nutrient regeneration The release of nutrients from organic matter by decomposers.

Noah's ark blot This is a species screening Southern Blotting technique. DNA samples from males and females of the species to be investigated are separated by electrophoresis and probed with the DNA of interest. This technique is useful in gene mapping and in the estimation of evolutionary distance between species.

Nocardioforms Bacteria that resemble members of the genus Nocardia; they develop a substrate mycelium that readily breaks up into rods and coccoid elements (a quality sometimes called fugacity).

Nociceptor A sensory receptor responding to potentially harmful stimuli; produces a sensation of pain.

Nocturnal Describing an animal that is active mainly or exclusively at night, rather than by day.

Node The region of a stem between two internodes, where there is branching of the vascular tissue into leaves or other appendages.

Nodes of Ranvier The small gaps in the myelin sheath between successive glial cells along the axon of a neuron; also, the site of high concentration of voltage-gated ion channels.

Nodule A tumorlike growth on the roots of certain higher plants that encloses a population of nitrogen-fixing bacteria.

Nodus The kink or notch on the costal margin of the dragonfly wing. The name is also used for the strong, short cross-vein just behind the notch.

Noma A highly destructive and usually fatal infection of the teeth and jaws which is a progression of ANUG Only found in malnourished children. - also called cancrum oris.

Nomadic A way of life in which there is no permanent residence site and the group moves from place to place according to the season, the available of food supply and other such factors; the route of travel often follows a traditional pattern.

Nomenclature The study of the naming of organisms in the fashion that reflects their evolutionary relationships.

Nomograph A graph that allows a third variable to be measured when the values of two related variables are known.

Nonamniote lineage The vertebrate lineage leading to modern amphibians.

Nonbiodegradable A chemical that cannot be broken down by bacteria or other organisms. It is said to be persistent.

Noncompetitive inhibitor A substance that reduces the activity of an enzyme by binding to a location remote from the active site, changing its conformation so that it no longer binds to the substrate.

Nonconservative substitution A mutation which results in the substitution of one *amino acid* within a polypeptide chain with an amino acid belonging to a different polarity/charge group.

Noncyclic electron flow A route of electron flow during the light reactions of photosynthesis that involves both photosystems and produces ATP, NADPH, and oxygen; the net electron flow is from water to $NADP^+$.

Noncyclic photophosphorylation The process in which light energy is used to make ATP when electrons are moved from water to NADP1 during photosynthesis; both photosystem I and photosystem II are involved.

Nondisjunction The failure of chromosomes to separate properly during cell division. The unequal segregation of chromosomes during meiosis. This forms cells with either too many (possibly one or more single or sets of chromosomes too many) or too few chromosomes. Thought to be a common cause for Down Syndrome, where sufferers often have an extra copy of chromosome 21.

Nongonococcal urethritis (NGU) Any inflammation of the urethra not caused by Neisseria gonorrhoeae.

Nongovernmental organizations (NGOs) A term referring collectively to pressure and research groups, advisory agencies, political parties, professional societies, and other groups concerned about environmental quality, resource use, and many other issues.

Noninterference A principle that requires us to refrain from interfering with the lives or destinies of other moral agents or moral subjects except to satisfy essential needs.

Nonmalfeasance Doing no harm to any being that has goods (values, interests, moral values) of their own.

Nonpoint sources Scattered, diffuse sources of pollutants, such as runoff from farm fields, golf courses, construction sites, etc.

Nonpolar covalent bond The type of bond that is formed when electrons spend as much time orbiting one nucleus as the other; thus, the distribution of charges is symmetrical.

Nonrenewable resource A resource that is in limited supply and can't be replenished by natural processes, at least not for thousands of years or more. Fossil fuels are nonrenewable resources.

Nonsense codon A codon that does not code for an amino acid but is a signal to terminate protein synthesis.

Nonsense mutation A change in the sequence of a nucleic acid that causes a *nonsense* (*stop* or *termination*) *codon* to replace a codon representing an amino acid.

Nonseptate Lacking cross walls; commonly used in reference to certain fungal hyphae. Also termed aseptate.

Nonseptate hypha Fungal hyphae that lack septa or cross walls.

Nonshattering Seeds and/or fruits do not split off and scatter from the fruiting head; a trait associated with domesticated plants.

Nonshivering thermogenesis The hormonal triggering of heat production.

Nonspecific resistance Refers to those general defense mechanisms that are inherited as part of the innate structure and function of each animal; also known as nonspecific, innate or natural immunity.

Nontranslated RNA (NTR) The segments located at the 5' and 3' ends of a mRNA molecule which do not encode any part of the polyprotein; may contain important translational control elements.

Nonvascular plants Plants lacking lignified vascular tissue (xylem), vascularized leaves, and having a free-living, photosynthetic gametophyte stage that dominates the life cycle. Common examples are the mosses and liverworts.

Noradrenaline A hormone, produced by the medulla of the adrenal gland, that increases the concentration of glucose in the blood, raises blood pressure and heartbeat rate, and increases muscular power and resistance to fatigue; also one of the principal neurotransmitters; also called norepinephrine.

Norepinephrine A hormone produced in the adrenal medulla and secreted under stress; contributes to the "fight or flight" response.

Norm of reaction The range of phenotypic possibilities for a single genotype, as influenced by the environment.

Northern blot A technique for analyzing mixtures of RNA, whereby the presence and rough size of one particular type of RNA (usually an mRNA) can be ascertained. After Dr. E. M. Southern invented the Southern blot, it was adapted to RNA and named the "Northern" blot.

Northern coniferous forest The biome characterized by cool summers, cold winters, and short growing seasons. Characteristic animals include insects, snowshoe hares, lynx, wolves, caribou, and moose.

Nosocomial infection An infection that develops within a hospital (or other type of clinical care facility) and is produced by an infectious organism acquired during the stay of the patient.

No-take reserves Geographic areas where by law no one is allowed to fish or collect biological specimens. Rules could apply to one or all species.

Notaulix One of a pair of longitudinal grooves on the mesonotum of certain hymenopterans, dividing the mesonotum into a central area and two lateral areas (plural notaulices)

Notochord In chordates, a cellular rod that runs the length of the body and provides dorsal support. Also, a structure of mesoderm in the embryo that will become the vertebrae of the spinal column. The stiff rod-like structure that all members of the Phylum Chordata develop at some stage during their life.

Notopleuron A triangular area on the thorax of certain flies, just behind the humeral callus and occupying parts of both dorsal and lateral surfaces.

Notopodium Lobe of parapodium nearer the dorsal side in polychaete annelids.

Notum The dorsal or upper surface of any thoracic segment: usually prefixed by pro-, meso-, or meta- to indicate the relevant segment.

Noxious Harmful to the health.

NT Abbreviation for nucleotide; i.e. the monomeric unit from which DNA or RNA are built. One can express the size of a nucleic acid strand in terms of the number of nucleotides in its chain; hence 'nt' can be a measure of chain length.

Nucellus Tissue in the ovule within which the embryo sac develops; integuments surround the nucellus.

Nuclear area In prokaryotic cells, a region containing the cell's genetic information. Unlike the nucleus in eukaryotic cells, it is not surrounded by a membrane.

Nuclear envelope The complex double-membrane structure forming the outer boundary of the eucaryotic nucleus. It is covered by pores through which substances enter and leave the nucleus.

Nuclear fission The radioactive decay process in which isotopes split apart to create two smaller atoms.

Nuclear fusion A process in which two smaller atomic nuclei fuse into one larger nucleus and release energy; the source of power in a hydrogen bomb.

Nuclear membrane The double membrane which surrounds the eukaryotic nucleus. It has many pores in its surface which regulate the flow of large compounds into and out of the nucleus.

Nuclear pores Openings in the membrane of a cell's nuclear envelope that allow the exchange of materials between the nucleus and the cytoplasm.

Nuclear run-on A method used to estimate the relative rate of transcription of a given gene, as opposed to the steady-state level of the mRNA transcript (which is influenced not just by transcription rates, but by the stability of the RNA). This technique is based on the assumption that a highly-transcribed gene should have more molecules of RNA polymerase bound to it than will the same gene in a less-active state. If properly prepared, isolated nuclei will continue to transcribe genes and incorporate 32P into RNA, but only in those transcripts that were in progress at the time the nuclei were isolated. Once the polymerase molecules complete the transcript they have in progress, they should not be able to re-initiate transcription. If that is true, then the amount of radiolabel incorporated into a specific type of mRNA is theoretically proportional to the number of RNA polymerase complexes present on that gene at the time of isolation.

Nuclease An enzyme which degrades nucleic acids. A nuclease can be DNA-specific (a DNase), RNA-specific (RNase) or non-specific. It may act only on single stranded nucleic acids, or only on double-stranded nucleic acids, or it may be non-specific with respect to strandedness. A nuclease may degrade only from an end (an exonuclease), or may be able to start in the middle of a strand (an endonuclease). To further complicate matters, many enzymes have multiple functions; for example, Bal31 has a 3'-exonuclease activity on double-stranded DNA, and an endonuclease activity specific for single-stranded DNA or RNA.

Nucleic acid A polymeric molecule composed of nucleotide monomers joined by phosphodiester bonds. These link the 5' ribose carbon atom of one nucleotide residue with the 3' ribose carbon atom of the neighbouring nucleotide residue, through a phosphate group. Each phosphate group ionises by losing a hydrogen ion thus becoming negatively charged, and giving the molecule its acidic character. The negative charge of nucleic acids is made use of in electrophoretic separations.

Nucleic acid hybridization Coming together (annealling) of single-stranded nucleic acid sequences by hydrogen bonding of complementary bases to form double-stranded molecules; this process is the basis for molecular biological techniques in which a labeled probe sequence is used to detect another identical or similar sequence (e.g., Southern hybridization, Northern hybridization).

Nucleic acid probe In DNA technology, a labeled single-stranded nucleic acid molecule used to tag a specific nucleotide sequence in a nucleic acid sample. Molecules of the probe hydrogen-bond to the complementary sequence wherever it occurs; radioactive or other labeling of the probe allows its location to be detected.

Nucleocapsid The nucleic acid and its surrounding protein coat or capsid; the basic unit of virion structure.

Nucleoid Region in prokaryotes where the DNA is concentrated. Unlike a nucleus, it is not bound by a membrane.

Nucleoid region The region in a prokaryotic cell consisting of a concentrated mass of DNA.

Nucleolar organizer An area on certain chromosomes associated with the formation of the nucleolus.

Nucleolus A deeply staining body within the nucleus of a cell and containing RNA; nucleoli are specialized portions of certain chromosomes that carry multiple copies of the information to synthesize ribosomal RNA.

Nucleon The nucleus of an atom is built up of protons and neutrons, collectively called a nucleon.

Nucleoplasm Protoplasm of nucleus, as distinguished from cytoplasm.

Nucleoprotein. A molecule composed of nucleic acid and protein; occurs in the nucleus and cytoplasm of all cells.

Nucleoside A chemical structure where a nitrogenous base is attaches to a sugar. In nucleic acids (RNA and DNA), nucleosides are joined together with phosphate residues to form a long chain.

A chemical structure where a nucleoside is attached to a phosphate residue. The sugar molecule is in the 'middle' of the structure, having both the nitrogenous base and the phosphate attached to it.

Nucleosome The basic, beadlike unit of DNA packaging in eukaryotes, consisting of a segment of DNA wound around a protein core composed of two copies of each of four types of histone.

Nucleotides Are made up of three components, a pentose sugar (ribose or deoxyribose), a phosphate group and an organic base which may either be a purine (adenine or guanine) or a pyrimidine (cytosine thymine or uracil).This basic structure is found in many important cell molecules such as in ADP, ATP and coenzymes. Nucleotides also form the subunits from which nucleic acids (DNA and RNA)are built.

Nucleus Cell nucleus; a spheroid body within a cell, contained in a double membrane, the nuclear envelope, and containing chromosomes and one or more nucleoli. The genetic control center of a eukaryotic cell. The cell bodies of nerves within the central nervous system.

Nuda The class of ctenophorans whose members lack tentacles and have a flattened body with a highly branched gastrovascular cavity.

Nudibranchs (or sea slugs) Gastropods that lack a shell and have exposed gills.

Nuées ardentes Deadly, denser-than-air mixtures of hot gases and ash ejected from volcanoes.

Nuisance bloom A rapid increase of one or only a few species of phytoplankton, resulting in densities high enough to cause discoloration of the surface water, possible increase of toxins, and degradation of water quality aspects such as dissolved oxygen.

Null hypothesis In a statistical sense, to the statement that there will not be any difference between effects of experimental treatments.

Numbers pyramid A diagram showing the relative population sizes at each trophic level in an ecosystem; usually corresponds to the biomass pyramid.

Numerical aperture The property of a microscope lens that determines how much light can enter and how great a resolution the lens can provide.

Numerical taxonomy A system of classification in which there is no attempt to distinguish true and false similarities.

Nuptial flight The mating flight of insects, especially that of the queen with male or males.

Nurse cells. Single cells or layers of cells surrounding or adjacent to other cells or structures for which the nurse cells provide nutrient or other molecules (for example, for insect oocytes or Trichinella spp. juveniles).

Nut A dry, indehiscent one-seeded fruit with a hard pericarp (the shell).

Nutrient Any element or simple compound necessary for the health and survival of an organism. This includes air and water, as well as food.

Nutrient cycling All the processes by which nutrients are transferred from one organism to another. For instance, the carbon cycle includes uptake of carbon dioxide by plants, ingestion by animals, and respiration and decay of the animal.

Nyctinasty The "sleep movements" of leaves in response to change in turgor pressure of cells at the base of the petiole.

Nymph Name given to the young stages of those insects which undergo a partial metamorphosis. The nymph is usually quite similar to the adult except that its wings are not fully developed. It normally feeds on the same kind of food as the adult.

Nymphochrysalis Nonfeeding, prenymph stage in the life cycle of a chigger mite.

Nystatin A polyene antibiotic from Streptomyces noursei that is used in the treatment of Candida infections of the skin, vagina, and alimentary tract.

O

O antigen A polysaccharide antigen extending from the outer membrane of some gram-negative bacterial cell walls; it is part of the lipopolysaccharide.

Oacemaker enzyme The enzyme in a metabolic pathway that catalyzes the slowest or rate-limiting reaction; if its rate changes, the pathway's activity changes.

Obligate aerobe An organism that requires oxygen for cellular respiration and cannot live without it.

Obligate anaerobes Microorganisms that cannot tolerate the presence of oxygen and die when exposed to it.

Obligate symbiont Organism that is physiologically dependent on establishing a symbiotic relationship with another.

Observability Not all of the members of a group are always seen the same proportion of time. Animals of different ages, sexes, or of different dominance status may be seen more or less than other animals.

Observational learning The tendency to perform an appropriate action or response as the result of having observed another animal's performance of similar actions in the same situation.

Observational method In an historical context, the notion, developed by C. Lloyd Morgan, that only data gathered by direct experiment and watching the animals under study could provide the basis for generalizations. This was in contrast to Romanes's notion that much about animal behavior could be gained by inference from what was seen. In the modern sense, this refers to any of a number of techniques used to watch and record the actions and interactions of animals.

Observations Objectively measuring or recording the results of testing a theory; watching and evaluating.

Obtect pupa A pupa in which the legs and other appendages are closely appressed to the rest of the body and not capable of free movement - as in the butterfly chrysalis.

Occipital Suture A groove running round the posterior region of the head of some insects and separating the vertex from the occiput. On the sides of the head the same groove marks the posterior boundary of the cheeks or genae.

Occipital lobe The lobe of the cerebral cortex located at the rear of the head; is responsible for receiving and processing visual information.

Occiput Hindmost region of the top of the head, just in front of the neck membrane. In some insects it is separated from the vertex by the occipital suture, but it is not usually present as a distinct plate or sclerite.

Occlusal interference Contact between opposing teeth during chewing, which prevents the other teeth touching. In extreme lateral and protrusive positions of the jaw, this would happen in most dentitions and be of no concern. When occlusal interference occurs close to the area of maximum tooth contact, it may be troublesome.

Occlusal wear Loss of enamel on the biting surface of the teeth due to the abrasive action of chewing natural unprocessed food.

Ocean desert The concept of a region in the ocean in which lack of nutrients is reflected in a lack of plants and animals; a region of low biological productivity.

Ocean shorelines Rocky coasts and sandy beaches along the oceans; support rich, stratified communities.

Ocean thermal electric conversion (OTEC) Energy derived from temperature differentials between warm ocean surface waters and cold deep waters. This differential can be used to drive turbines attached to electric generators.

Oceanic Associated with sea-water environments seaward of the shelf-slope break

Oceanic crust The Earth's crust which is formed at mid-oceanic ridges, typically 5 to 10 kilometers thick with a density of 3.0 grams per centimeter cubed.

Oceanic ecosystems The marine ecosystem that extends from the continental shelves into the unproductive open ocean.

Oceanic islands Islands in the ocean; formed by breaking away from a continental landmass, volcanic action, coral formation, or a combination of sources; support distinctive communities.

Oceanic ridge A sinuous ridge rising from the deep-sea floor

Oceanic trench Deep steep-sided depression in the ocean floor caused by the subduction of oceanic crust beneath either other oceanic crust or continental crust.

Oceanic zone The region of water lying over deep areas beyond the continental shelf.

Ocellar bristles Bristles arising around or between the ocelli in various flies.

Ocellar triangle A triangular area, usually quite distinct from the rest of the head, on which the ocelli of true flies are carried.

Ocelli Simple eyes on some adult and larval insects. Typically there are three, which form an inverted triangle dorsally, the function of which is obscure.

Octomerous Eight parts, specifically, symmetry based on eight.

Ocyst Cystic form in the Apicomplexa, resulting from sporogony; the oocyst may be covered by a hard, resistant membrane (as in Eimeria), or it may not (as in Plasmodium).

Odontoblasts Cells lining the dental pulp, derived from the dental papilla wich form the dentine of the tooth crown and root. New odontoblasts may become differentiated from less specialised pericytes in the pulp.

Odontoblast process The extension of the cytoplasm of an odontoblast which remains surrounded by dentine during tooth formation. The process is still an active part of the cell and contributes to the production of intratubular dentine in response to ageing, tooth wear or arrested caries.

Odontopathogens Dental pathogens.

Odontophore Tooth-bearing organ in molluscs, including the radula, radular sac, muscles, and cartilages.

Oesophagus The narrow part of the alimentary canal immediately posterior to the pharynx and mouth.

Off-surround The photoreceptor cells in the outer portion of a receptive field that have an inhibitory effect on ganglion cells if stimulated by light.

Oils Triglycerides that are liquid at room temperature.

Okazaki fragments Short stretches of

polynucleotides produced during discontinuous DNA replication.

Olfactory Pertaining to the sense of smell.

Olfactory sacs Structures on both sides of the head of fishes that are sensitive to chemical stimuli.

Oligochaeta The class of annelids whose members are characterized by having few setae and no parapodia. Monoecious with direct development. The earthworm (Lumbricus) and Tubifex.

Oligodeoxyribonucleotide A short, single-stranded DNA molecule, generally 15-50 *nucleotides* in length, which may be used as a *primer* or a *hybridization probe*. Oligodeoxyribonucleotides are synthesized chemically under automated conditions.

Oligonucleotide A short nucleic acid molecule; normally refers to molecules between 5 and 200 nucleotide residues(bases) long.

Oligonucleotide primer short sequence of nucleotides which is needed for the functioning of the enzyme polymerase. In PCR, this sequence is chosen so that it initiates the polymerase chain reaction.

Oligopod larva Usual larva in Coleoptera and Neuroptera, with a well-developed head and thoracic legs.

Oligotrophic Condition of rivers and lakes that have clear water and low biological productivity (oligo 5 little; trophic 5 nutrition); are usually clear; cold, infertile headwater lakes and streams.

Oligotrophic lake A nutrient-poor, clear, deep lake with minimum phytoplankton.

Omasum The third compartment of the stomach of a ruminant mammal.

Omentum A folded peritoneal membrane from the stomach to the dorsal body wall.

Ommatidia The functional units of the invertebrate compound eye, each one of which has a lens, focusing cone, photoreceptive cells, and forms an image.

Omnivore Literally, an organism that will eat anything. Refers to animals who do not restrict their diet to just plants or other animals.

Omnivory Being able to feed in more than one distinct way (e.g., an organism capable of carnivory and herbivory)

On-center The inner circle of cells of a receptive field that stimulate ganglion cells if they are stimulated by light.

On-center, off-surround cell A ganglion cell that generates an action potential when light shines on the center of its receptive field and is inhibited when light shines on the periphery of its receptive field.

Onchocercoma Subcutaneous nodule containing masses of the nematode Onchocerca volvulus.

Onchosphere The larva of the tapeworm contained within the external embryonic envelope and armed with six hooks and cilia. Typically referred to as a coracidium when released into the water.

Oncogene A gene in a tumor virus or in cancerous cells which, when transferred into other cells, can cause transformation (note that only certain cells are susceptible to transformation by any one oncogene). Functional oncogenes are not present in normal cells. A normal cell has many "proto-oncogenes" which serve normal functions, and which under the right circumstances can be activated to become oncogenes. The prefix "v-" indicates that a gene is derived from a virus, and is generally an oncogene (like *v-src* , *v-ras*, *v-myb* , etc).

Oncomiracidium A ciliated larva of a monogenetic trematode.

Oncosphere Rounded larva common to all cestodes, bears hooks.

One gene one polypeptide hypothesis A revision of the one gene, one enzyme hypothesis. Some proteins are composed of different polypeptide chains encoded by separate genes, so the hypothesis now holds that mutation in a gene encoding a specifc polypeptide can alter the ability of the encoded protein to function and thus produce an altered phenotype.

One gene one enzyme hypothesis Holds that a single gene controls the production, specificity, and activity of each enzyme in a metabolic pathway. Thus, mutation of such a gene changes the ability of the cell to carry out a particular reaction and disrupts the entire pathway.

One-step growth experiment An experiment used to study the reproduction of lytic phages in which one round of phage reproduction occurs and ends with the lysis of the host bacterial population.

Ontogenetic niche The multitude of inherited ecological and social traits that are, in effect, passed on from generation to generation and that play integral roles, in concert with genetic inheritance, influencing behavior development.

Ontogeny The embryonic development of an organism.

Onychomycosis A fungal infection of the nail plate producing nails that are opaque, white, thickened, friable, and brittle. Also called ringworm of the nails and tinea unguium. Caused by Trichophyton and other fungi such as C. albicans.

Onychophora A phylum of terrestrial animals with 14 to 43 pairs of unjointed legs, oral papillae, and two large antennae. Onycophorans live in humid tropical areas of the world. Their ancestors may have been an evolutionary transition between annelids and arthropods. Velvet worms or walking worms.

Oocyte Stage in formation of ovum, just preceding first meiotic division (primary oocyte) or just following first meiotic division (secondary oocyte).

Oecium Brood pouch; compartment for developing embryos in ectoprocts.

Oogamy A condition in which male and female gametes differ, such that a small, flagellated sperm fertilizes a large, nonmotile egg.

Oogenesis The production of ova. The development of a diploid cell into a haploid ovum or egg cell.

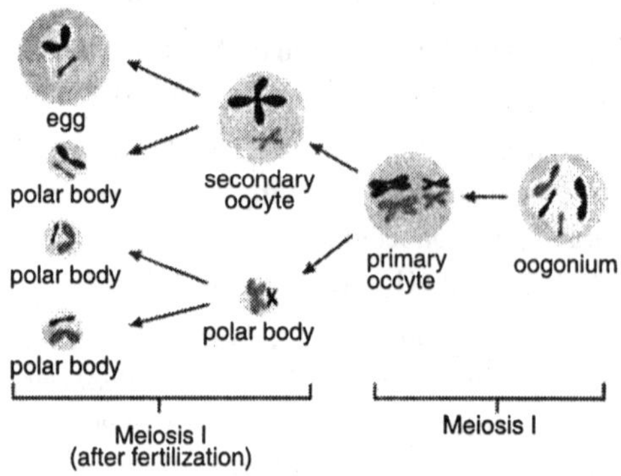

Fig. *Oogenesis*

Oogenotop Female genital complex of a flatworm, including oviduct, ootype, Mehlis' glands, common vitelline duct, and upper uterus.

Oogonia Mitotically dividing female structures that produce primary oocytes and gametes.

Oogonium In algae and fungi, a specialized egg-containing cell; a single-celled female gametangium, containing one or more eggs.

Ookinete The motile zygote of malarial parasites.

Oomycetes An assemblage of organisms, previously grouped with the fungi, but currently assigned to the kingdom Protista based in part on the presence of cellulose cell walls, rather than chitinous walls as found in the fungi.

Oontogeny The course of development of an individual from egg to senescence.

Oostegites Modified thoracic epipods in females of the crustacean superorder Peracarida. They form a pouch for brooding embryos.

Ootheca An egg case formed by the secretions of accessory genital glands or oviducts, such as the purse-like structure carried around by cockroaches or the spongy mass in which mantids lay their eggs.

Ootid Stage of formation of ovum after second meiotic division following expulsion of second polar body.

Ootype Part of oviduct in flatworms that receives ducts from vitelline glands and Mehlis' gland.

Opalinata The protozoan subphylum where members are cylindrical; covered with cilia. Examples: Opalina, Zelleriella.

Open access system A commonly-held resource for which there are no management rules.

Open access The concept of international law that permits free access by any nation to marine resources existing outside national jurisdictions.

Open canopy A forest where tree crowns cover less than 20 percent of the ground; also called woodland.

Open circulatory system A circulatory system in which blood is not confined to vessels in a part of its circuit within an animal; blood bathes tissues in blood sinuses.

Open community A community in which the populations have different density peaks and range boundaries and are distributed more or less randomly.

Open range Unfenced, natural grazing lands; includes woodland as well as grassland.

Open reading frame Any region of DNA or RNA where a protein could be encoded. In other words, there must be a string of nucleotides (possibly starting with a Met codon) in which one of the three reading frames has no stop codons.

Operant conditioning A type of learning that involves the animal's forming a variety of associations during which it learns to associate its behavior with the consequences of that behavior. That is, the sequence of events is dependent upon the behavior of the animal.

Operational sex ratio The number of sexually mature males and females in a population.

Operator A segment of DNA that interacts with a repressor protein to regulate the transcription of the structural genes of an operon.

Opercular chamber The closed cavity which covers the internal gills of the frog embryo, opening to the outside through a spiracle .

Operculum A cover. 1. The cover of a gill chamber of a bony fish (Chordata).

2. The cover of the genital pores of a horseshoe crab (Meristomata, Arthropoda).

3. The cover of the aperature of a snail shell (Gastropoda, Mollusca).

Operon A complete unit of bacterial gene expression and regulation, including the structural gene or genes, regulator gene(s), and control elements in DNA recognized by regulator gene products(s). A genetic unit consisting of a cluster of genes under the control of other genes, found in prokaryotes.

Ophiuroidea The class of echinoderms whose members have arms sharply set off from the central disk. Tube feet without suction disks. Brittle stars.

Ophthalmia neonatorum A gonorrheal eye infection in a newborn, which may lead to blindness. Also called conjunctivitis of the newborn.

Ophthalmic Pertaining to the eye.

Opisthaptor Posterior attachment organ of a monogenetic trematode.

Opisthomastigote A form of Trypanosomatidae with the kinetoplast at the posterior end. The flagellum runs through a long reservoir to emerge at the anterior. There is no undulating membrane. An example is Herpetomonas.

Opportunistic microorganism or pathogen A microorganism that is usually free-living or a part of the host's normal microbiota, but which may become pathogenic under certain circumstances, such as when the immune system is compromised.

Opportunistic species Species characterized by high reproduction rates, rapid development, early reproduction, small body size, and uncertain adult survival.

Opposable The capability of being placed against

the remaining digits of a hand or foot; e.g., the ability of the thumb to touch the tips of the fingers on that hand.

Opposite arrangement Two leaves borne per node and arranged across the stem from each other.

Opposite Leaf arrangement in which they are attached at a node directly across from one another on the stem.

Opsins Molecules in cone cells that bind to pigments, creating a complex that is sensitive to light of a given wavelength.

Opsonization Modification of the surface characteristics of an invading particle or organism by binding with antibody or a nonspecific molecule in such a manner as to facilitate phagocytosis by host cells.

Optical tweezer The use of a focused laser beam to drag and isolate a specific microorganism from a complex microbial mixture.

Optimal foraging theory A theory designed to predict the foraging behavior that maximizes food intake per unit time

Optimality model A model that attempts to predict the particular combination of costs and benefits of a behavior that will maximize an individual's inclusive fitness.

Oral Having to do with the mouth. The end of an animal containing the mouth.

Oral cavity The cavity within the mouth.

Oral sucker The sucker on the anterior end of a tapeworm, fluke, or leech.

Oral vibrissae The pair of large bristles just above the mouth in certain flies: usually simply called vibrissae.

Orbital In the current model of atomic structure, the volume of space surrounding the atomic nucleus in which an electron will be found 90 percent of the time.Discrete pathways or bands surrounding an atom through which electrons move.

Orchitis Inflammation of the testis.

Order A subdivision of a class or subclass containing a group of related families. Organophosphates. Organic compounds containing phosphorous; an important group of synthetic insecticides belong to this class of chemicals.

Ordovician A geological period beginning around 500 million years ago, believed to be the time when plants may have begun to migrate onto land.

Ordovician extinction Paleozoic-aged mass extinction possibly related to glaciation in the southern-hemisphere supercontinent Gondwana.

Ordovician period Geologic period of the Paleozoic Era after the Cambrian Period between 500 and 435 million years ago. Major advances during this period include the bony fish and possibly land plants (during the late Ordovician).

ORF Open reading frame. A continuous sequence of nucleotide triplets (codons) in which each triplet codes for an amino acid.

Organ Collection of tissues which performs a particular function or set of functions in an animal or plant's body. The heart, brain, and skin are three organs found in most animals. The leaf, stem, and root are three organs found in most plants. Organs are composed of tissues, and may be organized into larger organ systems.

Organ of corti The actual hearing organ of the vertebrate ear, located in the floor of the cochlear canal in the inner ear; contains the receptor cells (hair cells) of the ear.

Organ of ruffini Sensory receptor in the skin believed to be a sensor for touch-pressure, position sense of a body part, and movement. Also known as corpuscle of Ruffini.

Organ system A set of interconnected or interdependent parts that function together in a common purpose or produce results that cannot be achieved by one of them acting alone.

Organelle A structure within or on a cell that

performs specific functions and is related to the cell in a way similar to that of an organ to the body.One of several formed bodies with a specialized function, suspended in the cytoplasm and found in eukaryotic cells.Cell components that carry out individual functions; e.g., the cell nucleus and the endoplasmic reticulum. Subcellular structures (usually membrane-bound and unique to eukaryotes) that perform some function, e.g. chloroplast, mitochondrion, nucleus.

Organic Pertaining to (1) organisms or living things generally, or

(2) compounds formed by living organisms, or

(3) the chemistry of compounds containing carbon.

Organic chemistry The study of carbon compounds (organic compounds).

Organic compound A chemical comound containing the element carbon and usually synthesized by cells.

Organic evolution The change in an organism over time; a change in the sum of all genes in a population.

Organic molecule A molecule that contains one or more carbon atoms.

Organic nutrients Nutrients in the form of molecules synthesized by or originating from other organisms

Organ-identity gene A plant gene in which a mutation causes a floral organ to develop in the wrong location.

Organism An individual living thing, such as a bacterium, fungus, protist, plant or animal.

Organization, level of The extent of specialization and organization of the cells of an organism. Organization may be at the cellular, tissue, or organ level.

Organizational effects of hormones Changes resulting from the presence of hormones at critical time periods such that specific developmental pathways for specific brain regions and developing gonadal tissues are influenced to become either female or malelike.

Organizational effects Refers to the effects of exposure to certain hormones during critical periods in development that affect the central nervous system and other structures, producing morphological and behavioral changes in adulthood. Contrast with activational effects.

Organizer Area of an embryo that directs subsequent development of other parts.

Organogenesis Formation of organs in the developing embryo.

Organotrophs Organisms that use reduced organic compounds as their electron source.

Orgasm Rhythmic muscular contractions of the genitals (sex organs) combined with waves of intense pleasurable sensations; in males, results in the ejaculation of semen.

Oribatid mite A mite belonging to the Oribatei, a large unit of mites containing about 35 families in the suborder Sarcoptiformes.

Oriental sore Disease caused by Leishmania tropica. Also called Jericho boil, Delhi boil, Aleppo boil, or cutaneous leishmaniasis.

Orientation The way an organism positions itself in relation to external cues.

Origin A site within a DNA sequence of a chromosome, plasmid, or non-integrated virus at which replication of the DNA is initiated.

Origin of replication (*ori*) Section of DNA sequence which is recognised by a cell's DNA replication proteins, allowing initiation of new DNA synthesis. DNA molecules which do not carry an *ori* recognised by the host cell will be eventually lost from a growing population unless they are incorporated into the cell's genome.

Orogeny The tectonic processes of folding, faulting, and uplifting of the earth's crust that result in the formation of mountains.

Oroya fever Clinical form of Carrion's disease, caused by the bacterium Bartonella bacilliformis and transmitted by sand flies.

Orthogenesis A unidirectional trend in the evolutionary history of a lineage as revealed by the fossil record; also, a now discredited, anti-Darwinian evolutionary theory, popular around 1900, postulating that genetic momentum forced lineages to evolve in a predestined linear direction that was independent of external factors and often led to decline and extinction.

Oscillator The internal mechanism that is the clock in a biological rhythm.

Osculum The main opening through which filtered water is discharged. Found in sponges.

Osmoconformer An organism whose body fluids have the same or very similar osmotic pressure as that of its aquatic environment; a marine organism that does not utilize energy in osmoregulation.

Osmolarity Solute concentration expressed as molarity.

Osmole. Molecular weight of a solute, in grams, divided by the number of ions or particles into which it dissociates in solution. Adj., osmolar.

Osmophilic microorganisms Microorganisms that grow best in or on media of high solute concentration.

Osmoregulation The regulation of the movement of water by osmosis into and out of cells; the maintenance of water balance within the body.

Osmoregulator An animal whose body fluids have a different osmolarity than the environment, and that must either discharge excess water if it lives in a hypotonic environment or take in water if it inhabits a hypertonic environment.

Osmosis Diffusion of water molecules across a membrane in response to differences in solute concentration. Water moves from areas of high-water/low-solute concentration to areas of low-water/high-solute concentration. Diffusion of water across a semi-permeable barrier such as a cell membrane, from high water potential (concentration) to lower water potential (concentration).

Osmotic conformers Organisms that tolerate large variations of internal ionic concentrations without serious damage by remaining isosmotic with the water around them.

Osmotic potential The tendency of water to move across a selectively permeable membrane into a solution; it is determined by measuring the pressure required to stop the osmotic movement of water into the solution.

Osmotic potential. Osmotic pressure.

Osmotic pressure A measure of the tendency of a solution to take up water when separated from pure water by a selectively permeable membrane.

Osmotic pressure In hypoosmotic conditions, the internal fluid pressure that develops from the osmotic inflow of water.

Osmotic pressure Pressure generated by water moving by osmosis into or out of a cell.

Osmotic pressure Water tends to move towards dense concentrations of ions. Sugar solutions on the surface of exposed dentine cause water to be drawn out of the dentinal tubules causing which distorts the odontoblast causing pain.

Osmotic shock Alternating seawater and dilute freshwater in intertidal zones causes erratic osmotic balance in plants growing in these areas.

Osmotolerant Organisms that grow over a fairly wide range of water activity or solute concentration.

Osmotroph A heterotrophic organism that absorbs dissolved nutrients.

Osphradium A sense organ in aquatic snails and bivalves that tests incoming water.

Osseous integration A term used to describe the desired adhesion between an implant and the bone which holds it in place.

Ossicles Small separate pieces of echinoderm endoskeleton. Also, tiny bones of the middle ear of vertebrates.

Ossification The process by which embryonic cartilage is replaced with bone.

Ossified skull Bony skull.

Osteichthyes The class of fishes whose members are characterized by the presence of a bony skeleton, a swim bladder, and an operculum. Bony fishes.

Osteoarthritis A degenerative condition associated with the wearing away of the protective cap of cartilage at the ends of bones. Bone growths or spurs develop, restricting movement and causing pain.

Osteoblast A cell which produces bone matrix. An osteoblast will become an osteocyte once it becomes embedded within the matrix of the bone.

Osteocalcin A calcium binding protein, synthesised by the osteoblast and secreted into the matrix at the time of bone mineralisation. Mice bread without the osteocalcin gene develop heavy bones suggesting that osteocalcin is a negative regulator of bone formation

Osteoclast A multi-nucleate cellcapable of removing both the organic and mineral component of bone. Osteoclasts activity is controlled by nearby osteoblasts.

Osteocytes Bone cells that lay down new bone; found in the concentric layers of compact bone. Bone cell, a type of connective tissue.

Osteoderm A bony deposit in the form of a plate or scale found in the dermal layers of the skin.

Osteoid The extracellular matrix inwhich bone forms. It is high in collagen and other bone proteins but lacks any crystal formation.

Osteon Unit of bone structure; Haversian system.

Osteonectin A bone glycoprotein which has the property of binding to both collagen fibres and the hydroxyapatite crystals, and thus may be important in initiating bone mineralisation by acting as a template for nucleation. Osteonectin is also produced by endothelial cells and plateletes and is able to bindfibrinogen.

Osteopontin An adhesive glycoprotein related to sialoproteins, which is secreted by osteoclasts to assist in their adhesion to the bone surface. After bone resorption it may then act as a signal to stimulate osteoblast activity

Osteoporosis A reduction in bone mass which occurs commonly in post menopausal females, but also in older men. It is due to a reduction in the activity of the ovaries and a decreased secretion of oestrogen. Bone formation and bone healing are not affected but more bone is resorbed by osteoclasts than is replaced. Lack of exercise is also a factor in bone loss.

Osteostracans A group of Paleozoic (Upper Silurian to Upper Devonian) agnathans belonging to the order Cephalaspidiformes.

Ostium Opening.

Ostracoderm An extinct agnathan; a fishlike creature encased in an armor of bony plates.

Otoacariasis Infestation of the external ear canal by ticks or mites.

Otolith Calcareous concretions in the membranous labyrinth of the inner ear of lower vertebrates, or in the auditory organ of certain invertebrates.

Out of Africa hypothesis Holds that modern human populations (*Homo sapiens*) are all derived from a single speciation event that took place in a restricted region in Africa.

Outbreak The sudden, unexpected occurrence of a disease in a given population.

Outbreeding depression A reduction in the fitness of offspring resulting from matings among individuals from different populations or demes, possibly caused by the breaking up of coadapted gene complexes.

Outcrop Any place where bedrock is visible on the surface of the Earth.

Outcrosser A flower that must be cross-pollinated to successfully complete the reproductive process.

Outer membrane A special membrane located outside the peptidoglycan layer in the cell walls of gram-negative bacteria.

Outgroup. In phylogenetic systematic studies, a species or group of species closely related to but not included within a taxon whose phylogeny is being studied, and used to polarize variation of characters and to root the phylogenetic tree.

Outwelling The export of detritus and other organic matter from estuaries to other ecosystems.

Ovarian cycle The cycle in the ovary during which the oocyte matures and ovulation occurs.

Ovarian follicle The layers of epithelial cells that surround a developing mammalian oocyte. The process begins with the primary follicle and ends with the Graafian follicle, which contains the mature ovum ready for release during ovulation.

Ovarian stroma The structural framework of the ovary that is formed by connective tissue and smooth muscle.

Ovaries (1) In animals, the female gonads, which produce eggs (ova) and female sex hormones.

(2) In flowers, part of the female reproductive structure in the carpel; contain the ovules, where egg development occurs. The lower part of the carpel that contains the ovules within which the female gametophyte develops.

Overburden Overlying layers of noncommercial sediments that must be removed to reach a mineral or coal deposit.

Over-consumption The use of resources at a rate that exceeds the ability of natural processes to replace them.

Overdispersion Nonrandom dispersion of individuals in a habitat, such as when a minority of host individuals bears a majority of parasites.

Overdominance Selection favoring heterozygotes.

Overhang Single-stranded section of DNA at the end of a double-stranded fragment, for example in a sticky end. Overhangs may be 3' or 5' i.e. the end of the single-stranded section may carry a free 3' or 5' end. This will depend on how the overhang was created.

Overkill The shooting, trapping, or poisoning of certain populations, usually for sport or economic reasons.

Overnutrition Receiving too many calories.

Overpopulation Too many organisms (especially humans) for the available resources to support in a sustainable way.

Overshoot The extent to which a population exceeds the carrying capacity of its environment.

Overturn The sinking of surface water that has become more dense than the water below .

Overwinter A period of rest or hibernation by which insects survive the winter.

Ovicapt Sphincter on the oviduct of a flatworm.

Oviducts Tubes that connect the ovaries and the uterus; transport sperm to the ova, transport the fertilized ova to the uterus, and serve as the site of fertilization; also called the fallopian tubes or uterine tubes.

Oviger Leg that carries eggs in pycnogonids.

Oviparity Reproduction in which eggs are released by the female; development of offspring occurs outside the maternal body.

Oviparous Referring to a type of development in which young hatch from eggs laid outside the mother's body.

Ovipositor The tubular or valved egg-laying apparatus of a female insect: concealed in many insects, but extremely large among the bush-crickets and some parasitic hymenopterans.

Ovisac External sac attached to the somite that bears openings of gonoducts in females of many Copepoda. Fertilized eggs pass into the ovisacs for embryonation.

Ovovitellarium Mixed mass of ova and vitelline cells; found in the monogenean genus Gyrodactylus and in a few tapeworms.

Ovoviviparity Reproduction in which eggs develop within the maternal body without additional nourishment from the parent and hatch within the parent, or immediately after laying.

Ovoviviparous Describes reproduction in which embryos develop within the maternal body without additional nourishment from the parent and hatch within the parent or immediately after emerging.

Ovulation The discharge of an ovum from a vesicular follicle of the ovary; this usually occurs on the 14th day after the first day of the last menstrual period.

Ovulation The release of the oocyte onto the surface of the ovary; occurs at the midpoint of the ovarian cycle. The release of the ovum (egg) from the ovary after the peaking of luteinizing hormone concentration in the blood during the menstrual cycle.

Ovule In seed plants, a protective structure in which the female gametophyte develops, fertilization occurs, and seeds develop; contained within the ovary. Structures inside the ovary of the flower within which the female gametophyte develops after megasporogenesis has produced a megaspore inside each ovule.

Ovuliferous scale In conifers, an axillary, scalelike shoot that bears one or more ovules. Each strobilus may contain one or many ovuliferous scales, depending on the species.

Ovum The female gamete; the haploid, unfertilized egg, which is usually a relatively large, nonmotile cell.

Oxidation The loss of an electron by an atom or molecule; sometimes addition of oxygen chemically to a substance. Opposite of reduction, in which an electron is accepted by an atom or molecule.

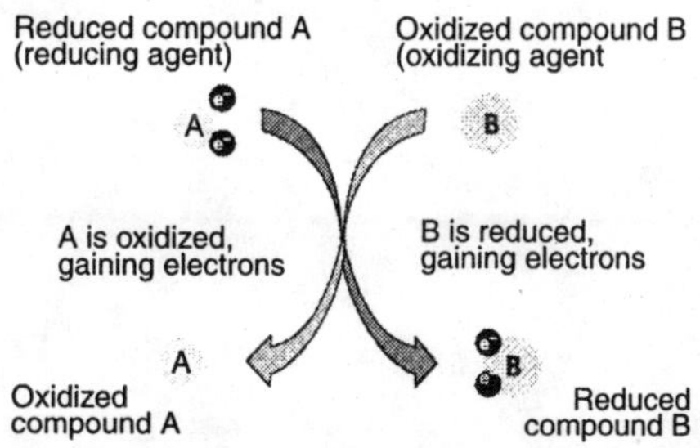

Fig. *Oxidation*

Oxidation-reduction reactions Reactions involving electron transfers; the reductant donates electrons to an oxidant.

Oxidative phosphorylation The electron transport system associated with aerobic respiration and mitochondria. In the release of energy through a series of cytochromes, three molecules of ATP are made.

Oxidizing agent or oxidant The electron acceptor in an oxidation-reduction reaction.

Oxygen An element that is one of the constituents of water, organic matter, and many other chemicals. Oxygen gas (O_2), composed of two oxygen atoms, is needed for respiration and is produced by photosynthesis.

Oxygen cycle The circulation and reutilization of oxygen in the biosphere.

Oxygen debt In muscle, the cumulative deficit of oxygen that develops during strenuous exercise when the supply of oxygen is inadequate for the demand; ATP is produced anaerobically by glycolysis, and the resulting pyruvic acid is converted to lactic acid, which is subsequently metabolized when adequate oxygen is available.

Oxygen dissociation curve A curve showing the percent saturation of a blood pigment, such as hemoglobin, as a function of oxygen concentration of the fluid

Oxygen minimum layer A depth zone, usually below the thermocline, in which dissolved oxygen is minimal

Oxygen minimum zone the ocean layer below the photic zone where dissolved oxygen concentration is lowest

Oxygen sag Oxygen decline downstream from a pollution source that introduces materials with high biological oxygen demands.

Oxygen technique The estimation of primary productivity by the measurement of the rate of oxygen increase

Oxygenic photosynthesis Photosynthesis that oxidizes water to form oxygen; the form of photosynthesis characteristic of eucaryotic algae and cyanobacteria.

Oxyhemoglobin Compound formed when oxygen combines with hemoglobin.

Oxytalin fibres Are related to elastic fibres, though they have a smaller core of elastin. They are found in the periodontal ligament and in the epidermis of thin skin, but not in the oral mucosa.

Oxytocin A peptide hormone secreted by the posterior pituitary that stimulates the contraction of the uterus during childbirth.

Oyster reef A dense oyster bed present in some estuaries and other marine environments.

Ozone A highly reactive molecule containing three oxygen atoms; a dangerous pollutant in ambient air. In the stratosphere, however, ozone forms an ultraviolet absorbing shield that protects us from mutagenic radiation.

Ozone layer Ozone (O_3) in the atmosphere that deflects ultraviolet radiation, which is harmful to life.

P

P element Mobile sequence of DNA (transposon/transposable element) found in Drosophila that can 'hop' in and out of fly chromosomes; can be used with transgenic techniques.

P53 protein. A tumor suppressor protein with critical functions in normal cells. A mutation in the gene that encodes it, p53, can result in loss of control over cell division and thus cancer.

P680 The reaction center for Photosystem II; a chlorophyll a molecule that is bound to a membrane protein and has a peak absorbance at 680 nm.

P700 The reaction center for Photosystem I; a chlorophyll a molecule that is bound to a membrane protein and has a peak absorbance at 700 nm.

Pacemaker A specialized region of the right atrium of the mammalian heart that sets the rate of contraction; also called the sinoatrial (SA) node.

Pachytene The third stage in prophase 1 of meiosis in which there is further shortening and thickening of the bivalents. The chromosomes split into chromatids, allowing crossing over to occur and hence the formation of chiasmata. This continues in the following stage of diplotene.

Pacinian corpuscle A sensory receptor in skin, muscles, body joints, body organs, and tendons that is involved with the vibratory sense and firm pressure on the skin; also called a lamellated corpuscle.

Package In recombinant DNA procedures, refers to the step of incorporation of *cosmid* or other lambda *vector* DNA with an *insert* into a *phage* head for transduction of DNA into host.

Paddy A flooded field used to cultivate lowland rice.

Paedogenesis Reproduction by immature or larval animals caused by acceleration of maturation. Progenesis.

Paedomorphosis Condition in which a larva becomes sexually mature without attaining the adult body form. Paedomorphosis may be achieved by neotony or by progenesis.

Pair bond. An affiliation between an adult male and an adult female for reproduction. Characteristic of monogamous species.

Palatal complex Bones and soft tissue that make up the palate.

Palea One of two bracts around the grass flower.

Paleoherb Any member of a group of basal flowering herbs which may be the closest relatives of the monocots. They include the water lilies, Piperales, and Aristolochiales.

Paleolithic Old Stone Age; a cultural period during

which early humans obtained food solely by foraging; ending in some areas approximately 10,000 years ago.

Paleontologist A person who studies fossils and all aspects of extinct life.

Paleontology The study of ancient life by collection and analysis of fossils.

Paleosol Soil horizon from the geologic past.

Paleo-tethys ocean A large ocean that originated between eastern Gondwana, Siberia, Kazakhstan, and Baltica in the Ordovician and finally closed in the Jurassic; replaced by the Tethys Ocean as eastern Pangea was assembled.

Paleozoic era The period of time beginning 570 million years ago ending 245 million years ago; falls between the Proterozoic and Mesozoic Eras and is divided into the Cambrian, Ordovician, Silurian, Devonian, Carboniferous, and Permian Periods.

Palindromic sequence A nucleotide sequence which is the same when read in either direction, usually consisting of adjacent inverted repeats. *Restriction endonuclease recognition sites* are palindromes:

5'—>3'

GAATTC *Eco*RI recognition site.

Palisade Layer of mesophyll cells in leaves that are closely placed together under the epidermal layer of the leaf. Palisade parenchyma: Columnar cells located just below the upper epidermis in leaves the cells where most of the light absorbtion in photosynthesis occurs.

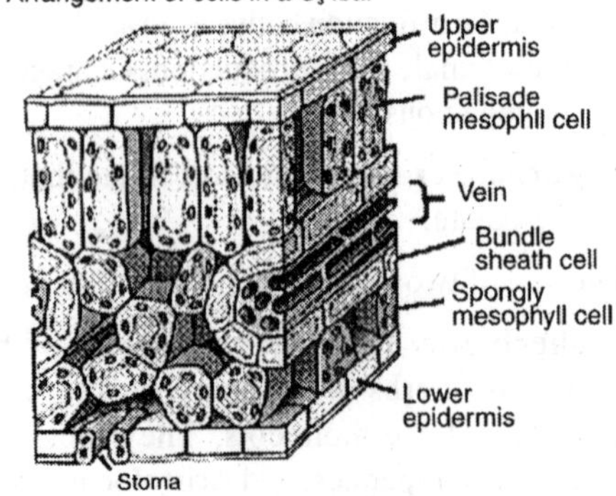

Fig. *Palisade*

Palisade parenchyma Parenchyma cells in the leaf mesophyll characterized by uniform rows of tightly packed cells with many chloroplasts beneath the upper epidermis.

Pallium Mantle of a mollusc or brachiopod.

Palmate An arrangement of leaflets (or lobes on a simple leaf), each originating from a common point, usually the axial end

Palmately compound leaf Leaflets radiating from a common point.

Palp A segmented leg-like structure arising on the maxilla or labium. Palps have a sensory function and play a major role in tasting food.

Palpebrae Eyelid.

Palynology The study of palynomorphs and other acid-resistant microfossils usually produced by plants, protists, and fungi.

Palynomorph Generic term for any object a palynologist studies.

Pampiniform A threadlike plexus.

Pancreas A gland in the abdominal cavity that secretes digestive enzymes into the small intestine and also secretes the hormones insulin and glucagon into the blood, where they regulate blood glucose levels. A digestive organ that produces trypsin, chymotrypsin and other enzymes as a pancreatic juice, but which also has endocrine functions in the production of the hormones somatostatin, insulin, and glucagon.

Pancreatic islets Clusters of endocrine cells in the pancreas that secrete insulin and glucagon; also known as islets of Langerhans.

Pandemic An increase in the occurrence of a disease within a large and geographically widespread population (often refers to a worldwide epidemic).

Paneth cell The granular cell located at the base of glands in the small intestine; it produces the enzyme lysozyme.

Pangaea The name proposed by German

meteorologist Alfred Wegener for a supercontinent that existed at the end of the Paleozoic Era and consisted of all the Earth's landmasses.

Pangenesis Darwin's hypothesis that hereditary characteristics are carried by individual body cells that produce particles that collect in the germ cells.

Panicle A branched inflorescence with the branches bearing loose flower clusters.

Panning A selection method by which a transgenic virus expressing a desired protein may be selected from a mixture of viruses.

Pannotia A supercontinent that existed in the Late Precambrian and gave rise to the continents of Gondwana, Laurentia, Siberia, and Baltica in the Cambrian.

Pannus A superficial vascularization of the cornea with infiltration of granulation tissue.

Pansporoblast Myxosporidean sporoblast that gives rise to more than one spore. Also called a sporoblast mother cell.

Panthalassic ocean A vast ocean that existed from the Late Precambrian to the Jurassic, circling the globe and connecting to smaller oceans that developed throughout the Phanerozoic; also known as the Panthalassa.

Panzootic The wide dissemination of a disease in an animal population.

Papatasi fever Virus disease transmitted by sand flies. Also called sand fly fever.

Papilla A small nipplelike projection. A vascular process that nourishes the root of a hair, feather, or developing tooth.

Papula Respiratory processes on skin of sea stars; also, pustules on skin.

Parabasal body Golgi body located near the basal body (kinetosome) of some flagellate protozoa, from which the parabasal filament runs to the basal body.

Parabasal filament Fibril, with periodicity visible in electron micrographs, that courses between the parabasal body and a kinetosome.

Parabiosis The fusion of two individuals, resulting in mutual physiological intimacy.

Parabronchi The tiny air tubes within the lung of a bird across which gas exchange occurs.

Paracrine Cell messengers also called cytokines which are locally acting, produced by neighbouring cells or the extracellular matrix, as distinct from as distinct from endochrine or hormonal messengers.

Paraglossa One of a pair of lobes at the outer edges of the tip of the labium: with the central glossae, the paraglossae make up the ligula.

Parakeratinised An epithelium in which the superficial cells have **not** lost their nuclei , but have become filled with keratin.

Parallel evolution The development of similar characteristics in organisms that are not closely related (not part of a monophyletic group) due to adaptation to similar environments and/or strategies of life.

Parallel venation A type of venation in which the main veins of a leaf are parallel or nearly so, but converge at the apex and base.

Paralytic shellfish poisoning Dinoflagellates (Gonyaulax spp.) produce a powerful neurotoxin called saxitoxin. Shellfish accumulate saxitoxin and are poisonous when consumed by animals and humans. Saxitoxin paralyzes the striated respiratory muscles by inhibiting sodium transport. Paralytic shellfish poisoning is characterized by numbness of the mouth, lips, face, and extremities.

Paramastigote Form of trypanosomatid in which the kinetosome and kinetoplast are beside the nucleus.

Paramere Copulatory appendage in male cimicid bugs.

Paramylon bodies Organelles containing the starch-like substance paramylon; in some algae and flagellates.

Parapatric speciation Speciation that occurs in small, local populations, called demes.

Parapatric speciation The differentiation into distinct species of populations experiencing some gene flow.

Paraphyletic Term applied to a group of organisms which includes the most recent common ancestor of all of its members, but not all of the descendants of that most recent common ancestor.

Paraphyly The condition that a taxon or other group of organisms contains the most recent common ancestor of all members of the group but excludes some descendants of that ancestor; contrasts with monophyly and polyphyly.

Parapodia A sort of "false foot" formed by extension of the body cavity. Polychaetes and some insect larvae have parapodia in addition to their legs, and these provide extra help in locomotion.

Parapolar cells Cells making up the ciliated somatoderm immediately behind the calotte of a mesozoan.

Parasite An organism that lives on or within a host (another organism); it obtains nutrients from the host without benefiting or killing (although it may damage) the host; parasitic- adj.; parasitism- a type of symbiotic relationship in which one organism benefits and the other does not.

Parasitic castration Condition in which a parasite causes retardation in development or atrophy of host gonads, often accompanied by failure of secondary sexual characteristics to develop.

Parasitic Denoting an association where one living organism benefits at the expense of another.

Parasitism A form of symbiosis in which the population of one species benefits at the expense of the population of another species; similar to predation, but differs in that parasites act more slowly than predators and do not always kill the host. A type of symbiosis in which one organism benefits at the expense of the other, for example the influenza virus is a parasite on its human host. Viruses, are obligate intracellular parasites.

Parasitoid Organism that is a typical parasite early in its development but that finally kills the host during or at the completion of development; often used in reference to many insect parasites of other insects.

Parasitologist Quaint person who seeks truth in strange places; a person who sits on one stool, staring at another.

Parasitology The study of the biology of parasites. For example, the study of the biology of the parasite that causes malaria, *Plasmodium falciparum*.

Parasitophorous vacuole Vacuole within a host cell that contains a parasite.

Parasphenoid teeth In amphibians, teeth actually borne by the parasphenoid bone itself.

Parasympathetic division One of two divisions of the autonomic nervous system; generally enhances body activities that gain and conserve energy, such as digestion and reduced heart rate.

Parasympathetic nervous system Part of the autonomic nervous system that regulates the involuntary activity of glands, smooth muscle, and cardiac muscle. It is derived from the vagal and sacral neural crest cells.

Paratenic host Host in which a parasite survives without undergoing further development. Also known as a transport host.

Parathyroid glands Four endocrine glands, embedded in the surface of the thyroid gland, that secrete parathyroid hormone and raise blood calcium levels.

Parenchyma One of the three major cell types in plants. Parenchyma cells have thin, usually multisided walls, are unspecialized but carry on photosynthesis and cellular respiration and can store food; form the bulk of the plant body; found in the fleshy tissue of fruits and seeds, photosynthetic cells of leaves, and the vascular system. Generalized plant cells whose numerous functions include photosynthesis, storage, bulk of herbaceous stem tissues, lateral transport in woody stems. Parenchyma are variously shaped but are characterized by thin walls and remain alive at functional maturity.

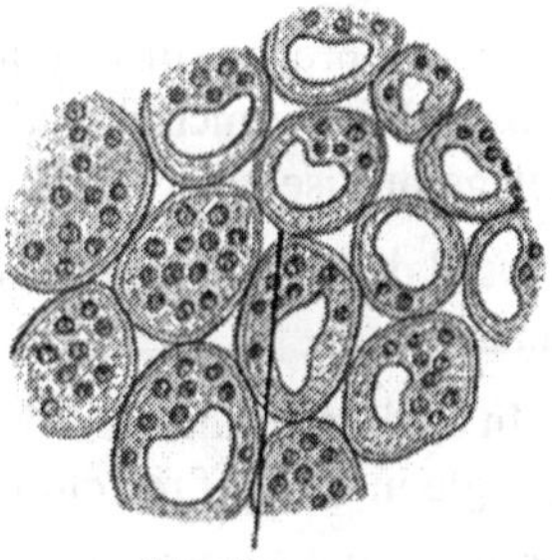

Fig. *Parenchyma*

Parent material Undecomposed mineral particles and unweathered rock fragments beneath the subsoil; weathering of this layer produces new soil particles for the layers above.

Parental generation In an experimental genetic cross, the parents of the F1 generation; homozygous for the trait(s) being studied.

Parental investment Any investment in an offspring that increases its chances of survival and reproduction at the expense of the parents' ability to invest in other offspring.

Parental manipulation The selective providing of care to some offspring at the expense of other offspring so as to maximize the parents' reproductive success.

Parenteral route A route of drug administration that is nonoral (e.g., by injection).

Parfocal A microscope that retains proper focus when the objectives are changed.

Parietal cell Cell of a gastric gland that secretes hydrochloric acid and intrinsic factor.

Parietal eye A sensory structure capable of light reception, located on the dorsal side of the diencephalon, and opening to the outside through the parietal foramen.

Parietal lobe The lobe of the cerebral cortex that lies at the top of the brain; processes information about touch, taste, pressure, pain, and heat and cold.

Parity The condition or fact of having borne offspring.

Paronychia Inflammation involving the folds of tissue surrounding the nail; usually caused by Candida albicans.

Parotoid gland Large, swollen, glandular area lying behind the eye on the head, and extending in some species well onto the neck, in some species of anurans. Prominently developed in bufonid (toad) species, in many of which the glandular secretions are potent poisons.

Pars prostatica Dilation of the ejaculatory duct of a flatworm, surrounded by unicellular prostate cells.

Parsimony Refers to a rule used to choose among possible cladograms, which states that the cladogram implying the least number of changes in character states is the best.

Parthenocarpy Fruit development without fertilization.

Parthenogenesis Special reproductive strategy in which unfertilized eggs undergo cell division and embryogenesis to develop into viable adult individuals ("virgin birth"). The embryo develops without a genetic contribution from the sperm, although in some species fertilization is necessary for egg activation.

Partial pressures The concentration of gases; a fraction of total pressure.

Particulate material Atmospheric aerosols, such as dust, ash, soot, lint, smoke, pollen, spores, algal cells, and other suspended materials; originally applied only to solid particles but now extended to droplets of liquid.

Particulate organic matter Particulate material in the sea derived from the decomposition of the nonmineral constituents of living organisms.

Parts per billion (ppb) Number of parts of a chemical found in one billion parts of a particular gas, liquid, or solid mixture.

Parts per million (ppm) Number of parts of a chemical found in one million parts of a particular gas, liquid, or solid mixture.

Parts per trillion (ppt) Number of parts of a chemical found in one trillion (1012) parts of a particular gas, liquid, or solid mixture.

Parturition The process of childbirth.

Paruterine organ Fibromuscular organ in some cestodes that replaces the uterus.

Passive avoidance learning A form of operant conditioning, in which the animal does not make an overt response, but rather is trained to avoid making a particular response by being exposed to some type of noxious stimulus each time the response occurs.

Passive continental margin A continental margin that is located at the "trailing edge" of a continent and as a result shows little geological activity . Compare active continental margin.

Passive diffusion The process in which molecules move from a region of higher concentration to one of lower concentration as a result of random thermal agitation.

Passive heat absorption The use of natural materials or absorptive structures without moving parts to gather and hold heat; the simplest and oldest use of solar energy.

Passive immunity Immunisation, usually of relatively short duration, induced by the introduction of preformed antibodies into an individual. This may occur naturally, as with the transfer of antibodies between mother and baby, either via the placenta or via breast milk. Or the antibodies may be acquired via injection.

Passive immunization Immune state in an animal created by inoculation with serum (containing antibodies) or lymphocytes from an immune animal, rather than by exposure to the antigen.

Passive transport Diffusion across a plasma membrane in which the cell expends no energy.

Pasteur effect The decrease in the rate of sugar catabolism and change to aerobic respiration that occurs when microorganisms are switched from anaerobic to aerobic conditions.

Pasteurization The process of heating milk and other liquids to destroy microorganisms that can cause spoilage or disease.

Pasture Enclosed domestic meadows or managed grazing lands.

Patagium 1. In birds, a feathered web of skin that spans the angle in front of the elbow.

2. a fold of skin in flying squirrels, flying lizards, and other arboreal gliding animals that encloses the limbs on both sides from neck to tail, enabling the animal to glide.

Patch A nonlinear habitat type that differs from the surrounding vegetation.

Patch model A type of optimal foraging model that assumes that prey occur in discrete clumps and that seeks to predict where and when organisms will forage for these prey.

Patchiness A condition in which organisms occur in aggregations

Patchiness Within a larger ecosystem, the presence of smaller areas that differ in some physical conditions and thus support somewhat different communities; a diversity-promoting phenomenon.

Patella Knee bone.

Patent A legal document, protecting the inventor/discoverer of a new process which is deemed to be 'novel, inventive and innovative' from having her/his work copied and sold by others for a profit.

Pathogen An organism that produces disease in a host organism, disease being an alteration of one or more metabolic functions in response to the presence of the organism.

Pathogenesis Production and development of disease.

Pathogenic Giving origin to disease.

Pathogenic potential The degree that a pathogen causes morbid signs and symptoms.

Pathogenicity The condition or quality of being pathogenic, or the ability to cause disease.

Pathogenicity island A large segment of DNA in some pathogens that contains the genes responsible for virulence; often it codes for the type III secretion system that allows the pathogen to secrete virulence proteins and damage host cells. A pathogen may have more than one pathogenicity island.

Pathway architecture The analysis, design, and modification of biochemical pathways to increase process efficiency.

Pattern formation The ordering of cells into specific three-dimensional structures, an essential part of shaping an organism and its individual parts during development.

Paurometabolous metamorphosis A form of insect development in which immatures resemble parents, and molting is restricted to the immature stages.

Pauropoda A class of arthropods whose bodies are small, soft, 11 segmented, and have 9 pairs of legs.

PBR322 A common plasmid. Along with the obligatory origin of replication, this plasmid has genes which make the *E. coli* host resistant to ampicillin and tetracycline. It also has several restriction sites (BamHI, PstI, EcoRI, HindIII etc.) into which DNA fragments could be spliced in order to clone them.

PCR Polymerase Chain Reaction. A method for amplifying specific DNA sequences in a laboratory. Many millions of copies of the DNA can be produced in just a few hours.

Peanut worms Burrowing invertebrates with an unsegmented body and an anterior end that can be pulled into the body.

Peat Deposits of moist, acidic, semidecayed organic matter.

Peat moss A relatively sterile, inert medium composed of partially decomposed plants of the genus Sphagnum. Exceptionally high water-holding capacity.

Pebrine An infectious disease of silkworms caused by the protozoan Nosema bombicis.

Pecten Any of several types of comblike structures on various organisms, for example, a pigmented, vascular, and comblike process that projects into the vitreous humor from the retina at a point of entrance of the optic nerve in the eyes of all birds and many reptiles.

Pectin A complex polysaccharide in the middle lamella and primary walls of plant cells.

Pectinate Having branches which arise from the main axis like the teeth of a comb: usually applied to antennae.

Pectines Sensory appendage on abdomens of scorpions.

Pectoral Of or pertaining to the breast or chest; to the pectoral girdle; or to a pair of horny shields of the plastron of certain turtles.

Pectoral fin Each of the pair of fins just behind the head of fishes.

Pectoral girdle In humans, the bony arch by which the arms are attached to the rest of the skeleton; composed of the clavicle and scapula.

Ped A natural soil aggregate, formed partly through bacterial and fungal growth in the soil.

Pedal locomotion The type of locomotion exhibited by flatworms, some cnidarians, and the gastropod molluscs. This locomotion involves waves of activity in the muscular system, which is applied to the substratum.

Pedalium Flattened blade at the base of the tentacles in cubozoan medusae.

Pedicel The 2nd antennal segment: the name is also given to the narrow waist of an ant.

Pedicellariae Pincerlike structures found on the body wall of many echinoderms. They are used in cleaning and defense.

Pedigree A family tree describing the occurrence of heritable characters in parents and offspring across as many generations as possible.

Pedigree analysis A type of genetic analysis in which a trait is traced through several generations of a family to determine how the trait is inherited. The information is displayed in a pedigree chart using standard symbols.

Pedipalps The second pair of appendages of cheliceromorphs. In many arachnids, such as spiders, the pedipalps are enlarged in the male and used for copulation.

Pelagic Refers to the plants and animals that live in the water column or in the open waters of the ocean rather than the ocean floor. Life is found throughout the pelagic zone, however is more concentrated at shallower depths. Pelagic organisms can be further divided into the *plankton* and *nekton.*

Pelagic organism An organism that lives above the bottom in the open sea.

Pelagic zone One of the two basic subdivisions of the marine biome; consists of the water above the sea floor and its organisms.

Pellagra Lassitude, torpor, dermatitis, diarrhea, dementia, and death brought about by a diet deficient in tryptophan and niacin.

Pellets Compacted aggregations of particles resulting either from egestion (fecal pellets) or from burrow-constructing activities of marine organisms

Pellicle A thin, frequently noncellular covering of an animal (e.g., the protective and supportive pellicle of protists occurs just below the plasma membrane); may be composed of a cell membrane, cytoskeleton, and other organelles.

Pelt The hide or skin of an animal.

Pelvic Situated at or near the pelvis, as applied to girdle, cavity, fins, and limbs.

Pelvic fin Each of the second pair of ventral fins of fishes.

Pelvic girdle In humans, the bony arch by which the legs are attached to the rest of the skeleton; composed of the two hipbones.

Pelvic inflammatory disease (PID) A severe infection of the female reproductive organs. The disease that results when gonococci and chlamydiae infect the uterine tubes and surrounding tissue.

Pelycosaur Any of a group of carnivorous Permian synapsids distinguished by powerful jaws, stabbing teeth, and a large skin-covered sail on the back.

Pen in squids, a thin, chitinous structure extending the length of the mantle tissue

Penetrance In genetics, the proportion of individuals with a particular genotype that show the phenotype ascribed to that genotype.

Penetration anchor In hydraulically burrowing organisms, any device used to penetrate and gain an initial purchase on the sediment so that the body can be thrust in farther

Penicillin An antibiotic produced by certain species of Penicillium. An ascomycete fungus.

Pennsylvanian A geological period of the Carboniferous beginning about 320 million years ago.

Pentadactyl With five digits, or five fingerlike parts, to the hand or foot.

Pentamerous symmetry A radial symmetry based on five or multiples thereof.

Pentaradial symmetry A form of radial symmetry found in the echinoderms in which body parts are arranged in fives around an oral-aboral axis.

Pentastomida A phylum of worms that are all endoparasites in the lungs or nasal passageways of carnivorous vertebrates. Tongue worms.

Penton or pentamer A capsomer composed of five protomers.

Pentose phosphate pathway The pathway that oxidizes glucose 6-phosphate to ribulose 5-phosphate and then converts it to a variety of three to seven carbon sugars; it forms several important products (NADPH for biosynthesis, pentoses, and other sugars) and also can be used to degrade glucose to CO_2.

Penultimate phalange Next to last digit.

PEP carboxylase The enzyme responsible for CO_2 fixation in the primary fixation of C_4 metabolism.

Peplomer or spike A protein or protein complex that extends from the virus envelope and often is important in virion attachment to the host cell surface.

Pepo A berry with the outer wall formed from receptacle tissue fused to the exocarp; the fleshy interior is mesocarp and endocarp.

Pepsin An enzyme produced from pepsinogen that initiates protein digestion by breaking down protein into large peptide fragments. An enzyme, produced by the stomach, that chemically breaks down peptide bonds in polypeptides and proteins.

Pepsinogen An inactive form of pepsin; synthesized and stored in cells lining the gastric pits of the stomach.

Peptic ulcer Damage to the epithelial layer of the stomach lining; generally caused by bacterial infection.

Peptidase An enzyme that breaks down simple peptides, releasing amino acids.

Peptide A molecule formed by peptide bonds covalently linking two or more *amino acids*. Short peptides (generally less than 60 amino acid residues, and usually only half that length) can be chemically synthesized by one of several different methods; larger peptides (more correctly, *polypeptides*) are usually *expressed* from recombinant DNA.

Peptide bonds A covalent bond made between the carbon atom of the carboxyl group of one amino acid and the nitrogen atom from the amine group of another. In the process a molecule of water is removed. Peptide bonds allow chains of amino acids to form polypeptides and proteins When peptide bonds are broken apart they need water to reform the amino acids. This process is known as hydrolysis, and occurs during cooking and in digestion.

Peptide interbridge A short peptide chain that connects the tetrapeptide chains in some peptidoglycans.

Peptidoglycan A large polymer composed of long chains of alternating N-acetylglucosamine and N-acetylmuramic acid residues. The polysaccharide chains are linked to each other through connections between tetrapeptide chains attached to the N-acetylmuramic acids. It provides much of the strength and rigidity possessed by bacterial cell walls.

Peptidyl or donor site (P site) The site on the ribosome that contains the peptidyl-tRNA at the beginning of the elongation cycle during protein synthesis.

Peptidyl transferase The enzyme that catalyzes the transpeptidation reaction in protein synthesis; in this reaction, an amino acid is added to the growing peptide chain.

Peptones Water-soluble digests or hydrolysates of proteins that are used in the preparation of culture media.

Perception The interpretation of sensations by the brain.

Perceptual world The manner in which an organism's brain analyzes and interprets all incoming stimuli. Each species has a different view of the world, its Umwelt.

Percolation The movement of water by gravitational pull down through soil particles.

Pereiopod Thoracic appendage of a crustacean.

Perennial A plant that overwinters and continues to grow for many years. It may reproduce every year, or only on rare occasions.

Perennibranchiate Having permanent gills, relating especially to certain paedomorphic salamanders.

Perfect flower A flower having both stamens and carpels.

Perfect stage The phase during the life cycle of a fungus when sexual fusion occurs, producing characteristic sexual spores.

Perforin pathway The cytotoxic pathway that uses perforin protein, which polymerizes to form membrane pores that help destroy cells during cell-mediated cytotoxicity. Perforin is produced by cytotoxic T cells and NK cells and stored in granules that are released when a target cell is contacted.

Perianth The sepals and petals of a flower are together called the perianth; literally "around the anthers".

Periaquaductal grey Is an integrative centre for inputs form the autonomic nervous system, the limbic system and from sensory and motor pathways. It has an inhibitory affect on pain transmission due to descending connections through the raphe nucleus along the corticospinal tract to the cells of the dorsal horn.

Pericardial coelom The cavity that surrounds the heart and is bounded by the pericardium.

Pericardium Area around heart; membrane around heart.

Pericarp The fruit wall that develops from the ovary wall.

Perichondrium A layer of connective tissue that forms around the cartilage during bone formation. Cells in the perichondrium lay down a peripheral layer that develops into compact bone.

Pericycle Root tissue sandwiched between the endodermis and the phloem; it is the outermost layer of the stele; meristematic region that gives rise to branch roots.

Pericytes Small cells lying next to the endothelial cells of capillaries which have the capacity to differentiate into osteoblasts.

Periderm The protective coat that replaces the epidermis in plants during secondary growth, formed of the cork and cork cambium.

Peridinin Carotenoid pigment found in dinoflagellates.

Perigynous A flower in which the base of the sepals, petals, and stamens form a cup around the ovary.

Period The fundamental unit in the hierarchy of time units; a part of geologic time during which a particular sequence of rocks designated as a system was deposited. Units of geological time that are the major subdivisions of Eras.

Period of infectivity Refers to the time during which the source of an infectious disease is infectious or is disseminating the pathogen.

Periodontal disease A disease located around the teeth or in the periodontium-the tissue investing and supporting the teeth, including the cementum, periodontal ligament, alveolar bone, and gingiva.

Periodontal pocket Loss of epithelial attachment to the tooth, producing an increase in gingival sulcus depth beyond the normal 1-2 mm.

Periodontitis An inflammation of the periodontium.

Periodontosis A degenerative, noninflammatory condition of the periodontium, which is characterized by destruction of tissue.

Periosteum A fibrous membrane that covers bones and serves as the site of attachment for skeletal muscles; contains nerves, blood vessels, and lymphatic vessels.

Periostracum Outer horny layer of a mollusc shell.

Peripheral nervous system The division of the nervous system that connects the central nervous system to other parts of the body. Components of the nervous system that transmit messages to the central nervous system.

Peripheral protein A protein on the surface of a biological membrane.

Periphyton Dense strands of algal growth that cover the water surface between the emergant aquatic plants. *Spirogyra* is commonly responsible for this growth.

Periplasm The substance that fills the periplasmic space.

Periplasmic flagella The flagella that lie under the outer sheath and extend from both ends of the spirochete cell to overlap in the middle and form the axial filament. Also called axial fibrils and endoflagella.

Periplasmic space The space between the plasma membrane and the outer membrane in gram-negative bacteria, and between the plasma membrane and the cell wall in gram-positive bacteria.

Periproct Region of aboral plates around the anus of echinoids.

Perisarc Sheath covering the stalk and branches of a hydroid.

Perissodactyl Pertaining to an order of ungulate mammals with an odd number of digits.

Peristalsis Involuntary contractions of the smooth muscles in the walls of the esophagus, stomach, and intestines that propel food along the digestive tract. Waves of muscle contraction in the esophagus that propel food from the oral cavity to the stomach.

Peristome A set of cells or cell parts which surround the opening of a moss sporangium. In many mosses, they are sensitive to humidity, and will alter their shape to aid in spore dispersal.

Peristomium The segment of the body of an annelid that surrounds the mouth.

Perithecium A flask-shaped ascocarp.

Peritoneum A membrane that lines the body cavity and forms the external covering of the visceral organs.

Peritreme Elongated sclerite extending forward from the stigma of certain mites, mainly in the suborder Mesostigmata.

Peritrichous A cell with flagella evenly distributed over its surface.

Peritrophic membrane Noncellular, delicate membrane lining an insect's midgut.

Peritubular capillaries In the vertebrate kidney, the capillaries that surround the renal tubule; water and solutes are reabsorbed into the bloodstream through the peritubular capillaries and some substances are secreted from them into the renal tubule.

Perivitelline space The fluid-filled area formed between an external membrane of the egg (e.g. fertilization membrane, ascaroside layer, zona pellucida, etc.) and the plasma membrane.

Permafrost The permanently frozen soil in polar regions.

Permanent parasite Parasite that lives its entire adult life within or on a host.

Permanent retrievable storage Placing waste storage containers in a secure building, salt mine, or bedrock cavern where they can be inspected periodically and retrieved, if necessary, for repacking or for transfer if a better means of disposal or reuse is developed.

Permanent wilting point The soil moisture content at the point when a given plant's root system can no longer absorb water.

Permeability A property of membranes allowing all substances to pass freely.

Permease. A transporter molecule; a molecule in the cell membrane that makes it possible for another molecule (to which the membrane is not otherwise permeable) to be transported across the membrane, that is, mediated transport.

Permian The last geological period of the Paleozoic era beginning about 286 million years ago and ending 251 million years ago. The end of the Permian was a time of abrupt extinction in the diversity of flora and fauna, especially of sea life.

Permian period The last geologic time period of the Paleozoic Era, noted for the greatest mass extinction in earth history, when nearly 96% of species died out.

Permineralization Fossilization process that occurs when minerals, carried by ground water, enter and harden in the pores of an organism's structures.

Peroxisome A microbody containing enzymes that transfer hydrogen from various substrates to oxygen, producing and then degrading hydrogen peroxide.

Pertussis An acute, highly contagious infection of the respiratory tract, most frequently affecting young children, usually caused by Bordetella pertussis or B. parapertussis. Consists of peculiar paroxysms of coughing, ending in a prolonged crowing or whooping respiration; hence the name whooping cough.

Pes A foot.

Pest An organism that interferes with human activities, property, or health, or is objectionable.

Pest resurgence Rebound of pest populations due to acquired resistance to chemicals and nonspecific destruction of natural predators and competitors by broadscale pesticides.

Pesticides Chemical products used to reduce or eliminate unwanted organisms, regarded as "pests". Pesticides are often used on gardens, agricultural land, roadsides, and golf courses to eliminate species considered undesirable or damaging.

Pesticide rain Long-range transport of pesticides by air currents and deposition through precipitation in sites far from its origin; analogous to acid rain.

Pesticide treadmill A need for constantly increasing doses or new pesticides to prevent pest resurgence.

Pest-resistant crops Crops that possess attributes which minimize damage by pests.

Petals Usually brightly colored elements of a flower that may produce fragrant oils; nonreproductive structures that attract pollinators. Sterile leaf-like (white, colorless, but usually colored) structures in flowers that serve to attract pollinators.

Petaloids Describes flowerlike arrangement of respiratory podia in irregular sea urchins.

Petiole The generally non-leafy part of the leaf that attaches the leaf blade to the stem; celery and rhubarb are examples of a leaf petiole that we use as food. The stalk connecting the leaf blade to the stem.

Petri dish A shallow dish consisting of two round, overlapping halves that is used to grow microorganisms on solid culture medium; the top is larger than the bottom of the dish to prevent contamination of the culture.

Petrifaction A fossil formed when plant parts are infiltrated or replaced by mineral substances such that the structure is preserved but the fossil is actually rock.

Petroleum products Oil and the materials made from crude natural petroleum.

Peyer's patches Lymphoid tissue in the wall of the intestine; not circumscribed by a tissue capsule.

PGA A three-carbon molecule formed when carbon dioxide is added to ribulose biphosphate (RuBP) during the dark reaction of photosynthesis (Calvin, or Calvin-Benson Cycle). PGA is converted to PGAL, using ATP and NADPH.

PGAL (phosphoglyceraldehyde) A substance formed from PGA during the dark reaction of photosynthesis. Some PGAL leaves the cycle and can be converted to glucose, while other PGAL molecules are used to reform ribulose biphosphate (RuBP) to continue the dark reaction.

PH. A symbol referring to the relative concentration of hydrogen ions in a solution; pH values are from 0 to 14, and the lower the value, the more acid or hydrogen ions in the solution. Equal to the negative logarithm of the hydrogen ion concentration.A measure of how acid or alkali a solution is. As the pH gets lower, the solution is more acid. At a pH of 7 the solution is neither acid nor alkali. pH is the inverse of the logarithm of the concentration of hydrogen ions.

PH scale The numerical scale that measures acidity and alkalinity, ranging from 0 (most acidic) to 14 (most basic); pH stands for potential hydrogen and refers to the concentration of hydrogen ions ($H+$).

Phage (fage) A virus that infects bacteria; also called a bacteriophage.

Phage display Phage display vectors express the inserted DNA as a protein at a prominent position on their capsid. This allows capture and isolation of recombinant phage clones by immobilised interacting proteins (eg. antibodies). This screening method is called biopanning.

Phagemid A plasmid-based vector which carries a filamentous phage *ori*, leading to production of single-stranded copies of the phagemid when the cell is infected with the relevant helper phage. Under normal circumstances the phagemid behaves in exactly the same way as a plasmid.

Phagocytes White blood cells that can engulf (by phagocytosis) and destroy microorganisms including viruses and bacteria; cells in this category include neutrophils and monocytes.

Phagocytic vacuole A membrane-delimited vacuole produced by cells carrying out phagocytosis. It is formed by the invagination of the plasma membrane and contains solid material.

Phagocytosis The endocytotic process in which a cell encloses large particles in a membrane-delimited phagocytic vacuole or phagosome and engulfs them.

Phagolysosome The vacuole that results from the fusion of a phagosome with a lysosome.

Phagotroph A heterotrophic organism that ingests solid particles for food.

Phalanges The bones of the fingers or toes.

Phalanx Any one of the bones in the fingers or toes.

Phanerozoic The geologic eon that includes the interval of time from approximately 543 million years ago to the present, comprising the Paleozoic, Mesozoic, and Cenozoic eras.

Pharyngeal arches Columns of mesenchyme found in the neck of the developing vertebrate embryo derived from cranial neural crest. In lower vertebrates, blood vessels formed here become part of the gills; in higher vertebrates derivatives include portions of the jaw and middle ear. Also known as branchial arches, gill arches, or visceral arches.

Pharyngeal slits Characteristic of chordates, pharyngeal slits are openings through which water is taken into the pharynx, or throat. In primitive chordates the pharyngeal slits are used to strain water and filter out food particles; in fishes they are modified for respiration. Most terrestrial vertebrates have pharyngeal slits only in the embryonic stage.

Pharyngitis Inflammation of the pharynx, often due to a S. pyogenes infection.

Pharynx The part of the digestive tract between the mouth cavity and the esophagus that, in vertebrates, is common to both digestive and respiratory tracts. In cephalochordates the gill slits open from it.

Phase A specified, recognizable portion of the activity cycle.

Phase-contrast microscope A microscope that converts slight differences in refractive index and cell density into easily observed differences in light intensity.

Phasmid One of a pair of glands or sensory structures found in the posterior end of certain nematodes.

Phellem Also termed cork; the outer, protective tissue of stems and roots composed of nonliving cells with suberized cell walls, formed centrifugally by the cork cambium or phellogen.

Phelloderm A tissue laid down to the inside of the phellogen; the inner part of the periderm.

Phellogen The cork cambium.

Phenetic Refers to the use of a criterion of overall similarity to classify organisms into taxa; contrasts with classifications based explicitly on a reconstruction of phylogeny.

Phenetic system A classification system that groups organisms together based on the similarity of their observable charactistics.

Phenol coefficient test A test to measure the effectiveness of disinfectants by comparing their activity against test bacteria with that of phenol.

Phenology The seasonal life history of an insect population.

Phenotype matching A mechanism by which kin may recognize one another; individuals use kin as a referent whose phenotypes are learned by association.

Phenotype The physical appearance of an individual that is the result of that individual's genotype and the interaction of the genotype with the environment during development. Hence, individuals with the same genotype may have different phenotypes in different environments. Compare *genotype*.

Phenotypic gradualism. The hypothesis that new traits, even those that are strikingly different from ancestral ones, evolve by a long series of small, incremental steps.

Phenotypic plasticity The capacity of an individual to produce different phenotypes under different conditions. Non-genetic potential variability within the range of a single individual.

Phenotypic ratio The relative numbers of progeny in each phenotypic category produced by a genetic cross.

Pheromone A substance secreted by an animal which when released externally in small amounts causes a specific reaction, such as stimulation to mate with or supply food to a receiving individual of the same species.

Phi scale Scale used for measuring the grain size of sediments. = -log2 (grain diameter)

Philopatric Of or relating to species or groups that remain in or habitually return to their native regions or territories.

Phloem Tissue in the vascular system of plants that moves dissolved sugars and other products of photosynthesis from the leaves to other regions of the plant. Phloem tissue consists of cells called sieve tubes and companion cells. Cells of the vascular system in plants that transport food from leaves to other areas of the plant.

Phloem rays The part of the vascular ray which is located in the secondary phloem.

Phonoreceptor A specialized nerve ending that responds to sound.

Phoresis Form of symbiosis when the symbiont, the phoront, is mechanically carried about by its host. Neither is physiologically dependent on the other.

Phoronida A phylum of marine animals whose members live in permanent chitinous tubes in muddy, sandy, or solid substrates. Feed via an anterior lophophore with two parallel rings of long tentacles.

Phosphagen A term for creatine phosphate and arginine phosphate, which store and may be sources of high-energy phosphate bonds.

Phosphatase, alkaline An enzyme which catalyzes the hydrolysis of phosphomonoesters of the 5' nucleotides. Used to dephosphorylate (remove phosphate groups from) the 5' ends of DNA or RNA molecules, to facilitate 5' end-labeling with 32P added back by T4 polynucleotide kinase; or to dephosphorylate the 5' ends of DNA molecules to prevent unwanted *ligation* reactions during cloning.

Phosphatasing = dephosphorylating. Removal of 5' phosphate from an RNA or DNA molecule, using alkaline phosphatase.

Phosphate A salt in which the negatively charged part is a phosphorus molecule joined to four oxygen molecules as PO_4.

Phosphate group A chemical group composed of a central phosphorous bonded to three or four oxygens. The net charge on the group is negative.

Phosphate group transfer potential A measure of the ability of a phosphorylated molecule such as ATP to transfer its phosphate to water and other acceptors. It is the negative of the DGo_ for the hydrolytic removal of phosphate.

Phosphatide A lipid with phosphorus, such as lecithin. A complex phosphoric ester lipid, such as lecithin, found in all cells. Phospholipid.

Phosphodiester bond The covalent bond between the 3' hydroxyl in the sugar ring of one nucleotide and the 5' phosphate group of the sugar ring of the adjacent nucleotide residue within a nucleic acid:

5'-Ribose- 3' - O - P(O)2 - O - 5' -Ribose - 3' - etc.

Phospholipids Asymmetrical lipid molecules with a hydrophilic head and a hydrophobic tail. Lipids with a phosphate group in place of one of the three fatty acid chains. Phospholipids are the building blocks of cellular membranes. Phospholipids have hydrophilic heads (glycerol and phosphate) and hydrophobic tails (the non-polar fatty acids).

Phospho-proteins Proteins which contain available phosphate groups.

Phosphorus cycle The movement of phosphorus atoms from rocks through the biosphere and hydrosphere and back to rocks.

Phosphorylate To add phosphate; e.g., to a protein to alter its function.

Phosphorylation The addition of a phosphate monoester to a macromolecule, catalyzed by a specific kinase enzyme. With respect to proteins, certain *amino acid* side chains (serine, threonine, tyrosine) are subject to phosphorylation catalyzed by protein kinases; altering the phosphorylation status of a protein may have dramatic effects on its biologic properties, and is a common cellular control mechanism. With respect to DNA, 5' ends must be phosphorylated for *ligation*.

Photic zone The surface layer of the ocean that is penetrated by sunlight. The photic zone is the layer of the ocean that has been explored the most as it is relatively easy to access with conventional diving equipment. In the photic zone phytoplankton flourish and it is where the fish, marine mammals, and marine invertebrates that most people are familiar with are found. Light can penetrate down to approximately 200m which marks the end of the photic zone. Also referred to as the Sunlight Zone or the Epipelagic Zone.

Photoautotroph An organism that harnesses light energy to drive the synthesis of organic compounds from carbon dioxide.

Photochemical oxidants Products of secondary atmospheric reactions.

Photodegradable plastics Plastics that break down when exposed to sunlight or to a specific wavelength of light.

Photoinhibition reduction of photsynthetic rates due to too much light

Photolithotrophic autotrophs Organisms that use light energy, an inorganic electron source (e.g., H2O, H2, H2S), and CO2 as a carbon source.

Photolysis The splitting of a molecule of water during the light reactions of photosynthesis; the Hill reaction.

Photon A unit of light energy.

Photoperiod The amount of time per day that an organism is exposed to light

Photoperiodism The ability of certain plants to sense the relative amounts of light and dark in a

24-hour period; controls the onset of flowering in many plants.The system within organisms that causes certain events, including the onset of reproduction, to be related to the length of day.

Photopigment A molecule in visual receptor cells (e.g., rhodopsin) that responds to light energy.

Photoreactivation The process in which blue light is used by a photoreactivating enzyme to repair thymine dimers in DNA by splitting them apart.

Photoreceptors Sensory cells that contain photopigments and respond specifically to light energy.

Photorespiration A metabolic pathway that consumes oxygen, releases carbon dioxide, generates no ATP, and decreases photosynthetic output; generally occurs on hot, dry, bright days, when stomata close and the oxygen concentration in the leaf exceeds that of carbon dioxide.

Photosynthate A substance synthesized in the process of photosynthesis

Photosynthesis Carbon and nitrogen atoms in atmospheric CO_2 and N_2 are in an oxidized and inert state; to make them more reactive they have to be given a larger number of electrons. In photosynthesis, electrons are removed from a poor electron donor and transferred to a strong electron donor by chlorophyll in a reaction driven by sunlight; the strong electron donor is then used to reduce CO_2 or N_2.

Photosynthetic efficiency The percentage of available light captured by plants and used to make useful products.

Photosynthetic pigment A molecule such as chlorophyll that is responsible for capturing solar energy in photosynthesis.

Photosynthetic quotient In photosynthesis, the moles of oxygen produced, divided by the moles of carbon dioxide assimilated

Photosynthetic rate The rate of conversion of dissolved carbon dioxide and bicarbonate ion to photosynthetic product

Photosynthetic unit A group of associated chlorophyll molecules, including antenna molecules and a central chlorophyll a collector molecule.

Photosystem I The photosystem in eucaryotic cells that absorbs longer wavelength light, usually greater than about 680 nm, and transfers the energy to chlorophyll P700 during photosynthesis; it is involved in both cyclic photophosphorylation and noncyclic photophosphorylation.The second part of the Z-scheme in which a chlorophyll a molecule absorbs most effectively at 700 nm.The first part of the Z-scheme in which the chlorophyll a molecule absorbs most effectively at 680 nm.

Photosystem II The photosystem in eucaryotic cells that absorbs shorter wavelength light, usually less than 680 nm, and transfers the energy to chlorophyll P680 during photosynthesis; it participates in noncyclic photophosphorylation.

Phototactic Moving in response to light

Phototaxis A taxis in which light is the orienting stimulus. An involuntary tendency for an organism to turn toward (positive) or away from (negative) light.

Phototrophs Organisms capable of using CO2 in the presence of light as a source of metabolic energy.Organisms that use sunlight to synthesize organic nutrients as their energy source; e.g., cyanobacteria, algae, and plants.

Phototropism The reaction of plants to light in which the plants bend toward the light. Plant response to light by unequal growth caused by concentration of the plant hormone Indole Acetic Acid (IAA, an auxin) on the darker side of the plant shoot.

Photovoltaic cell An energy-conversion device that captures solar energy and directly converts it to electrical current.

Phragmoplast A system of microtubules and vesicles that arises between two daughter nuclei at telophase and forms the cell plate.

Phreatophyte A desert shrub with a long trap root that enables the plant to avoid reliance on rainwater by tapping into groundwater.

Phycobilins A group of photosynthetic pigments that includes phycocyanin, a bluish pigment in cyanobacteria (blue-green algae), and phycoerythrin, a red pigment in red algae.

Phycobiliproteins Photosynthetic pigments that are composed of proteins with attached tetrapyrroles; they are often found in cyanobacteria and red algae.

Phycobilisomes Special particles on the membranes of cyanobacteria that contain photosynthetic pigments and electron transport chains.

Phycobiont The algal or cyanobacterial partner in a lichen.

Phycocolloid One of several starch-like chemicals found in some seaweeds. They are of significant commercial importance.

Phycocyanin An accessory pigment found in cyanobacteria and the chloroplasts of red algae.Blue, water-soluble pigment found in the cyanobacteria and the red algae.

Phycoerythrin A red photosynthetic phycobiliprotein pigment used to trap light energy.A red photosynthetic pigment found in cyanobacteria and red algae.An accessory pigment found in cyanobacteria and the chloroplasts of red algae.

phyletic change The changes taking place in a single lineage of organisms over a long period of time; one of the principal patterns of evolutionary change.

Phyletic gradualism. A model of evolution in which morphological evolutionary change is continuous and incremental and occurs mainly within unbranched species or lineages over long periods of geological time; contrasts with punctuated equilibrium.

Phyletics Phylogenetic systematics, cladistics.

Phyllopodium Leaflike swimming appendage of branchiopod crustaceans.

Phyllosphere The surface of plant leaves.

Phylogenetic Pertaining to a phylogeny.

Phylogenetic species concept. An irreducible (basal) cluster of organisms, diagnosably distinct from other such clusters, and within which there is a parental pattern of ancestry and descent.

Phylogenetic systematics The study of the phylogenetic relationships among organisms in which true and false similarities are differentiated; cladistics.

Phylogenetic tree A graph made of nodes and branches, much like a tree in shape, that shows phylogenetic relationships between groups of organisms and sometimes also indicates the evolutionary development of groups.

Phylogenetics Field of biology that deals with the relationships between organisms. It includes the discovery of these relationships, and the study of the causes behind this pattern.

Phylogeny (1) the study of evolutionary relationships within a monophyletic group.

(2) evolutionary hypotheses represented as a dendrogram or branching diagram. The evolutionary relationships among organisms; the patterns of lineage branching produced by the true evolutionary history of the organisms being considered.The origin and diversification of any taxon, or the evolutionary history of its origin and diversification, usually presented in the form of a dendrogram.

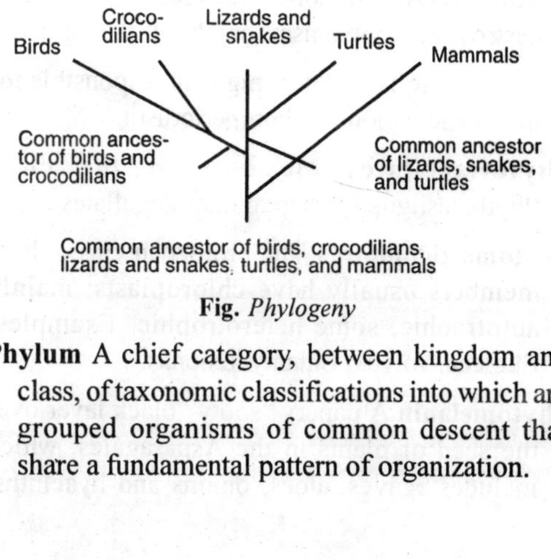

Fig. *Phylogeny*

Phylum A chief category, between kingdom and class, of taxonomic classifications into which are grouped organisms of common descent that share a fundamental pattern of organization.

Physical control Control of pests by physical means such as heat, cold, sound waves, etc.

Physical map A map of the locations of identifiable landmarks on DNA (e.g., *restriction enzyme cutting sites, genes*), regardless of inheritance. Distance is measured in *base pairs*. For the human *genome*, the lowest- resolution *physical map* is the banding patterns on the 24 different *chromosomes*; the highest- resolution map would be the complete *nucleotide* sequence of the chromosomes.

Physical or abiotic factors Nonliving factors, such as temperature, light, water, minerals, and climate, that influence an organism.

Physiological dependence The condition in which there is a physical need for a drug to avoid withdrawal symptoms.

Physiological limits The tolerance range of given species for temperature induced rates of metabolic activity.

Physiological race A geographically defined population of a species that is physiologically distinct from other populations

Physiology A branch of biology dealing with the organic processes and phenomena of an organism or any of its parts or of a particular bodily process.

Physostomous swim bladder In fish, a swim bladder with an air passage or duct to the esophagus.

Phytoalexin An antibiotic, produced by plants, that destroys microorganisms or inhibits their growth.

Phytochrome The protein pigment responsible for the phenomenon of photoperiodism.

Phytoflagellates Members of the class Phytomastigophorea, plantlike flagellates.

Phytomastigophorea The protozoan class where members usually have chloroplasts; mainly autotrophic, some heterotrophic. Examples: Euglena, Volvox, Chlamydomonas.

Phytomelanin A papery "sooty" black layer over the seed of plants in the Asparagales, which includes agaves, aloes, onions and hyacinths. It is an important character for defining the group.

Phytophagous Organisms that feed on plants.

Phytoplankton A community of floating photosynthetic organisms, largely composed of algae and cyanobacteria.A floating layer of photosynthetic organisms, including algae, that are an important source of atmospheric oxygen and form the base of the aquatic food chain.

Phytoremediation The use of plants and their associated microorganisms to remove, contain, or degrade environmental contaminants.

Phytotoxic Poisonous to plants.

Pia mater The innermost meninx that is in direct contact with the brain and spinal cord.

Picoplankton The component of the plankton that consists of extremely small organisms, mostly bacterioplankton, 0.2 to 2 mm (0.0002 to 0.002 mm) in size they are too small to catch in a standard plankton net.

Piedra A fungal disease of the hair in which white or black nodules of fungi form on the shafts.

Pigment Any colorful compound, used by living things to absorb or block sunlight, and in sexual displays.

Pilidium Free-swimming, hat-shaped larva of nemertine worms.

Pilidium larva Free-swimming, hat-shaped larva of nemertean worms characterized by an apical tuft of cilia.

Pillow lava Lava extruded beneath water characterised by pillow-type shapes.

Pilose Densely clothed with hair.

Piloting The use of familiar landmarks to find a direction or area.

Pilus Projection from surface of a bacterial cell (F+) that can donate genetic material to another (F-).

Pinacocyte Thin, flat cell covering the outer surface, and some of the inner surface, of poriferans.

Pinacoderm The layer of pinacocytes in sponges.

Pineal gland A small gland located between the cerebral hemispheres of the brain that secretes melatonin.An endocrine gland located near the midline of the brain that produces melatonin, a hormone involved in biological rhythms, particularly in annual cycles.

Pinkeye Bacterial conjunctivitis, sometimes transmitted by flies of the genus Hippolates.

Pinna The external ear. Also a feather, wing, or fin or similar part.

Pinnate Denoting an arrangement of leaflets (or lobes on a simple leaf) along a main central unit.

Pinnately compound Leaves which are divided up like a feather are said to be pinnately compound.

Pinnipeds Mammals with paddle-shaped flippers: seals, eared seals (sea lions and fur seals), and the walrus.

Pinocytosis The endocytotic process in which a cell encloses a small amount of the surrounding liquid and its solutes in tiny pinocytotic vesicles or pinosomes.

Pioneer community The first community to become established in an area.

Pioneer organism An organism that first colonizes soil or pioneer species.In primary succession on a terrestrial site, the plants, lichens, and microbes that first colonize the site.

Pioneer species A species that is an early occupant of newly created or disturbed areas. A member of the early stage communities in ecological succession.

Pipestem fibrosis Thickening of the walls of a bile duct as the result of the irritating presence of a parasite.

Piroplasm Any of the class Piroplasmea, while in a circulating erythrocyte.

Pistil Female reproductive structures in flowers, consisting of the stigma, style, and ovary. Also known as a carpel in some books.

Pistillate Denoting a unisexual flower having a pistil but no stamens (carpellate).

Pistillate flower A flower having carpels but no stamens.

Pit A pore in a secondary cell wall.

Pit organ Receptor of infrared radiation (heat) on the head of some snakes (pit vipers).

Pitched Pertaining to inoculation of a nutrient medium with yeast, for example, in beer brewing.

Pith Parenchymal tissue in the center of a stem located interior to the vascular bundles.

Pith ray The region or tissue located between vascular bundles in a stem and connecting the pith and cortex; also termed interfascicular region or medullary ray.

Pituitary gland A small gland located at the base of the brain; consists of an anterior and a posterior lobe and produces numerous hormones. The master gland of the endocrine system, the pituitary releases hormones that have specific targets as well as those that stimulate other glands to secrete hormones. Part of the pituitary is nerve tissue, the rest is glandular epithelium.

Pituitary pituitar = mucous secretion. An endocrine gland at the base of the brain.

Placenta Embryonic/maternal organ that serves nutritional and respiratory functions of the mammalian fetus. It is composed of the embryonic chorion and maternal uterine endometrium, allowing provision of oxygen and nutrients to the fetus and removal of carbon dioxide and other waste products.

Placental mammal A member of a group of mammals, including humans, whose young complete their embryonic development in the uterus, joined to the mother by a placenta. One of three groups of mammals that carry their young in the mother's body for long periods during which the fetus is nourished by the placenta. Humans are placental mammals.

Placental A mammal whose young form a placenta as they develop in the mother's uterus.

Placentome The structure involved in connecting

the fetal and maternal tissues consisting of a cotyledon and a caruncle in the cotyledonary placenta. The cotyledons or chorionic villi are of fetal origin and "plug into" the caruncles or receptacles in the maternal uterine wall.

Placid Plates on Kinorhyncha.

Placode Localized, plate-like thickening of vertebrate head ectoderm from which a specialized structure develops; such structures include eye lens, special sense organs, and certain neurons.

Placoderm A member of an extinct class of fishlike vertebrates that had jaws and were enclosed in a tough, outer armor.

Placoid scale Type of scale found in cartilaginous fishes, with basal plate of dentin embedded in the skin and a backward-pointing spine tipped with enamel.

Placozoa A phylum of small, flattened, marine animals that feed by forming a temporary digestive cavity. Tricoplax adherans.

Plague An acute febrile, infectious disease, caused by the bacillus Yersinia pestis, wnicn has a high mortality rate; the two major types are bubonic plague and pneumonic plague.

Planaria Small free-living flatworms (Phylum Platyhelminthes) with bilateral symmetry and cephalization. The freshwater type is often used as an experimental organism.

Planidium First instar of hypermetamorphic, parasitic Diptera and Hymenoptera, which is apodous but moves actively by means of thoracic and caudal setae.

Planktivore An animal that feeds on plankton.

Plankton Pelagic organisms that float through the water column, not attached to any substrate and unable to move against the currents and tides. Plankton can be further divided into *phytoplankton* and *zooplankton, meroplankton* and *holoplankton.* Compare *nekton.*

Planktonic A form of life style in which an organisms floats freely in a fluid without significant attachment or association with other living forms.

Planktonic organisms "Floaters"; one of the two main types of organisms in the pelagic zone of the marine biome.

Planktotrophic larva Planktonic-dispersing larva that derives its nourishment by feeding in the plankton

Plant growth regulator Any molecule that exhibits hormone-like effects in a plant, whether synthesized or naturally occurring.

Plantae The plant kingdom; nonmobile, autotrophic, multicellular eukaryotes. Kingdom of the plants, autotrophic eukaryotes with cellulose in their cell walls and starch as a carbohydrate storage product, with chlorophylls a and b as photosynthesis pigments.

Plantigrade Pertaining to animals that walk on the whole surface of the foot (for example, humans and bears); compares with digitigrade.

Plants Members of the kingdom Plantae, which consists of photosynthetic, eukaryotic, and mostly multicellular organisms.

Planula Free-swimming, ciliated larval type of cnidarians; usually flattened and ovoid, with an outer layer of ectodermal cells and an inner mass of endodermal cells.

Planuloid ancestor Hypothetical form representing ancestor of Cnidaria and Platyhelminthes.

Plaque 1. A clear area in a lawn of bacteria or a localized area of cell destruction in a layer of animal cells that results from the lysis of the bacteria by bacteriophages or the destruction of the animal cells by animal viruses.

2. The term also refers to dental plaque, a film of food debris, polysaccharides, and dead cells that cover the teeth. It provides a medium for the growth of bacteria (which may be considered a part of the plaque), leading to a microbial plaque ecosystem that can produce dental decay.Is a film of bacteria in a matrix of salivary and bacterial

polymers. It can be called a biofilm as it has a complex population of organisms which when mature, reach an ecological balance with one another.

Plaque lift Blot taken from plaques (usually bacteriophage on a bacterial lawn) growing on an agar plate. Normally used for detection of a plaque containing a phage with a specific inserted sequence (screening).

Plasma The liquid matrix of blood in which the cells are suspended.

Plasma cells Antibody secreting cells that develop from B-cells in response to the sequence of events occurring in the humoral immune system upon detection of certain types of antigens.

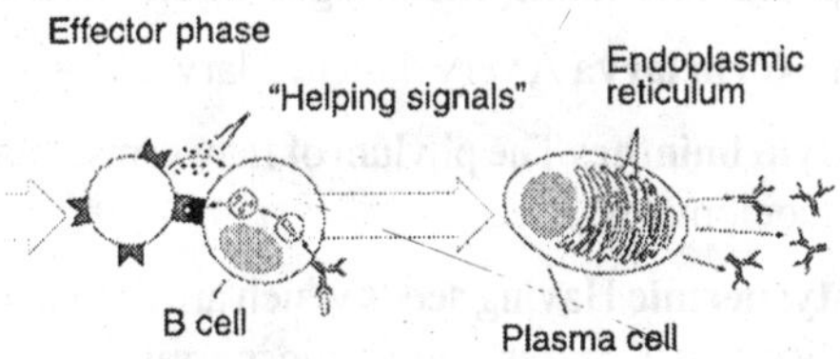

Fig. *Plasma cells*

Plasma membrane Outermost membrane of a cell; its surface has molecular regions that detect changes in external conditions and act as a selective barrier to the passage of ions and molecules between the cell and its environment. The external covering of a protozoan.

Plasmid A double-stranded DNA molecule that can exist and replicate independently of the chromosome or may be integrated with it. A plasmid is stably inherited, but is not required for the host cell's growth and reproduction.A circular piece of DNA present in bacteria or isolated from bacteria. *Escherichia coli*, the usual bacteria in molecular genetics experiments, has a large circular genome, but it will also replicate smaller circular DNAs as long as they have an "origin of replication". Plasmids may also have other DNA inserted by the investigator. A bacterium carrying a plasmid and replicating a million-fold will produce a million identical copies of that plasmid. Common plasmids are pBR322, pGEM, pUC18.Autonomously replicating, extrachromosomal circular DNA molecules, distinct from the normal bacterial *genome* and nonessential for cell survival under nonselective conditions. Some plasmids are capable of integrating into the host genome. A number of artificially constructed plasmids are used as *cloning vectors*. Double stranded, circular DNA molecule found in bacteria and some fungi. Size range from ~5 - 200kbp, contains genes which may be useful to host, but are not essential. Plasmids can be eliminated from bacterial populations by growth under selective conditions ("curing"). Carry an origin of replication (*ori*) which allows replication in a specific host species. Plasmids used in recombinant DNA work are generally smaller than 10kbp in size and have a copy number of up to 500/cell depending on plasmid type.

Plasmid fingerprinting A technique used to identify microbial isolates as belonging to the same strain because they contain the same number of plasmids with the identical molecular weights and similar phenotypes.

Plasmodesma A cytoplasmic strand that connects adjacent plant cells through pores in the cell wall.

Plasmodesmata Junctions in plants that penetrate cell walls and plasma membranes, allowing direct communication between the cytoplasm of adjacent cells.

Plasmodial (acellular) slime mold A member of the division Myxomycota that exists as a thin, streaming, multinucleate mass of protoplasm, which creeps along in an amoeboid fashion. A stage in the life cycle of myxomycetes (plasmodial slime molds); a multinucleate mass of protoplasm surrounded by a membrane. Also, a parasite of the genus Plasmodium.

Plasmogamy The fusion of the cytoplasm of cells from two individuals; occurs as one stage of syngamy.

Plasmolysis A phenomenon in walled cells in which the cytoplasm shrivels and the plasma membrane pulls away from the cell wall when the cell loses water to a hypertonic environment.

Plasmotomy Division of a multinucleate cell into multinucleate daughter cells, without accompanying mitosis.

Plastic song One of the stages of song development in birds during which the bird has begun to sing using the species-typical pattern, but with a certain degree of variation.

Plasticity Cell wall "stretchability" induced by auxin.

Plastids Membrane-bound organelles in plant cells that function in storage (of food or pigments) or food production. Term for any double membrane-bound organelle. Chloroplasts contain the chemicals for photosynthesis, amyloplasts store starch, chromoplasts contain colorful pigments such as in the petals of a flower or epidermis of a fruit.

Plastron The ventral part of the turtle shell, consisting of a series of paired bones, overlain by series of laminae alternating with bones.

Plate Rigid parts of the Earth's crust and part of the Earth's upper mantle that move and adjoin each other along zones of seismic activity. The theory that the crust and part of the mantle are divided into plates that interact with each other causing seismic and tecotnic activity is called plate tectonics.

Plate tectonics The movement of the plates that make up the surface of the Earth. The revolutionary paradigm in geology that the earth's crust is composed of rigid segments (plates) in constant (although considered slow in a human-scale time frame) motion (tectonics) relative to each other.

Platelet Are small colourless discs of cytoplasm found in blood. When platelets come into contact with a damaged vessels surface they change in several important ways. They begin to swell, their shape becomes irregular with protruding processes, they become sticky and they release an enzyme which causes the formation of thromboxane, one of the precursors of thrombin . Thromboxane also activates nearby platelets, thus starting a positive feedback which rapidly increases the mass of sticky platelets which form a platelet plug. This process accounts for daily damage to capillary walls. Damage on a larger scale requires other mechanism for haemostasis. Platelets also release serotonin which acts as a powerful vasoconstrictor. In vertebrates, cell fragments that bud off from the megakaryocytes in the bone marrow; carry chemicals needed for blood clotting. Cell fragment functioning in blood clotting.

Platelet- derived growth factor A cytokine found especially in platelets. It stimulates cell proliferation and encourages wound healing.

Platyform larva A very flattened larva.

Platyhelminthes The phylum of flatworms; bilateral acoelomates.

Platyspermic Having seeds which are flattened and disc-like. Contrast with radiospermic.

Play behavior Certain locomotor, social, and manipulative behavior patterns exhibited by young and some adult mammals and birds.

Pleated sheet One form of the secondary structure of proteins in which the polypeptide chain folds back and forth, or where two regions of the chain lie parallel to each other and are held together by hydrogen bonds.

Pleiotropic A term describing a genotype with multiple phenotypic effects. For example: sickle-cell anemia produces a multitude of consequences in those it affects, such as heart disease, jidney problem, etc.

Pleistocene A geological period, which began about 2 million years ago, characterized by a series of ice ages. The first geologic epoch of the Quaternary Period of the Cenozoic Era that ended 10,000 years ago with the retreat of the last glaciers.

Pleistocene age The geologic epoch of the

Quartenary period extending from the end of the Pliocene to the beginning of the Holocene, and the rocks formed during that time.

Pleomorphic Refers to bacteria that are variable in shape and lack a single, characteristic form.

Pleopod One of the swimming appendages on the abdomen of a crustacean.

Plerocercoid larva Metacestode that develops from a procercoid larva; it usually shows little differentiation.

Plerocercoid Metacestode that develops from a procercoid. It usually shows little differentiation.

Plerocercus Tapeworm metacestode in the order Trypanorhyncha in which the posterior forms a bladder, the blastocyst, into which the rest of the body withdraws.

Plesiomorphic Ancestral characters; characters possessed by members of both ingroup and outgroup.

Plesiomorphic character A primitive phenotypic character possessed by a remote ancestor.

Plesiomorphy A primitive character state for the taxa under consideration.

Pleura A thin sheet of epithelium that covers the inside of the thoracic cavity and the outer surface of the lungs.

Pleurae In trilobites and other arthropods, pleurae are elongated flat outgrowths from each body segment, that overlie and protect the appendages.

Pleural cavity The space between the sheets of pleura (one covering the inside of the thoracic cavity, the other covering the outside of the lungs).

Pleural Concerning the side walls of the body.

Pleural suture A vertical or diagonal groove on each of the thoracic pleura, separating the episternum at the front from the epimeron at the back.

Pleurite Lateral sclerite of a somite in an arthropod.

Pleuron The side wall of a thoracic segment.

Pleuston Refers to plankton that have a float protruding above the sea surface, such as the Portugese man-of-war.

Plexus A network, especially of nerves or blood vessels.

Plica polonica Develops in untreated head louse (Pediculus humanus capitis) infection, hair matted together with exudate, fungal growth, fetid odor.

Plicate Folded like a paper fan, as in the leaves of palms, cyclanthoids, and some orchids.

Plumage The feather covering of birds, sometimes colored for purposes of camouflage, courtship display, etc., and divided in certain stages of a bird's growth to indicate age or maturity, such as natal plumage, juvenile plumage, first winter plumage, first nuptial plumage, and the like.

Plumes The conspicuous feather or feathers of a bird.

Plumose With numerous feathery branches: applied especially to antennae.

Plumule The shoot apex of a seedling, including embryonic leaves.

Pluripotent Of a cell or nucleus, having the ability to produce many, but not all of the cell types in an organism (Contrast with totipotent). Note: In mammalian development, cells (e.g., embryonic or other stem cells) capable of giving rise to all cell types of the embryo, but not extraembryonic tissues (derivatives of trophoblast) are called 'pluripotent.' [For example, in a recent issue of Nature on ES cells, 'pluripotent' was defined in one article as "able to give rise to all cell types found in the embryo and adult animal."] This usage is different from that used with other organisms and is a potential source of confusion. In mammalian systems, the word "multipotent" is used for a cell capable of producing some, but not all cell types of the embryo.

Pluripotent stem cell A cell within bone marrow that is a progenitor for any kind of blood cell.

Plus strand or positive strand The virus nucleic-acid strand that is equivalent in base sequence to the viral mRNA.

Pluteus Echinoid or ophiuroid larva with elongated processes like the supports of a desk; originally called "painter's easel larva."

Pluton Any body of igneous rock that solidified below the earth's surface.

Plutonic Applies to igneous rocks formed beneath the surface of the Earth; typically with large crystals due to the slowness of cooling. Synonym of intrusive. Antonym of volcanic.

Plywood A building material consisting of two or more thin sheets of wood bonded together.

Pnctuated equilibrium A theory of evolution advocating spurts of relatively rapid change followed by long periods of stasis.

Pneumatic duct in fish, the connection between the esophagus and swim bladder

Pneumatic sacs Gas-filled sacs that arise from the esophagus, or another part of the digestive tract, of fishes. Pneumatic sacs are used in buoyancy regulation (swim bladders) or gas exchange (lungs).

Pneumatocyst A gas-filled bladder in seaweeds.

Pneumatophore A specialized root that grows upward into the air from roots growing in the mud; aids in gaseous exchange.

Pneumostome The opening of the mantle cavity (lung) of pulmonate gastropods to the outside.

Poachers Those who hunt wildlife illegally.

Pod 1. A dry dehiscent fruit that splits along two seams; a legume.

2. A school of cetaceans.

Podium A footlike structure, for example, the tube foot of echinoderms.

Podomere More or less cylindrical segment of a limb of an arthropod, generally articulated at both ends.

Poecilostome Describes mouthparts borne by members of the copepod order Poecilostomata (buccal cavity large, somewhat slitlike, with sickle-shaped mandibles); also, a member of the order Poecilostomata.

Pogonophora A phylum of protostomate, marine animals that are distributed throughout the world's oceans. Live in secreted, chitinous tubes in cold water at depths exceeding 100 m. Lack a mouth and digestive tract. Nutrients absorbed across the body wall and from endosymbiotic bacteria that they harbor. Beard worms.

Poikilotherm An organism whose body temperature is identical to that of the external environment

Poikilothermic Pertaining to animals whose body temperature is variable and fluctuates with that of the environment; cold blooded; compares with ectothermic.

Point mutations A change in the structure of a gene that usually arises from the addition, deletion, or substitution of one or more nitrogenous bases.

Point sources Specific locations of highly concentrated pollution discharge, such as factories, power plants, sewage treatment plants, underground coal mines, and oil wells.

Polar Having parts or areas with opposed or contrasting properties, such as positive and negative charges, head and tail.

Polar body Minute, nonfunctioning cell produced during those meiotic divisions that lead to egg cells; contains a nucleus but very little cytoplasm.

Polar capsule Compartment bearing the polar filaments in myxozoans.

Polar covalent bond A covalent bond in which atoms share electrons in an unequal fashion. The resulting molecule has regions with positive and negative charges. The presence of polar covalent bonds allows other polar molecules to surround molecule: example: glucose sugar in water.

Polar easterlies Variable winds that blow at high latitudes.

Polar filament Threadlike organelles in Myxozoa and Microspora.

Polar flagellum A flagellum located at one end of an elongated cell.

Polar granule Refractile granule within a coccidian oocyst.

Polar ice cap The portions of the globe close to the poles that are permanently covered with ice.

Polar molecule A molecule (such as water) with opposite charges on opposite sides.

Polar nuclei In angiosperms, the two nuclei of the central cell of the female gametophyte; they fuse with a sperm nucleus to form the triploid ($3n$) endosperm nucleus.Two of the eight haploid nuclei of a megagametophyte that migrate, one from each end, to the middle of the embryo sac. These two nuclei fuse with a sperm nucleus during double fertilization.

Polar ring Electron-dense organelles of unknown function, located under the cell membrane at the anterior tip of sporozoites and merozoites.

Polarity In systematics, the ordering of alternative states of a taxonomic character from ancestral to successively derived conditions in an evolutionary transformation series. In developmental biology, the tendency for the axis of an ovum to orient corresponding to the axis of the mother. Also, condition of having opposite poles; differential distribution of gradation along an axis.

Polarity of characters The states of characters used in a cladistic analysis, either original or derived.Original characters are those acquired by an ancestor deeper in the phylogeny than the most recent common ancestor of the taxa under consideration. Derived characters are those acquired by the most recent common ancestor of the taxa under consideration.

Polarization The arrangement of positive electrical charges on one side of a surface membrane and negative electrical charges on the other side (in nerves and muscles).

Polarized A description for a membrane that has a potential difference due to an unequal distribution of ions across the membrane.

Polaroplast Organelle, apparently a vacuole, near the polar filament of a microsporidean.

Polian vesicles Vesicles opening into ring canal in most asteroids and holothuroids.

Poliomyelitis An acute, contagious viral disease that attacks the central nervous system, injuring or destroying the nerve cells that control the muscles and sometimes causing paralysis; also called polio or infantile paralysis.

Polishing *In vitro* digestion of the single-stranded section of an overhanging end (by exonuclease) to produce a blunt end.

Pollen An immature male gametophyte that develops within the anthers of stamens in a flower.

Pollen basket The pollen-carrying region on the hind leg, of a bee: also known as the corbicula.

Pollen grains The containers for male gametophytes of seed plants produced in a microsporangium by meiosis. Microspores produced by seed plants that contain the male gametophyte.

Pollen tube Structure produced by the tube nucleus in the pollen grain through which the sperm nucleus (or nuclei in angiosperms) proceed to travel through to reach the egg.

Pollex The thumb.

Pollinate To transfer pollen grains from a stamen to a stigma or ovule of a plant.

Pollination Process of transferring the pollen from its place of production to the place where the egg cell is produced. This may be accomplished by the use of wind, water, insects, birds, bats, or other means. Pollination is usually followed by fertilization, in which sperm are released from the pollen grain to unite with the egg cell.

Pollinator Animal which carries pollen from one seed plant to another, unwittingly aiding the plant in its reproduction. Common pollinators include insects, especially bees, butterflies, and moths, birds, and bats.

Pollinia A mass of fused pollen produced by many orchids.

Pollution charges Fees assessed per unit of pollution based on the "polluter pays" principle.

Pollution The introduction of harmful substances or heat energy into the environment by humans.

Poly-A tail - polyadenosine tail A sequence of repeatingadenosine ribonucleotides added to the 3' end of a newly transcribed pre- mRNA before it exits the nucleus. This modification can be up to 200-300 bases long and functions to increase the stability and translatability of the mRNA. After an mRNA is transcribed from a gene, the cell adds a stretch of A residues (typically 50-200) to its 3' end. It is thought that the presence of this "polyA tail" increases the stability of the mRNA (possibly by protecting it from nucleases). Note that not all mRNAs have a polyA tail; the histone mRNAs in particular do not.

Poly-a track A lengthy adenylic acid polymer (RNA) which is covalently linked to the 3' end of newly synthesized *mRNA* molecules in the nucleus. Function not polymerase chain reaction (PCR) - A DNA amplification reaction involving multiple (30 or more) cycles of *primer annealing, extension*, and denaturation, usually using a heat-stable DNA *polymerase* such as *Taq polymerase*. Paired primers are used, which are complementary to opposing strands of the DNA and which flank the area to be amplified. Under optimal conditions, single DNA sequence can be amplified a million-fold.

Polyacrylamide gel (PAGE) Used to separate proteins and smaller DNA fragments and *oligonucleotides* by electrophoresis. When run under conditions which denature proteins (i.e., in the presence of 2-mercaptoethanol, SDS, and possibly urea), molecules are separated primarily on the basis of size.

Polyandrous Having more than one male mate. Polyandry is advantageous when food is plentiful but, because of predation or other factors, the chances of successfully rearing young are low.

Polyandry Condition of having more than one male mate at one time.

Poly-b-hydroxybutyrate A linear polymer of b-hydroxybutyrate used as a reserve of carbon and energy by many bacteria.

Polycarpic Denoting a plant that flowers more than once during its lifetime.

Polycentric complex Cities with several urban cores surrounding a once dominant central core.

Polychaeta The class of annelids whose members are mostly marine and are characterized by a head with eyes and tentacles and a body with parapodia. Parapodia bear numerous setae. Examples: Nereis, Arenicola.

Polychaets A class of mostly marine worms in the Phylum Annelida, with anterior tentacles and palps and most segments bearing parapodia with bristles; free-swimming or sessile in tubes or burrows; often brightly colored, most 5-10 cm in length.

Polychlorinated biphenyls (PCBs) A group of nonbiodegradable pollutants.

Polyembryonic (eggs) A single egg that divides to form two or more (often hundreds) identical embryos.

Polyembryony The production of several embryos from a single egg, as in some chalcids.

Polyestrus Having more than one estrus cycle in a year. Many small mammals that have multiple litters each year are polyestrus.

Polygamy Condition of having more than one mate at a time.

Polygenes Genes at multiple loci that influence a trait in a quantitative fashion.

Polygenic disorders Genetic disorders resulting from the combined action of *alleles* of more than one *gene* (e.g., heart disease, diabetes, and some cancers). Although such disorders are inherited, they depend on the simultaneous presence of several alleles; thus the hereditary patterns are usually more complex than those of *single- gene disorders*. Compare single- gene disorders.

Polygenic inheritance. Inheritance of traits influenced by multiple alleles; traits show continuous variation between extremes; offspring are usually intermediate between the two parents; also known as blending and quantitative inheritance.

Polygenic trait Traits that are influenced by many genes and that are usually continuously distributed within a population.

Polygynous Having more than one female mate. Polygyny tends to occur in species whose young are relatively independent at birth or hatching.

Polygyny threshold The point at which a female will benefit more by joining an already mated male possessing a good territory rather than an unmated male on a poor territory.

Polygyny When a male mammal mates with more than one female during a breeding season.

Polyinvagination islands Disconnected clusters of cells of the primary hypoblast that are the first to migrate beneath the epiblast of the avian blastoderm.

Polykinetid Rows or fields of kinetids in ciliates linked by fibrous networks.

Polymer Organic molecule composed of smaller units known as monomers. A large molecule composed of smaller subunits, for example starch is a polymer of glucose, proteins are polymers of amino acids.

Polymerase An enzyme which links individual nucleotides together into a long strand, using another strand as a template. There are two general types of polymerase — DNA polymerases (which synthesize DNA) and RNA polymerase (which makes RNA). Within these two classes, there are numerous sub-types of polymerase, depending on what type of nucleic acid can function as template and what type of nucleic acid is formed. A DNA-dependant DNA polymerase will copy one DNA strand starting from a primer, and the product will be the complementary DNA strand. A DNA-dependant RNA polymerase will use DNA as a template to synthesize an RNA strand.

Polymerase chain reaction (PCR) A method for amplifying a DNA *base sequence* using a heat-stable *polymerase* and two 20- base *primers*, one *complementary* to the (+)- strand at one end of the sequence to be amplified and the other complementary to the (-)- strand at the other end. Because the newly synthesized DNA strands can subsequently serve as additional templates for the same primer sequences, successive rounds of primer annealing, strand elongation, and dissociation produce rapid and highly specific amplification of the desired sequence. PCR also can be used to detect the existence of the defined sequence in a DNA sample. Is the basis for a number of extremely important methods in molecular biology. It can be used to detect and measure vanishingly small amounts of DNA and to create customized pieces of DNA. It has been applied to clinical diagnosis and therapy, to forensics and to vast numbers of research applications. It would be difficult to overstate the importance of PCR to science.A technique for replicating a specific piece of DNA *in-vitro* , even in the presence of excess non-specific DNA. Primers are added (which initiate the copying of each strand) along with nucleotides and Taq polymerase. By cycling the temperature, the target DNA is repetitively denatured and copied. A single copy of the target DNA, even if mixed in with other undesirable DNA, can be amplified to obtain billions of replicates. PCR can be used to amplify RNA sequences if they are first converted to DNA via reverse transcriptase. This two-phase procedure is known as 'RT-PCR'.

Polymerase, DNA or RNA *Enzymes* that catalyze the synthesis of *nucleic acids* on preexisting nucleic acid templates, assembling RNA from ribonucleotides or DNA from deoxyribonucleotides.

Polymerization The process of forming a polymer or polymeric compound.

Polymers Polymers are a subclass of macromolecules.

Polymerization is the chemical process in which a series of simple structural units (monomers) combine to form a large chainlike molecule (polymers).

Chains of amino acids (polypeptides or proteins) or nucleotides (polynucleotides) are polymers.

Polymorphic A locus that contains two or more alleles within a population.

Polymorphism Difference in DNA sequence among individuals. Genetic variations occurring in more than 1% of a population would be considered useful polymorphisms for genetic *linkage* analysis. Compare *mutation.*

Polymorphonuclear leukocyte A leukocyte that has a variety of nuclear forms.

Polynomial A scientific name for an organism composed of more than two words.

Polynucleotides Polynucleotides are either RNA or DNA.

They guide the formation of exact copies of their own sequence through complementary pairing of nucleotides subunits (U with A and G with C).

Catalytic functions that polymerize nucleotides and insure polynucleotides replication are provided by highly specialized proteins called enzymes.

Nucleotide pairing between different regions of the same polynucleotide chain causes the molecule to fold up a certain way

The 3-D shape adopted by the polynucleotide:

Depends entirely on the sequence of its nucleotide.

Affects its stability, its actions on other molecules, and its ability to replicate, so that not all polynucleotide shapes will be equally successful in a replicating mixture.

RNA molecules can be selected for their ability to bind to almost any other molecule. Such RNA molecules are retained and used as templates to direct production of multiple copies of their own sequences.

Polynucleotide chains Attachment of one nucleotide to another in a linear fashion.

Polynucleotide kinase Enzyme which catalyzes the transfer of the terminal phosphate of ATP to 5' hydroxyl termini of polynucleotides, either DNAor RNA. Usually derived from T4 bacteriophage.

Polyp The sessile form of life history in cnidarians; e.g., the freshwater hydra.

Polypeptides Chains of amino acids joined by peptide bonds . They are not the size of proteins, but may be biologically very active. Some hormones are peptides, such as insulin which has 51 amino-acid residues. Peptides may also be powerful neurotransmitter substances.

Polyphagous Feeding on a variety of plants and or animals.

Polyphasic taxonomy An approach in which taxonomic schemes are developed using a wide range of phenotypic and genotypic information.

Polyphyletic group An assemblage of organisms that includes multiple evolutionary lineages. Polyphyletic assemblages usually reflect insufficient knowledge regarding the phylogeny of a group of organisms.

Polyphyletic Term applied to a group of organisms which does not include the most recent common ancestor of those organisms; the ancestor does not possess the character shared by members of the group.

Polyphyly The condition that a taxon or other group of organisms does not contain the most recent common ancestor of all members of the group, implying that it has multiple evolutionary origins; such groups are not valid as formal taxa and are recognized as such only through error. Contrasts with monophyly and paraphyly.

Polyphyodont Continuous replacement of teeth with many generations (from "poly" =many; "phylo" = generation and "dont" =teeth).

Polypide An individual or zooid in a colony, specifically in ectoprocts, which has a lophophore, digestive tract, muscles, and nerve centers.

Polyplacophora The class of molluscs whose members are elongate, dorsoventrally flattened, and have a shell consisting of eight dorsal plates.

Polyploid An organism possessing more than two full homologous sets of chromosomes.

Polyploidy Abnormal variation in the number of chromosome sets. The condition when a cell or organism has more than the customary two sets of chromosomes. This is an especially effective speciation mechanism in plants since the extra chromosomes will establish reproductive isolation with the parental population(s), an essential for speciation.

Polypod larva Caterpillar type of larva found in Lepidoptera and some Hymenoptera. It has thoracic appendages and abdominal locomotory processes (prolegs). Also called cruciform.

Polyribosome An aggregation of several ribosomes attached to one messenger RNA molecule.

Polysaccharide A carbohydrate composed of many monosaccharide units, for example, glycogen, starch, and cellulose.Long chains of monosaccharide units bonded together; e.g., glycogen, starch, and cellulose.

Polysome Two or more ribosomes connected by a molecule of messenger RNA.

Polyspermy Entry of more than one sperm into the egg during fertilization.

Polytene chromosome Large chromosome resulting from repeated DNA replication and alignment of homologs in the absence of mitosis; found in salivary glands and some other tissues of dipteran insects.

polyunsaturated fat A fat having several to many double bonds between carbon atoms.

Polyzoic A tapeworm forming a strobila of several to many proglottids; also, a colony of many zooids.

Pome A fleshy fruit derived from a compound inferior ovary; the fleshy edible part is the ripened tissue surrounding the ovary (derived from receptacle and perianth tissue); the ovary matures into the core and contains the seed.example: apple or pear.

Pongid Of or relating to the primate family Pongidae, comprising the anthropoid apes (gorillas, chimpanzees, gibbons, orangutans).

Pons The region that, with the medulla oblongata, makes up the hindbrain, which controls heart rate, constriction and dilation of blood vessels, respiration, and digestion.

Pontiac fever A bacterial disease caused by Legionella pneumophila that resembles an allergic disease more than an infection. First described from Pontiac, Michigan.

Population The individuals of a given species that occupy the same locality and form the interbreeding group in that location. A group of two or more populations that regularly exchange genes is known as a *metapopulation.*.

Population bottleneck Type of genetic drift that occurs as the result of a population being drastically reduced in numbers by an event having little to do with the usual forces of natural selection.

Population crash A sudden population decline caused by predation, waste accumulation, or resource depletion; also called a dieback.

Population density The number of individuals of a population per unit area or volume of living space.

Population dynamics The study of the factors that affect the growth, stability, and decline of populations, as well as the interactions of those factors.

Population explosion Growth of a population at exponential rates to a size that exceeds environmental carrying capacity; usually followed by a population crash.

Population genetics The study of hereditary at the population level. This may include measurement of allele frequencies and the genetic relationships between members of the group.

Population hurdle The need for investment in infrastructure and social services in a rapidly growing population that prevents the capital investment necessary for real economic development.

Population momentum A potential for increased population growth as young members reach reproductive age.

Population viability analysis (PVA) A method of predicting whether or not a species will persist in a particular environment.

Populational gradualism. The observation that new genetic variants become established in a population by increasing their frequencies across generations incrementally, initially from one or a few individuals and eventually characterizing a majority of the population.

Pore Any opening into or through a tissue or body structure.

Pore cell The tube-like cell of sponges that forms a pore.

Pore space Space between soil particles.

Porifera The animal phylum whose members are sessile and either asymmetrical or radially symmetrical. Body organized around a system of water canals and chambers. Cells are not organized into tissues or organs. Sponges.

Porifera The phylum comprising the sponges.

Porin proteins Proteins that form channels across the outer membrane of gram-negative bacterial cell walls. Small molecules are transported through these channels.

Porocyte Type of cell found in asconoid sponges through which water enters the spongocoel.

Porose area Sunken areas on the basis capituli of certain mites and ticks.

Porosity The relative amount of pore space versus soil particulate space in soil.

Porrect Extending horizontally forward: applied especially to antennae.

Portal porta = gate. A pathway or entrance.

Portal system System of large veins beginning and ending with a bed of capillaries; for example, hepatic portal and renal portal system in vertebrates.

Positional information Signals, to which genes regulating development respond, indicating a cell's location relative to other cells in an embryonic structure.

Positive feedback A physiological control mechanism in which a change in some variable triggers mechanisms that amplify the change.

Positive feedback control Occurs when information produced by the feedback increases and accelerates the response.

Postanal tail A tail that extends posterior to the anus; one of the four unique characteristics of chordates.

Posterior neuropore Posterior open end of the neural tube that must close during the development of the embryo; failure to close results in spina bifida

Posterior Situated at or toward the rear of the body; situated toward the back; in human anatomy the upright posture makes posterior and dorsal identical.

Posterior station Development of a protozoan in the hindgut or posterior midgut of its insect host, such as in the section Stercoraria of the Trypanosomatidae.

Postharvest physiology The study of the storage and perishability of fruits following harvest.

Postherpetic neuralgia The severe pain after a herpes infection.

Post-kala-azar dermal leishmanoid Disfiguring dermal condition developing about one to two years after inadequate treatment of kala-azar.

Postmatcrialist values A philosophy that emphasizes quality of life over acquisition of material goods.

Postmating isolation Isolation that occurs when fertilization is prevented even though mating has occurred.

Postmentum The basal region of the labium.

Postscutellum A small division of the mesonotum just behind the scutellum: usually very small or absent, but well developed in certain flies.

Postsynaptic membrane The surface of the cell on the opposite side of the synapse from the synaptic terminal of the stimulating neuron that contains receptor proteins and degradative enzymes for the neurotransmitter.

Post-transcriptional regulation Any process occurring after transcription which affects the amount of protein a gene produces. Includes RNA processing efficiency, RNA stability, translation efficiency, protein stability. For example, the rapid degradation of an mRNA will reduce the amount of protein arising from it. Increasing the rate at which an mRNA is translated will increase the amount of protein product.

Post-translational modification Modifications made to a polypeptide molecule after its initial synthesis, this includes proteolytic cleavages, *phosphorylation, glycosylation*, carboxylation, addition of fatty acid moieties, etc.

Post-translational processing The reactions which alter a protein's covalent structure, such as phosphorylation, glycosylation or proteolytic cleavage.

Post-vertical bristles A pair of bristles - divergent, parallel, or crossing - on the back of the head of various flies, some way behind the ocelli.

Postzygotic barrier Any of several species-isolating mechanisms that prevent hybrids produced by two different species from developing into viable, fertile adults.

Potable Refers to water suitable for drinking.

Potential energy Stored energy that is latent but available for use. A rock poised at the top of a hill or water stored behind a dam are examples of potential energy.

Potocytosis Endocytosis of certain small molecules and ions bound to specific receptors limited to small areas on the cell surface. The areas of the receptors are invaginated and pinch off to form tiny vesicles.

Pour plate A petri dish of solid culture medium with isolated microbial colonies growing both on its surface and within the medium, which has been prepared by mixing microorganisms with cooled, still liquid medium and then allowing the medium to harden.

Power The rate of energy delivery; measured in horsepower or watts.

Ppain receptor A modified nerve ending that, when stimulated, gives rise to the sense of pain.

Ppeck order A hierarchy of social privilege in a flock of birds.

Ppelagic Pertaining to the open ocean.

Ppolyembryony Development of a single zygote into more than one offspring.

Ppolygenic Multiple genetic influences controlling a single trait, causes the trait to display continuous variation.

Ppolygyny Condition of having more than one female mate at one time.

Ppropodeum First abdominal segment of hymenopterans, fused to the thorax.

PPT A measure of the salt content of sea water in terms of kg salt per kg of water, reckoned in parts per thousand. A conductivity version of this measure is the PSU, which differs from PPT by very little, on the order of 0.02 PSU or less.

Prairie An area of land dominated by grasses with occasional shrubby plants and small trees occurring where the grass cover is broken and with herbaceous perennials during certain seasons.

Praniza Parasitic larva of the isopod suborder Gnathiidea. It parasitizes fishes and feeds on blood.

PRE progesterone response element A binding site in a promoter to which the activated progesterone receptor can bind. The progesterone receptor is essentially a transcription factor which is activated only in the presence of progesterone . The activated receptor will bind to a PRE, and transcription of the adjacent gene will be altered.

Preadaptation Occurs when a structure or a process present in members of a species proves useful in promoting reproductive success when an individual encounters new environmental situations.

Pre-apical Arising just before the tip: many flies. for example, have pre-apical bristles just before the tip of the tibia.

Prebiotic synthesis. The chemical synthesis that occurred before the emergence of life.

Precambrian Informal term describing 7/8 of geologic time from the beginning of the earth to the beginning of the Cambrian Period of the Paleozoic Era. During this time the atmosphere and oceans formed, life originated (or possibly "colonized" Earth), eukaryotes and simple animals evolved and by the end of the precambrian they began to accumulate hard preservable parts, the common occurrence of which marks the beginning of the Cambrian.

Prechordal plate A mass of chordamesodermal cells located anterior to the notochord which will form head mesenchyme.

Precious corals Gorgonians that secrete a red or pink skeleton consisting of fused calcareous spicules.

Precipitation Any form of water particles, such as frozen water in snow or ice crystals, or liquid water in raindrops or drizzle.

Precision Precision is the repeatability of a measurement. A measurement can be precise, but not accurate.

Precocial young Young that are born or hatched at a relatively advanced state of development and are capable of a more independent existence beginning at birth or hatching. Compare to altricial.

Precocious Appearing, developing, or maturing earlier than is usual.

Precostal area The area in front of , or to the fore of the costa.

Precycling Making environmentally sound decisions at the store and reducing waste before we buy.

Predaceous Preying upon other organisms, predatory.

Predaceous, predacious Living by killing and consuming other animals; predatory.

Predation The derivation of an organism of elements essential for its existence from organisms of other species that it consumes and destroys. The ingestion of prey by a predator for energy and nutrients.

Predator An organism that consumes another living organism (carnivores and herbivores are both predators by this definition)

Predator mutualism Two or more species of predators working together to capture prey

Predatory release Occurs when a predator species is removed from a prey species such as by great reduction in the predator's population size or by the migration of the prey species to an area without major predators. The removal of the predator releases the prey from one of the factors limiting its population size.

Predentine The extracellular matrix produced by odontoblasts, which becomes mineralised to form dentine. Similar in structure to osteoid .

Predictive model a somewhat simplified representation of a physical process that yields information with predictive value

Preformation The erroneous idea that gametes contain miniaturized versions of all of the elements present in an adult.

Preformationism 17th century theory of inheritance that hypothesized that all the organs of an adult were prefigured in miniature within either the sperm or the ovum.

Prehensile Adapted for grasping.

Prehensile movement The ability to seize or grasp.

Premating isolation When behaviors or other factors prevent animals from mating.

Prementum The distal region of the labium, from which spring the labial palps and the ligula.

Pre-Mrna An RNA molecule which is *transcribed* from chromosomal DNA in the nucleus of eukaryotic cells, and subsequently processed through *splicing* reactions to generate the mRNA which directs protein synthesis in the cytoplasm.

Premunition Resistance to reinfection or superinfection, conferred by a still existing infection, that does not destroy the organisms of the infection already present.

Prenatal testing Testing to detect the presence of a genetic disorder in an embryo or fetus; commonly done by amniocentesis or chorionic villi sampling.

Prenymph Nonfeeding, quiescent stage in the life cycle of a chigger mite.

Preoptic nuclei Neurons in the hypothalamus that produce various neuropeptides.

Preovipositional period The period between the emergence of an adult female and the start of its egg laying.

Preparedness The genetically based predisposition to learn. Depending upon the learning situation, animals of a particular species may be prepared, unprepared, or contraprepared to learn.

Prepatent Developmental stage in an infection before agents produce evidence of their presence.

Prepupa The last larval instar after it ceases to feed; often it takes on a distinctive appearance becoming quiescent and rather shrunken, and often looks dead.

Presoma The proboscis, neck, and attached muscles and organs of an acanthocephalan.

Pressure flow hypothesis The theory that organic solutes move along a concentration gradient from source to sink through the phloem.

Pressure potential The water force created by a real pressure against a membrane.

Pressure The weight exerted over a unit area of surface. This is equal to 1 atmosphere (14.7 lbs per square inch) at the sea surface and 1 atmosphere plus the pressure exerted by the water column, which is 1 atmosphere per 10 m (33 ft) of depth.

Pressure-flow hypothesis A hypothesis accounting for sap flow through the phloem system. According to this hypothesis, the solution containing nutrient sugars moves through the sieve tubes by bulk flow, moving into and out of the sieve tubes by active transport and diffusion.

Presumptive organization Arrangement of cells in the embryo into groups which in normal development become a particular organ or tissue.

Presymptomatic screening Testing to detect genetic disorders that only become apparent later in life. The tests are done before the condition actually appears, such as with Huntington disease.

Pretarsus In insects the terminal segment of the leg bearing the pretarsal claws.

Prevalance In epidemiology, the number of cases of a disease at a given time; that is, a static measurement. Contrast with incidence.

Prevalence rate Refers to the total number of

individuals infected at any one time in a given population regardless of when the disease began.

Prey An animal that is used by other animals for food.

Prey model A type of optimal foraging model that addresses the types of organisms an individual should try to catch and eat (i.e., its diet).

Prey switching The tendency of predators to switch to a more readily available prey when one prey species becomes rare; allows the first prey population to rebound and helps prevent its extinction.

Prezygotic barrier A reproductive barrier that impedes mating between species or hinders fertilization of ova if interspecific mating is attempted.

Priapulida A phylum of aschelminths commonly called priapulids.

Pribnow box A special base sequence in the promoter that is recognized by the RNA polymerase and is the site of initial polymerase binding.

Price elasticity A situation in which supply and demand of a commodity respond to price.

Primary (frank) pathogen Any organism that causes a disease in the host by direct interaction with or infection of the host.

Primary amebic meningoencephalitis An infection of the meninges of the brain by the free-living amoebae Naegleria or Acanthamoeba.

Primary bilateral symmetry. Usually applied to a radially symmetrical organism descended from a bilateral ancestor and developing from a bilaterally symmetrical larva.

Primary body Those parts of a plant produced by the shoot and root apical meristems.

Primary cell wall The cell wall outside the plasma membrane that surrounds plant cells; composed of the polysaccharide cellulose.

Primary compounds Chemicals made by plants and needed for the plant's own metabolism.

Primary consumer A heterotrophic, herbivorous organism that feeds directly on a primary producer. Urchins are a primary consumer as they feed on kelp.

Primary endosperm nucleus The product of the fusion of a sperm and two polar nuclei in the embryo sac of angiosperms; double fertilization.

Primary germ layers Blocks or layers of embryonic cells that give rise to tissues and organs of animals.

Primary growth Cells produced by an apical meristem. The growth a plant by the actions of apical meristems on the shoot and root apices in producing plant primary tisues.

Primary immune response The initial immune response to an antigen, which appears after a lag of several days.

Primary macronutrients Elements that plants require in relatively large quantities: nitrogen, phosphorus, and potassium.

Primary meristems The apical meristems on the shoot and root apices in plants that produce plant primary tissues.

Primary metabolites Microbial metabolites produced during the growth phase of an organism.

Primary oocyte During oogenesis, the immature egg cell in which the first meiotic division begins, leading to the production of the first polar body and the secondary oocyte.

Primary pit fields Regions within the primary cell wall in which plasmodesmata traverse the cell wall.

Primary pollutants Chemicals released directly into the air in a harmful form.

Primary producer An autotrophic organism that makes complex organic compounds from inorganic compounds through the process of photosynthesis or chemosynthesis. For example, phytoplankton synthesize sugars (organic compounds) from carbon dioxide and water(inorganic compounds) using energy from the sun. Phytoplankton is one example of a marine primary producer.

Primary production The production of living matter by photosynthesizing organisms or by chemosynthesizing organisms. Usually expressed as grams of carbon per square meter per year

Primary productivity Synthesis of organic materials (biomass) by green plants using the energy captured in photosynthesis.

Primary radial symmetry .Usually applied to a radially symmetrical organism that did not have a bilateral ancestor or larva, in contrast to a secondarily radial organism.

Primary reproductives Those members of a social group of insects whose primary role is reproduction, (often the founders of the colony). Compared to secondary reproductive who may produce some young but are primarily involved in some other activity.

Primary root The first root formed by a plant.

Primary standards Regulations of the 1970 Clean Air Act; intended to protect human health.

Primary structure Refers to the sequence of *amino acid residues* or *nucleotides* within protein or nucleic acid molecules, respectively.

Primary succession Plant successional events occurring in a pristine or newly forming habitat.

Primary tissue Any tissue derived from the apical meristem, either shoot or root.

Primary transcript When a gene is transcribed in the nucleus, the initial product is the primary transcript, an RNA containing copies of all exons and introns. This primary transcript is then processed by the cell to remove the introns, to cleave off unwanted 3' sequence, and to polyadenylate the 5' end. The mature message thus formed is then exported to the cytoplasm for translation.

Primary treatment A process that removes solids from sewage before it is discharged or treated further.

Primary wall The wall layer of a plant cell deposited during cell expansion, generally thin and elastic.

Primates The taxonomic order of mammals that includes prosimians (lemurs and tarsiers), monkeys, apes, and humans; characteristics include large brain, stereoscopic vision, and grasping hand.

Primer A small oligonucleotide (anywhere from 6 to 50 nt long) used to prime DNA synthesis. The DNA polymerases are only able to extend a pre-existing strand along a template; they are not able to take a naked single strand and produce a complementary copy of it *de-novo*. A primer which sticks to the template is therefore used to initiate the replication. Primers are necessary for DNA sequencing and PCR.

Primer extension A reaction in which DNA is *reverse transcribed* from an RNA *template* to which a specific oligonucleotide *primer* has been *annealed*. The new cDNA product is an extension of the primer, which is synthesized at the 3' end of the primer in a direction extending toward the 5' end of the RNA. This reaction is useful for exploring the extreme 5' end of RNA molecules.This is a method used to figure out how far upstream from a fixed site the start of an mRNA is.

Priming pheromone A chemical signal that alters the physiology of another organism, eventually causing a change in its behavior. Contrast with releasing pheromone.

Primite Anterior member of a pair of gregarines in syzygy.

Primitive Describes a character state that is present in the common ancestor of a clade. A primitive character state is inferred to be the original condition of that character within the clade under consideration. For example, "presence of hair" is a primitive character state for all mammals, whereas the "hairlessness" of whales is a derived state for one subclade within the Mammalia.

Primitive folds Thickened ridges of the primitive streak; formed by convergent flow of epiblast.

Primitive streak A medial thickening along the dorsal margin of an amniote embryo that forms during the migration of endodermal and mesodermal cells into the interior of the embryo.

Primordial follicle An immature ovarian follicle that will develop into successively more mature follicles as the oocyte matures.

Principle of allocation The concept that each organism has an energy budget, or a limited amount of total energy available for all of its maintenance and reproductive needs.

Principle of competitive exclusion A result of natural selection whereby two similar species in a community occupy different ecological niches, thereby reducing competition for food.

principle of independent assortment One of Mendel's observations on the behavior of hereditary units during gamete formation. A modern interpretation of this principle is that genes carried in one chromosome are distributed to gametes without regard to the distribution of genes in nonhomologous chromosomes.

Principle of segregation One of Mendel's observations on the behavior of hereditary units during gamete formation. A modern interpretation of the principle of segregation is that genes exist in pairs, and during gamete formation, members of a pair of genes are distributed into separate gametes.

Prion An infectious form of protein that may increase in number by converting related proteins to more prions.Infectious agents composed only of one or more protein molecules without any accompanying genetic information.

Pristine Denoting a natural and undisturbed state.

Proamnion Region anterior to developing avian embryo, surrounding the head; contains only endoderm and ectoderm. It will later become invaded with mesoderm and become incorporated into the amnion.

Probe A fragment of DNA or RNA which is labeled in some way (often incorporating 32P or 35S), and which is used to hybridize with the nucleic acid in which we are interested. For example, if we want to quantitate the levels of alpha subunit mRNA in a preparation of pituitary RNA, we might make a radiolabeled RNA *in-vitro* which is complementary to the mRNA, and then use it to probe a Northern blot of the pit RNA. A probe can be radiolabeled, or tagged with another functional group such as biotin. A probe can be cloned DNA, or might be a synthetic DNA strand.

Probiotic (1) The oral administration of either living microorganisms or substances to promote the health and growth of an animal or human.

(2) A living organism that may provide health benefits beyond its nutritional value when it is ingested.

Proboscis A snout or trunk. Also, tubular sucking or feeding organ with the mouth at the end as in planarians, leeches, and insects. Also, the sensory and defensive organ at the anterior end of certain invertebrates.Name given to various kinds of sucking mouths in which some of the mouth-parts are drawn out to form tubes.

Procambium A primary meristem of roots and shoots that forms the vascular tissue.

Procaryote A cell in which the genetic material is dispersed throughout the cell in distinction to a eucaryote which has a nucleus and other organelles. Bacteria and blue-green algae are procaryotes

Procaryotic cells Prokaryotes include the kingdoms of Monera (simple bacteria) and Archaea.

They are characterized by:

The lack of a nucleus;

The possession of a single double-stranded DNA molucule

The possession of a very small range of organelles

Procaryotic cells replicate by dividing in two by means of binary fission, a very fast division process. The rapid division process allows for rapid adaptation to the changes in their environment.

Procaryotic species A collection of strains that share many stable properties and differ significantly from other groups of strains.

Procercoid Cestode metacestode developing from a coracidium in some orders. It usually has a posterior cercomer.

Procercoid larva Cestode developing from a coracidium in some orders; it usually has a posterior cercomer. Developmental stage between oncosphere and plerocercoid.300

Process To project from.

Processing The reactions occurring in the nucleus which convert the primary RNA transcript to a mature mRNA. Processing reactions include capping, splicing and polyadenylation. The term can also refer to the processing of the protein product, including proteolytic cleavages, glycosylation, etc.

Processivity The extent to which an RNA or DNA *polymerase* adheres to a *template* before dissociating, determines the average length (in kilobases) of the newly synthesized nucleic acid strands. Also applies to the action of *exonucleases* in digesting from the ends to the middle of a nucleic acid.

Proctodeal valve In insects, a valve in the anterior end of the hindgut that serves as an occlusor mechanism.

Procuticle Thicker layer beneath the epicuticle of arthropods that lends mass and strength to the cuticle. It contains chitin, sclerotin, and also inorganic salts in Crustacea. The layers within the procuticle vary in structure and composition.

Prodromal stage The period during the course of a disease in which there is the appearance of signs and symptoms, but they are not yet distinctive and characteristic enough to make an accurate diagnosis.

Producer An organism that synthesizes food molecules from inorganic compounds by using an external energy source; most producers are photosynthetic. The first level in a food pyramid; consist of organisms that generate the food used by all other organisms in the ecosystem; usually consist of plants making food by photosynthesis.

Producer, in ecological systems An autotrophic organism, usually a photosynthesizer, that contributes to the net primary productivity of a community.

Production frontier The maximum output of two competing commodities at different levels of production.

Production. In ecology, the energy accumulated by an organism that becomes incorporated into new biomass.

Productivity The amount of plant product resulting from a combination of water, nutrients and environmental factors.

Profile A graph that shows changes in temperature, salinity, or any other parameter with depth

Profundal zone The deepest portion of a lake.

Progametangia Swollen hyphae of fungi that fuse at the point of contact and eventually form gametangia.

Progametangium The cell that gives rise to a gametangium and a proximal suspensor during the early stages of sexual reproduction in zygomycetous fungi.

Progenesis Accelerated development of reproductive organs relative to somatic tissue, leading to paedomorphosis.

Progesterone A female sex hormone secreted by the corpus luteum that, in conjunction with estradiol, functions to prepare and maintain the uterine mucosa for the reception, implantation, and development of a fetus if fertilization occurs.

Proglottid One set of reproductive organs in a tapeworm strobila; usually corresponds to a segment. One of the linearly arranged segmentlike sections that make up the strobila of a tapeworm.

Prognathic A prominent lower jaw which may bring the lower teeth ahead of the upper teeth

Prognathous Having a more or less horizontal head, with the mouth-parts at the front.

Progymnosperm An extinct group of plants, the Progymnospermophyta, which included woody plants bearing large plannated braching systems; the prototypes of the gymnospems (e.g., Archaeopteris).

Prohaptor Collective adhesive and feeding organs at the anterior end of a monogenetic trematode.

Prohormone A precursor of a hormone, especially a peptide hormone.

Prokaryote Type of cell that lacks a membrane-bound nucleus and has no membrane organelles; a bacterium. Prokaryotes are more primitive than eukaryotes. Cells lacking membrane-bound organelles and having a single circular chromosome, and ribosomes surrounded by a cell membrane. Prokaryotes were the first forms of life on earth, evolving over 3.5 billion years ago.

Prokaryotic cell A type of cell lacking a nucleus and membrane-bound organelles; found in the Kingdom Monera.

Prolactin A hormone produced by the anterior pituitary; secreted at the end of pregnancy when it activates milk production by the mammary glands.

Proleg One of the fleshy, stumpy legs on the hind region of a caterpillar. Unjointed abdominal appendage in the larva of Lepidoptera and some other insects.

Proliferate To grow or multiply by rapidly producing new tissue, parts, cells, buds, or offspring.

Proliferative kidney disease A protozoan disease caused by an unclassified myxozoan in salmonids throughout the world.

Proline One of 20 amino acids common in proteins. It is a common amino acid of collagen and like lycine must be hydroxylated by ascorbic acid in order to allow the formation of hydrogen bonds which will hold the triple helix together.

Proline-rich proteins A group of proteins in saliva which have the ability to bind to calcium. They provide the protective layer of pellicle on the tooth surfaces by binding to the calcium in enamel. They also bind onto microorganism, providing a link between organism and the tooth surface. Proline rich proteins help to de-toxify tanins, which are potentially poisonous plant substances found in tea, and unripe fruit.

Promastigote Form of Trypanosomatidae with the free flagellum and the kinetoplast anterior to the nucleus, as in Leptomonas.

Prometaphase The phase of mitosis in which the nuclear envelope breaks into fragments. Some of the spindle fibers reach the chromosomes and attach to protein structures at the centromeres, called kinetochores, while others make contact with microtubules coming from the opposite pole. The opposing spindle fibers move the chromosomes toward the metaphase plate, an imaginary plane equidistant from the poles.

Promethean environmentalism A form of technological optimism that predicts that human ingenuity and enterprise will find cures for all our problems.

Promiscuity A mating system in which there is no prolonged association between the sexes and multiple mating by at least one sex.

Promiscuous A mating behavior in which the male and female do not form lasting pair bonds; one male may mate with several females, or one female with several males.

Promoter A specific sequence within a double-stranded DNA molecule that is recognized by an RNA *polymerase*, which binds to it and uses it to begin transcribing the DNA *template* into a new RNA. The location and orientation of the *promoter* within a DNA molecule determines the start site of the new RNA. Other proteins (e.g. transcriptional activators such as *sigma factor*) are usually required for an RNA polymerase to recognize a promoter. The first few hundred nucleotides of DNA "upstream" (on the 5' side) of a gene, which control the transcription of that gene. The promoter is part of the 5' flanking DNA, i.e. it is not transcribed into RNA, but without the promoter, the gene is not functional. Note that the definition is a bit hazy as far as the size of the region encompassed, but the "promoter" of a gene starts with the nucleotide immediately upstream from the cap site, and includes binding sites for one or more transcription factors which can not work if moved farther away from the gene.

Pronatalist pressures Influences that encourage people to have children.

Pronephros Most anterior of three pairs of embryonic renal organs of vertebrates, functional only in adult hagfishes and larval fishes and amphibians, and vestigial in mammalian embryos. Adj., pronephric.

Pronotal comb A row of stout spines on the hind margin of the pronotum of certain fleas.

Pronotum The upper, often shield-like, hardened body-wall plate, located just behind the head of an insect.

Pronucleus Haploid, gametic nucleus (from sperm or egg) in a fertilized egg, prior to fusion to form a zygote nucleus

Prop roots Adventitious roots arising on a stem above the ground and imparting some mechanical support to plants. The angled roots may provide for absorption of water and nutrients.

Propagated epidemic An epidemic that is characterized by a relatively slow and prolonged rise and then a gradual decline in the number of individuals infected. It usually results from the introduction of an infected individual into a susceptible population, and the pathogen is transmitted from person to person.

Propagation The process of increasing in number.

Propagule In animals, the minimum number of individuals of a species capable of colonizing a new area. This may be fertilized eggs, a mated female, a single male and a single female, or a whole group of organisms depending upon the biological and behavioral requirements of the species. In plants, a propagule is whatever structure functions to reproduce the species: a seed, spore, stem or root cutting, etc.

Prophase (1) The first stage of mitosis during which chromosomes condense, the nuclear envelope disappears, and the centrioles divide and migrate to opposite ends of the cell.

(2) The first stage of mitosis and meiosis (although in meiosis this phase is denoted with either a roman numeral I or II) where the chromatin condenses to form chromosomes, nucleolus dissolves, nuclear envelope dissolves, and the spindle begins to form.

Proplastids Membrane-bound particles that develop some internal structure; may subsequently develop into chloroplasts, chromoplasts, or leucoplasts.

Propodeum The I st abdominal segment in the hymenopteran group known as the Apocrita: it is completely fused with the thorax.

Propodosoma Portion of the podosoma that bears the first and second pairs of legs of a tick or mite.

Propolar cells Anterior tier of cells in the calotte of a dicyemid mesozoan.

Proprioceptor Sensory receptor located deep within the tissues, especially muscles, tendons, and joints, that is responsive to changes in muscle stretch, body position, and movement.

Propupa In thrips, the next to the last nymphal instar in which the wing pads are present and the legs short and thick. Also in male scale insects.

Prosencephalon The most anterior brain region of developing embryos which later divides into the telencephalon (most anterior brain region) and diencephalon. These regions later differentiate into the cerebral hemispheres, thalamus and hypothalamus and optic vesicles. Synonym: forebrain.

Prosimian Any member of a group of arboreal primates including lemurs, tarsiers, and lorises, but excluding monkeys, apes, and humans.

Prosoma Anterior part of an invertebrate in which primitive segmentation is not visible; fused head and thorax of arthropod; cephalothorax.

Prosopyle Connections between the incurrent and radial canals in some sponges.

Prostaglandins Are members of a class of hormones known as the eicosanoids. They are released by cells which have been damaged and have a powerful ability to sensitise nerve endings causing tenderness to the damaged area, and to cause vasoconstriction by contracting the smooth muscle of arterioles. They belong to a group of compounds which have a similar effect on nervous tissue, known asneuropeptides.

Prostate gland A gland that is located near and empties into the urethra; produces a secretion that enhances sperm viability. Gland involved in the reproductive system in males, the prostate secretes a sperm activating chemical into the semen during the arousal/ejaculation response.

Prostate The prostate gland is "before" the bladder.

Prosternum Ventral surface of the 1st thoracic segment.

Prostheca An extension of a bacterial cell, including the plasma membrane and cell wall, that is narrower than the mature cell.

Prosthetic group Nonprotein groups that are attached to an enzyme or other protein and necessary for its function.

Prostomium Anterior closure of a metameric animal, anterior to the mouth.

Prostration Extreme physical weakness or exhaustion.

Protamine Protein that binds DNA in sperm, replacing histones and allowing chromosomes to become more highly condensed than possible with histones.

Protandric hermaphroditism A pattern of sexuality in which a single individual functions as male and then female in sequence.

Protandrous Condition of hermaphroditic animals and plants in which male organs and their products appear before the corresponding female organs and products, thus preventing self-fertilization.

Protandry The condition in a monoecious organism in which male gonads mature before female gametes; prevents self-fertilization.

Protease An enzyme that hydrolyzes proteins to their constituent amino acids. Also called a proteinase.

Proteasome A large, cylindrical protein complex that degrades ubiquitin-labeled proteins to peptides in an ATP-dependent process.

Protective coloration Coloration that benefits the individual by providing concealment from predators.

Protein A large molecule composed of one or more chains of *amino acids* in a specific order; the order is determined by the *base sequence* of *nucleotides* in the *gene* coding for the protein. Proteins are required for the structure, function, and regulation of the bodys cells, tissues, and organs, and each protein has unique functions. Examples are hormones, *enzymes*, and antibodies. Class of biochemical compounds constructed from amino acids. Proteins may be structural, such as those that make up hair and cartilage, or they may be reactive, such as the enzymes.

Protein engineering The rational design of proteins by constructing specific amino acid sequences through molecular techniques, with the objective of moaifying protein characteristics.

Protein kinase An enzyme that transfers phosphate groups from ATP to a protein.

Protein phosphatase An enzyme that removes phosphate groups from proteins, often functioning to reverse the effect of a protein kinase.

Protein polymorphism Presence of several variants of a protein of a given type (e.g., a certain enzyme, such as carboxylase) in a population

Protein primary structure The linear sequence of amino acids in a polypeptide or protein. This is the amino acid sequence for which the triplet code of bases in the mRNA codes.

Protein purification Chemical methods, usually based on a combination of chemical properties and physical properties, by which a desired protein is isolated from the other proteins with which it normally occurs.

Protein sequencing The determination of the primary structure of a protein. This is usually done using a machanised version of the Edman reaction, a chemical process which sequentially removes amino acids from the protein and identifies the amino acid which is removed.

Protein splicing The post-translational process in which part of a precursor polypeptide is removed before the mature polypeptide folds into its final shape; it is carried out by self-splicing proteins that remove inteins and join the remaining exteins.

Protein synthesis The assemblage of protein as determined by the nucleotide sequence of a messenger RNA and the assistance of transfer RNA aligning amino acids in the proper arrangement.

Protein trafficking The movement of proteins around a cell or within organelles so that they reach the correct place to carry out their activity.

Proteinaceous Describes any structure which is composed of protein.

Proteinoids Polymers of amino acids formed spontaneously from inorganic molecules; have enzyme-like properties and can catalyze chemical reactions.

Proton The positively charged elements of the nucleus of an atom. A hydrogen atom without its electron amounts to a single proton charge.

Protelean parasite Organism parasitic during its larval or juvenile stages and free living as an adult, usually changing form with each stage.

Proteobacteria A large group of bacteria, primarily gram-negative, that 16S rRNA sequence comparisons show to be phylogenetically related; proteobacteria contain the purple photosynthetic bacteria and their relatives and are composed of the a, b, g, d, and e subgroups.

Proteoglycans A glycoprotein in the extracellular matrix of animal cells, rich in carbohydrate.

Proteome The complete collection of proteins that an organism produces.

Proterosoma In mites, the anterior part of the body when there is a demarcation of the body between the second and third pair of legs.

Protests Single-celled organisms; a type of eukaryote. Protista

Prothallial cell Sterile cells present in the pollen grain of gymnosperms, believed to represent the last remnant of the vegetative male gametophyte thallus, which was, ancestrally,

Prothallus In ferns, a small heart-shaped bisexual gametophyte.

Prothoracic gland An invertebrate endocrine gland located in the prothorax region directly behind the head, which secretes ecdysone, a steroid hormone that promotes molting.

Prothorax The first of the three thoracic segments of an insect; usually contains the first pair of walking appendages.

Prothrombin A constituent of blood plasma that is changed to thrombin by a catalytic sequence that includes thromboplastin, calcium, and plasma globulins; involved in blood clotting.

Protista The taxonomic Kingdom from which the other three eukaryotic kingdoms (Fungi, Animalia and Plantae) are thought to have evolved. The earliest eukaryotes were single-celled organisms that would today be placed in this admittedly not monophyletic group. The endosymbiosis theory suggests that eukaryotes may have evolved independently several times.

Protochordates Chordates that lack a backbone.

Protocoel The anterior coelomic compartment in some deuterostomes, corresponds to the axocoel in echinoderms.

Protocooperation. A mutually beneficial interaction between organisms in which the interaction is not physiologically necessary to the survival of either.

Protoderm The outermost primary meristem, which gives rise to the epidermis of roots and shoots.

Protogynous Hermaphroditism in which the female gonads mature before the male gonads.

Protogyny A sequentially hermaphroditic species in which individuals change from females to males.

Protomer An individual subunit of a viral capsid; a capsomer is made of protomers.

Protomerite Anterior half of a cephaline gregarine protozoan.

Protomers Subunits from which a larger structure is built. Thus the tubulin heterodimer is the protomer for microtubule assembly, G-actin the protomer for F-actin. Because it avoids the difficulty that arises with, for example, dimers that serve as subunits for assembly, it is a useful term that deserves wider currency.

Proton A subatomic particle in the nucleus of an atom that carries a positive charge. The positively charged (+1) subatomic particle located in the atomic nucleus and having a mass slightly less than that of a neutron. Elements differ by the number of protons in their atoms.A subatomic particle with a single positive electrical charge, found in the nucleus of the atom.

Proton motive force (PMF) The force arising from a gradient of protons and a membrane potential that is thought to power ATP synthesis and other processes.

Proton pump An active transport mechanism in cell membranes that consumes ATP to force hydrogen ions out of a cell and, in the process, generates a membrane potential.

Protonema The early, filamentous growth of the gametophyte in bryophytes and ferns.

Protonephridium Primitive osmoregulatory or excretory organ composed of a tubule terminating internally with a flame bulb or solenocyte; the unit of a flame bulb system. Protonephridia are specialized for ultrafiltration.

Proton-motive force The potential energy stored in the form of an electrochemical gradient, generated by the pumping of hydrogen ions across biological membranes during chemiosmosis.

Protonymph Early, bloodsucking stage in the life cycle of some mesostigmatid mites.

Proto-oncogene A cellular *oncogene*-like sequence which is thought to play a role in controlling normal cellular growth and differentiation. A gene present in a normal cell which carries out a normal cellular function, but which can become an oncogene under certain circumstances. The prefix "c-" indicates a cellular gene, and is generally used for proto-oncogenes (examples: *c-myb* , *c-myc* , *c-fos* , *c-jun* , etc).

Protoplasm Organized living substance; cytoplasm and nucleoplasm of the cell.

Protoplast All of a plant cell excluding the wall.

Protoplast fusion The technique of enzymatically digesting the cell wall of two distinctly different cells, then treating the plasma membrane so that protoplasts of the two cells fuse. The resulting hybrid may be difficult or impossible to achieve with traditional plant breeding.

Protopod (protopodite)Coxa and basis together.

Protopod larva Larva found in some parasitic Hymenoptera and Diptera; limbs are rudimentary or absent; internal organs are incompletely differentiated; requires highly nutritive and sheltered environment for further development.

Protopodite The basal segment of a biramous appendage of a crustacean.

Protoscolex Juvenile scolex budded within a coenurus or a hydatid metacestode of a taeniid cestode.

Protostele When a plant's vascular tissue develops in a solid central bundle, it is said to have a protostele.

Protostomes Broad classification of organisms that tend to share certain embryological traits; among these the formation of the "mouth first" (hence the name) during gastrulation, before the future anus. (The site of gastrulation initiation, the blastopore, becomes the mouth.) Major protostome phyla include molluscs, annelids and arthropods.

Protostomia A group of phyla in which cleavage is determinate, the coelom (in coelomate forms) is formed by proliferation of mesodermal bands (schizocoelic formation), the mesoderm is formed from a particular blastomere (called 4d), and the mouth is derived from or near the blastopore. Includes the Annelida, Arthropoda, Mollusca, and a number of minor phyla. Compares with Deuterostomia.

Protostyle A rotating mucoid mass into which food is incorporated in the gut of a gastropod (phylum Mollusca).

Protothecosis A disease of humans and animals produced by the green alga Prototheca moriformis.

Prototroph A microorganism that requires the same nutrients as the majority of naturally occurring members of its species.

Protoxylem The first xylem cells formed in the primary xylem.

Protozoa The Subkingdom of the Kingdom Animalia, in older classification systems, that includes all unicellular organisms that lack differentiated tissues and that primarily reproduce through asexual means, although conjugation does occur.Single-celled protists grouped by their method of locomotion. This group includes *Paramecium*, *Amoeba*, and many other commonly observed protists.

Protozoan or protozoon A microorganism belonging to the Protozoa subkingdom. A unicellular or acellular eucaryotic protist whose organelles have the functional role of organs and tissues in more complex forms. Protozoa vary greatly in size, morphology, nutrition, and life cycle. A microscopic, single-celled organism that is largely aquatic and includes many parasitic forms.Animal-like protists; the various groups of unicellular and eukaryotic organisms that are mostly heterotrophic.

Protozoologist A person who studies protozoa.

Protozoology The study of protozoa.

Protract A muscle action that draws an element anteriorly.

Proventriculus In birds the glandular stomach between the crop and gizzard. In insects, a muscular dilation of foregut armed internally with chitinous teeth.

Province A geographically defined area with a characteristic set of species or characteristic percentage representation by given species

Proviral DNA Viral DNA that has been integrated into host cell DNA. In retroviruses it is the double-stranded DNA copy of the RNA genome.

Proviral state The condition of a host cell after having been transformed.

Provirus Viral DNA that inserts into a host genome.

Proximal Situated toward or near the point of attachment; opposite of distal, distant.

Proximal tubule The winding section of the renal tubule where most reabsorption of water, sodium, amino acids, and sugar takes place.

Proximate causation (factors)Mechanistic explanations for how behavior occurs, including, in particular, hormones, the nervous system, and behavior development.The factors that underlie the functioning of a biological system at a particular place and time, including those responsible for metabolic, physiological, and behavioral functions at the molecular, cellular, organs, and population levels.

Pruinose Covered with a powdery deposit, usually white or pale blue: especially applied to Odonata.

Pruning The selective removal of parts of a plant, usually woody shrubs or trees.

Pseudocoel A body cavity not lined with peritoneum and not a part of the blood or digestive systems, embryonically derived from the blastocoel.

Pseudocoelom In nematodes, a closed fluid-containing cavity that acts as a hydrostatic skeleton to maintain body shape, circulate nutrients, and hold the major body organs.

Pseudocoelomates Animals that have a body cavity that is in direct contact with the outer muscular layer of the body and does not arise by splitting of the mesoderm; e.g., roundworms.

Pseudoconditioning The strengthening of a response to a previously neutral stimulus by repeatedly eliciting the response with another stimulus without pairing the presentation of the two stimuli.

Pseudocyst Pocket of protozoa within a host cell but not surrounded by a cyst wall of parasite origin.

Pseudoelaters Moisture-sensitive cells produced in the sporangium of hornworts.

Pseudoextinction The apparent disappearance of a taxon. In cases of pseudoextinction, this disappearance is not due to the death of all members, but the evolution of novel features in one or more lineages, so that the new clades are not recognized as belonging to the paraphyletic ancestral group, whose members have ceased to exist. The Dinosauria, if defined so as to exclude the birds, is an example of a group that has undergone pseudoextinction.

Pseudofeces Material rejected by suspension feeders or deposit feeders as potential food before entering the gut

Pseudogene Inactive but stable components of the genome which derived by duplication and mutation of an ancestral, active gene. Pseudogenes can serve as the donor sequence in *gene conversion* events.

Pseudoknot A feature of RNA *tertiary structure*; best visualized as two overlapping *stem-loops* in which the loop of the first stem-loop participates as half of the stem in the second stem-loop.

Pseudolabia Bilateral lips around the mouth of many nematodes of the order Spirurata; they are not homologous to the lips of most other nematodes but develop from the inner wall of the buccal cavity.

Pseudomurein A modified peptidoglycan lacking D-amino acids and containing N-acetyltalosaminuronic acid instead of N-acetylmuramic acid; found in methanogenic archaea.

Pseudomyiasis Presence within a host of a fly not normally parasitic.

Pseudoplasmodium A sausage-shaped amoeboid structure consisting of many myxamoebae and behaving as a unit; the result of myxamoebal aggregation in the cellular slime molds; also called a slug.

Pseudopodia Temporary cytoplasmic extensions from a cell that enables it to move (sing.: pseudopodium).

Pseudopodium or pseudopod A nonpermanent cytoplasmic extension of the cell body by which amoebae and amoeboid organisms move and feed.

Pseudoreplication Results from either repeated testing of the same animal without proper statistical correction using a repeated measures design, or counting members of an interacting social group as separate data points. In both cases, problems with independence of data points arises with pseudoreplication.

Pseudorevertant A mutant virus or organism which has recovered a wildtype phenotype due to a second-site mutation (potentially located in a different region of the genome, or involving a different polypeptide) which has eliminated the effect of the initial mutation.

Pseudoscorpions Small arachnids, seldom over 5 mm. long, scorpion-like in general appearance but without sting.

Pseudovipositor The slender tube to which the posterior part of the abdomen is reduced in the female of certain insects.

Psilophyta A division of leafless, rootless, vascular, homosporous plants, as exemplified by the genus Psilotum.

Psittacosis A disease due to a strain of Chlamydia psittaci, first seen in parrots and later found in other birds and domestic fowl (in which it is called ornithosis). It is transmissible to humans.

Psoas The loin.

PSU practical salinity units A measure of the salt content of seawater, based upon electrical conductivity of reference sea water, which now happens to be a reference set of diluted sea water samples from the North Atlantic of known salt content.

Psychoactive drug A drug that affects the central nervous system by influencing the release of neurotransmitters or mimicking their actions.

Psychoactive plants Plants that contain compounds that

Psychobiology The study of the mechanism and function of the central nervous system from both psychological and biological perspectives.

Psychological dependence A condition marked by the strong desire to repeat the use of a drug to reexperience the feelings of well-being induced by the drug.

Psychopharmacology The study of the brain's behavior in terms of chemical, physiological, and psychological parameters.

Psychrophile A microorganism that grows well at 0°C and has an optimum growth temperature of 15°C or lower and a temperature maximum around 20°C.

Psychrotroph A microorganism that grows at 0°C, but has a growth optimum between 20 and 30°C, and a maximum of about 35°C.

Pteridophyte Plant in which the sporophyte generation is the larger phase and in which the gametophyte lives an existence independent of its parent sporophyte. Pteridophytes are almost all vascular plants, and include the lycophytes, trimerophytes, sphenophytes, and ferns.

Pteridosperm An extinct group of seed plants which bore fern-like leaves.

Pterobranchia The class of hemichordates whose members lack gill slits and have two or more arms. Colonial, living in externally secreted encasements.

Pterophyta The division of vascular plants commonly referred to as the ferns.

Pteropods Pelagic gastropods in which the foot is modified for swimming and the shell is reduced or absent.

Pterostigma A small coloured area near the wing-tip of dragonflies, bees, and various other clear-winged insects: also called the stigma.

Pterygoid bones In the skull of most lower vertebrates, four large, medially situated bones of the palatal complex that is located on the lower surface of the palatal cartilage.

Pterygoid pterygion = wing; oid = like.

Pterygote Any member of the sub-class Pterygota, which includes all winged and some secondarily wingless insects.

Ptilinum In Diptera an organ that can be inflated to a bladder-like structure and thrust out through a frontal suture of the head at the time of emergence from the puparium.

Puberty The age at which an organism can first reproduce.

Pubescence Having hairs or trichomes on the surface.

Pubescent Covered with short, soft hair

Pubis The hair that appears at puberty.

Public trust A doctrine obligating the government to maintain public lands in a natural state as guardians of the public interest.

Pudendo Pudenda = to be ashamed. External female genitalia.

Puerperal fever An acute, febrile condition following childbirth; it is characterized by infection of the uterus and/or adjacent regions and is caused by streptococci.

Puff Strands of DNA spread apart at certain locations on giant chromosomes of some flies where that DNA is being transcribed.

Pull factors Conditions that draw people from the country into the city.

Pulmonary Pertaining to the lungs.

pulmonary anthrax A form of anthrax involving the lungs. Also known as woolsorter's disease.

Pulmonary artery The artery that carries blood from the right ventricle of the vertebrate heart to the lungs. Artery carrying oxygen-poor blood from the heart to the lungs.

Pulmonary circuit The system of blood vessels from the right ventricle of the heart to the lungs, transporting deoxygenated blood and returning oxygenated blood from the lungs to the left atrium of the heart.

Pulmonary Relating to or associated with lungs.

Pulmonary vein In birds and mammals, a vein that carries oxygenated blood from the lungs to the left atrium of the heart, from which blood is pumped into the left ventricle and from there to the body tissues.

Pulmonate lung The gas exchange structure in the Pulmonata-the land snails and slugs.

Pulp The dental pulp is a connective tissue trapped inside the fully formed tooth with just one entry and exit for nerves and vessels, at the apex of the root. The characteristic cell of the pulp, is the odontoblast, which lines the walls of the pulp chamber and is able to form dentine throughout life.

Pulping The process of partially digesting and breaking up wood fibers to make paper.

Pulpitis Inflammation of the dental pulp caused by irritation from chemical, physical or bacterial injury, usually transmitted to the pulp via the dentine. It is important for the clinical management of the tooth, to decide whether the pulpitis is reversible, that is will it resolve if the irritation is removed, or whether it has been damaged beyond its capacity to repair.

Pulse A measurement of heart rate; distention of an artery that can be felt each time the heart contracts.

Pulsed field gel electrophoresis (PFGE) A gel technique which allows size-separation of very large fragments of DNA, in the range of hundreds of kb to thousands of kb. As in other gel electrophoresis techniques, populations of molecules migrate through the gel at a speed related to their size, producing discrete bands. In normal electrophoresis, DNA fragments greater than a certain size limit all migrate at the same rate through the gel. In PFGE, the electrophoretic voltage is applied alternately along two perpendicular axes, which forces even the larger DNA fragments to separate by size.

Pulvillus The little pad beneath each claw on the foot of a fly.

Pumice A lightweight white, yellow, or gray stone formed from volcanic glass; used in polishing and cleaning.

Punctate Covered with tiny pits or depressions, like the elytra of many beetles and the thoraxes of many hymenopterans.

Punctuated equilibria The hypothesis that evolution occurs in relatively rapid bursts, interspersed with long periods of stasis.

Punctuated equilibrium A model that holds that the evolutionary process is characterized by long periods with little or no change interspersed with short periods of rapid speciation.

Punctuated equilibrium model The idea that evolutionary change can occur rapidly over periods of thousands of years and that these periods of rapid change are interrupted by periods of constancy (stasis).

Punnett square A tool used by geneticists to help predict the results of a genetic cross. Different kinds of gametes produced by each parent are placed on each axis of the square. Combining gametes in the interior of the square gives the results of random fertilization.

Pupa The 3rd stage in the life history of butterflies and other insects undergoing a complete metamorphosis during which the larval body is rebuilt into that of the adult insect a non-feeding and usually inactive stage. A developmental stage of some insects, in which the organism is nonfeeding, immotile, and sometimes encapsulated or in a cocoon; the pupal stage occurs between the larval and adult phases.In metamorphozing insects, a stage between the larva and adult during which the organism undergoes major developmental changes. Inactive quiescent stage of the holometabolous insects. It follows the larval stages and precedes the adult stage.

Pupariation Formation of a puparium by the third-stage larvae of certain families of Diptera.

Puparium The barrel-shaped case which conceals the pupa of the house-fly and many other true flies. It is formed from the skin of the last larval instar.

Puparium/ puparia A case formed by the hardening of the last larval skin, in which the pupa is formed; usually of flies.

Pupate To turn into and exist as a pupa.

Pupiparous Insects which give birth to fully-grown larvae which pupate almost immediately are said to be pupiparous. The main examples are various blood-sucking flies.

Pure culture A population of cells that are identical because they arise from a single cell.

Purine A basic, heterocyclic, nitrogen-containing molecule with two joined rings that occurs in nucleic acids and other cell constituents; most purines are oxy or amino derivatives of the purine skeleton. The most important purines are adenine and guanine.A nitrogen- containing, single- ring, basic compound that occurs in nucleic acids. The purines in DNA and RNA are adenine and guanine.

Purine bases Adenine (A) or Guanine (G).

Purple membrane An area of the plasma membrane of Halobacterium that contains bacteriorhodopsin and is active in photosynthetic light energy trapping.

Push factors Conditions that force people out of the country and into the city.

Putrefaction The microbial decomposition of organic matter, especially the anaerobic breakdown of proteins, with the production of foul-smelling compounds such as hydrogen sulfide and amines.

Pycnocline the ocean layer, usually near the bottom of the photic zone, marked by a sharp change in density

Pycnogonida The class of chelicerate arthropods whose members have a reduced abdomen and four to six pairs of walking legs. Without special respiratory or excretory structures. Sea spiders.

Pycnoxylic Wood in which there is little or no parenchyma tissue among the xylem is called pycnoxylic. Conifers and flowering plants have pycnoxylic wood. Contrast with mannoxylic.

Pygidium In trilobites, the posterior division of the body, formed by fusion of the telson with one or more posterior pleurae.

Pygostyle The fused posterior caudal vertebrae of a bird; helps support tail feathers that are important in steering.

Pyloric caecum Each of the slender blind tubes found in the intestine of many bony fishes.

Pyloric sphincter The ring of muscle at the junction of the stomach and small intestine that regulates the movement of food into the small intestine.

Pyogenic Pus producing.

Pyramid of biomass A representation showing the total mass of living matter in each trophic level of an ecosystem.

Pyramid of energy A diagram of the energy flow between the trophic levels of an ecosystem; plants or other autotrophs (at the base of the pyramid) represent the greatest amount of energy, herbivores next, then primary carnivores, secondary carnivores, etc.

Pyramid of numbers The decrease in the number of individuals that is observed in each higher level of a food chain.

Pyrenoid The differentiated region of the chloroplast that is a center of starch formation in green algae and stoneworts.

Pyrimidine A basic, heterocyclic, nitrogen-containing molecule with one ring that occurs in nucleic acids and other cell constituents; pyrimidines are oxy or amino derivatives of the pyrimidine skeleton. The most important pyrimidines are cytosine, thymine, and uracil.

Pyrimidine bases Cytosine (C), Thymine (T) or Uracil (U).

Pyrogenic Substance that causes a rise in body temperature; causes fever.

Pyrophosphate Inhibitors of mineralisation, they may offer up phospate ions in the presence of alkaline phosphatase. Crystals of calcium pyrophosphates are found in abnormal calcification of soft tissue, such as the disc of the TM Joint.

Pyrrhophyta A division of protists commonly termed the dinoflagellates.

Pyruvic acid The end product 3-carbon compound resulting from glycolysis of glucose which enters the Krebs cycle in aerobic respiration.

PŽbrine An infectious disease of silkworms caused by the protozoan Nosema bombycis.

Q

Q fever An acute zoonotic disease caused by the rickettsia Coxiella burnetii.

Q10 Increase of metabolic rate with an increase of 10ºC

Quadrate One of the bones which together with the articular bones and the dentary, made/make up a reptile's jaw. In mammals the quadrate bone is incorporated into the middle ear as the incus.

Quadriceps quadr = four; caput = head.

Quadrilateral A cell near the base of the damselfly wing, whose shape is important in separating the families.

Quantitative character A heritable feature in a population that varies continuously as a result of environmental influences and the additive effect of two or more genes (polygenic inheritance).

Quantitative genetics The study of the genetic basis of traits, usually explained in terms of the interaction of a group of genes with the environment

Quantitative trait A genetic trait that is determined by multiple interacting loci, and for which there is a range of phenotypes between phenotypic extremes.

Quantitative trait locus (QTL) The set of genes that governs the quantity of a trait that is not completely determined by any one gene acting alone.

Quantum models of speciation Models of evolution that hold that speciation sometimes occurs rapidly as well as over long periods, as the classical theory proposed.

Quark A subatomic particle found in the nucleus of the atom.A fundamental constituent of matter generally found in protons and neutrons. To date, six quarks are known and termed; up, down, strange, charm, bottom, and top.

Quartan malaria Malaria with fevers recurring every 72 hours. Caused by Plasmodium malariae.

Quaternary period The most recent geologic period of the Cenozoic Era, the Quaternary began 2 million years ago with the growth of northern hemisphere continental glaciers and the ice age.

Quaternary structure In some proteins, a fourth structural level created by interactions with other proteins. Aspect of protein structure determined by the number and arrangement of polypeptides in a large protein such as hemoglobin.

Queen cell The special cell in which a queen honey bee develops from egg to the adult stage.

Queen. In entomology, the single fully developed female in a colony of social insects such as bees, ants, and termites, distinguished from workers, nonreproductive females, and soldiers.

Quellung reaction The increase in visibility or the swelling of the capsule of a microorganism in the presence of antibodies against capsular antigens.

Quiescent Being at rest; quiet; still; inactive or motionless.

Quiescent center A region located within the zone of cell division in plant roots, containing meristematic cells that divide very slowly.

Quorum sensing The process in which bacteria monitor their own population density by sensing the levels of signal molecules that are released by the microorganisms. When these signal molecules reach a threshold concentration, the population density has attained a critical level or quorum, and quorum-dependent genes are expressed.

Quotidian malaria Malaria with fevers recurring every 24 hours. Found in cases of overlapping infections.

R

R plasmid A bacterial plasmid carrying genes that confer resistance to certain antibiotics.

R selection Selection favoring rapid rates of reproduction and growth, especially among species that specialize in colonizing short-lived habitats.

Rabies An acute infectious disease of the central nervous system, which affects all warm-blooded animals (including humans). It is caused by an ssRNA virus belonging to the genus Lyssavirus in the family Rhabdoviridae.

Race A subdivision of a species that is capable of interbreeding with other members of the species.

Raceme A vertical inflorescence with stalked flowers.

Rachis Central, longitudinal, supporting structure in the ovary of some nematodes.

Racking The removal of sediments from wine bottles.

Radappertization The use of gamma rays from a cobalt source for control of microorganisms in foods.

Radial A spoke or ray.

Radial canals. Canals along the ambulacra radiating from the ring canal of echinoderms; also choanocyte-lined canals in syconoid sponges.

Radial cleavage. Embryonic development in which early cleavage planes are symmetrical to the polar axis, each blastomere of one tier lying directly above the corresponding blastomere of the next layer; indeterminate cleavage.

Radial cut A longitudinal section cut along a radius.

Radial micellation Reinforcement of the cell wall of guard cells in specific regions such that the cells curve outward when fully turgid.

Radial sector The posterior of the two main branches of the radius, usually abbreviated to Rs. It usually has several branches of its own.

Radial symmetry. A morphological condition in which the parts of an animal are arranged concentrically around an oral-aboral axis, and more than one imaginary plane through this axis yields halves that are mirror images of each other.In animals, refers to organisms with their body parts arranged around a central axis. Such animals tend to be circular or cylindrical in shape.

Radiata Group of simple animals with radial symmetry, including the cnidarians (or coelenterates) and ctenophores.

Radiation Event of rapid cladogenesis, believed to occur under conditions where a new feature permits a lineage to move into a new niche or new habitat, and is then called an adaptive radiation.

Radicle The primary root of a germinating seed.

Radioactive An unstable isotope that decays spontaneously and releases subatomic particles or units of energy.

Radioactive dating A method of determining the age of fossils and rocks using half-lives of radioactive isotopes.

Radioactive decay A change in the nuclei of radioactive isotopes that spontaneously emit high-energy electromagnetic radiation and/or subatomic particles while gradually changing into another isotope or different element.

Radioactive isotope An unstable form of an element that spontaneously emits either high- energy electromagnetic radiation or subatomic particles (or both). The decay rate describes the time necessary for a radioactive isotope to change to a stable element.

Radioactivity The emission of radiation by unstable atoms in the form of particles and rays.

Radio-carbon dating Method for determining the age of an organic substance by measuring the amount of the carbon isotope, carbon-14, remaining in the substance; useful for determining ages in the range of 500 to 70,000 years.

Radiocarbon technique The estimation of primary productivity by the measurement of radiocarbon uptake

Radioimmunoassay A very sensitive assay technique that uses a purified radioisotope-labeled antigen or antibody to compete for antibody or antigen with unlabeled standard and samples to determine the concentration of a substance in the samples.

Radioisotope An isotope that is radioactive, i.e., one having an unstable nucleus, which gives it the property of decay by one or more of several processes. Also called a radioactive isotope.

Radioisotope dating A technique for determining the age of fossil materials based on the rate of disintegration of the radioisotopes.

Radiolarian Members of the classes Acantharea, Phaeodarea, and Polycystinea (phylum Sarcomastigophora) with actinopodia and beautiful tests.

Radiolarian ooze A type of biogenous sediment that consists mostly of the silica shells of radiolarians.

Radioles Featherlike processes from the head of many tubicolous polychaete worms (phylum Annelida), used primarily for feeding.

Radiometric dating A method paleontologists use for determining the ages of rocks and fossils on a scale of absolute time, based on the half-life of radioactive isotopes.

Radionucleides Isotopes that exhibit radioactive decay.

Radiospermic Having seeds which are round or ovoid. Contrast with platyspermic.

Radiotelemetry A method of tracking the movements of a specific species.

Radius One of the main longitudinal veins, running near the front of the wing and usually the 3rd and abbreviated to R. It gives off a posterior branch - the radial sector - and the smaller branches of these veins are numbered R1, R2, etc.

Radula The rasping, tonguelike structure of most molluscs that is used for scraping food; composed of minute chitinous teeth that move over a cartilaginous odontophore.

Rain forest A forest with high humidity, constant temperature, and abundant rainfall (generally over 380 cm [150 in] per year); can be tropical or temperate.

Rain shadow The dry region on the leeward side of a mountain range, where rainfall is noticeably less than on the windward side. For example, the White Mountains in east central California are in the rain shadow of the Sierra Nevada.

Ram ventilation The movement of water across gills as a fish swims through the water with its mouth open.

Ramus The vertical part of the mandible which

supports the coronoid and the condylar processes.

Random motion Molecular movement that is always present but nondirectional.

Random spatial distribution Situation in which individuals are randomly distributed in a space; probability of an individual's being located at any given point is the same irrespective of location in the space

Range of optimum The range of values for a condition in the environment that is best able to support survival and reproduction of an organism.

Rangeland Grasslands and open woodlands suitable for livestock grazing.

Rank In traditional taxonomy, taxa are ranked according to their level of inclusiveness. Thus a genus contains one or more species, a family includes one or more genera, and so on.

RAPD random amplification of polymorphic DNA A method for identifying differences between genomes of different individuals by PCR with a single short (usually 10-base) primer, which will anneal with complementary sequence at undetermined positions in the genome. The products form a type of "genetic fingerprint".

Raphe a groove in the frustules of pennate diatoms through which cytoplasm extends for locomotion

Raprenox A process to reduce nitrogen oxide emissions by mixing combustion gases with isocyanic acid, which converts nitrogen oxides to dimolecular nitrogen gas.

Raptorial Adapted for seizing and grasping prey, like the -front legs of a mantis.

Ras protein. A protein that initiates a cascade of reactions leading to cell division when a growth factor is bound to the cell surface. The gene encoding Ras becomes an oncogene when a mutation produces a form of Ras protein that initiates the cascade even in the absence of the growth factor.

Rathke's pocket or pouch An evagination of the stomodeum that forms part of the pituitary gland (the adenohypophysis or anterior lobe of the pituitary gland). Synonym: hypophysis.

Ratite Referring to birds having an unkeeled sternum; compares with carinate.

Ray-finned Taxonomic group of fish, such as trout, tuna, salmon, and bass, that have thin, bony supports holding the fins away from the body and an internal swim bladder that changes the buoyancy of the body; one of the two main types of bony fishes.

Rays Parenchymal cells found in secondary xylem and phloem in both woody angiosperms and conifers, which provide for lateral (radial) transport.

RbcL A gene which is located in the chloroplast of photosynthetic organisms. It codes for the large subunit of the protein rubisco, and its sequence has been useful in plant phylogenies.

Reabsorption The return to the blood of most of the water, sodium, amino acids, and sugar that were removed during filtration; occurs mainly in the proximal tubule of the nephron.

Reactant A starting material in a chemical reaction.

Reaction wood Wood produced in response to a tree that has lost its vertical position, causing the tree to straighten.

Reading frame Refers to a polypeptide sequence potentially encoded by a single-stranded *mRNA*. Because *codons* are nucleotide triplets, each mRNA has 3 reading frames (each nucleotide can participate in 3 codons, at the 1st, 2nd, and 3rd base position). Duplex DNA strands have 6 reading frames, 3 in each strand:

AlaSerProLeuVal . . 1st reading frame

ProAlaProTERTrp . . 2nd reading frame: TER = Stop

GlnProProSerGly . . 3rd reading frame

GCCAGCCCCCUAGTGGG... Nucleotide sequence of Mrna. When mRNA is translated by the cell, the nucleotides are read three at a time.

By starting at different positions, the groupings of three that are produced can be entirely different. The following example shows a DNA sequence and the three reading frames in which it could be read. Not only is an entirely different amino acid sequence specified by the different reading frames, but two of the three frames have stop codons, and thus are not open reading frames (asterisks indicate a stop codon).

A DNA open reading frame:

...ATG ACA TGT AAA GAT AGA CTA ACC TTT TGG..

...Met Thr Cys Lys Asp Arg Leu Thr Phe Trp...

Same bases, different 'frame':

...A TGA CAT GTA AAG ATA GAC TAA CCT TTT GG..

... *** His Val Lys Ile Asp *** Pro Phe Gly..

Same sequence, the last of the 3 possible frames:

...AT GAC ATG TAA AGA TAG ACT AAC CTT TTG G.

... Asp Met *** Arg *** Thr Asn Leu Leu ...

If we shift the grouping again, we will just get the first reading frame again. The reading frame that is actually used is determined by the first methionine codon (the initiation codon). Once that first AUG is recognized, the pattern of triplet groupings follows unambiguously.

Reagin Antibody that mediates immediate hypersensitivity reactions. IgE is the major reagin in humans.

Realized niche The multidimensional space that an animal actually occupies in the presence of competitors, predators, pathogens, and limited food. Contrast with fundamental niche.

Rebound v. To spring back after a weight has been removed.

Recapitulation. Summing up or repeating; hypothesis that an individual repeats its phylogenetic history in its development.

Receptacle The expanded tip of a pedicel or peduncle to which the floral organs are attached.

Receptive field The area of a group of sensory cells that, when stimulated, influences the activity of a given neuron.

Receptor On or in a cell, a specific protein molecule whose shape fits that of a specific molecular messenger, such as a hormone.

Receptor potential An initial response of a receptor cell to a stimulus, consisting of a change in voltage across the receptor membrane proportional to the stimulus strength. The intensity of the receptor potential determines the frequency of action potentials traveling to the nervous system.

Receptor-mediated endocytosis The movement of specific molecules into a cell by the inward budding of membranous vesicles containing proteins with receptor sites specific to the molecules being taken in; enables a cell to acquire bulk quantities of specific substances.

Recessive A gene whose expression is masked when it is present in combination with a dominant allele. Refers to an allele of a gene that is expressed when the dominant allele is not present. An allele expressed only in homozygous form, when the dominant allele is absent.

Recessive gene Gene that is masked by its dominant allelic partner, having the recessive phenotype expressed only when both alleles for a given trait are recessive.

Recessive trait A trait caused by an allele, the phenotype of which is suppressed when it occurs with the dominant allele. Hence, two recessive alleles are necessary for the recessive phenotype to be displayed.

Recharge zone Area where water infiltrates into an aquifer.

Reciprocal altruism The trading of altruistic acts by individuals at different times, i.e., the payback to the altruist occurs some time after the receipt of the act.

Recognition sequence A specific *palindromic*

sequence within a double-stranded DNA molecule which is recognized by a *restriction endonuclease*, and at which the restriction endonuclease specifically cleaves the DNA molecule.

Recognition species concept The idea that specific mating adaptations become fixed in a population and form the basis of species identification.

Recognition surface The three-dimensional structure of a biological membrane surface that gives specificity due to macromolecules of various sizes and shapes that extend above the lipid bilayer.

Recolonization Revegetation; the reestablishment of a natural community following a natural or unnatural event that removed the existing community.

Recombinant A DNA or protein molecule produced as a result of assembling and joining DNA sequences from different sources. Sometimes used to refer to an organism carrying such a gene.

Recombinant clones *Clones* containing *recombinant DNA molecules.*

Recombinant DNA A DNA from a plasmid into which has been inserted a foreign gene. The plasmid is then introduced into a host cell, often the bacterium E.coli. The host cell may then express the foreign gene and secrete the desired protein. This process, commonly known as genetic engineering, has been used to great effect in synthesising proteins such as insulin and interferon.

Recombinant DNA technologies Procedures used to join together DNA segments in a cell- free system (an environment outside a cell or organism). Under appropriate conditions, a recombinant DNA molecule can enter a cell and replicate there, either autonomously or after it has become integrated into a cellular *chromosome.*

Recombinant-vector vaccine The type of vaccine that is produced by the introduction of one or more of a pathogen's genes into attenuated viruses or bacteria. The attenuated virus or bacterium serves as a vector, replicating within the vertebrate host and expressing the gene(s) of the pathogen. The pathogen's antigens induce an immune response.

Recombination A natural cellular process through which DNA molecules of similar or identical sequence can be exchanged. This process is the molecular basis of crossing over during meiosis and some DNA repair mechanisms. Another form of recombination, site-specific recombination, is found where transfer, rearrangement or insertion of specific sections of DNA occurs e.g. insertion of viral DNA into chromosomes, gene switching etc.

Recombination-repair A mode of filling a gap in one strand of duplex DNA by retrieving a homologous single strand from another duplex. Usually the underlying mechanism behind *homologous recombination* and *gene conversion.*

Recruitment The residue of those larvae that have:

(1) dispersed;

(2) settled at the adult site;

(3) made some final movements toward the adult habitat;

(4) metamorphosed successfully, and

(5) survived to be detected by the observer

Rectal gland The excretory organ of elasmobranchs and the coelacanth located near the rectum; it excretes a hyperosmotic salt solution.

Rectum In insects, the posterior expanded part of the hindgut, typically pear shaped.

Recycling Reprocessing of discarded materials into new, useful products; not the same as reuse of materials for their original purpose, but the terms are often used interchangeably.

Red algae Common name for the algae placed in the division Rhodophyta.

Red bed Sedimentary layers composed primarily of

sandstone, siltstone, and shale, that are predominantly red in color due to the presence of iron oxides; often used in reference to the Permian or Triassic sediments of the western U.S.

Red blood cell The type of blood cell that contains hemoglobin and no nucleus. During their formation in mammals, erythrocytes lose their nuclei; those of other vertebrates retain the nuclei.

Red clay A fine sediment that is the most common type of lithogenous sediment in the open ocean floor.

Red queen hypothesis The hypothesis that sexual reproduction has evolved because the genetic variation that results is adaptive in the evolutionary "arms race" between hosts and their pathogens and parasites as well as between predators and prey.

Red tides Red tides occur frequently in coastal areas and often are associated with population blooms of dinoflagellates. Dinoflagellate pigments are responsible for the red color of the water. Under these conditions, the dinoflagellates often produce saxitoxin, which can lead to paralytic shellfish poisoning.

Redfield ratio The carbon-nitrogen-phosphorus ratio of aquatic microorganisms. This ratio is important for predicting limiting factors for microbial growth.

Redox reaction A chemical reaction involving the transfer of one or more electrons from one reactant to another; also called oxidation-reduction reaction.Oxidation-reduction reaction; a chemical reaction involving the transfer of electrons from one molecule to another.

Redox-potential discontinuity That depth below the sediment-water interface marking the transition from chemically oxidative to reducing processes

Reduced enamel epithelium (REE) The epithelium produced by the combination of the external and internal enamel epithelium. The REE remains covering the enamel crown until the tooth erupts when it fuses with the oral epithelium. The REE remaining on the enamel surface becomes the junctional epithelium.

Reduced tillage systems Systems, such as minimum till, conserve-till, and no-till, that preserve soil, save energy and water, and increase crop yields.

Reducing agent or reductant The electron donor in an oxidation-reduction reaction.

Reduction The gaining of electrons by a substance involved in a redox reaction. The gain of an electron or a hydrogen atom. The gain of electrons or hydrogens in a chemical reaction.

Reduction division The first nuclear division of meiosis during which the paired homologues migrate to opposite poles, resulting in cells with a reduction from a diploid to a haploid number of chromosomes.

Reductive dehalogenation The cleavage of carbon-halogen bonds by anaerobic bacteria that creates a strong electron-donating environment.

Reef A large ridge or mound-like structure within a body of water that is built by calcareous organisms such as corals, red algae, and bivalves; barrier reef- A reef growing offshore from a land mass and separated by a lagoon or estuary, e.g, the Great Barrier Reef of Australia; patch reef- A discontinuous reef growing in small areas, separated by bare areas of sand or debris, often part of a larger reef complex.

Reef crest The shallow outer edge of the reef slope of a coral reef.

Reef flat The wide and shallow upper surface of a coral reef..

Reef slope The outer, steep margin of a coral reef.

Refined carbohydrates Natural carbohydrates from which other bulk such as fibres have been removed. Granulated sugar is a refined form of sugar cane.

Reflex A response to a stimulus that occurs without conscious effort; one of the simplest forms of behavior.

Reflex arc Pathway of neurons, effector(s) and sensory receptors that participate in a reflex.

Refraction The deflection of a light ray from a straight path as it passes from one medium (e.g., glass) to another (e.g., air).

Refractive index The ratio of the velocity of light in the first of two media to that in the second as it passes from the first to the second.

Refractory period The short time immediately after an action potential in which the neuron cannot respond to another stimulus, owing to an increase in potassium permeability.

Refuge A device by which an individual can avoid predation

Refuse-derived fuel Processing of solid waste to remove metal, glass, and other unburnable materials; organic residue is shredded, formed into pellets, and dried to make fuel for power plants.

Regeneration The ability of an organism to grow a body part that has been lost.

Regenerative farming Farming techniques and land stewardship that restore the health and productivity of the soil by rotating crops, planting ground cover, protecting the surface with crop residue, and reducing synthetic chemical inputs and mechanical compaction.

Region of division The area of cell division in the tip of a plant root.

Region of elongation The area in the tip of a plant root where cells grow by elongating, thereby increasing the length of the root.

Region of maturation (differentiation) The area where primary tissues and root hairs develop in the tip of a plant root.

Regression (n) A drop in sea level that causes an area of the land to be uncovered by seawater.

Regular symmetry Radial symmetry; having two or more planes through which the structure (flower) could be cut to result in mirror-image halves; actinomorphy.

Regulation The ability of cells to change their developmental fates according to their location within the whole organism. Regulation can be manifested experimentally as an ability for a cell or cells to develop normally following removal or addition of other cells or tissue.

Regulative development. Progressive determination and restriction of initially totipotent embryonic material.

Regulator An organism that can maintain constant some aspect of its physiology (e.g., body temperature) constant despite different and changing properties of the external environment

Regulator genes Genes that control the activity of other genes that code for enzymes or structural proteins.

Regulator T cell Regulator T cells control the development of effector T cells. Two types exist: T-helper cells (CD41 cells) and T-suppressor cells. There are three subsets of T-helper cells: TH1, TH2, and TH0. TH1 cells produce IL-2, IFN-g, and TNF-b. They effect cell-mediated immunity and are responsible for delayed-type hypersensitivity reactions and macrophage activation. TH2 cells produce IL-4, IL-5, IL-6, IL-10, IL-13. They are helpers for B-cell antibody responses and humoral immunity; they also support IgE responses and eosinophilia. TH0 cells exhibit an unrestricted cytokine profile.

Regulatory mutants Mutant organisms that have lost the ability to limit synthesis of a product, which normally occurs by regulation of activity of an earlier step in the biosynthetic pathway.

Regulon A collection of genes or operons that is controlled by a common regulatory protein.

Rehabilitate land A utilitarian program to make an area useful to humans.

Reinforcement In learning, anything that alters the probability of behavior.

Relatedness Two clades are more closely related when they share a more recent common ancestor between them than they do with any other clade.

Relative fitness The contribution of one genotype

to the next generation compared to that of alternative genotypes for the same locus.

Relative humidity At any given temperature, a comparison of the actual water content of the air with the amount of water that could be held at saturation.

Relative time Type of geologic time (absolute time being the other) that places events in a sequence relative to each other.

Relativists Those who believe moral principles are always dependent on the particular situation.

Relay neuron Neuron that transmits signals between different regions of the central nervous system.

Releaser A signal stimulus that functions as a communication signal between individuals of the same species.

Releaser gland A gland in turbellarians that secretes a chemical that dissolves the attachment of the organism from a substrate.

Releasing factors Hormones or neurosecretions from the hypothalamus that travel either via the hypothalamic-pituitary portal system or along axons between the hypothalamus and the pituitary where they exert their effect in terms of production and release of hormones.

Releasing hormone A hormone produced by neurosecretory cells in the hypothalamus of the vertebrate brain that stimulates or inhibits the secretion of hormones by the anterior pituitary.

Remineralisation The replacement of mineral salts lost by de-mineralisation of a solid salt.

Remipedia A class of crustaceans whose members possess about 30 body segments and uniform, biramous appendages. This class contains a single species of cave-dwelling crustaceans from the Bahamas.

Remodelling Of bone refers to the constant removal by osteoclasts and rebuilding by osteoblasts The mass of bone can be controlled constantly by altering the balance between removal and rebuilding. The shape of a bone can also be altered by removing in one place and building somewhere else, without necessarily changing the total mass of a bone mass.

Renal Pertaining to the kidney.

Renal corpuscle The filtration unit found in the cortex of the kidneys and made up of Bowman's capsule and a glomerulus.

Renal tubule The portion of the nephron where urine is produced.

Renette An excretory structure found in some worms.

Renewable resources Resources normally replaced or replenished by natural processes; resources not depleted by moderate use; examples include solar energy, biological resources such as forests and fisheries, biological organisms, and some biogeochemical cycles.

Renin An enzyme secreted by the kidneys that converts angiotensinogen into angiotensin II.

Rennin A milk-clotting endopeptidase secreted by the stomach of some young mammals, including bovine calves and human infants.

Repetitive DNA A surprising portion of any genome consists not of genes or structural elements, but of frequently repeated simple sequences. These may be short repeats just a few nt long, like CACACA etc. They can also range up to a few hundred nt long. Examples of the latter include Alu repeats, LINEs, SINEs. The function of these elements is often unknown. In shorter repeats like di- and tri-nucleotide repeats, the number of repeating units can occasionally change during evolution and descent. They are thus useful markers for familial relationships and have been used in paternity testing, forensic science and in the identification of human remains.

Replacement Fossilization process that occurs when an organism is completely decomposed and replaced by mineral.

Replica plating A technique for isolating mutants

from a population by plating cells from each colony growing on a nonselective agar medium onto plates with selective media or environmental conditions, such as the lack of a nutrient or the presence of an antibiotic or a phage; the location of mutants on the original plate can be determined from growth patterns on the replica plates.

Replication fork The Y-shaped structure where DNA is replicated. The arms of the Y contain template strand and a newly synthesized DNA copy.

Replicative form A double-stranded form of nucleic acid that is formed from a single-stranded virus genome and used to synthesize new copies of the genome.

Repolarized The reestablishment of polarity, especially the return of plasma membrane potential to resting potential after depolarization.

Reporter gene The use of a functional enzyme, such as beta-galactosidase, luciferase, or chloramphenicol acetyltransderase, downstream of a gene, promoter, or translational control element of interest, to more easily identify successful introduction of the gene into a host and to measure transcription and/or translation.

Repositioning Of a tooth refers to its movement within the entire dentition which involves the remodelling of the tooth socket. Repositioning of teeth occurs naturally due to continued eruption and mesial (or distal) drift.

Repressible enzyme An enzyme whose level drops in the presence of a small molecule, usually an end product of its metabolic pathway.

Repression Inhibition of transcription (or translation) by the binding of a repressor protein to a specific site on DNA (or *mRNA*).

Repressor protein DNA regulatory protein that binds the operator (a specific DNA sequence) in the absence of the inducing substance (in the case of the lac operon, lactose).

Reproductive effort The fraction of assimilated nutrients that are devoted to reproductive behavior and gamete production.

Reproductive isolation/ Reproductive barrier A form of genetic isolation where gene transfer between different populations of a species is prevented. The isolating mechanisms may be ecological (such as water barriers), seasonal (eg breeding seasons at different times of the year), behavioural (eg different courtship behaviours in each group) or physiological (some aspect of body form or function prevents mating between individuals from each group). In time, reproductive isolation may lead to the evolution of new sub-species or new species.

Reproductive isolating mechanism Biological or behavioral characteristics that reduce or prevent interbreeding with other populations; e.g., the production of sterile hybrids. Establishment of reproductive isolation is considered essential for development of a new species.

Reproductive potential The number of offspring a female of a given age can be expected to produce.

Reproductive strategy The reproductive patterns that are followed by a particular species.

Reproductive success The number of progeny born, or surviving progeny produced by an organism.

Reproductive system One of eleven major body organ systems in animals; is responsible for reproduction and thus the survival of the species.

Reptiles/ Reptilia Vertebrates with scales on their skin and leathery eggs that are laid on land. Marine reptiles include sea turtles, sea snakes, the marine iguana, and the saltwater crocodile.

Repugnatorial glands Glands secreting a noxious substance for defense or offense, for example, as in the millipedes.

Reservoir A site, alternate host, or carrier that normally harbors pathogenic organisms and serves as a source from which other individuals can be infected.

Reservoir host An organism other than a human that is infected with a pathogen that can also infect humans.

Residence time The length of time a component,

such as an individual water molecule, spends in a particular compartment or location before it moves on through a particular process or cycle.

Residue As applied to proteins, what remains of an *amino acid* after its incorporation into a peptide chain, with subsequent loss of a water molecule.

Residuesphere The region surrounding organic matter such as a seed or plant part in which microbial growth is stimulated by increased organic matter availability.

Resilience The ability of a community or ecosystem to recover from disturbances.

Resilin The elastic protein found in some arthropod joints that stores energy and functions in jumping.

Resin canal Long intercellular spaces present in the longitudinal system of vascular cells of pine, spruce, larch, and Douglas fir. Adjacent parenchymal cells secrete the resin into the canals.

Resistance The ability of a community to resist being changed by potentially disruptive events.

Resolution Degree of molecular detail on a *physical map* of DNA, ranging from low to high.

Resolving power A measure of the clarity of an image; the minimum distance that two points can be separated and still be distinguished as two separate points.

Resonance A vibration of electrons from one atom to another within chlorophyll.

Resource A commodity that is required by an organism and is potentially in short supply

Resource defense polygyny Males control access to females indirectly by monopolizing critical resources.

Resource partitioning In a biological community, various populations sharing environmental resources through specialization, thereby reducing direct competition.

Resource scarcity A shortage or deficit in some resource.

Respiration (1) In aerobic organisms, the intake of oxygen and the liberation of carbon dioxide.

(2) In cells, the oxygen-requiring stage in the breakdown and release of energy from fuel molecules.

Respiratory burst The respiratory burst occurs when an activated phagocytic cell increases its oxygen consumption to support the increased metabolic activity of phagocytosis. The burst generates highly toxic oxygen products such as singlet oxygen, superoxide radical, hydrogen peroxide, hydroxyl radical, and hypochlorite.

Respiratory fibrotic agents Special class of irritants, including chemical reagents and particulate materials, that damages the lungs, causing scar tissue formation that lowers respiratory capacity.

Respiratory pigment Organic compounds that have either metallic copper or iron with which oxygen combines.

Respiratory quotient The ratio of moles of carbon dioxide produced to oxygen consumed in respiration

Respiratory surface A thin, moist, epithelial surface that oxygen can cross to move into the body and carbon dioxide can cross to move out of the body.

Respiratory syncytial virus A member of the family Paramyxoviridae and genus Pneumovirus; it is a negative-sense ssRNA virus that causes respiratory infections in children.

Respiratory system One of eleven major body organ systems in animals; moves oxygen from the external environment into the internal environment and removes carbon dioxide from the body.

Respiratory tree A pair of tubules attached to the rectum of a sea cucumber that branch through the body cavity and function in gas exchange.

Response element By definition, a "response element" is a portion of a gene which must be present in order for that gene to respond to some hormone or other stimulus. Response elements are binding sites for transcription factors. Certain transcription factors are activated by stimuli such as hormones or heat shock. A gene may respond to the presence of that hormone because the gene has in its promoter region a binding site for hormone-activated transcription factor. Example: the glucocorticoid response element (GRE).

Rest position A position the jaw adopts when at rest with the lips lightly together.

Resting membrane potential/ Resting potential The potential difference that results from the separation of charges along the plasma membrane of a neuron.

Restoration ecology Seeks to repair or reconstruct ecosystems damaged by human actions.

Restore ecosystems To reinstate an entire community of organisms to naturally occurring association.

Restricted transduction A transduction process in which only a specific set of bacterial genes are carried to another bacterium by a temperate phage; the bacterial genes are acquired because of a mistake in the excision of a prophage during the lysogenic life cycle.

Restriction endonuclease A bacterial enzyme which recognizes a specific *palindromic sequence (recognition sequence)* within a double-stranded DNA molecule and then catalyzes the cleavage of both strands at that site. Also called a restriction enzyme. Restriction endonucleases may generate either *blunt* or *sticky ends* at the site of cleavage.

Restriction enzyme A class of enzymes ("restriction endonucleases") generally isolated from bacteria, which are able to recognize and cut specific sequences ("restriction sites") in DNA. For example, the restriction enzyme BamHI locates and cuts any occurrence of:

5'-GGATCC-33'-CCTAGG-5'

Note that both strands contain the sequence GGATCC, but in antiparallel orientation. The recognition site is thus said to be palindromic, which is typical of restriction sites. Every copy of a plasmid is identical in sequence, so if BamHI cuts a particular circular plasmid at three sites producing three "restriction fragments", then a million copies of that plasmid will produce those same restriction fragments a million times over. There are more than six hundred known restriction enzymes.

Bacteria produce restriction enzymes for protection against invasion by foreign DNA such as phages. The bacteria's own DNA is modified in such a way as to prevent it from being clipped.

Restriction enzyme cutting site A specific *nucleotide sequence* of DNA at which a particular *restriction enzyme* cuts the DNA. Some sites occur frequently in DNA (e.g., every several hundred *base pairs*), others much less frequently (*rare- cutter*; e.g., every 10,000 base pairs).

Restriction fragment The piece of DNA released after restriction digestion of plasmids or genomic DNA. One can digest a plasmid and isolate one particular restriction fragment (actually a set of identical fragments). The term also describes the fragments detected on a genomic blot which carry the gene of interest.

Restriction fragment length polymorphism (RFLP) Variation between individuals in DNA fragment sizes cut by specific *restriction enzymes; polymorphic sequences* that result in RFLPs are used as *markers* on both *physical maps* and genetic *linkage maps*. RFLPs are usually caused by *mutation* at a cutting site.

Restriction map A "cartoon" depiction of the locations within a stretch of known DNA where restriction enzymes will cut.

The map usually indicates the approximate length of the entire piece (scale on the bottom), as well as the position within the piece at which designated enzymes will cut. This map happens to be of a plasmid, and the two ends are joined together with about 25 nt between the EcoRI and HindIII sites.

Restriction site A specific sequence on a DNA strand that is recognized as a "cut site" by a restriction enzyme.

Resurgence (pest) The development of large populations of pests that had previously been suppressed.

Rete mirabile A network of small blood vessels so arranged that the incoming blood runs countercurrent to the outgoing blood and thus makes possible efficient exchange between the two bloodstreams. Such a mechanism serves to maintain the high concentration of gases in the fish swim bladder.

Rete system Highly branched system of tubules in acanthocephalans, lying on longitudinal muscles or between longitudinal and circular muscles; thought to assist in contraction stimuli.

Reticular fibres Are fine type III collagen fibres forming a net-like supporting framework or reticulum. They are found around small blood vessels, nerve cells, muscle fibres and in particular beneath epithelial membranes as part of thebasal lamina.

Reticular Resembling a net in appearance or structure.

Reticular formation In the central core of the medulla, it consists of several structures, including the periaquaductal grey. The reticular formation integrates information from many sources and influences sensory motor and autonomic activity. It is involved in aversive drive (behaviour which is an instinctive turning away from the unpleasant).

Reticulate body (RB) The form in the chlamydial life cycle whose role is growth and reproduction within the host cell.

Reticulate venation Netted venation; the arrangement of veins in a leaf that resembles a net; characteristic of dicot leaves.

Reticulation Joining of separate lineages on a phylogenetic tree, generally through hybridization or through lateral gene transfer. Fairly common in certain land plant clades; reticulation is thought to be rare among metazoans.

Reticulocyte lysate A lysate of rabbit reticulocytes, which has been extensively digested with micrococcal nuclease to destroy the reticulocyte *mRNA*s. With the addition of an exogenous, usually synthetic, mRNA, *amino acids* and a source of energy (ATP), the translational machinery of the reticulocyte (*ribosomes*, eukaryotic translation factors, etc.) will permit in vitro *translation* of the added mRNA with production of a new *polypeptide*. This is only one of several available in vitro translation systems.

Reticuloendothelial (RE) system Total complement of fixed macrophages in the body, especially reticular connective tissue and the lining epithelium of the blood vascular system. Some authorities also include the phagocytic white blood cells.

Reticulopodia Long thread-like pseudopodia that branch apart and rejoin, forming a fine network. They are characteristic of forams.

Reticulum A fine network (e.g., endoplasmic reticulum).

Retina The innermost layer of the vertebrate eye, containing photoreceptor cells (rods and cones) and neurons; transmits images formed by the lens to the brain via the optic nerve.

Retinal The light-absorbing pigment in rods and cones of the vertebrate eye.

Retinoic acid A product of retinol(Vitamin A) which binds onto cell membranes and controls cell division and differentiation through gene expression.

Retrofection Process of reinfection, whereby juvenile nematodes hatch on the skin and reenter the body before molting to third-stage larvae.

Retroperitoneal External to the peritoneum.

Retrotransposon A transposable DNA element (transposon) which is replicated through an RNA intermediate via reverse transcriptase. Within their characteristic long terminal repeats (LTRs), some retrotransposons encode retroviral-like proteins (e.g., gag, pol) for reverse transcriptase and integrase. These are likely derived from retroviruses; other retroransposons lack viral genes. Retrotransposons are ubiquitous elements in eukaryotic genomes. Examples include yeast Ty elements, Drosophila copia elements, and human LINE1 elements (which comprise 17% of the human genome).

Retroviral vector Artificial DNA construct derived from a retrovirus, used to insert sequences into an organism's chromosomes.

Retroviruses Viruses that contain a single strand of RNA as their genetic material and reproduce by copying the RNA into a complementary DNA strand using the enzyme reverse transcriptase. The single-stranded DNA is then copied, and the resulting double-stranded DNA is inserted into a chromosome of the host cell.

Retting The process that frees flax fibers by allowing microbial decomposition to break down the outer part of the stem.

Revegetate The natural or induced replacement of plants into a cleared area; the recurrence of the same plant community that existed prior to clearing.

Reverse bohr effect Effect that occurs when lactate builds up in the blood of certain invertebrates and pH decreases, increasing the affinity of hemocyanin for oxygen

Reverse transcriptase (RT) An RNA-dependent DNA polymerase that uses a viral RNA genome as a template to form a DNA copy; this is a reverse of the normal flow of genetic information, which proceeds from DNA to RNA.

Reverse transcription Process of transcribing a single-stranded DNA from a single-stranded RNA (the reverse of transcription); used by retroviruses as well as in biotechnology.

Reversible covalent modification A mechanism of enzyme regulation in which the enzyme's activity is either increased or decreased by the reversible covalent addition of a group such as phosphate or AMP to the protein.

Reversing thermometer A thermometer that permits the measurement of temperature at depth, since the mercury column breaks when the thermometer is turned upside down.

Reye's syndrome An acute, potentially fatal disease of childhood that is characterized by severe edema of the brain and increased intracranial pressure, vomiting, hypoglycemia, and liver dysfunction. The cause is unknown but is almost always associated with a previous viral infection (e.g., influenza or varicella-zoster virus infections).

Reynolds number A number that represents the relative importance of viscous forces and inertial forces in a fluid. As Re increases, inertial forces become more important. In sea water, Re increaseswith increasing water velocity and with the size of the object in the water

RFLP Restriction fragment length polymorphism The acronym is pronounced "riflip". Although two individuals of the same species have almost identical genomes, they will always differ at a few nucleotides. Some of these differences will produce new restriction sites (or remove them), and thus the banding pattern seen on a genomic Southern will thus be affected. For any given probe (or gene), it is often possible to test different restriction enzymes until we find one which gives a pattern difference between two individuals - a RFLP. The less related the individuals, the more divergent their DNA sequences are and the more likely we are to find a RFLP.

Rhabdite A rodlike structure in the cells of the epidermis or underlying parenchyma in certain tubellarians that are discharged in mucous secretions.

Rheoreceptor A sensory organ of aquatic animals that responds to water current.

Rheumatic fever An autoimmune disease

characterized by inflammatory lesions involving the heart valves, joints, subcutaneous tissues, and central nervous system. The disease is associated with hemolytic streptococci in the body. It is called rheumatic fever because two common symptoms are fever and pain in the joints similar to that of rheumatism.

Rheumatoid arthritis A crippling form of arthritis that begins with inflammation and thickening of the synovial membrane, followed by bone degeneration and disfigurement.

Rhinencephalon The part of the vertebrate brain that comprises the septum, cingulate gyrus, hypothalamus, amygdala, hippocampus, and the fornix; also called the limbic system.

Rhinophore Chemoreceptive tentacles in some molluscs (opisthobranch gastropods).

Rhizoid A cellular outgrowth of a plant that usually aids in anchoring to the surface and increasing surface area to acquire water or nutrients; found in mosses, liverworts, and hornworts.Filamentous structures in the plants group known as bryophytes that attach to a substrate and absorb moisture. The term is also applied to similar structures found outside the bryophytes.

Rhizosphere A region around the plant root where materials released from the root increase the microbial population and its activities.

RhO factor The protein that helps RNA polymerase dissociate from the terminator after it has stopped transcription.

Rhodophyta The division of protists commonly called the red algae.

Rhodopsin A visual pigment consisting of retinal and opsin. When rhodopsin absorbs light, the retinal changes shape and dissociates from the opsin, after which it is converted back to its original form.

Rhombencephalon The posteriormost of 3 primary vesicles (bulges) formed from the anterior neural tube. The hindbrain further subdivides into a posterior myelencephalon which will give rise to the medulla oblongata, and an anterior metencephalon, which will give rise to the cerebellum and pons.

Rhombogen Stage in the life cycle of a dicyemid mesozoan.

Rhopalium A sensory structure at the margin of the scyphozoan medusa. It consists of a statocyst and a photoreceptor (phylum Cnidaria).

Rhoptry/Rhoptries Club-shaped bodies in Apicomplexa composing one of the structures of the apical complex; open at anterior and apparently functioning in penetration of host cell.

Rhynchocoel In nemerteans, the dorsal tubular cavity that contains the inverted proboscis. It has no opening to the outside.

Rhyniophyta An extinct division of early, vascular land plants characterized by their small stature, dichotomous branching, lack of leaves, roots, and possession of terminal sporangia (e.g., Rhynia, Cooksonia).

Rhyolite Highly felsic igneous volcanic rock, typically light in color; rough volcanic equivalent of granite.

Ribbon (or nemertean) worms (phylum Nemertea)Invertebrates with a complete digestive tract, a true circulatory system, and a long proboscis to capture prey.

Ribonuclease (RNase) An enzyme which catalyzes the hydrolysis of RNA. There are many different RNases, some of the more important include:

RNase A Cleaves ssRNA 3' of pyrimidines

RNase T1 Cleaves ssRNA at guanasine nucleotides

RNase V1 Cleaves dsRNA (helical regions)

RNase H Degrades the RNA part of RNA:DNA hybrids.

Ribonucleic acid (RNA) A type of nucleic acid consisting of nucleotide monomers with a ribose sugar and the nitrogenous bases adenine (A), cytosine (C), guanine (G), and uracil (U); usually single-stranded; functions in protein synthesis and as the genome of some viruses.Nucleic acid containing ribose sugar and the base Uracil; RNA functions in protein synthesis. The single starnded molecule transcribed from one strand of the DNA. There are three types of RNA, each is involved in protein synthesis. RNA is made up nucleotides containing the sugar ribose, a phosphate group, and one of four nitrogenous bases (adenine, uracil, cytosine or guanine).

Riboprobe A strand of RNA synthesized *in-vitro* (usually radiolabeled) and used as a probe for hybridization reactions. An RNA probe can be synthesized at very high specific activity, is single stranded (and therefore will not self anneal), and can be used for very sensitive detection of DNA or RNA.

Ribose A 5-carbon sugar important in RNA and many other compounds.

Ribosome A complex ribonucleoprotein particle (eukaryotic ribosomes contain 4 RNAs and at least 82 proteins) which is the "machine" which translates *mRNA* into protein molecules. In eukaryotic cells, ribosomes are often in close proximity to the endoplasmic reticulum.Part of protein synthetic machinery, made of ribosomal RNAs and proteins.Subcellular structure composed of protein and ribonucleic acid. May be free in the cytoplasm or attached to the membranes of the endoplasmic reticulum; functions in protein synthesis.

Ribosomal binding sequence In prokaryotic organisms, part or all of the polypurine sequence AGGAGG located on *mRNA* just upstream of an AUG *initiation codon*; it is complementary to the sequence at the 3' end of 16S rRNA; and involved in binding of the ribosome to *mRNA*. The *internal ribosomal entry site* found in some viruses may be an analogous eukaryotic genetic element.

Ribosomal RNA One of the three types of RNA; rRNA is a structural component in ribosomes.

Ribosomal subunits Two units that combine with mRNA to form the ribosomal-mRNA complex at which protein synthesis occurs.

Ribotyping Ribotyping is the use of E. coli rRNA to probe chromosomal DNA in Southern blots for typing bacterial strains. This method is based on the fact that rRNA genes are scattered throughout the chromosome of most bacteria and therefore polymorphic restriction endonuclease patterns result when chromosomes are digested and probed with rRNA.

Ribozyme A catalytically active RNA. A good example is the hepatitis delta virus RNA which is capable of self-cleavage and self-ligation in the absence of protein enzymes.

Ribulose-1,5-bisphosphate carboxylase The enzyme that catalyzes the incorporation of CO_2 in the Calvin cycle.

Rickettsia A very small and specialized pathogenic bacterium transmitted by ticks, mites, and fleas.

Ridge and rise system the interconnecting chain of seafloor mountains that trace the edges of crustal plates and the sites of new oceanic crust production

Rift A long, narrow crack in the entire thickness of the Earth's crust, which is bounded by normal faults on either side and forms as the crust is pulled apart; v. To split the Earth's crust.

Rift basin or **rift valley** The long, and fairly wide trough that has formed as a section of the Earth's crust has dropped down along faults, e.g., African Rift Valley in East Africa.

Rift zone The area on continents where a trough bounded by normal faults is forming; the site of crustal extension, similar to that which occurs at mid-oceanic ridges

Rill erosion The removing of thin layers of soil as little rivulets of running water gather and cut small channels in the soil.

Rimrock 1. A ledge of cliff overlooking lower ground and formed by the outcropping of a horizontal layer of resistant rock on an elevated area.

2. a cliff or vertical face of an outcrop or rock in the canyon wall.

Ring-porous wood Secondary xylem characterized by larger vessels and tracheids being produced early in the season (or following favorable growing conditions). Thus each term of growth activity is seen as a ring.

Ringworm The common name for a fungal infection of the skin, even though it is not caused by a worm and is not always ring-shaped in appearance.

Rip current A concentrated rapid current moving offshore from a beach fronting a longshore current

Riparian A general classification of habitat along streams, ponds and lakes or any water courses.

Rise Bottom of low relief at the base of the continental slope

Rise period or burst The period during the one-step growth experiment when host cells lyse and release phage particles.

Risk Probability that something undesirable will happen as a consequence of exposure to a hazard.

Risk assessment Evaluation of the short-term and long-term risks associated with a particular activity or hazard; usually compared to benefits in a cost-benefit analysis.

Risk-sensitive foraging A model that incorporates variation in prey encounter rates to predict where predators forage. Depending on their internal state and other factors, predators may be risk-prone, foraging where encounter rates are variable, or risk-averse, foraging where encounter rates are relatively constant.

Ritualization. In ethology, the evolutionary modification, usually intensification, of a behavior pattern to serve communication.

RNA "ribonucleic acid". The nucleic acid which carries the DNA message into parts of the cell where it is interpreted and used. The 18S ribosomal RNA sequence has been used in many groups of organisms to reconstruct phylogeny.

RNA polymerase During transcription, an enzyme that attaches to the promoter region of the DNA template, joins nucleotides to form the synthesized strand of RNA and detaches from the template when it reaches the terminator region.

RNA processing Modification of RNA before it leaves the nucleus, a process unique to eukaryotes.

RNA splicing A complex and incompletly understood series of reactions occuring in the nucleus of eukaryotic cells in which *pre-mRNA transcribed* from chromosomal DNA is processed such that noncoding regions of the pre-mRNA (*introns*) are excised, and coding regions (*exons*) are covalently linked to produce an *mRNA* molecule ready for transport to the cytoplasm. Because of splicing, eukaryotic DNA representing a gene encoding any given protein is usually much larger than the mRNA from which the protein is actually *translated.*

RNA transcript Term applied to RNA transcribed in the nucleus.

RNA world. Hypothetical stage in the evolution of life on earth in which both catalysis and replication were performed by RNA, not protein enzymes and DNA.

RNAi 'RNA interference' (a.k.a. 'RNA silencing') Is the mechanism by which small double-stranded RNAs can interfere with expression of any mRNA having a similar sequence. Those small RNAs are known as 'siRNA', for short interfering RNAs. The mode of action for siRNA appears to be via dissociation of its strands, hybridization to the target RNA, extension of those fragments by an RNA-dependent RNA polymerase, then fragmentation of the target. Importantly, the remnants of the target molecule appears to then act as an siRNA itself; thus the effect of a small amount of starting siRNA is effectively amplified and can have long-lasting effects on the recipient cell.

The RNAi effect has been exploited in numerous research programs to deplete the call of specific messages, thus examining the role of those messages by their absence.

Rnase protection assay This is a sensitive method to determine (1) the amount of a specific mRNA present in a complex mixture of mRNA and/or

(2) the sizes of exons which comprise the mRNA of interest. A radioactive DNA or RNA probe (in excess) is allowed to hybridize with a sample of mRNA (for example, total mRNA isolated from tissue), after which the mixture is digested with single-strand specific nuclease. Only the probe which is hybridized to the specific mRNA will escape the nuclease treatment, and can be detected on a gel. The amount of radioactivity which was protected from nuclease is proportional to the amount of mRNA to which it hybridized. If the probe included both intron and exons, only the exons will be protected from nuclease and their sizes can be ascertained on the gel.

Rnase Ribonuclease An enzyme which degrades RNA. It is ubiquitous in living organisms and is exceptionally stable. The prevention of RNase activity is the primary problem in handling RNA.

Rock cycle The process whereby rocks are broken down by chemical and physical forces; sediments are moved by wind, water, and gravity, sedimented and reformed into rock, and then crushed, folded, melted, and recrystallized into new forms.

Rockweeds Brown algae such as Fucus that are inhabitants of rocky shores in temperate zones.

Rocky mountain spotted fever An acute, infectious, and sometimes fatal tick-borne disease caused by *Rickettsia rickettsii*; characterized by fever, bone and muscle pain, headache, and rash. The disease occurs only in North and South America.

Rod/ Rod cell A class of vertebrate photoreceptors that, unlike cone cells, is usually insensitive to color.

Rodinia A supercontinent that existed during the Late Precambrian before the supercontinent Pannotia; the oldest supercontinent for which we have a good record; Russian for "homeland."

Rolling-circle mechanism A mode of DNA replication in which the replication fork moves around a circular DNA molecule, displacing a strand to give a tail that is also copied to produce a new double-stranded DNA.

Romana's sign Symptoms of recent infection by Trypanosoma cruzi, consisting of edema of the orbit and swelling of the preauricular lymph node.

Romanovsky stain Complex stain, based on methylene blue and eosin, used to stain blood cells and hemoparasites. Wright's and Giemsa's stains are two common examples.

Roost 1. A perch upon which birds or fowls rest at night.

2. a large cage, house, or place for fowls or birds to roost in.

Root Organs, usually occurring underground, that absorb nutrients and water and anchor the plant; one of the three major plant organ systems.

Root cap A thimble-shaped group of cells found at the tip of roots; it functions to protect the meristem.

Root hair A tiny projection growing just behind the root tips of plants, increasing surface area for the absorption of water and minerals.

Root nodule Gall-like structures on roots that contain endosymbiotic nitrogen-fixing bacteria (e.g., Rhizobium or Bradyrhizobium is present in legume nodules).

Root pressure A positive pressure in the xylem due to a negative solute potential and closed stomata.

Root resorption Resorption of cementum and underlying root dentine by osteoclasts. Temporary zones of root resorption may occur during orthodontic tooth repositioning. More extensive and irreversible root resorption may occur if the root becomes ankylosed .

Root system Plant organ systems that anchors the plant in place, stores excess sugars, and absorbs water and mineral nutrients. That part of the plant below ground level.

Rootbound The condition of a restricted root system caused by excess growth in too small a volume of soil; sometimes called potbound.

Root-leaf-vascular system axis Refers to the arrangement in vascular plants in which the roots anchor the plant and absorb water and nutrients, the leaves carry out photosynthesis, and the vascular system connects the roots and leaves, carrying water and nutrients to the leaves and carrying sugars and other products of photosynthesis from the leaves to other regions of the plant.

Rorquals Blue, fin, and the other baleen whales that display long grooves on the underside.

Roseola infantum A skin eruption that produces a rose-colored rash in infants. Caused by the human herpesvirus 6. The disease is short-lived and characterized by a high fever of 3 to 4 days' duration.

Rosette A growthform in which the leaves are arranged in concentric circles or whorls around a central bud.

Rostellum Projecting structure on scolex of tapeworm, often with hooks.

Rostrum A beak or snout, applied especially to the piercing mouth-parts of bugs and the elongated snouts of weevils.

Rotifera The phylum of aschelminths that has members with a ciliated corona surrounding a mouth; muscular pharynx (mastax) present with jawlike features; nonchitinous cuticle; parthenogenesis common; both freshwater and marine species.

Rough ER A portion of the endoplasmic reticulum containing ribosomes.

ROV Abbreviation for remotely operated vehicle, usually a submersible tethered to a ship, with facilities for video, remote sampling by grabbing arms, and precise navigation

R selection The concept that in certain (r-selected) populations, a high reproductive rate is the chief determinant of life history.

R strategist Species of organism that uses a survival and reproductive "strategy" characterized by high fecundity, high mortality, short longevity. Populations are controlled by density-independent factors.

RT/PCR reaction A series of reactions which result in RNA being copied into DNA and then amplified. A single *primer* is used to make single-stranded *cDNA* copies from an RNA *template* under direction of *reverse transcriptase*. A second primer *complementary* to this "first strand" cDNA is added to the reaction mix along with *Taq polymerase*, resulting in synthesis of double-stranded DNA. The reaction mix is then cycled (denaturation, *annealing* of primers, *extension*) to amplify the DNA by conventional PCR.

Rubella A moderately contagious skin disease that occurs primarily in children 5 to 9 years of age that is caused by the rubella virus, which is acquired by droplet inhalation into the respiratory system; German measles.

Rubisco Protein which fixes carbon in photosynthetic organisms. It binds molecules of carbon dioxide to a five-carbon molcule. Rubisco is the most common protein on earth.

Rubp carboxylase The enzyme that fixes CO2 in the Calvin cycle of photosynthesis.

Rubp Ribulose biphosphate; the 5-carbon chemical that combines with carbon dioxide at the beginning of the Calvin Cycle.

Rubpribulose 1, 5-bisphosphate, the 5-carbon compound that combines with CO2 in the Calvin cycle.

Rudimentary Poorly or imperfectly developed.

Ruffles Slender projections of the exterior surface of a dicyemid mesozoan.

Rugae Raised ridges of epithelium, each with its core of lamina propria, found on the anterior wall of the hard palate.

Rumen The expanded upper portion or first compartment of the stomach of ruminants.

Ruminant Mammals that have 4 stomachs and chew their cud are ruminants. They are herbivores and the plant material they eat is difficult to digest. In ruminants the 4 digestive chambers accommodate a large number of bacteria which help digest the plant material. Members of the deer family and mountain sheep and goats are all ruminants.

Rumination The process of digestion of ruminants, whereby the animal swallows food quickly, and then regurgitates and chews it more thoroughly at a later time until digestion is completed.

Run The straight line movement of a bacterium.

Runaway selection Selection for male ornaments that happens due to the genetic correlation, and resulting positive feedback relationship, between the trait and the preference for the trait.

Runner A horizontal stem that grows along the surface of the ground; also known as a stolon.

Runoff The excess of precipitation over evaporation; the main source of surface water and, in broad terms, the water available for human use.

Runoff transcript RNA which has been synthesized from plasmid DNA (usually by a bacteriophage *RNA polymerase* such as *T7* or *SP6*) and which terminates at a specific 3' site because of prior cleavage of the plasmid DNA with a *restriction endonuclease*.

Run-of-the-river flow Ordinary river flow not accelerated by dams, flumes, etc. Some small, modern, high-efficiency turbines can generate useful power with run-of-the-river flow or with a current of only a few kilometers per hour.

Rural area An area in which most residents depend on agriculture or the harvesting of natural resources for their livelihood.

Rut Is a general term that refers to the breeding period of mammals, especially the deer, elk and moose. During the rut, males exhibit specific behaviors to establish harems or to attract females to mate with.

S

S curve A curve that depicts logistic growth; called an S curve because of its shape.

S phase That period of interphase when new DNA is synthesized as part of replication of the chromatin.

S1 end mapping A technique to determine where the end of an RNA transcript lies with respect to its template DNA (the gene). Can't be described in a short paragraph.

S1 nuclease An enzyme which digests only single-stranded nucleic acids.

SA (sinoatrial) node The pacemaker of the heart, located in the wall of the right atrium. At the base of the wall separating the two atria is another patch of nodal tissue called the atrioventricular node.

Saccule Small chamber of the membranous labyrinth of the inner ear.

Sacrum Bone formed by fused vertebrae to which pelvic girdle is attached; pertaining to the sacrum.

Saefftigen's pouch Internal, muscular sac near the posterior end of a male acanthocephalan. It contains fluid that aids in manipulating the copulatory bursa.

Sagittal Pertaining to the median anteroposterior plane that divides a bilaterally symmetrical organism into right and left halves.

Salicylic acid A compound with pain-relieving characteristics, found in willow bark and other plants; the basic ingredient of aspirin.

Salience In the context of classical conditioning, the fact that animals can learn to make certain associations, depending upon which stimuli are involved, more rapidly than other associations.

Salinity Number of kg of dissolved salts in one kg g of seawater, measured in parts per thousand. Actually this definition stands alongside another definition based upon water standards whose electrical conductivities are measured.

Salinization A process in which mineral salts accumulate in the soil, killing plants; occurs when soils in dry climates are irrigated profusely.

Saliva The enzyme-containing secretion of the salivary glands.

Salivarium Chamber in buccal cone of acarines into which salivary ducts open.

Salivary amylase An enzyme secreted by the salivary glands that begins the breakdown of complex sugars and starches.

Salivary gland Any of the glands in molluscs, vertebrates, and other animals that release digestive enzymes into the mouth.

Salmon ranching Practice in which cultured juvenile salmon are released into fresh water and allowed to migrate to sea so they can be harvested later when they return as adults.

Salmonellosis An infection with certain species of the genus Salmonella, usually caused by ingestion of food containing salmonellae or their products. Also known as Salmonella gastroenteritis or Salmonella food poisoning.

Salps A group of pelagic tunicates (phylum Urochordata), either colonial or solitary, with buccal and atrial siphons on opposite sides of the body

Salt gland An orbital gland of many reptiles and birds that secretes a hyperosmotic NaCl or KCl solution an important osmoregulatory organ, especially for marine species.

Salt The reaction product of an acid and a base; dissociates in water solution to negative and positive ions, but not H+ or OH_.

Salt lick A natural or artificial deposit of exposed salt that animals lick for nutrients.

Salt marsh A coastal habitat consisting of salt-resistant plants residing in an organic-rich sediment accreting toward sea level

Salt wedge A layer of denser, saltier seawater that flows along the bottom in estuaries .

Saltatorial Refers to a form of locomotion in which the animal hops as it moves along. This is found in jackrabbits, the western jumping mouse, and even mule deer appear to utilize a form of saltatorial locomotion where all four feet are off the ground at the same time and they run in a springy fashion.

Saltatory conduction Rapid transmission of a nerve impulse along an axon resulting from the action potential jumping from one node of Ranvier to another, skipping the myelin-sheathed regions of membrane.

Saltwater intrusion Movement of saltwater into freshwater aquifers in coastal areas where groundwater is withdrawn faster than it is replenished.

Samara A simple, dry indehiscent fruit with the pericarp bearing winglike outgrowths; winged fruit of maple.

Sampling Estimating the density of organisms (pests or natural enemies) or damage by examining a defined portion of the crop.

Sand dollars Sea urchins with a flat, round test and short spines that live partly buried in soft sediments.

Sand fly Member of the dipteran subfamily Phlebotominae, family Psychodidae; sometimes also applied to Simuliidae (New Zealand) and Ceratopogonidae (Caribbean).

Sand Soil inorganic particles in the .02—2.0 mm size range.

Sandstone A sedimentary rock composed of sand.

Sanitization Reduction of the microbial population on an inanimate object to levels judged safe by public health standards; usually, the object is cleaned.

Saponin A glycoside with a steroid molecule as the active component, such as diosgenin from yams.

Saprobe An organism deriving its food from the dead body or nonliving products of another organism.

Saprophagous Feeding on decaying matter; saprobic; saprozoic.

Saprophyte A plant lacking chlorophyll that obtains its nutrients from dead organic matter. The bacteria and fungi of decay are examples, but there are also flowering plants like the white Indian pipes of eastern US Temperate Broadleaf Deciduous Forests or the giant *Rafflesia* of the Indo-Malyasian formation of the Tropical Braodleaf Evergreen Forest that are saprophytes.

Saprophytic Living on dead or decaying organic matter.

Saprozoic nutrition Nutrition of an animal by absorption of dissolved salts and simple organic nutrients from surrounding medium. Also refers to feeding on decaying organic matter.

Sapwood Layers of secondary xylem that are still functional in older woody plants; visible as the outer lighter areas in the cross section of a tree trunk.

Sarcocystin Powerful toxin produced by zoitocysts of Sarcocystis.

Sarcodina The protozoan subphylum where members have pseudopodia for movement and food gathering; naked or with shell or test; mostly free living.

Sarcolemma The specialized plasma membrane surrounding a muscle cell (muscle fiber); capable of propagating action potentials.

Sarcomastigophora The protozoan phylum where members possess flagella, pseudopodia, or both for locomotion and feeding; single type of nucleus.

Sarcomere The contractile unit of a myofibril. The repeating units, delimited by the Z bands along the length of the myofibril. The functional units of skeletal muscle; consist of filaments of myosin and actin.

Sarcoplasm The clear, semifluid cytoplasm between the fibrils of muscle fibers.

Sarcoplasmic reticulum A modified form of endoplasmic reticulum in striated muscle cells that stores calcium used to trigger contraction during stimulation.

Sarcoptic mange Disease caused by mites of the genus Sarcoptes. Also called scabies.

Sargasso Sea The region of the Atlantic Ocean north of the West Indies that is characterized by floating masses of Sargasso weed, a brown alga.

Satellite Posterior member of a pair of gregarines in syzygy.

Satellite DNA A genomic DNA fraction which has a different density to the main body of genomic DNA and therefore forms a separate band (satellite) during density gradient ultracentrifugation. The difference in density is due to a bias in base content caused by the presence of repeat sequence DNA.

Saturated fat/ Saturated fatty acid A fat in which all the carbons in the fatty acids are connected by single bonds, thereby having the maximum number of hygrogen atoms. A fatty acid in which all carbons in the hydrocarbon tail are connected by single bonds, thus maximizing the number of hydrogen atoms that can attach to the carbon skeleton.

Saturated solutions Salts such as the apatites do not readily become ionized and dissolve in water. When no more ions can dissolve the solution is said to be saturated. The concentration of ions in a saturated solution, its solubility product, is constant for each salt, at a neutral pH . If the solution becomes more acidic, more ions can dissolve from the solid. Saliva is a super saturated solution of calcium phosphate.

Saturation light intensity The light intensity that maximizes the photosynthetic rate.

Saturation point The maximum concentration of water vapor the air can hold at a given temperature.

Sauropterygians Mesozoic marine reptiles.

Savanna A tropical grassland biome with scattered individual trees, large herbivores, and three distinct seasons based primarily on rainfall, maintained by occasional fires and drought.

Saxitoxin a paralytic toxic produced by dinophytes that accumulates in the butter clam (Saxadoma)

Scabies Disease caused by mites of the genus Sarcoptes. Also called sarcoptic mange.

Scaffolding proteins Special proteins that are used to aid procapsid construction during the assembly of a bacteriophage capsid and are removed after the completion of the procapsid.

Scale A platelike organic structure found on the surface of some cells (chrysophytes).

Scalid A set of complex spines found on the kinorhynchs, loriciferans, priapulans, and larval nematomorphs, with sensory, locomotor, food capture, or penetrant function.

Scan sampling A method for recording behavior observations that involves looking briefly at each animal at prescribed time intervals (e.g., every minute or every five minutes) to record their activity at the time of the sample.

Scanning electron microscope (SEM) An electron microscope that scans a beam of electrons over the surface of a specimen and forms an image of the surface from the electrons that are emitted by it.

scanning probe microscope A microscope used to study surface features by moving a sharp probe over the object's surface (e.g., the scanning tunneling microscope).

Scanning tunneling microscope A type of scanning probe microscope used to image a surface by moving a fine probe over it at a constant height, which is maintained by keeping a constant electron flow (tunneling current) between the tip and surface.

Scaphopoda A class of molluscs whose members have a tubular shell that is open at both ends. Possess tentacles but no head. Dentalium.

Scapula The shoulder blade.

Scarabaeiform A grub like larva having a thick, soft body with a well-developed head and strong thoracic legs but with no legs on the hind region: often permanently curved into a C. The larvae of the lamellicorn beetles are of this type.

Scarification A mechanical or chemical degradation of a hard surface, such as seed coats, so that oxygen and water can penetrate the hard layers.

Scarlet fever A disease that results from infection with a strain of Streptococcus pyogenes that carries a lysogenic phage with the gene for erythrogenic (rash-inducing) toxin. The toxin causes shedding of the skin. This is a communicable disease spread by respiratory droplets.

Scavenger An organism that feeds upon dead and dying organisms.

Scent gland A gland located around the feet, face, or anus of many mammals; secretes pheromones, which may be involved with defense, species and sex recognition, and territorial behavior. Musk gland.

Schistosomule Juvenile stage of a blood fluke, between a cercaria and an adult; a migrating form taking the place of a metacercaria in the life cycle.

Schizeckenosy System of waste elimination found in some mites with a blindly ending midgut; the lobe breaks free from the ventriculus and is expelled through a split in the posterodorsal cuticle.

Schizocarp A dry indehiscent fruit that splits into two one-seeded halves at maturity.

Schizocoel A coelom formed by the splitting of embryonic mesoderm. Noun, schizocoelomate, an animal with a schizocoel, such as an arthropod or mollusc. Adj., schizocoelous.

Schizocoelous mesoderm formation Embryonic formation of the mesoderm as cords of cells between ectoderm and endoderm; splitting of these cords results in the coelomic space.

Schizogony Form of asexual reproduction in which multiple mitoses take place, followed by simultaneous cytokineses, resulting in many daughter cells at once.

Schizont Cell undergoing schizogony, in which nuclear divisions have occurred but cytokinesis is not completed; in its late phase sometimes called a segmenter.

School a well-defined social organization of marine animals consisting of a single species with all members of a similar size

Schüffner's dots Small surface invaginations that appear as stippling on the membrane of an erythrocyte infected with Plasmodium vivax after Romanovsky staining.

Schwann cells Members of a family of nerve-supporting (neuroglial) cells. The Schwann cell has an extensive cytoplasm which allows it to wrap a sheath around nerve axons.

Science A methodical, precise, objective way to study the natural world.

Scientific method Systematic apporach of observation, hypothesis formation, hypothesis testing and hypothesis evaluation that forms the basis for modern science.The objective process of approaching a problem. Involves hypothesis establishment, testing and observing the results, reevaluation of the hypothesis in light of new knowledge, and retesting to seek repeatability and thus validity of the hypothesis.

Scientific name A Latinized binomial (genus and species) unique to each identified organism.

Scion A small twig or bud that is grafted to a stock.

Scleractinia Order of coelenterates, usually producing calcareous skeletons with hexameral symmetry

Sclereid A sclerenchyma cell with a thick, lignified secondary wall having many pits; variable in form but not usually elongated.

Sclerenchyma One of the three major cell types in plants; have thickened, rigid, secondary walls that are hardened with lignin; provide support for the plant. Sclerenchyma cells include fibers and sclereids. Plant tissue type consisting of elongated cells with thickened secondary walls for support of the plant.

Sclerite A hard chitinous or calcareous plate or spicule; one of the plates making up the exoskeleton of arthropods, especially insects.

Scleroblast An amebocyte specialized to secrete a spicule, found in sponges.

Sclerocyte An amebocyte in sponges that secretes spicules.

Sclerophyllous Vegetation characterized by thick leathery leaves with abundant sclerenchyma cells; vegetation of a chaparral.

Sclerosponges Sponges with a massive calcareous skeleton.

Sclerotin Highly resistant and insoluble protein occurring in the cuticle of arthropods; also thought to occur in structures secreted by various other animals, such as in the eggshells of some trematodes, in which stabilization of the protein is achieved by orthoquinone crosslinks between free imino or amino groups of the protein molecules.

Sclerotium A fungal resting body resistant to unfavorable conditions; a firm, hardened mass of hyphae (or a hardened plasmodium of a slime mold) that will germinate on the return of favorable conditions.

Sclerotization The hardening and darkening processes in the cuticle (involves the epicuticle and exocuticle with a substance called sclerotin).

Scolex The attachment or holdfast organ of a tapeworm, generally considered the anterior end; it is used to adhere to the host.

Scoliosis Lateral curvature of the spine.

Scopa The pollen-collecting apparatus of a bee, whether it be the pollen basket on the leg or a brush of hairs on the abdomen.

Scope for growth The surplus of energy available for growth beyond that required for maintenance

Scopula A small tuft of hairs.

Scorpion Any member of the arachnid order Scorpionidae; they have an elongate body and a poison sting at the end of abdomen.

Scramble polygyny A mating system in which males actively search for mates without overt competition.

Screening by antibody Is an option if the bacteria and plasmid are designed to express proteins from the cDNA inserts. The principle is similar to hybridization, in that we lift replica filters from bacterial plates, but then we use the antibody (perhaps generated after olde tyme protein purification rituals) to show which colony expresses the desired protein.

Screening by hybridization Involves spreading the mixture of bacteria out on a dozen or so agar plates to grow several ten thousand isolated colonies. Membranes are laid onto each plate, and some of the bacteria from each colony stick, producing replicas of each colony in their original growth position. The membranes are lifted and the adherent bacteria are lysed, then hybridized to a radioactive piece of alpha DNA (the source of which is a story in itself). When X-ray film is laid on the filter, only colonies carrying alpha sequences will "light up". Their position on the membranes show where they grew on the original plates.

Scrotum In mammals, a pouch of skin located outside the body cavity into which the testes descend; provides proper temperature for the testes.

Scuba (self-contained underwater breathing apparatus) The use of tanks of compressed air for breathing underwater.

Scrub typhus Rickettsial disease transmitted by certain chigger mites.

Scutellum The 3rd of the major divisions of the dorsal surface of a thoracic segment: usually obvious only in the mesothorax and very large in some bugs.

Scutes Large scales. Usually means much the same as plate, lamina, shield.

Scutum The middle of the three main divisions of the dorsal surface of a thoracic segment. Also, in ticks, the sclerotized plate covering all or most of the dorsum in males, and the anterior portion in females, nymphs, and larvae of the Ixodidae.

Scyphistoma The polyp stage of a scyphozoan (phylum Cnidaria); develops from a planula and produces ephyrae by budding.

Scyphozoa A class of cnidarians whose members have prominent medusae. Gametes are gastrodermal in origin and are released to the gastrovascular cavity. Nematocysts are present in the gastrodermis. Polyps are small. Aurelia.

Sea A wave that has a sharp peak and a relatively flat trough. Seas are found in areas where waves are generated by the wind.

Sea anemones Anthozoans that consist of one large polyp.

Sea cucumbers Echinoderms with a soft, elongate body that lacks spines.

Sea spiders Arthropods that have a reduced body and four pair of legs .

Sea squirts Tunicates with a sac-like, attached body as adults.

Sea stars Echinoderms with five or more radiating arms and tube feet that are used in locomotion.

Sea urchins Echinoderms with a round or flattened test and movable spines.

Sea-floor spreading The process by which new sea floor is formed as it moves away from spreading centers in mid-ocean ridges.

Seagrasses Grass-like flowering plants such as eelgrass that are adapted to live at sea.

Seamount A submarine volcano in the abyssal plain.

Search image The hypothetical mental image of a prey species that a predator forms as it improves in its ability to capture that particular species.

Seasonal Fluctuations in environmental factors such as temperature over an annual cycle.

Seasonal estuary An estuary in which salinity at any one geographic point changes seasonally (e.g., decreases during the spring melt)

Seaward Side of an island that faces the direction of wave action generated either by winds or by currents generated by more indirect forces

Seaweed Any large photosynthetic protist, including rhodophytes and kelps. Seaweeds are not true plants, but like plants they can make their own food.

Sebaceous (oil) gland Gland of the skin that secretes sebum; oil gland.

Sebum Oily secretion from the sebaceous gland.

Secernentea The class of nematodes formerly called Phasmidea. Examples: Ascaris, Enterobius, Necator, Wuchereria.

Second law of thermodynamics The principle whereby every energy transfer or transformation increases the entropy of the universe. Ordered forms of energy are at least partly converted to heat, and in spontaneous reactions, the free energy of the system also decreases.

Second messenger The mechanism by which nonsteroid hormones work on target cells. A hormone binds to receptors on the cell's plasma membrane activating a molecule&emdash;the second messenger&emdash;that activates other intracellular molecules that elicit a response. The second messenger can be cyclic AMP, cyclic GMP, inositol triphosphate, diacrylglycerol, or calcium.

Second World The industrialized, socialist, centrally planned economy nations of Eastern Europe and the former Soviet Union and its allies.

Secondary (lateral) meristems Plant meristems that produce secondary growth from a cambium.

Secondary caries - caries which has occurred after a primary lesion has been restored. It is most commonly due to failure of the restoration at its margins, which have broken down and allowed a leak to develop between the restoration and the wall of the cavity. Caries bacteria which have been left behind during cavity preparation are unable to produce secondary caries if the margins of the restoration have achieved a proper seal against the tooth.

Secondary cell wall In woody plants, a second wall inside the primary cell wall; contains alternating layers of cellulose and lignin.

Secondary compounds Organic molecules synthesized by certain species of plants and not thought to be directly involved in essential metabolism.

Secondary consumer A heterotrophic, carnivorous organism that feeds on a primary consumer. Herring feeding on zooplankton are an example of a secondary consumer.

Secondary extinction The death of one population due to the extinction of another, often a food species.

Secondary growth Cells in a plant that are produced by a cambium. Increase in girth of a plant due to the action of lateral meristems such as the vascular cambium. The main cell produced in secondary growth is secondary xylem, better known as wood.

Secondary immune response The immune response elicited when an animal encounters the same antigen at some later time. The secondary immune response is more rapid, of greater magnitude, and of longer duration than the primary immune response.

Secondary immunity Resistance to an antigen the second time it appears. Because of the presence of B and T memory cells produced during the first exposure to the antigen, the second response is faster and more massive and lasts longer than the primary immune response.

Secondary lamella a small extension of a gill filament containing blood capillaries for gas exchange

Secondary macronutrients Elements that plants require in relatively small quantities: calcium, magnesium, and sulfur.

Secondary metabolites Products of metabolism that are synthesized after growth has been completed.

Secondary palate A plate of bone that separates the nasal and oral cavities of mammals and some reptiles.

Secondary parasite A parasite on another parasite.

Secondary phloem All phloem tissue formed by the vascular cambium in woody tissues.

Secondary pollutants Chemicals modified to a hazardous form after entering the air or that are formed by chemical reactions as components of the air mix and interact.

Secondary product Chemical compounds synthesized by plants or fungi but not critical for the basic metabolic functions of that organism; often functioning to deter predators or attract pollinators; a secondary metabolite.

Secondary production The production of living material per unit area (or volume) per unit time by herbivores. Usually expressed as grams carbon per meter square per year

Secondary recovery technique Pumping pressurized gas, steam, or chemical- containing water into a well to squeeze more oil from a reservoir.

Secondary structure Local structure within a protein which is conferred by the nature of the side chains of adjacent *amino acids* (e.g., alpha helix, beta sheet, random coil); local structure within an RNA molecule which is conferred by *base pairing* of *nucleotides* which are relatively closely positioned within the sequence (e.g., hairpins, stem-loop structures).

Secondary succession Revegetation of cleared land; return to previous community structure.

Secondary treatment The biological degradation of dissolved organic matter in the process of sewage treatment; the organic material is either mineralized or changed to settleable solids.

Secondary wall The innermost layer of a cell wall formed after cell elongation has ceased; often characterized by the deposition of lignin.

Secondary xylem Xylem produced by the vascular cambium in a woody plant stem or root; wood.

Secretin A hormone produced in the duodenum that stimulates alkaline secretions by the pancreas and inhibits gastric emptying.

Secretion (1) The discharge of molecules synthesized by the cell.

(2) In the vertebrate kidney, the discharge of wastes from the blood into the filtrate from the nephron tubules.

Secretory IgA The primary immunoglobulin of the secretory immune system.

Secretory vacuole In protists and some animals, these organelles usually contain specific enzymes that perform various functions such as excystation. Their contents are released to the cell exterior during exocytosis.

Secure landfill A solid waste disposal site lined and capped with an impermeable barrier to prevent leakage or leaching. Drain tiles, sampling wells, and vent systems provide monitoring and pollution control.

Sedatives Compounds that produce a relaxed, euphoric state—opium and its derivatives, morphine and heroin, are sedatives.

Sedentary Living in a fixed location, as with most plants, tunicates, sponges, etc. Contrast with motile.

Sedge Any member of the Cyperacaeae, a family of flowering plants that in their growthform resemble grasses.

Sediment Loose aggregate of solids derived from preexisting rocks, or solids precipitated from solution by inorganic chemical processes or extracted from solution by organisms.

Sedimentary rock Any rock composed of sediment, i.e., solid particles and dissolved minerals. Examples include rocks that form from sand or mud in riverbeds or on the sea bottom.

Sedimentation The deposition of organic materials or minerals by chemical, physical, or biological processes. Sediments can be transported from their source to their place of deposition by gravity, wind, water, or ice. If subjected to sufficient heat, pressure, or chemical reactions, sediments can solidify into sedimentary rock.

Seed A structure produced by seed plants which encapsulates the embryo. The seed often provides nourishment during germination, but may lie dormant for many years first.

Seed bank Storage facilities for seeds of domesticated plants and wild relatives; facilities for preserving genetic diversity.

Seed coat The outer layer of a seed that is developed from the integuments of the ovule; the testa.

Seed ferns Fossil plants with fernlike foliage but producing seeds belonging to the division Pteridospermophyta.

Seed plant Common term for gymnosperms and angiosperms.

Seeding In mariculture, the release of cultured juvenile individuals to enrich a natural stock.

Seedling The embryonic product of the germination of a seed; the young shoot and root axis.

Seed-scale complex The spirally arranged scales on a female strobilus in gymnosperms.

Seessel's pocket The anterior projecting portion of the foregut anterior to the stomodeum which will later form part of the mouth. Synonym: preoral gut.

Segment One of the rings or divisions of the body, or one of the sections of a jointed limb.

Segmentation (1) In many animal species, a series of body units that may be externally similar to, or quite different from, one another.

(2) The oscillating back-and-forth movement in the small intestine that mixes food with digestive secretions and increases the efficiency of absorption.

Segmented genome A virus genome that is divided into several parts or fragments, each probably coding for the synthesis of a single polypeptide; segmented genomes are very common among the RNA viruses.

Segmented worms Invertebrates that display an elongate body with distinct segmentation and a digestive tract that lies in a coelom.

Segments Repeating units in the body parts of some animals.

Segregation Separation of replicated chromosomes to opposite sides of the cell. Distribution of alleles on chromosomes into gametes during meiosis.

Seisonidea A class of rotifers containing members that are commensals of crustaceans; large and elongate body with rounded corona. Example: Seison.

Seive cell The organic solute-conducting cell of the phloem in gymnosperms.

Selectable marker A gene which is usually constitutively expressed and allows the selection of cells which carry it through growth on a selective medium. The most common example is the use of the ß-lactamase gene in plasmid vectors to confer ampicillin resistance on the host cell.

Selectins A family of cell adhesion molecules that are displayed on activated endothelial cells; examples include P-selectin and E-selectin. Selectins mediate leukocyte binding to the vascular endothelium.

Selection Process which favors one feature of organisms in a population over another feature found in the population. This occurs through differential reproduction—those with the favored feature produce more offspring than those with the other feature, such that they become a greater percentage of the population in the next generation.

Selection coefficient The difference between two fitness values, representing a relative measure of selection against an inferior genotype.

Selection differential The difference in mean phenotypic value for a selected group of breeding animals and the mean value in the population from which they are drawn.

Selection pressure A measure of the effectiveness of natural selection in altering the genetic composition of a population.

Selective breeding The selection of individuals with desirable traits for use in breeding. Over many generations, the practice leads to the development of strains with the desired characteristics.

Selective cutting Harvesting only mature trees of certain species and size; usually more expensive than clear-cutting, but it is less disruptive for wildlife and often better for forest regeneration.

Selective media Culture media that favor the growth of specific microorganisms; this may be accomplished by inhibiting the growth of undesired microorganisms.

Selective permeability The ability of the plasma membrane to let some substances in and keep others out.

Selective pressure An environmental factor that favors the survival and reproduction of those genetic variants within a population that are better adapted to the environment.

Selective toxicity The ability of a chemotherapeutic agent to kill or inhibit a microbial pathogen while damaging the host as little as possible.

Selectively permeable/ Selectively permeable membrane Permeable to small particles, such as water and certain inorganic ions, but not to larger molecules. A membrane that is permeable to small molecules, usually H2O, O2, and CO2, but not permeable to larger molecules or ions.

Self-assembly The spontaneous formation of a complex structure from its component molecules without the aid of special enzymes or factors.

Self-fertilization The union of egg and sperm produced by a single hermaphroditic organism.

Self-incompatibility Condition of a flower that cannot successfully complete the reproductive process with pollen produced by its own stamens.

Self-pollination The transfer of pollen from anther to stigma in the same flower or to another flower of the same plant, leading to self-fertilization.

Self-pollinating plant A plant that has its own pollen fall on its own stigma.

Self-propagating A description of the events occurring during an action potential, with each regional depolarization by sodium voltage-gated channels causing a similar event at an adjacent area downstream.

Self-regulating population A population with a growth rate that is dependent on its own numbers.

Self-shading Reduction in the amount of light available to phytoplankton that live below other phytoplankton.

Self-stimulation A situation in which a subject animal is able to introduce stimuli to its nervous system.

Semantic communication Of or relating to meaning of signals; specifically used to denote the use of different alarm signals to warn about different predators.

Semelparity The production of offspring by an organism once in its life.

Semen The thick, whitish secretion of the reproductive organs in the male; composed of sperm and secretions from the prostate, seminal vesicles, and various other glands and ducts.

Semicircular canals Bony canals in the vertebrate inner ear; the hair cells within the canals detect changes in the direction of movement of the head.

Semiconservative replication Process of DNA replication in which the DNA helix is unwound and each strand serves as a template for the synthesis of a new complementary strand, which is linked to the old strand. Thus, one old strand is retained in each new molecule.

Semidiurnal tide A tidal pattern with two high and two low tides each day.

Semilunar valve A valve between each ventricle of the heart and the artery connected to that ventricle.

Seminal receptacle A structure in the female reproductive system that stores sperm received during copulation (e.g., many insects and annelids).

Seminal vesicle 1. One of the paired accessory glands of the reproductive tract of male mammals. It secretes the fluid medium for sperm ejaculation (phylum Chordata).

2. A structure associated with the male reproductive tract that stores sperm prior to its release (e.g., earthworms_phylum Annelida).

Seminiferous Pertains to the tubules that produce or carry semen in the testes.

Seminiferous tubules The tightly coiled tubules found within the testes of vertebrates containing germ cells that are differentiated into spermatozoa.

Semipermeable Permeable to small particles, such as water and certain inorganic ions, but not to larger molecules.

Senescence The aging process, usually characterized by the loss of some functional capacity, including reproduction.

Sensation An impulse sent to the brain from activated receptors and sensory neurons.

Sense strand A gene has two strands: the sense strand and the anti-sense strand. The Sense strand is, by definition, the same 'sense' as the mRNA; that is it can be translated exactly as the mRNA sequence can. Given a sense strand with the following sequence: 5' - ATG GGG CCA CGG CTG TGA - 3'Met Gly Pro Arg Leu stop

The anti-sense strand will read as follows (note that the strand has been reversed and complemented): 5' - TCA CAG CCG TGG CCC CAT - 3'

The duplex DNA will pair as follows: 5' - ATGGGGCCACGGCTGTGA - 3'3' - TACCCCGGAGCCGACACT - 5'

Note however that when the RNA is transcribed from this sequence, the ANTI-SENSE strand is used as the template for RNA polymerization. After all, the RNA must base-pair with its template strand, so the process of transcription produces the complement of the anti-sense strand. This introduces some confusion about terminology:

Some people use the term 'coding strand' and 'non-coding strand' to refer to the sense and antisense strands, respectively.

Sensilla Modifications of the exoskeleton of an arthropod that, along with nerve cells, form sensory receptors.

Sensillum, A small sense organ, especially in the arthropods.

Sensitive period The time interval when an animal can develop an imprinting attachment. More broadly, in behavior development, this refers to time intervals when particular events must occur for proper ontogenetic sequencing.

Sensitization (1) Enhanced responsiveness to a repeated stimulus.

(2) Strengthening of a response that was initially produced via a CS resulting from a pairing with a US and UR.

(3) A stimulus priming the animal to pay particular attention to what follows.

Sensor In a closed system, the element that detects change and signals the effector to initiate a response.

Sensory adaptation A process that occurs at the level of the sensory receptors and that consists of a slowing down or cessation of nerve impulses transmitted to the central nervous system.

Sensory cortex A region of the brain associated with the parietal lobe.

Sensory deprivation Withholding all or a specified portion of the sensory input that an animal would normally be receiving.

Sensory filter Neural circuits that selectively transmit some features of a sensory input and ignore other features.

Sensory input Stimuli that the nervous system receives from the external or internal environment; includes pressure, taste, sound, light, and blood pH.

Sensory neuron or nerve A neuron or nerve that conducts an impulse from a receptor organ to the central nervous system. Neurons that carry signals from receptors and transmit information about the environment to processing centers in the brain and spinal cord. Neurons carrying messages from sensory receptors to the spinal cord. Sometimes referred to as an afferent neuron.

Sensory pathways The portion of the peripheral nervous system that carries information from the organs and tissues of the body to the central nervous system.

Sensory receptor A specialized structure that responds to specific stimuli from an animal's external or internal environment; transmits the information of an environmental stimulus to the animal's nervous system by converting stimulus energy to the electrochemical energy of action potentials.

Sepals Modified leaves that protect a flower's inner petals and reproductive structures. Small, leaf-like structures in flowers that enclose and protect the developing flower. These are often green, but in many monocots they are the same color as the petals (in which case the term tepal is applied since sepals and petals look so much alike).

Sepsis Systemic response to infection. This systemic response is manifested by two or more of the following conditions as a result of infection: temperature .38 or ,36°C; heart rate .90 beats per min; respiratory rate .20 breaths per min, or pCO2 ,32 mm Hg; leukocyte count .12,000 cells per ml3 or .10% immature (band) forms. Sepsis also has been defined as the presence of pathogens or their toxins in blood and other tissues.

Septa The thin plate of bone between the roots of teeth (Latin septum = a wall). Also the fibrous walls which separate sections of a gland. Septa separate sections of an orange or grapefruit.

Septate Divided by a septum or cross wall; also with more or less regular occurring cross walls.

Septic shock Sepsis associated with severe hypotension despite adequate fluid resuscitation, along with the presence of perfusion abnormalities that may include, but are not limited to, lactic acidosis, oliguria, or an acute alteration in mental status. Gram-positive bacteria, fungi, and endotoxin-containing gram-negative bacteria can initiate the pathogenic cascade of sepsis leading to septic shock.

Septic tank A tank used to process small quantities of domestic sewage. Solid material settles out and is partially degraded by anaerobic bacteria as sewage slowly flows through the tank. The outflow is further treated or dispersed in aerobic soil.

Septicemia Blood poisoning caused by pathogenic organisms.

Septum A partition or cross-wall that occurs between two cells in a bacterial (e.g., actinomycete) or fungal filament, or which partitions off fungal structures such as spores. Septa also divide parent cells into two daughter cells during bacterial binary fission.

Sequence As a noun, the sequence of a DNA is a buzz word for the structure of a DNA molecule, in terms of the sequence of bases it contains. As a verb, "to sequence" is to determine the structure of a piece of DNA; i.e. the sequence of nucleotides it contains.

Sequence database A computer-based collection of nucleic acid or amino acid sequences containing all published and some unpublished sequences. These can be searched to check for similarities with newly determined sequences. Examples are Genbank and Swissprot.

Sequence tagged site DNA sequence that has a single occurrence in the human genome and whose location and base sequence are known. Detectable by *polymerase chain reaction*, STSs are useful for localizing and orienting the mapping and sequence data reported from many different laboratories and serve as landmarks on the developing *physical map* of the human genome. Expressed sequence tags (ESTs) are STSs derived from cDNAs.

Sequential hermaphrodite An individual that sequentially produces male and then female gametes or vice versa.

Sequential hermaphroditism The type of hermaphroditism that occurs when an animal is one sex during one phase of its life cycle and an opposite sex during another phase.

Seral stage One stage of the communities in an ecological succession.

Sere An entire successional sequence in an

ecosystem (e.g., the sequence of stages in the succession of a lake to a climax forest).

Serial homolog Series of segments in which each repeats the genetic expression of the genes in the segment (somite). Metameric structures that have evolved from a common form; the biramous appendages of crustaceans are serially homologous.

Seriously undernourished Those who receive less than 80 percent of their minimum daily caloric requirements; are likely to suffer permanently stunted growth, mental retardation, and other social and developmental disorders.

Serology The branch of immunology that is concerned with in vitro reactions involving one or more serum constituents (e.g., antibodies and compiement).

Serosa In histology, the term is used to mean a serous membrane such as the peritoneum, pleura, and pericardium. A serous membrane is composed of a layer of epithelium with a thin layer of connective tissue. Synonym: tunica serosa. 2) In embryology, the term is used to mean the most exterior extraembryonic membrane. It consists of ectoderm and somatic mesoderm (somatopleure) and encloses all other extraembryonic membranes including the amnion, allantois, and yolk sac. Synonym: chorion.

Serotonin Is present throughout the body, especially in blood platelets and in the intestines. Its release from blood platelets contributes to the pain, vasoconstriction and inflammation after injury. In nervous tissue it functions as a neurotransmitter, mainly in the midbrain in clusters of cells called the raphe, and in the medulla. The fibres of these cells connect with the forebrain, cerebellum and spinal cord. It therefore exerts a strong influence over arousal, sensory perception, emotion and thought. Drugs which slow down the removal of serotonin can reduce depression and pain.

Serotyping A technique or serological procedure that is used to differentiate between strains (serovars or serotypes) of microorganisms that have differences in the antigenic composition of a structure or product.

Sertoli cell A type of cell found within seminiferous tubules of the testes, a nutritive cell to which spermatids attach for the duration of spermatogenesis until the mature spermatozoa is formed. Sertoli cells form the blood-testis barrier. Synonym: nurse cell.

Serum The clear, fluid portion of blood lacking both blood cells and fibrinogen. It is the fluid remaining after coagulation of plasma, the noncellular liquid faction of blood.

Serum resistance The type of resistance that occurs with bacteria such as Neisseria gonorrhoeae because the pathogen interferes with membrane attack complex formation during the complement cascade.

Sesamoid bone A small bone which appears at the age of thirteen, adjacent to the carpo-metacarpal joint of the thumb and is of use in determining the skeletal age of a child.

Sessile Attached; not free to move about.Attached at the base; fixed to one spot, not able to move about.

Seston Particulate matter suspended in seawater

Seta Hairlike modifications of an arthropod's exoskeleton that may be set into a membranous socket. Displacement of a seta initiates a nerve impulse in an associated cell.

Setaceous Bristle-like, applied especially to antennae.

Settling basin A basin used during water purification to chemically precipitate out fine particles, microorganisms, and organic material by coagulation or flocculation.

Setules Chitinous projections from copepod maxillipeds that trap food particles

Severe combined immunodeficiency (SCID) A genetic disorder in which afflicted individuals have no functional immune system and are prone to infections. Both the cell-mediated immune response and the antibody-mediated response are absent.

Sex chromosomes The chromosomes that determine the sex of an organism. In humans, females have two X chromosomes, and males have one X chromosome and one Y chromosome. Chromosome that determines the gender (sex) of the individual. Human males have a large X and a smaller Y sex chromosomes, while human females have two X sex chromosomes.

Sex hormones A group of steroid hormones produced by the adrenal cortex. Hormones that are produced in the gonads and promoted development and maintainence of the secondary sex characteristics and structures, prepare the female for pregnancy, and aid in development of gametes. Males produce testosterone, while females produce estrogen and progesterone.

Sex linkage The condition in which the inheritance of a sex chromosome is coupled with that of a given gene; e.g., red-green color blindness and hemophilia in humans. Traits located on the X-chromosome.

Sex pilus A thin protein appendage required for bacterial mating or conjugation. The cell with sex pili donates DNA to recipient cells.

Sex-influenced traits Traits that behave as if they are dominant in one sex but recessive in the other sex.

Sex-linked genes Genes located on one sex chromosome but not the other.

Sex-linked trait An inherited trait, such as color discrimination, determined by a gene located on a sex chromosome and that therefore shows a different pattern of inheritance in males and females.

Sexual Pertaining to the fusion of gametes; sexual reproduction.

Sexual dimorphism Is found in mammals when the male and female have distinct differences such as size or other characteristics that make them appear different.

Sexual imprinting The process by which young of many species establish an attachment to opposite sex conspecifics; the effect manifests itself later in life during the process of finding mates.

Sexual reproduction A type of reproduction in which two parents give rise to offspring that have unique combinations of genes inherited through the gametes of the two parents. Sexual reproduction involves meiosis and syngamy.

Sexual selection Selection in relation to mating. It is composed of intrasexual competition among members of one sex (usually males) for access to the other sex and intersexual choice of members of one sex by members of the other sex (usually females).

Shale A sedimentary rock composed of mud.

Shallow ecology A critical term applied to superficial environmentalists who claim to be green but are quick to compromise and who do little to bring about fundamental change. Many of those accused of being shallow ecologists counter that they are merely pragmatic, progressive reformers who prefer to work within the established social contract rather than try to overthrow it.

Shantytowns Settlements created when people move onto undeveloped lands and build their own shelter with cheap or discarded materials; some are simply illegal subdivisions where a landowner rents land without city approval; others are land invasions.

Shaping The procedure used in operant conditioning wherein the experimenter can reinforce particular responses or behavioral actions by the subject and not others, thus enhancing the frequency of some actions and extinguishing other actions.

Shard A piece of broken clay or ceramic pot placed over the drainage hole of a pot to prevent the loss of soil during watering.

Sharpey's fibres. Are collagen fibres which have been trapped in bone or cementum in order to anchor them.

Shattering A trait found in wild plants in which the fruiting head breaks apart to scatter the seeds over a wide area.

Shear boundary The boundary between two plates that move past each other on the earth's surface.

Sheath A hollow tubelike structure surrounding a chain of cells and present in several genera of bacteria.

Sheet erosion Peeling off thin layers of soil from the land surface; accomplished primarily by wind and water.

Shelf break The section of the continental shelf where the slope abruptly becomes steeper, usually at a depth of 120 to 200 m (400 to 600 ft).

Shell The calcium carbonate outer layer of cnidarians, molluscs, and other animals. Produced by mucous glands.

Shigellosis The diarrheal disease that arises from an infection with a member of the genus Shigella. Often called bacillary dysentery.

SHIME system (simulated human intestinal microbial ecosystem) A set of connected chemostat-like reactors that provide a sequence of environments similar to the human digestive system.

Shine-Dalgarno sequence A segment in the leader of procaryotic mRNA that binds to a special sequence on the 16S rRNA of the small ribosomal subunit. This helps properly orient the mRNA on the ribosome.

Shingles A reactivated form of chickenpox caused by the varicella-zoster virus.

Shivering thermogenesis The generation of heat by shivering, especially within the animal body.

Shoot The plant stem; provides support for the leaves and flowers; one of the three major plant organs; also referred to as the shoot system.

Shoot system Collective term for the stem and its leaves; generally the aboveground portion of the plant.

Short grass prairie A grassland biome characterized by short grass and low rainfall; also known as the plains.

Short-day plant A plant that flowers when the length of day is shorter than some critical value.

Shotgun cloning The practice of randomly clipping a larger DNA fragment into various smaller pieces, cloning everything, and then studying the resulting individual clones to figure out what happened. For example, if one was studying a 50 kb gene, it "may" be a bit difficult to figure out the restriction map. By randomly breaking it into smaller fragments and mapping those, a master restriction map could be deduced. A way of determining the sequence of a large DNA fragment which requires little brainpower but lots of late nights. The large fragment is shotgun cloned, and then each of the resulting smaller clones ("subclones") is sequenced. By finding out where the subclones overlap, the sequence of the larger piece becomes apparent. Note that some of the regions will get sequenced several times just by chance.

Shrub A plant that is shrublike in habitat; usually a short, perennial plant without strong apical dominance.

Shuttle vector A small *plasmid* capable of *transfection* into both prokaryotic and eukaryotic cells.

Sialoproteins A family of adhesion molecules which include osteopontin. Bone sialoprotein is formed by cells lining the root surface and influences cementoblast differentiation which encourages mineralisation. Dentine sialoprotein appears to inhibit mineralisation. Osteoclasts adhere both to bone sialoprotein and osteopontin.

Siberia A separate continental plate that existed from the Latest Precambrian to the Carboniferous, composed of a large part of central Russia, namely Siberia.

Siblicide The killing of one sibling by another.

Sibling species Closely related species that are so similar that they are nearly indistinguishable morphologically

Sickle cell anemia A condition that causes the red blood cells to collapse (sickle) under oxygen stress. The condition becomes manifest when an individual is homozygous for the gene for hemoglobin-S (HbS).

Siderite Also called ironstone, that is a concretion of iron carbonate. Common in the Mazon Creek fossil beds.

Siderophore A small molecule that complexes with ferric iron and supplies it to a cell by aiding in its transport across the plasma membrane.

Sieve elements Tubular, thin-walled cells that form a system of tubes extending from the roots to the leaves in the phloem of plants; lose their nuclei and organelles at maturity, but retain a functional plasma membrane.

Sieve plate The part of the wall of sieve-tube members bearing one or more highly differentiated sieve areas or modified primary pit fields.

Sieve pores Holes or modified plasmodesmata found within a sieve area either on a sieve plate or sieve cell.

Sieve tube A long tube specialized for the conduction of food materials (products of photosynthesis) and consisting of several-

Sieve-tube member One of the cells of a sieve tube; found primarily in flowering plants and typically associated with a companion cell; also called sieve-tube element.

Sigma Parameter expressing the seawater density: and equal to 1 minus the density of seawater, measured at a given temperature and at a pressure of l atmosphere

Sigma factor Certain small ancillary proteins in bacteria that increase the binding affinity of RNA polymerase to a promoter. Different sigma factors recognize different promoter sequences.

Sign An objective change in a diseased body that can be directly observed (e.g., a fever or rash).

Sign stimuli Specific external stimuli that trigger stereotyped responses from conspecifics, usually called fixed action patterns.

Signal Any behavior pattern that conveys information. Patterns modified through evolution to convey information are called displays.

Signal peptidase An enzyme present within the lumen of the endoplasmic reticulum which proteolytically cleaves a secreted protein at the site of a *signal sequence.*

Signal peptide A stretch of amino acids on polypeptides that targets proteins to specific destinations in eukaryotic cells.

Signal sequence A hydrophobic amino acid sequence which directs a growing peptide chain to be secreted into the endoplasmic reticulum.

Signaling pheromone A chemical signal that is quickly perceived and causes an immediate response. Also called releasing pheromone. Contrast with priming pheromone.

Signal-transduction pathway A mechanism linking a mechanical or chemical stimulus to a cellular response.

Silage Fermented plant material with increased palatability and nutritional value for animals, which can be stored for extended periods.

Silent mutation A nucleotide substitution (never a single deletion or insertion) which does not alter the amino acid sequence of an encoded protein due to the *degeneracy* of the genetic code. Such mutations usually involve the third base (*wobble* position) of *codons*.

Silica (SiO_3) A mineral similar to glass that is the major component of the cell wall, shell, or skeleton of many marine organisms.

Siliceous Containing silica.

Siliceous ooze A type of biogenous sediment that consists mostly of the silica shell and skéletons of marine organisms.

Silicification Process whereby silica replaces the original material of a substance. For example, silicified wood.

Silicoflagellates Unicellular and eukaryotic members of the phytoplankton that have a star-shaped silica skeleton.

Silique A simple fruit that develops from a two-carpeled ovary; at maturity the two halves fall away, leaving the seeds attached to the persistent central wall.

Sill A sheet-like igneous intrusion that parallels the plane of the surrounding rock.

Silt Soil inorganic particles in the 0.002-0.02 size range.

Siltation Deposition of earthy matter, fine sand, or the like by moving or running water.

Silurian period The geological time period of the Paleozoic Era following the Ordovician, between 435 and 395 million years ago, when plants colonized the land.

Simian Pertaining to monkeys or apes.

Simphyla A class of arthropods whose members are characterized by having long antennae, 10 to 12 pairs of legs, and centipedelike bodies. Occupy soil and leaf mold.

Simple diffusion The process of molecules spreading out randomly from where they are more concentrated to where they are less concentrated until they are evenly distributed.

Simple fruit A fruit that develops from a single ovary.

Simple leaf A leaf having a single blade portion; may be highly lobed or dissected.

Simple matching coefficient (SSM) An association coefficient used in numerical taxonomy; the proportion of characters that match regardless of whether or not the attribute is present.

Simple metamorphosis Metamorphosis in which the wings (when present) develop externally during the immature stage and there is no prolonged resting stage (i.e. pupa) preceding the last molt; stages included are the egg, nymphal, and adult. Also called gradual or partial metamorphosis, and paurometabolous development.

Simple pistil A pistil that contains a single carpel.

Simple sugar A sugar, such as glucose, that cannot be broken down into simpler sugar molecules.

Single gene disorder Hereditary disorder caused by a *mutant* allele of a single *gene* (e.g., Duchenne muscular dystrophy, retinoblastoma, sickle cell disease).

Single nucleotide polymorphism (SNP) A position in a genomic DNA sequence that varies from one individual to another. It is thought that the primary source of genetic difference between any two humans is due to the presence of single nucleotide polymorphisms in their DNA. Furthermore, these SNPs can be extremely useful in genetic mapping to follow inheritance of specific segments of DNA in a lineage. SNP-typing is the process of determining the exact nucleotide at positions known to be polymorphic.

Single radial immunodiffusion (RID) assay An immunodiffusion technique that quantitates antigens by following their diffusion through a gel containing antibodies directed against the test antigens.

Sinistral Pertaining to the left; in gastropods, shell is sinistral if opening is to left of columella when held with spire up and facing observer.

Sink A body or process that acts as a storage device or disposal mechanism; e.g., plants and the oceans act as sinks absorbing atmospheric carbon dioxide. Also, a location in a plant where sugar is being consumed, either in metabolism or by conversion to starch.

Sink habitat A habitat where mortality exceeds reproduction.

Sinkhole A natural depression in the surface of the land caused by the collapse of the roof of a cavern or subterranean passage, generally occurring in limestone regions.

Sinoatrial (SA) node A region of modified muscle cells in the right atrium that sends timed impulses to the heart's other muscle cells, causing them to contract; the heart's pacemaker.

Sinus A curved out hollow space in side the skull which is lined by respiratory epithelium and drains into the back of the throat. For example, maxillary s., ethmoid, s., sphenoids.

Sinus terminalis The vein encircling the area vasculosa.

Siphon Opening in molluscs or in urochordates which draws water into the body cavity. In many molluscs, the siphon may be used to expel water forcibly, providing a means of propulsion.

Siphonogamy A reproductive process in seed plants in which a pollen tube carries the sperm cells to the egg located within the integumented megasporangium.

Siphonoglyph Ciliated furrow in the gullet of sea anemones.

Siphonophores A group of specialized hydrozoan cnidarians, consisting of large planktonic polymorphic colonies

Siphonostele When a plant's vascular tissue develops as a central cylinder, it is said to have a siphonostele.

Siphuncle Cord of tissue running through the shell of a nautiloid, connecting all chambers with body of animal.

Sipuncula A phylum of protostomate worms whose members burrow in soft marine substrates throughout the world's oceans. Range in length from 2 mm to 75 cm. Peanut worms.

Sirenians (sea cows; order Sirenia)Marine mammals with anterior flippers, no rear limbs, and a paddle-shaped tail.

Sister chromatid One of the two identical parts of a duplicated chromosome in a eukaryotic cell. Sister chromatids consist of exact copies of a long coiled DNA molecule with associated proteins. Sister chromatids are joined at the centromere of a duplicated chromosome.

Sister group. The relationship between a pair of species or higher taxa that are each other's closest phylogenetic relatives.

Site-directed mutagenesis The introduction of a mutation, usually a *point mutation* or an insertion, into a particular location in a cloned DNA fragment. This mutated fragment may be used to "*knock out*" a gene in the organism of interest by *homologous recombination.*

Site-specific recombination Occurs between two specific but not necessarily homologous sequences. Usually catalyzed by enzymes not involved in general or *homologous recombination.*

Skeletal muscle Muscle that is generally attached to the skeleton and causes body parts to move; consists of muscle fibers. Voluntary muscle cells that have a striated appearance. These muscles control skeletal movements and are normally under conscious control.

Skeletal system/ Skeleton One of eleven major body organ systems in animals; supports the body, protects internal organs, and, with the muscular system, allows movement and locomotion. Support structure in animals, against which the force of muscles acts. Vertebrates have a skeleton of bone or cartilage; arthropods have one made of chitin; while many other invertebrates use a hydrostatic skeleton, which is merely an incompressible fluid-filled region of their body.

Skin One of eleven major body organ systems in animals; the outermost layer protecting multicellular animals from the loss or exchange of internal fluids and from invasion by foreign microorganisms; composed of two layers: the epidermis and dermis.

Skin-associated lymphoid tissue (SALT) The lymphoid tissue in the skin that forms a first-line defense as a part of nonspecific immunity.

Slash-and-burn agriculture The cutting down and burning of tropical vegetation to make mineral nutrients available for use by introduced agricultural crops.

S-layer A regularly structured layer composed of protein or glycoprotein that lies on the surface of many bacteria. It may protect the bacterium and help give it shape and rigidity.

Sled A benthic sampling device designed to slide along the sediment surface, digging into the bottom to a depth of at most a few centimeters

Sleep movement In legumes, the movement of the leaves in response to daily rhythms of dark and light. The leaves are horizontal in daylight and folded vertically at night.

Sleeping sickness African trypanosomiasis and mosquito-borne, virus-induced encephalitis.

Sliding-filament model The theory explaining how muscle contracts, based on change within a sarcomere, the basic unit of muscle organization, stating that thin (actin) filaments slide across thick (myosin) filaments, shortening the sarcomere; the shortening of all sarcomeres in a myofibril shortens the entire myofibril.

Slime The viscous extracellular glycoproteins or glycolipids produced by staphylococci and Pseudomonas aeruginosa bacteria that allows them to adhere to smooth surfaces such as prosthetic medical devices and catheters. More generally, the term often refers to an easily removed, diffuse, unorganized layer of extracellular material that surrounds a bacterial cell.

Slime ball Mass of mucus-covered cercariae of dicrocoeliid flukes, released from land snails. Also a term of derogation applied to really disgusting persons.

Slime layer A layer of diffuse, unorganized, easily removed material lying outside the bacterial cell wall.

Slime mold A common term for members of the divisions Acrasiomycota and Myxomycota.

Slime mold Members of the Myxomycetes, characterized by a creeping, plasmodial stage.

Slope A steep-sloping bottom extending seaward from the edge of the continental shelf and downward toward the rise

Slot blot Similar to a dot blot, but the analyte is put onto the membrane using a slot-shaped template. The template produces a consistently shaped spot, thus decreasing errors and improving the accuracy of the analysis.

Slow sand filter A bed of sand through which water slowly flows; the gelatinous microbial layer on the sand grain surface removes waterborne microorganisms, particularly Giardia, by adhesion to the gel. This type of filter is used in some water purification plants.

Slow virus disease A progressive, pathological process caused by a transmissible agent (virus or prion) that remains clinically silent during a prolonged incubation period of months to years after which progressive clinical disease becomes apparent.

Sludge A general term for the precipitated solid matter produced during water and sewage treatment; solid particles composed of organic matter and microorganisms that are involved in aerobic sewage treatment (activated sludge).

Slums Legal but inadequate multifamily tenements or rooming houses; some are custom built for rent to poor people, others are converted from some other use.

Small intestine A coiled tube in the abdominal cavity that is the major site of chemical digestion and absorption of nutrients; composed of the duodenum, jejunum, and ileum.

Small nuclear ribonucleoprotein (snRNP) One of a variety of small particles in the cell nucleus, composed of RNA and protein molecules; functions are not fully understood, but some form parts of spliceosomes, active in RNA splicing.

Smallpox (variola) Once a highly contagious, often fatal disease caused by a poxvirus. Its most noticeable symptom was the appearance of blisters and pustules on the skin. Vaccination has eradicated smallpox throughout the world.

Smog The term used to describe the combination of smoke and fog in the stagnant air of London; now often applied to photochemical pollution products or urban air pollution of any kind.

Smolt Ayoung salmon just before it migrates downstream and out to sea.

Smooth ER The portion of endoplasmic reticulum that lacks ribosomes.

Smooth muscle Muscle that lacks striations; found around circulatory system vessels and in the walls of such organs as the stomach, intestines, and bladder. Involuntary, not striated cells that control autonomic functions such as digestion and artery contraction.

Snapping division A distinctive type of binary fission resulting in an angular or a palisade arrangement of cells, which is characteristic of the genera Arthrobacter and Corynebacterium.

Social Living in more or less organized communities of individuals.

Social behavior Behavior that takes place in a social context and results from the interaction between and among individuals.

Social deprivation Withholding or removal of contact with any form of stimulation from conspecifics.

Social dominance A hierarchical pattern of social organization involving domination of some members of a group by other members in a relatively orderly and long-lasting pattern.

Social ecology A socialist/humanist philosophy based on the communitarian anarchism of the Russian geographer Peter Kropotkin. It shares much with deep ecology except that it is more humanist in its outlook.

Social justice Equitable access to resources and the benefits derived from them; a system that recognizes inalienable rights and adheres to what is fair, honest, and moral.

Social organization The species-typical pattern of relationships among all members of a group. This would include spatial distribution patterns, interindividual relationships involving dominance hierarchies or territoriality, mating systems, parenting, and dispersal.

Societies The most highly organized type of social organization; consist of individuals that show varying degrees of cooperation and communication with one another; often have a rigid division of labor.

Sociobiology A study that involves the application of the principles of evolution to the study of the social behavior and social systems of animals.

Sodium-potassium ATPase pump The active transport mechanism that functions to concentrate sodium ions on the outside of a plasma membrane and potassium ions on the inside of the membrane.

Soft corals Colonial anthozoans with no hard skeleton.

Softwood General term for the wood (secondary xylem) of conifers.

Soil A complex mixture of weathered mineral materials from rocks, partially decomposed organic molecules, and a host of living organisms.

Soil horizons Horizontal layers that reveal a soil's history, characteristics, and usefulness.

Soil profile A vertical section of soil showing the zones of particle sizes from surface down to bedrock.

Soldier In termites, sterile males or females with large heads and mandibles; they function to protect the colony.

Solenia Channels through the coenenchyme connecting the polyps in an alcyonarian colony (phylum Cnidaria).

Solenocyte Special type of flame bulb in which the bulb bears a flagellum instead of a tuft of flagella.

Solenophage Blood-feeding arthropod that introduces its mouthparts directly into a blood vessel to feed.

Soleus The sole of the foot.

Solitary Occurring singly or in pairs, not in colonies.

Solubility The relative ability of a solute to be dissolved.

Solubility product A value found by multiplying the concentration of positive IONS by the concentration of negative ions in a solution of a salt, hence $[Ca]^+ \times [P0_4]^- = K_{sp}$(solubility product). The value for K_{sp} is constant when the solution of ions is saturated and in balance with its solid crystalline form. Acid helps increase the solubility of a weakly soluble salt.

Solute Any material dissolved in a solution.

Solute potential The water potential component caused by the presence of solutes in water.

Solution A homogeneous, liquid mixture of two or more substances.solvent The dissolving agent of a solution. Water is the most versatile solvent known.

Solution hybridization A method closely related to RNase protection. Solution hybridization is designed to measure the levels of a specific mRNA species in a complex population of RNA. An excess of radioactive probe is allowed to hybridize to the RNA, then single-strand specific nuclease is used to destroy the remaining unhybridized probe and RNA. The "protected" probe is separated from the degraded fragments, and the amount of radioactivity in it is proportional to the amount of mRNA in the sample which was capable of hybridization. This can be a very sensitive detection method.

Solvent The liquid matrix in which a solute is dissolved.

Soma The whole of an organism except the germ cells (germ plasm).

Somaclonal variant A plant showing a mutation that developed asexually during the tissue culture of a single callus.

Somatic Either pertaining to body cells (as opposed to gametes) or to the body wall (as opposed to the viscera).

Somatic cell Ordinary body cell; pertaining to or characteristic of a body cell. Any cell other than a germ cell or germ-cell precursor.

Somatic growth Growth of the body, exclusive of gametes

Somatic mesoderm The cellular layer on the external side of the coelom; formed from the delamination of lateral plate mesoderm. In conjunction with somites, it will later form body wall and limbs.

Somatic mutation A mutation that occurs in cells of leaves, stems, or roots; a mutation occurring in any cells that are not involved in gamete formation.

Somatic nervous system The branch of the motor division of the vertebrate peripheral nervous system composed of motor neurons that carry signals to skeletal muscles in response to external stimuli.

Somatic senses All senses except vision, hearing, taste, and smell; include pain, temperature, and pressure.

Somatocoel Posterior coelomic compartment of echinoderms; left somatocoel gives rise to oral coelom, and right somatocoel becomes aboral coelom.

Somatoplasm The living matter that makes up the mass of the body as distinguished from germ plasm, which makes up the reproductive cells. The protoplasm of body cells.

Somatopleure The membrane composed of ectoderm and somatic mesoderm. It forms part of body wall as well as the chorion and amnion.

Somatostatin Pancreatic hormone that controls the rate of nutrient absorption into the bloodstream.

Somatotropin A hormone, produced by the pituitary gland, that stimulates protein synthesis and promotes the growth of bone; also known as growth hormone.

Somite Block of dorsal mesodermal cells adjacent to the notochord during vertebrate organogenesis. These transient structures define the segmental pattern of the embryo, and subsequently give rise to vertebrae and ribs, dermis of the back, and skeletal muscles of the back, body wall and limbs.

Sonar or biosonar. A system that uses sound at sonic or ultrasonic frequencies to detect and locate objects.

Sorocarp The fruiting structure of the Acrasiomycetes.

Sorting (of a sediment) The range of scatter of particle sizes about the median grain size of a sediment

Sorting. Differential survival and reproduction among varying individuals; often confused with natural selection which is one possible cause of sorting.

Sorus A type of fruiting structure composed of a mass of spores or sporangia.

SOS repair A complex, inducible repair process that is used to repair DNA when extensive damage has occurred.

Sound window The use of frequencies for communication that are transmitted through the environment with little loss of strength (attenuation).

Source The location or object from which a pathogen is immediately transmitted to the host, either directly or through an intermediate agent.

Source habitat A habitat where reproduction exceeds mortality and from which excess individuals disperse.

Southern blot/ Southern blotting DNA is separated by electrophoresis (usually in *agarose gels*), then transferred to nitrocellulose paper or other suitable solid-phase matrix (e.g., nylon membrane), and denatured into single strands so that it can be *hybridized* with a specific *probe*. The Southern blot was developed by E.M. Southern, a molecular biologist in Edinburgh. *Northern* and *western* blots were given contrasting names to reflect the different target substances (RNA and proteins, respectively) that are subjected in these procedures to electrophoresis, blotting and subsequent detection with specific probes. A method of transfer of bands of DNA, separated on a gel electrophoresis, onto a membrane, from which the bands may be treated with chemicals to enable them to be seen. {Named after the inventor of the process, E.M Southern. Related methods have been named 'Northern Blotting' (RNA blotting from electrophoresis gels) and 'Western Blotting' (protein blotting) by scientists.}

Southwestern blot The binding of protein to a nucleic acid on a matrix similar to what is done for western, northern, and southern blots. This technique is used to identify DNA binding proteins and the recognition sites for these proteins.

Sp6 RNA polymerase A bacteriophage *RNA polymerase* which is commonly used to transcribe *plasmid* DNA into RNA. The plasmid must contain an SP6 *promoter* upstream of the relevant sequence.

Space limited Description of a situation in which space is a limiting resource

Sparganum Cestode plerocercoid of unknown identity.

Spatial autocorrelation A situation in which some parameter at any location (e.g., population density) can be predicted through a knowledge of the values of the parameter in other locations

Spatial distribution The arrangement of individuals in a space

Spawning The release of gametes or eggs into the water.

Special senses Vision, hearing, taste, and smell.

Specialist A species with a very narrow range in habitat or food requirements. For example, the

Marbled Murrelet nests in old-growth forests on thick branches high up in the forest canopy. Compare *generalist*.

Specialized (1) Of cells, having particular functions in a multicellular organism.

(2) Of organisms, having special adaptations to a particular habitat or mode of life.

Specialized niche A species occupying a niche with a narrow range of tolerance.

Speciation The process by which new species arise. The process by which discontinuities between populations occur due to the development of mechanisms creating the reproductive isolation of one population from the other.

Allopatric speciation: species formation that occurs during geographic isolation of populations. Generally believed to be the most common way in which new species arise, especially among the higher animals.

Sympatric speciation: the formation of new species without geographic isolation; the acquisition of reproductively isolating mechanism among individuals coexisting in the same area. Not infrequent in plants.

Species Species of higher organisms are groups of interbreeding or potentially interbreeding natural populations that are reproductively isolated. Bacterial species are collections of strains that have many stable properties in common and differ significantly from other groups of strains.

Species abundance The total number of individual of a species within a given area or community. Compare *species richness*.

Species diversity A measure of both species abundance and species richness. An area that has a large number of species and many representative individuals from each species is more diverse than an area that has only a single species.

Species packing The phenomenon in which present-day communities generally contain more species than earlier communities because organisms have evolved more adaptations over time.

Species recovery plan A plan for restoration of an endangered species through protection, habitat management, captive breeding, disease control, or other techniques that increase populations and encourage survival.

Species richness The number of different species that exist within a given area or community. Compare *species abundance*.

Species selection A theory maintaining that species living the longest and generating the greatest number of species determine the direction of major evolutionary trends.

Species-area effect A regular logarithmic relationship between the number of species in a confined geographic area (e.g., an island) and the area in which the species occur

Species-specific Characteristic of (and limited to) a particular species.

Species-typical behavior Actions and displays that are broadly characteristic of a species.

Specific heat The amount of heat required to raise the temperature of 1 gm of any substance 1 degree Celsius.

Specific-mate recognition system (SMRS) Signals that increase the likelihood that members of the same population will mate with one another. Contrast with species isolating mechanisms.

Spectrophotometer An instrument that measures the proportions of light of different wavelengths absorbed and transmitted by a pigment solution.

Speculum An iridescent patch of color on the wings of certain ducks and other birds.

Sperm A male gamete.

Sperm competition A situation in which one male's sperm fertilize a disproportionate number of eggs when a female copulates with more than one male.

Sperm nuclei Each pollen grain produces two sperm nuclei, which effect double fertilization in angiosperms.

Spermaceti organ a large organ in the forehead of sperm whales that is filled with a fine-quality liquid or waxy spermaceti oil.

Spermagonium A structure that produces spermatia in the rust fungi.

Spermalege Organ that receives the sperm in the female cimicid bug during copulation.

Spermatheca A sac in the female reproductive organs for the reception and storage of sperm.

Spermatid A growth stage of a male reproductive cell arising by division of a secondary spermatocyte; gives rise to a spermatozoon.

Spermatocytes The diploid ($2n$) cells formed by the enlargement and differentiation of the spermatogonia; they give rise by meiotic division to the spermatids.

Spermatodactyl Modification in some Acari of chelicera, which functions in transfer of sperm from male's gonopore to copulatory receptacles between third and fourth coxae of female.

Spermatogenesis The development of sperm cells from spermatocytes to mature sperm, including meiosis.

Spermatogonia The unspecialized diploid ($2n$) cells on the walls of the seminiferous tubules that, by enlargement, differentiation, and meiotic division, become spermatocytes, then spermatids, then sperm cells.

Spermatophore Formed "container" or packet of sperm that is placed in or on the body of a female, in contrast to the sperm in copulation which are conducted directly from male reproductive structures into the female's body.

Spermatophyte A seed plant.

Spermatozoon A mature haploid male gamete.

Spermiogenesis The process of differentiation by which haploid spermatids are transformed into mature spermatozoa.

Spermosphere The region surrounding a germinating seed where released organic matter stimulates microbial growth.

Sphenoid The sphenoid bone.

Sphenomandibular ligament A ligament which joins the lingula of the mandible to the spine of the sphenoid bone.

Sphenophyta The vascular plant division commonly termed the horsetails to which the genus Equisetum belongs.

Spheroplast A relatively spherical cell formed by the weakening or partial removal of the rigid cell wall component (e.g., by penicillin treatment of gram-negative bacteria). Spheroplasts are usually osmotically sensitive.

Sphincter A ringlike valve, consisting of modified muscles in a muscular tube, such as a digestive tract; closes off the tube like a drawstring.

Spicule Crystalline or mineral deposits found in sponges, sea cucumbers, or urochordates. They are structural components in many sponges, and may serve a protective function in other organisms.

Spike An inflorescence in which the main axis is elongated and the flowers are sessile.

Spikelet A small group of grass flowers; a unit of the inflorescence in grasses.

Spinal cord A cylinder of nerve tissue extending from the brain stem; receives sensory information and sends output motor signals; with the brain, forms the central nervous system. Nerve cell collections extending from the base of the brain to just below the last rib vertebrae.

Spindle An assemblage of microtubules that orchestrates chromosome movement during eukaryotic cell division.

Spindle apparatus Microtubule construction that aligns and segregates chromosomes during eukaryotic cell division.

Spindle fibers A collection of microtubules attached to a centromere during mitosis and meiosis that are responsible for the movement of the chromosomes to opposite poles.

Spine A multicellular, thorn like process or

outgrowth of the integument not separated from it by a joint.

Spiracle In insects and some other terrestrial arthropods, a small opening through which air is taken into the tracheae. Insects have several spiracles, arranged along the sides of the abdomen.

Spiracular plate A plate like sclerite next to or surrounding a spiracle.

Spiral cleavage A type of embryonic development in protostomes, in which the planes of cell division that transform the zygote into a ball of cells occur obliquely to the polar axis, resulting in cells of each tier sitting in the grooves between cells of adjacent tiers.

Spiral valve A spiral portion in the intestine of cartilaginous fishes.

Spirillum A spirally coiled bacterium.

Spirochete A flexible, spiral-shaped bacterium with periplasmic flagella.

Spittle In insects, a frothy fluid produced by the nymphs of spittlebugs (Cercopidae).

Splanchnic mesoderm The cell layer on the internal side of the coelom; formed by the delamination of lateral plate mesoderm. Later forms the heart, part of the circulatory system and many other structures.

Splanchnopleure The cellular layer consisting of splanchnic mesoderm and endoderm. It will later form part of the body wall in the embryo as well as the yolk sac and allantois.

Spleen A secondary lymphoid organ where old erythrocytes are destroyed and blood-borne antigens are trapped and presented to lymphocytes.

Spliceosome A complex assembly that interacts with the ends of an RNA intron in splicing RNA; releases an intron and joins two adjacent exons.

Split or interrupted gene A structural gene with DNA sequences that code for the final RNA product (expressed sequences or exons) separated by regions coding for RNA absent from the mature RNA (intervening sequences or introns).

Spondylosis Degeneration of a vertebra.

Sponges (phylum Porifera)Invertebrates that consist of a complex aggregation of cells, including collar cells, and have a skeleton of fibers and/or spicules.

Spongiform encephalopathies Degenerative central nervous system diseases in which the brain has a spongy appearance; they appear due to prions.

Spongin Fibrous, collagenous material making up the skeletal network of horny sponges.

Sponging Proteinacous compound of which the spicules in Demospongiae are composed.

Spongioblast Cell in a sponge that secretes spongin, a protein.

Spongocoel Central body cavity of sponges.

Spongocyte A cell in sponges that secretes spongin.

Spongy bone The inner layer of bone; found at the ends of long bones and is less dense than compact bone. Some spongy bone contains red marrow.

Spongy mesophyll Parenchyma cells found in plant leaves that are irregularly shaped and have large intracellular spaces.

Spongy parenchyma A tissue composed of loosely packed, irregular parenchymatous cells containing chloroplasts; commonly found in leaves.

Spontaneous generation The hypothesis that living organisms can arise from nonliving matter.

Spontaneous ovulators Species in which females release eggs whether they have copulated or not.

Spontaneous recovery A process in which a conditioned response has been extinguished and then is followed by some time interval (generally one minute up to one day or more depending upon the species and experimental conditions), after which the animal may immediately exhibit a nearly normal correct response rate upon reintroduction to the test situation.

Sporadic disease A disease that occurs occasionally and at random intervals in a population.

Sporadin Mature trophozoite of a gregarine protozoan.

Sporangia The structures in which spores are produced (sing.: sporangium).

Sporangiophore A stalk to which sporangia are attached.

Sporangiospore A spore that develops within a sporangium.

Sporangium A saclike structure or cell, the contents of which are converted into an indefinite number of spores. It is borne on a special hypha called a sporangiophore.

Spore A differentiated, specialized form that can be used for dissemination, for survival of adverse conditions because of its heat and dessication resistance, and/or for reproduction. Spores are usually unicellular and may develop into vegetative organisms or gametes. They may be produced asexually or sexually and are of many types.

Sporoblast Cell mass that will differentiate into a sporocyst within an oocyst.

Sporocyst (1) Stage of development of a sporozoan protozoan, usually with an enclosing membrane, the oocyst.

(2) An asexual stage of development in some digenean trematodes that arises from a miracidium and gives rise to rediae.

Sporocyst residuum Cytoplasmic material "left over" within a sporocyst after sporozoite formation; seen as an amorphous mass.

Sporogony Multiple fission that produces sporozoites after zygote formation. Occurs in the class Sporozoea.

Sporont Undifferentiated cell mass within an unsporulated oocyst.

Sporophyll Any leaf which bears sporangia is called a sporophyll.

Sporophyte The diploid stage in the life cycle of an organism undergoing an alternation of generations. The sporophyte is multicellular and develops from a zygote. The mature sporophyte meiotically produces haploid spores that later generate the gametophyte generation.

Sporoplasm Amebalike portion of a microsporan or myxosporan cyst that is infective to the next host.

Sporopollenin A secondary product, a polymer synthesized by a side branch of a major metabolic pathway of plants that is resistant to almost all kinds of environmental damage; especially important in the evolutionary move of plants onto land.

Sporotrichosis A subcutaneous fungal infection caused by the dimorphic fungus Sporothrix schenckii.

Sporozoans Members of the protists that are referred to as slime molds; may include organisms resembling the ancestors of fungi.

Sporozoite A stage in the life history of many sporozoan protozoa; released from oocysts.

Spout (blow)The water vapor and seawater that is observed when whales surface and exhale.

Spread plate A petri dish of solid culture medium with isolated microbial colonies growing on its surface, which has been prepared by spreading a dilute microbial suspension evenly over the agar surface.

Spring diatom increase The major rapid population increase of diatoms, occurring in the spring in temperate-boreal latitudes

Spring overturn Springtime lake phenomenon that occurs when the surface ice melts and the surface water temperature warms to its greatest density at 4C and then sinks, creating a convection current that displaces nutrient-rich bottom waters.

Spring tides The tides with a large tidal range; they

occur around the times of full or new moon. Compare neap tides.

Springwood The cells in the secondary xylem that are formed early in the season, usually with wide vessels (angiosperms) or wide tracheids (gymnosperms); also called

Spur A large and usually movable spine, normally found on the legs.

Spurious vein A false vein formed by a thickening of the wing membrane and usually unconnected with any of the true veins.

Sputum The mucous secretion from the lungs, bronchi, and trachea that is ejected (expectorated) through the mouth.

Squalene A liquid acyclic triterpene hydrocarbon found especially in the liver oil of sharks.

Squama Any of the membranous flaps that arise near the base of the wing in many true flies (plural: squamae).

Squamous epithelium Simple epithelium of flat, nucleated cells.

Stress Physical, chemical, or emotional factors that place a strain on an animal. Plants also experience physiological stress under adverse environmental conditions.

Stability One of the phases of a population's life cycle. The population's size remains roughly constant, fluctuating around some average density. Also, the ability of a community to persist unchanged.

Stability-time hypothesis Hypothesis that states that higher diversity occurs in habitats that are ancient and stable environmentally

Stabilizing selection A process of natural selection that tends to favor genotypic combinations that produce an intermediate phenotype; selection against the extremes in variation.

Stable runoff The fraction of water available year-round; usually more important than total runoff when determining human uses.

Stable transfection A form of transfection experiment designed to produce permanent lines of cultured cells with a new gene inserted into their genome. Usually this is done by linking the desired gene with a "selectable" gene, i.e. a gene which confers resistance to a toxin (like G418, aka Geneticin). Upon putting the toxin into the culture medium, only those cells which incorporate the resistance gene will survive, and essentially all of those will also have incorporated the experimenter's gene.

Stablizing selection Natural selection that favors intermediate variants by acting against extreme phenotypes.

Stadium The time interval between molts in a developing insect.

Stage A distinct, sharply differentiated period in the development of an insect, e.g., egg stage, larval stage, pupal stage, adult stage; in mites and ticks, each instar.

Stalk A nonliving bacterial appendage produced by the cell and extending from it. A leaf's petiole; the slender stem that supports the blade of a leaf and attaches it to a larger stem of the plant.

Stalk-and-rush Predators that approach prey as closely as possible, then close with a sudden burst of speed.

Stamens The male reproductive structures of a flower; usually consist of slender, thread-like filaments topped by anthers. The male reproductive structures in the flower, composed of a filament and anther.

Staminate flower A flower having stamens but no carpels.

Staminodium A sterile stamen; nonfunctional anthers and often with petaloid filaments.

Standard deviation A statistical measure of the degree of variation from the mean value among the individual measurements in a series of values.

Standard error of the mean (SEM) A statistical measure of variation most properly restricted to use with a group of means, though often reported as a measure of variation around the mean value in a series of individual measurements.

Standard free energy change The free energy change of a reaction at 1 atmosphere pressure when all reactants and products are present in their standard states; usually the temperature is 25°C.

Standard reduction potential A measure of the tendency of a reductant to lose electrons in an oxidation-reduction (redox) reaction. The more negative the reduction potential of a compound, the better electron donor it is.

Standing crop The amount of living material per unit area or volume; may be expressed as grams of carbon, total dry weight, and so on

Standing stock The total amount, or biomass, of an organism at a given time.

Stapes One of the three bones that function in hearing.

Staphylococcal food poisoning A type of food poisoning caused by ingestion of improperly stored or cooked food in which Staphylococcus aureus has grown. The bacteria produce exotoxins that accumulate in the food.

Staphylococcal scalded skin syndrome A disease caused by staphylococci that produce an exfoliative toxin. The skin becomes red (erythema) and sheets of epidermis may separate from the underlying tissue.

Starch A polysaccharide composed of a thousand or more glucose molecules; the chief food storage material of most plants.

Start codon The codon (AUG) on a messenger RNA molecule where protein synthesis begins.

Starter culture An inoculum, consisting of a mixture of carefully selected microorganisms, used to start a commercial fermentation.

Stasis A period of little or no discernible change in a lineage.

Stationary phase The phase of microbial growth in a batch culture when population growth ceases and the growth curve levels off.

Statistics The mathematics of the collection, organization, and interpretation of numerical data.

Statoblast Biconvex capsule containing germinative cells and produced by most freshwater ectoprocts by asexual budding. Under favorable conditions it germinates to give rise to new zooid.

Statocyst A type of mechanoreceptor that functions in equilibrium in invertebrates through the use of statoliths, which stimulate hair cells in relation to gravity.

Statolith Small calcareous body resting on tufts of cilia in the statocyst.

Steady-state economy Characterized by low birth and death rates, use of renewable energy sources, recycling of materials, and emphasis on durability, efficiency, and stability.

Stele The central vascular cylinder in roots where xylem and phloem are located.

Stem The aboveground part of the axis of vascular plants, as well as anatomically similar portions below ground (such as rhizomes).

Stem cells The common, self-regenerating cells in the marrow of long bones that give rise, by differentiation and division, to red blood cells and all of the different types of white blood cells.

Stem group All the taxa in a clade preceding a major cladogenesis event. They are often difficult to recognize because they may not possess synapomorpies found in the crown group.

Stem-loop A feature of RNA *secondary structure*, in which two complementary, inverted sequences which are separated by a short-intervening sequence within a single strand of RNA base pair to form a "“stem” with a “loop” at one end. Similar to a *hairpin*, but these usually have very small loops and longer stems.

Stemma The simple eye in holometabolous larvae. Also called lateral ocellus.

Stem-nodulating rhizobia Rhizobia (members of the

genera Rhizobium, Bradyrhizobium, and Azorhizobium) that produce nitrogen-fixing structures above the soil surface on plant stems. These most often are observed in tropical plants and produced by Azorhizobium.

Stenohaline Pertaining to aquatic organisms that have restricted tolerance to changes in environmental saltwater concentration.

Stenophagous Eating few kinds of foods.

Stenotopic Refers to an organism with

Steppe Refers to arid lands having vegetation that is adapted to dry conditions, and having extreme temperature variations between the hot of the summer season and the cold of the winter season. In Idaho it often has considerable sagebrush.

Stereoblastula Embryo produced by spiral cleavage, characterized by the absence of a blastocoel; formed by embryos of annelid worms, turbellarian flatworms, nemertean worms, and all molluscs except cephalopods.

Stereocilia The short, modified cilia at the apex of a hair cell.

Stereogastrula A solid type of gastrula, such as the planula of cnidarians.

Stereoisomer A molecule that is a mirror image of another molecule with the same molecular formula.

Stereom Meshwork structure of endoskeletal ossicles of echinoderms.

Stereoscopic vision Ability to perceive a single, three-dimensional image from the simultaneous but separate images delivered to the brain by each eye.

Stereotyped behavior A pattern of behavior repeated with little variation in performance.

Sterilization The process by which all living cells, viable spores, viruses, and viroids are either destroyed or removed from an object or habitat.

Sternite The plate or sclerite on the underside of a body segment.

Sternum Ventral plate of an arthropod body segment; breastbone of vertebrates.

Steroids A class of lipids characterized by a carbon skeleton consisting of four rings with various functional groups attached.

Sterol One of a class of organic compounds containing a molecular skeleton of four fused carbon rings; it includes cholesterol, sex hormones, adrenocortical hormones, and vitamin D.

Stewards Protectors, careful managers of land and nature's resources.

Stichosome Column of large, rectangular cells called stichocytes, supporting and secreting into the esophagus of most nematodes of the family Trichuridae.

Sticky ends After digestion of a DNA with certain restriction enzymes, the ends left have one strand overhanging the other to form a short (typically 4 nt) single-stranded segment. This overhang will easily re-attach to other ends like it, and are thus known as "sticky ends". For example, the enzyme BamHI recognizes the sequence GGATCC, and clips after the first G in each strand:

The overhangs thus produced can still hybridize ("anneal") with each other, even if they came from different parent DNA molecules, and the enzyme ligase will then covalently link the strands. Sticky ends therefore facilitate the ligation of diverse segments of DNA, and allow the formation of novel DNA constructs.

Stieda body Plug in the inner wall of one end of a coccidian oocyst.

Stigma The sticky tip of a pistil. Or, the dense region of pigments found in many photosynthetic protists which is sensitive to light, and thus functions somewhat like a miniature eye.

Stimulant A psychoactive compound that excites and enhances mental alertness and physical activity; often reduces fatigue and suppresses hunger.

Stimulus A physical or chemical change in the

environment that leads to a response controlled by the nervous system.

Stimulus filter The ability of the nervous system to block incoming stimuli that are unimportant for the animal.

Stipe A supporting stalk; such as those in mushrooms and brown algae.

Stippled A pattern which is made of small dots. Gingiva has a stippled appearance due to small depressions caused by the attachment of clumps of fibres in the lamina propria to the basement membrane of the epithelium..

Stipules Paired appendages found at the base of the leaves of many flowering plants.

Stock The size of a population.

Stock recruitment models Fishery models that predict the amount of juvenile recruitment as a function of the parent stock

Stolon A rootlike extension of the body wall giving rise to buds that may develop into new zooids, thus forming a compound animal in which the zooids remain united by the stolon. Found in some colonial anthozoans, hydrozoans, ectoprocts, and ascidians.

Stoma/stomata A microscopic pore surrounded by guard cells in the epidermis of leaves and stems that allows gas exchange between the environment and the interior of the plant.

Stomach The muscular organ between the esophagus and small intestine that stores, mixes, and digests food and controls the passage of food into the small intestine.

Stomach poison An insecticide that is lethal only after it has been ingested by an insect, entering the insect body through the gut.

Stomata Openings in the epidermis of a stem or leaf of a plant which permit gas exchange with the air. In general, all plants except liverworts have stomata in their sporophyte stage.

Stomatal apparatus The stomata and guard cells that control the size of the stoma.

Stomochord Anterior evagination of the dorsal wall of the buccal cavity into the proboscis of hemichordates; the buccal diverticulum.

Stomodeal valve In insects, the cylindrical or funnel-shaped invagination of the foregut into the midgut.

Stomodeum Anterior part of the embryonic alimentary canal formed as an invagination of the ectoderm; the future mouth.

Stoneworts A group of approximately 250 species of algae that have a complex growth pattern, with nodal regions from which whorls of branches arise; they are abundant in fresh to brackish waters.

Stony corals Anthozoans, often colonial, that secrete a calcareous skeleton.

Stop codon The codon on a messenger RNA molecule where protein synthesis stops.

Strain A population of organisms that descends from a single organism or pure culture isolate.

Strangler A plant that begins life as an epiphyte in the canopy of a forest and sends its roots down the trunk of a host tree and into the substrate. The roots may form a thick woody net around the host's trunk, but it is apparently the foliage of the strangler that eventually kills the host by blocking sunlight from the latter's leaves. A type of growthform found in tropical forests.

Strata Layers of sedimentary rock, the oldest rocks occurring at the bottom.

Strategic minerals Materials a country cannot produce itself but that it uses for essential materials or processes.

Strategy A group of related traits, evolved under the influence of natural selection, that solve particular problems encountered by living organisms; often includes anatomical, physiological, and behavioral characteristics.

Stratification The separation of the water column into layers, with the densest at the bottom and the least dense at the surface. A stratified water column is said to be stable. An unstable column results when the surface water becomes more dense than the water below.

Stratigraphy The study of rock layers, especially their distribution, environment of deposition, and age; stratigraphic, adj.

Stratosphere The zone in the atmosphere extending from the tropopause to about 50 km (30 mi) above the earth's surface; temperatures are stable or rise slightly with altitude; has very little water vapor but is rich in ozone.

Stratum corneum The outer layer of the epidermis of vertebrates. The layer is keratinized for protection from desiccation and foreign agents.

Stratum granulosum A layer of follicle cells that line the theca of a maturing ovarian follicle.

Streak plate A petri dish of solid culture medium with isolated microbial colonies growing on its surface, which has been prepared by spreading a microbial mixture over the agar surface, using an inoculating loop.

Streambank erosion Washing away of soil from banks of established streams, creeks, or rivers, often as a result of the removal of trees and brush along streambanks or cattle damage to the banks.

Streptavidin A bacterial analog of egg white *avidin.*

Streptococcal pneumonia An endogenous infection of the lungs caused by Streptococcus pneumoniae that occurs in predisposed individuals.

Streptococcal sore throat One of the most common bacterial infections of humans. It is commonly referred to as "strep throat." The disease is spread by droplets of saliva or nasal secretions and is caused by Streptococcus spp. (particularly group A streptococci).

Streptolysin-O (SLO) A specific hemolysin produced by Streptococcus pyogenes that is inactivated by oxygen (hence the "O" in its name). SLO causes beta-hemolysis of blood cells on agar plates incubated anaerobically.

Streptolysin-S (SLS) A product produced by Streptococcus pyogenes that is bound to the bacterial cell but may sometimes be released. SLS causes beta hemolysis on aerobically incubated blood-agar plates and can act as a leukocidin by killing leukocytes that phagocytose the bacterial cell to which it is bound.

Streptomycin A bactericidal aminoglycoside antibiotic produced by Streptomyces griseus.

Streptophytes The clade consisting of the plants plus their closest relatives, the charophytes.

Stress shock A loose set of physical, psychological, and/or behavioral changes thought to result from the stress of excess competition and extreme closeness to other members of the same species.

Stressed community A community that is disturbed by human activity, such as road building or pollution, and is inadvertently simplified. Some species become superabundant while others disappear.

Stretch receptor Sensory receptor that responds to stretch; found in muscle tissue, lungs, and other organs that undergo changes in position or size.

Striae Grooves running across or along the body: applied especially to the grooves on beetle elytra.

Striated ducts Ducts which carry saliva from the intercalated ducts to a series of main collecting ducts. Striated duct cells are actively involved in secretion and absorption. Their striated appearance is due to the many long folds of the cell membrane.

Striated muscle In the striated muscle cells it is well established that in the appropriate ionic environment, and using energy derived from ATP, thick ('myosin') and thin ('actin') filaments can associate to generate contractile force. To date, this is the only interaction between microfilamentous structures that has been shown, for certain, to be capable of force generation

Strict aerobe An organism that can survive only in

an atmosphere of oxygen, which is used in aerobic respiration.

Strict anaerobe An organism that cannot survive in an atmosphere of oxygen. Other substances, such as sulfate or nitrate, are the terminal electron acceptors in the electron transport chains that generate their ATP.

Stridulation The production of sounds by rubbing two parts of the body together: best known in grasshoppers and other orthopterans.

Strike Deposition of fly eggs or larvae on a living host. The direction or trend of a bedding plane or fault, as it intersects the horizontal.

Stringency A term used to describe the conditions of hybridization. By varying the conditions (especially salt concentration and temperature) a given probe sequence may be allowed to hybridize only with its exact complement (high stringency), or with any somewhat related sequences (relaxed or low stringency). Increasing the temperature or decreasing the salt concentration will tend to increase the selectivity of a hybridization reaction, and thus will raise the stringency.

Strip-farming Planting different kinds of crops in alternating strips along land contours; when one crop is harvested, the other crop remains to protect the soil and prevent water from running straight down a hill.

Strip-mining Removing surface layers over coal seams using giant, earth-moving equipment; creates a huge open-pit from which coal is scooped by enormous surface-operated machines and transported by trucks; an alternative to deep mines.

Strobila A stage in the development of the scyphozoan jellyfish. Also, the chain of proglottids of a tapeworm.

Strobilocercoid Cysticercoid that undergoes some strobilation; found only in Schistotaenia.

Strobilocercus Simple cysticercus with some evident strobilation.

Strobilus A number of sporangia- or ovule-bearing stuctures (sporophylls, sporangiophores, scales, etc.) grouped together on an axis.

Stroma The fluid of the chloroplast surrounding the thylakoid membrane; involved in the synthesis of organic molecules from carbon dioxide and water.

Stromatolite Dome-like microbial mat communities consisting of filamentous photosynthetic bacteria and occluded sediments (often calcareous or siliceous). They usually have a laminar structure. Many are fossilized, but some modern forms occur.

Stromatolites Massive calcareous skeletons formed by cyanobacteria (blue-green algae).

Structural color A color that results when light is reflected by a particular surface.

Structural formula A type of molecular notation in which the constituent atoms are joined by lines representing covalent bonds.

Structural genes Genes that code for enzymes or structural proteins rather than proteins that control the function of other genes.

Structural molecule A complex molecule, such as cellulose, that provides support and protection.

Structure The physical arrangement or spatial patterns of the components of an ecosystem, especially the plant life. Includes growthforms, number of canopy layers, degree of cover, distribution patterns of species within the ecosystem (patches, for example). Structure may also refer to the organization of the ecosystem in terms of trophic levels.

Structure functionalism The scientific tradition that stresses the relationship between a physical structure and its function, for example, the related disciplines of anatomy and physiology.

Style Part of the female reproductive structure in the carpel of a flower; formed from the ovary wall. The tip of the style carries the stigma to which pollen grains attach. Part of the pistil that separates the stigma from the ovary.

Stylet A needle-like object: applied to the various components of piercing mouthparts and also to a part of the sting of a bee or other hymenopteran.

Stylops Member of the insect order Strepsiptera.

Stylostome The tube formed by the host as a result of the feeding of a chigger; in secreting salivary fluids, the chigger partially digests skin tissues, which induces the host to form a proteinaceous tube walling off the injury.

Subacute sclerosing panencephalitis Diffuse inflammation of the brain resulting from virus and prion infections.

Sub apical Situated just before the tip or apex.

Subatomic particles The three kinds of particles that make up atoms: protons, neutrons, and electrons.

Subcephalic pocket The cavity formed beneath the embryonic head as the head fold develops.

Subchelate Condition of an arthropod appendage in which the terminal podomere can fold back like a pincer against the subterminal podomere.

Subcosta Usually the first of the longitudinal veins behind the front edge of the wing, although it is often missing or very faint: abbreviated to Sc.

Subduction The downward movement of a plate into the mantle that occurs in trenches, which are also known as subduction zones.

Subduction zone n A long narrow area in which subduction is taking place, e.g. the Peru-Chile trench, where the Pacific Plate is being subducted under the South American Plate.

Suberin Waxy, waterproof chemical in some plant cells, notably cork (in stems) and endodermis cells.

Subgingival plaque The plaque that forms at the dentogingival margin and extends down into the gingival tissue.

Sub-imago Found only among the mayflies, the sub-imago or dun is the winged insect which emerges from the nymphal skin. It is rather dull in colour, but very soon moults again - the only example of a winged insect undergoing a moult - to reveal the imago.

Sublimation The process by which water can move between solid and gaseous states without ever becoming liquid.

Sub-marginal cells Cells lying just behind the stigma in the hymenopteran forewing: important in the identification of bees and sphecid wasps.

Submarine canyon A narrow, deep depression in the continental shelf formed by the erosion of rivers or glaciers before the shelf was submerged.

Submergent plant community a marine plant community restricted to subtidal environments

Submetacentric Pertaining to a chromosome with a centromere located between the center and one end of the chromosome.

Submucosa A layer beneath the lamina propria of mucosa, which is loose and elastic. It may contain large blood vessels, nerves, glands and lymphatic tissue.

Subnivean Applied to environments beneath snow, in which snow insulates against a colder atmospheric temperature.

Subsidence A settling of the ground surface caused by the collapse of porous formations that result from withdrawal of large amounts of groundwater, oil, or other underground materials.

Subsoil A layer of soil beneath the topsoil that has lower organic content and higher concentrations of fine mineral particles; often contains soluble compounds and clay particles carried down by percolating water.

Subspecies A taxonomic subdivision of a species; a population of a particular region genetically distinguishable from other such populations and capable of interbreeding with them.

Substance P A neuropeptide with a particularly powerful ability to excite a post synaptic cell. When substance P is released into a synapse of a sensory neurone it causes severe pain.

Substiedal body Additional plug material underlying a Stieda body.

Substitution A type of mutation in which one base is substituted for another.

Substrate "Supporting surface" On which an organism grows. The substrate may simply provide structural support, or may provide water and nutrients. A substrate may be inorganic, such as rock or soil, or it may be organic, such as wood.

Substrate feeders Animals such as earthworms or termites that eat the soil or wood through which they burrow.

Substrate-level phosphorylation The synthesis of ATP from ADP by phosphorylation coupled with the exergonic breakdown of a high-energy organic substrate molecule.

Subsurface biosphere The region below the plant root zone where microbial populations can grow and function.

Subtidal zone The bottom above the continental shelf.

Subtropical Refers to the portion of the temperate zone closest to the equator

Succession A predictable ordering of a dominance of a species or groups of species following the opening of an environment to biological colonization.

Succulent A plant that is able to store water in its tissues and then withdraw it during times of drought. Water storage tissue may be found in the stem, leaves, or roots depending on the species. Stem succlulents, leaf succulents, and root succulents are types of growthforms.

Sucrose Cane sugar; a common disaccharide found in many plants; a molecule of glucose linked to a molecule of fructose.

Sudden infant death syndrome (SIDS) A disorder resulting in the unexpected death during sleep of infants, usually between the ages of two weeks and one year. The causes are not fully understood, but are believed to involve failure of automatic respiratory control.

Sudoriferous gland A sweat gland.

Sugar A monosaccharide; a carbohydrate with the general formula CnH2nOn.

Sulcus A furrow or groove.

Sulfate reduction The process of sulfate use as an oxidizing agent, which results in the accumulation of reduced forms of sulfur such as sulfide, or incorporation of sulfur into organic molecules, usually as sulfhydryl groups.

Sulfide One of the minerals that is abundant in the hot water that seeps through hydrothermal vents.

Sulfonamide A chemotherapeutic agent that has the SO2-NH2 group and is a derivative of sulfanilamide.

Sulfur dioxide A colorless, corrosive gas directly damaging to both plants and animals.

Sulfur oxide Produced from burning coal, this gas is a pollutant that can produce sulfuric acid in rain droplets.

Summation A phenomenon of neural integration in which the membrane potential of the postsynaptic cell in a chemical synapse is determined by the total activity of all excitatory and inhibitory presynaptic impulses acting on it at any one time.

Summerwood The cells in the secondary xylem that are formed late in the season, usually with few vessels (angiosperms) or narrow tracheids (gymnosperms); also called late wood.

Sun compass The mechanism in which animals use the sun and an internal clock that allows them to adjust for movement in the sun as they navigate.

Superantigen Superantigens are bacterial proteins that stimulate the immune system much more extensively than do normal antigens. They stimulate T cells to proliferate nonspecifically through simultaneous interaction with class II MHC proteins on antigen-presenting cells and variable regions on the b chain of the T-cell receptor complex. Examples include streptococcal scarlet fever toxins, staphylococcal toxic shock syndrome toxin-1, and streptococcal M protein.

Supercoil Double-stranded circular DNA which is twisted about itself. Commonly observed with *plasmids* and circular viral DNA genomes (such as that of hepatitis B virus). A nick in one strand of the plasmid may remove the twist, resulting in a *relaxed*, circular DNA molecule. A complete break in the DNA puts the plasmid in a linear form. Supercoils, relaxed circular DNA, and linear DNA all have different migration properties in agarose gels, even though they contain the same number of base pairs.

Superfamily A group of closely related families; superfamily names end in -oidea.

Superior Above a point of reference (e.g., the neck is superior to the chest of humans).

Superior ovary Attachment of the ovary to the receptacle above and free from the attachment of the other floral parts.

Superior vena cava Blood from the head returns to the heart through this main vein.

Superoxide dismutase An enzyme that protects many microorganisms by catalyzing the destruction of the toxic superoxide radical.

Super-saturated solutions When a solution is saturated and still more ions are added, they cannot be held in solution but precipitate as a solid deposit. The proline-rich proteins of saliva are capable of binding calcium. They hold a store of calcium ions which allows saliva to carry more ions in solution than is theoretically possible. Saliva is thus a super-saturated solution of calcium phosphate.

Supplementary reproductives In termites the caste of males and females with short wings, light pigmentation, and small compound eyes. The females lay eggs in the colony supplementing the work of the queen.

Supply The quantity of that product being offered for sale at various prices, other things being equal.

Suppressor mutation A mutation that overcomes the effect of another mutation and produces the normal phenotype.

Suppressor T cells T cells that slow down and stop the immune response of B cells and other T cells. Immune system cells that shut off the antibody production when an infection is under control.

Suprachiasmtic nucleus (SCN) A brain nucleus involved in the visual pathway that has been clearly identified for its involvement in biological rhythms mediated by photoperiod.

Suprapopulation of parasites All individuals of a single parasite species at all stages in the life cycle in all hosts in an ecosystem.

Surf A wave that becomes so high and steep as it approaches the shoreline that it breaks.

Surface (mixed) layer The upper layer of water that is mixed by wind, waves, and currents.

Surface current long-term directional flow of water at the sea surface

Surface mining Some minerals are also mined from surface pits.

Surface of a cell. Eukaryotic flagella are longer versions of cilia. Flagellar undulations drive a cell through a fluid medium.

Surface tension A measure of how difficult it is to stretch or break the surface of a liquid. Water has a high surface tension because of the hydrogen bonding of surface molecules.

Surface-to-volume ratio (S/V ratio) The amount of surface area relative to the total volume of an organism.

Surra Disease of large mammals caused by Trypanosoma evansi.

Survivorship The percentage of a population reaching a given age or the proportion of the maximum life span of the species reached by any individual.

Survivorship curve A plot of the number of members of a cohort that are still alive at each age; one way to represent age-specific mortality.

Suspension feeder The type of feeding whereby an animal obtains its nutrients by the removal of suspended food particles from the surrounding water by some sort of capturing, trapping, or filtering structure.

Sustainable A sustainable way of life is one in which human needs are met without diminishing the ability of other people, wild species, or future generations to survive.

Sustainable agriculture An ecologically sound, economically viable, socially just, and humane agricultural system. Stewardship, soil conservation, and integrated pest management are essential for sustainability.

Sustainable development A real increase in well-being and standard of life for the average person that can be maintained over the long-term without degrading the environment or compromising the ability of future generations to meet their own needs.

Sustained yield Utilization of a renewable resource at a rate that does not impair or damage its ability to be fully renewed on a long-term basis.

Suture A groove on the body surface which usually divides one plate or sclerite from the next: also the junction between the elytra of a beetle.

Svedberg unit The unit used in expressing the sedimentation coefficient; the greater a particle's Svedberg value, the faster it travels in a centrifuge.

Swab A wad of absorbent material usually wound around one end of a small stick and used for applying medication or for removing material from an area; also, a dacron-tipped polystyrene applicator.

Swamp Wetland with trees, such as the extensive swamp forests of the southern United States.

Swarm cell A flagellated cell; the term is usually applied to the motile cells of the Myxomycota.

Swarmer Daughter trophozoites resulting from multiple fissions of Ichythophthirius multifiliis and a few other protozoa.

Swarming The aggregation of individuals for spawning or other purposes.

Swash rider Invertebrate that can migrate up and down shore with the rising and falling tide, in order to maintain station at a level that is moist but not overly washed by the waves

Sweeper tentacle A type of tentacle in corals that is used to sting neighboring colonies.

Swell A wave with a flatter, rounded crest and trough. Swells are found away from the area where waves are generated by the wind.

Swim bladder A gas-filled sac, which is usually located along the dorsal body wall of bony fishes; it is an outgrowth of the digestive tract and regulates buoyancy of a fish.

Sycon A type of canal system in certain sponges. Sometimes called syconoid.

Sylvatic Existing normally in the wild, not in the human environment.

Symbiont The smaller participant in a symbiotic relationship, living in or on the host.

Symbiosis A relationship between two organisms that live in intimate contact with each other; includes mutualism (both organisms benefit, they rely on each other for survival), parasitism (one organism benefits at its host's expense) and commensalism (one partner benefits and the other is neither benefitted nor harmed).

Symbiosome The final nitrogen-fixing form of Rhizobium that is active within root nodule cells.

Symbiotic An association of two or more different organisms in which both benefit.

Symmetrogenic fission Mitotic fission between the rows of flagella of protozoa.

Symmetry A balanced arrangement of similar parts on either side of a common point or axis.

Sympathetic division One of two divisions of the autonomic nervous system of vertebrates; generally increases energy expenditure and prepares the body for action.

Sympathetic nerves The sympathetic nervous system is one of he two main divisions of the autonomic nervous system. Sympathetic nerves make synapses at ganglia close to the spinal cord and have long post -ganglionic axons which, in general place the body on a state of alert. Always acts as a balance to activity of the parasympathetic nervous system.

Sympathetic nervous system Portion of the autonomic nervous system that arises from the thoracic and lumbar regions of the spinal cord; also called thoracolumbar division. The subdivision of the autonomic nervous system that dominates in stressful or emergency situations and prepares the body for strenuous physical activity, e.g., causing the heart to beat faster.

Sympatric Having the same or overlapping regions of geographical distribution. Noun, sympatry.

Sympatric speciation A mode of speciation occurring as a result of a radical change in the genome that produces a reproductively isolated subpopulation in the midst of its parent population.

Sympatry Populations or species with overlapping geographic distributions. Contrast with allopatry.

Symplast In plants, the continuum of cytoplasm connected by plasmodesmata between cells.

Symplastic Pertaining to the movement of water and solutes through tissues by passing through biological membranes.

Symplesiomorphies Taxonomic characters that are common to all members of a group of organisms. These characters indicate common ancestry but cannot be used to describe relationships within the group.

Symptom A change during a disease that a person subjectively experiences (e.g., pain, bodily discomfort, fatigue, or loss of appetite). Sometimes the term symptom is used more broadly to include any observed signs.

Synangium A cluster of sporangia which have become fused in development.

Synanthropism Habit of an organism of living in or around human dwellings.

Synapomorphies Characters that have arisen within a group since it diverged from a common ancestor. Synapomorphies are used to indicate degrees of relatedness within a group. Also called shared, derived characters.

Synapse The point at which one nerve cell connects with another. The nerve impulse is transmitted by the release of chemical neurotransmitter substances from the pre-synaptic cell membrane. The neurotransmitter substance diffuses across the synaptic cleft to the post-synaptic membrane which it depolarises. When the post-synaptic cell is sufficiently excited by a number of incoming impulses and enough neurotransmitter substance, it discharges an electrical impulse along its axon membrane to the next neurone.

Synapsids An amniote lineage comprising the mammals and the ancestral mammal-like reptiles, having a skull with a single pair of temporal openings.

Synapsis The alignment of chromosomes during meiosis I so that each chromosome is beside its homologue.

Synaptic cleft A narrow gap separating the synaptic knob of a transmitting neuron from a receiving neutron to an effector.

Synaptic knob The relay point at the tip of a transmitting neuron's axon, where signals are sent to another neuron or to an effector.

Synaptic processes The threadlike extensions at the end of an axon.

Synaptic terminal A bulb at the end of an axon in which neurotransmitter molecules are stored and released.

Synaptic vesicles Vesicles at the synapse end of an axon that contain the neurotransmitters.

Synaptonemal complex The structure that holds homologous chromosomes together during synapsis in prophase of meiosis I.

Syncitiotrophoblast Highly invasive, multinucleated derivative of the cytotrophoblast near the inner cell mass in the implanting mammalian embryo.

Syncitium Multinucleate cell; may arise either from karyokinesis without cytokinesis (e.g., Drosophila syncitial blastoderm) or from fusion of mononucleate cells (e.g. vertebrate skeletal muscles).

Syncline A fold of rock layers that is convex downwards. Antonym of anticline.

Syncytial hypothesis The idea that multicellular organisms could have arisen by the formation of cell boundaries within a large multinucleate protist.

Syncytium A multinucleated cell.

Syndrome A collection or group of signs and symptoms that occur together and characterize a particular disease or abnormality.

Synergid Cells in the embryo sac of angiosperms that flank the egg cell. The pollen tube grows through one (usually the smaller) of the synergids.

Synergism The interaction of two or more agents or forces so that their combined effect is greater than the sum of their individual effects, as when two hormones combine to affect target tissues.

Synergist syzyg = yoked together. Acting together.

Synergistic effects When an injury caused by exposure to two environmental factors together is greater than the sum of exposure to each factor individually.

Syngamy The process of union of two gametes; sometimes called fertilization. It encompasses both plasmogamy and karyogamy.

Synkaryon Zygote nucleus resulting from fusion of pronuclei.

Synonym One of two or more names which have been given to a single species. The earliest name usually (should) takes precedence.

Synovial fluid The lubricating fluid containing glycosaminoglycans which is held in the capsule of a synovial joint.

Synovial joint The most movable type of joint. The bones are covered by connective tissue, the interior of which is filled with synovial fluid, and the ends of the bones are covered with cartilage.

Synsacrum The fused posterior thoracic vertebrae, all lumbar and sacral vertebrae, and anterior caudal vertebrae of a bird; helps maintain proper flight posture.

Syntax Information provided by the sequence in which signals are transmitted.

Synthesis The formation of a more complex substance from simpler ones.

Synthesis phase In the cell cycle, the phase in which the DNA of the chromosomes is replicated and DNA-associated proteins, such as histones, are synthesized.

Synthetic oligonucleotide A single strand of DNA made by machine and used for priming DNA synthesis in sequencing or PCR, or as a probe. The maximum achievable length is around 50 bases, but longer molecules can be constructed using oligonucleotides as building blocks.

Syntrophism The association in which the growth of one organism either depends on, or is improved by, the provision of one or more growth factors or nutrients by a neighboring organism. Sometimes both organisms benefit. This type of mutualism is also known as cross-feeding or the satellite phenomenon.

Syphilis A sexually transmitted disease caused by a bacterial infection that produces an ulcer on the genitals and can have potentially serious effects if untreated.

Syrinx The vocal organ of birds located at the base of the trachea.

System A group of organs specialized for one function.

Systematic epidemiology The field of epidemiology that focuses on the ecological and social factors that influence the development of emerging and reemerging infectious diseases.

Systematic variation In learning, a procedure that involves examining qualitative differences in learning by investigating a wide variety of species from diverse taxonomic groups.

Systematics The area of biology that deals with the diversity of living organisms, their relationships to each other through evolution, and their classification. Can also be referred to as *taxonomy*.The classification of organisms based on information from observations and experiments; includes the reconstruction of evolutionary relatedness among living organisms. Currently, a system that divides organisms into five kingdoms (Monera, Protista, Plantae, Fungi, Animalia) is widely used.

Systemic A condition or process that affects the whole body; many metabolic poisons are systemic.

Systemic acquired resistance (SAR) A defensive response in infected plants that helps protect healthy tissue from pathogenic invasion.

Systemic circuit The portion of the circulatory system concerned with blood flow from the left ventricle of the heart to the entire body and back to the heart via the right atrium.

Systemic insecticide An insecticide capable of absorption into plant sap or animal blood and lethal to insects feeding on or within the treated host.

Systemic lupus erythematosus An autoimmune, inflammatory disease that may affect every tissue of the body.

Systole Phase of the cardiac cycle during which a heart chamber wall is contracted; also systolic pressure.

Systolic pressure The portion of blood pressure measurement that represents the highest pressure reached during ventricular ejection; it is the first number shown in a blood pressure reading.

Syzygy Stage during sexual reproduction of some gregarines in which two or more sporadins connect end to end.

T

T cell A type of lymphocyte derived from bone marrow stem cells that matures into an immunologically competent cell under the influence of the thymus. T cells are involved in a variety of cell-mediated immune reactions; also known as a T lymphocyte.

T7 RNA polymerase A bacteriophage *RNA polymerase* which is commonly used to transcribe plasmid DNA into RNA. The plasmid must contain a T7 *promoter* upstream of the relevant sequence.

Tachyzoite Small, merozoitelike stages of Toxoplasma. They develop in the host cells' parasitophorous vacuole by endodyogeny.

Tactile (touch) receptor A sensory receptor in the skin that detects light pressure; formerly called Meissner's corpuscle.

Tadpole larva The larva of tunicates.

Tagma A compound body section of an arthropod resulting from embryonic fusion of two or more segments; for example, head, thorax, abdomen.

Tagmatization The specialization of body regions of a metameric animal for specific functions. The head of an arthropod is specialized for feeding and sensory functions, the thorax is specialized for locomotion, and the abdomen is specialized for visceral functions.

Taiga biome The region of coniferous forest extending across much of northern Europe, Asia, and North America; characterized by long, cold winters and short, cool summers and by acidic, thin soils.

Tailings Mining waste left after mechanical or chemical separation of minerals from crushed ore.

Taking Unconstitutional confiscation of private property.

Tall grass prairie A grassland biome characterized by many tall grasses up to 5 meters tall.

Talus A steep, concave, downward sloping formation, formed by the accumulation of coarse, angular rock debris at the base of the cliff or slope.

Tandem repeat sequences Multiple copies of the same *base sequence* on a *chromosome*; used as a marker in *physical mapping*.

Tangential cut A longitudinal section cut at right angles

Tannin A secondary product found in many plants that have been widely utilized as stains, dyes, inks, or tanning agents for leather; believed to function in plants by discouraging herbivores.

Tantulus Larva of a tantulocaridan (subphylum Crustacea).

Tap root A primary root that grows vertically downward and gives off small lateral roots; occurs in dicots. Root system in plants characterized by one root longer than the other roots. Example: carrot.

Tapetum The nutritive somatic tissue surrounding the microsporocyte.

Tapeworms Parasitic flatworms, typically consisting of a chain of repeated units.

Taphonomy The study of what happens to a fossil, from the time of its initial creation (e.g. the death of an organism or the imprint left by the movement of an organism) and the time that the fossil is discovered by a paleontologist. For example, shells or bones can be moved by running water, and later be compressed by overlying sediment. Taphonomy is often broken into two parts, biostratinomy and the study of diagenesis.

Taq polymerase A DNA polymerase isolated from the bacterium *Thermophilis aquaticus* and which is very stable to high temperatures. It is used in PCR procedures and high temperature sequencing.

Tardigrada A phylum of animals whose members live in marine and freshwater sediments and in water films on terrestrial lichens and mosses. Possess four pairs of unsegmented legs and a proteinaceous cuticle. Water bears.

Target cell A cell that a particular hormone effects by its direct action (either passing through the membrane or binding to a surface receptor).

Targeting The movement of proteins around a cell or within organelles so that they reach the correct place to carry out their activity.

Tarsals The bones that make up the ankle joint.

Tarsus The insect's foot: primitively a single segment but consisting of several segments in most living insects.

Taste buds Structures in the mouth and other locations of fishes that are sensitive to chemical stimuli.

TATA box A sequence found in the promoter (part of the 5' flanking region) of many genes. Deletion of this site (the binding site of transcription factor TFIID) causes a marked reduction in transcription, and gives rise to heterogeneous transcription initiation sites.

Tautomer A molecule which can convert relatively easily from one isomeric form to another.

Tawny 1. Of a dark yellowish or dull yellowish-brown color.

2. a shade of brown tinged with yellow; dull yellowish brown.

Taxis Directed reactions to a stimulus involving an orientation of the long axis of the body in line with the stimulus source.

Taxon Any one of the levels in the taxonomic hierarchy

Kingdom One of the major subdivisions of life; based upon basic similarities in cell structure. Five kingdoms are recognized: Monera, Protoctista, Fungi, Animalia, and Plantae.

Phylum A subdivision of a kingdom encompassing all forms of life with the same distinctive body plan. [plural = phyla].

Class A higher taxon consisting of one or more orders and distinct from other taxa of similar rank

Order A higher taxon consisting of one or more families and distinct from other taxa of similar rank.

Family A taxonomic level including one or more genera of common phylogenetic origin and distinct from taxa of the same rank. [The Latin names of animal families end in "-idae".]

Genus A taxonomic category including one or more species with a presumed recent common ancestor.

Species Groups of actually or potentially interbreeding populations that are reproductively isolated from other such groups. The lowest taxonomic rank and presumably only real taxonomic unit in nature.

Subspecies A geographically defined aggregate of local populations which differ phenotypically from other such subdivisions of a species in other geographic areas.

Primitive (character) In cladistics, a character shared among and defining members of a large group or clade and believed to have arisen early in the group's evolution.

Taxonomy A systematic method of classifying plants and animals. Classification of organisms based on degrees of similarity purportedly representing evolutionary (phylogenetic) relatedness.

TB skin test Tuberculin hypersensitivity test for a previous or current infection with Mycobacterium tuberculosis.

T cell antigen receptor (TCR) The receptor on the T cell surface consisting of two antigen-binding peptide chains; it is associated with a large number of other glycoproteins. Binding of antigen to the TCR, usually in association with MHC, activates the T cell.

T dependent antigen An antigen that effectively stimulates B-cell response only with the aid of T-helper cells that produce interleukin-2 and B-cell growth factor.

Technological optimists Those who believe that technology and human enterprise will find cures for all our problems. Also called Promethean environmentalism.

Technology transfer The process of converting scientific findings from research laboratories into useful products by the commercial sector.

Technopolis Also called a vertical city; this model of city development proposes that cities grow vertically instead of horizontally.

Tectonic Describing the forces that cause the movements and deformation of Earth's crust on a large scale, also describes the resulting structures or features from these forces.

Tectonic estuary An estuary that results from the sinking of land due to movements of the crust.

Tectonic plates Huge blocks of the earth's crust that slide around slowly, pulling apart to open new ocean basins or crashing ponderously into each other to create new, larger landmasses.

Tectum A rooflike structure, for example, dorsal part of capitulum in ticks and mites.

Tegmen (plural tegmina) The leathery forewing of a grasshopper or similar insect, such as a cockroach

Tegula A small lobe or scale overlying the base of the forewing like a shoulder-pad.

Tegument Surficial covering of a multicellular organism, an integument.

Teichoic acids Polymers of glycerol or ribitol joined by phosphates; they are found in the cell walls of gram-positive bacteria.

Telencephalon The most anterior vesicle of the brain; the anterior-most subdivision of the prosencephalon that becomes the cerebrum and associated structures.

Teleology The philosophical view that natural events are goal directed and are preordained, as opposed to the scientific view of mechanical determinism.

Teleomorphic The sexual phase, or perfect stage, in a fungal life cycle.

Teleplanic larva Larva capable of dispersal over long distances, such as across oceans

Teliospore A thick-walled spore found in the rust and smut fungi; karyogamy occurs within the teliospore and it gives rise to the basidium.

Telmophage Blood-feeding arthropod that cuts through skin and blood vessels to cause a small hemorrhage of blood from which it feeds.

Telocentric Chromosome with centromere at the end.

Telolecithal Having the yolk concentrated at one end of an egg.

Telome theory A theory that regards the primitive vascular plant as consisiting of upright, dichotomously branched axes, some of which bear terminal sporangia and from which, through a series of geometric, architectural modifications, the reproductive and vegetative organs of other vascular land plants evolved.

Telomerase An enzyme that catalyzes the lengthening of telomeres; the enzyme includes a molecule of RNA that serves as a template for new telomere segments.

Telomere The protective structure at each end of a eukaryotic chromosome. Specifically, the tandemly repetitive DNA at the end of the chromosome's DNA molecule.

Telophase The flnal stage of mitosis in which the chromosomes migrate to opposite poles, a new nuclear envelope forms, and the chromosomes uncoil. The last phase of nuclear division in eukaryotes when the segregated chromosomes uncoil and begin to reform nuclei. This is immediately followed (in most cases) by cytokinesis.

Telson The last segment of the abdomen in many arthropods. May be flat and paddlelike, buttonlike, or long and spiny, as in the horseshoe crabs.

Temnospondyls A large lineage of amphibians that extended from the Carboniferous to the Triassic.

Temperate bacteriophage A bacterial virus that may become incorporated into the host-cell chromosome.

Temperate Characterized by a mild or moderate temperature. Region in which the climate undergoes seasonal change in temperature and moisture. Temperate regions of the earth lie primarily between 30 and 60 degrees latitude in both hemispheres.

Temperate deciduous forest A biome located throughout midlatitude regions where there is sufficient moisture to support the growth of large, broad-leaf deciduous trees.

Temperate forest biome Extends across regions of the northern hemisphere with abundant rainfall and long growing seasons. Deciduous, broad-leaved trees are the dominant plants.

Temperate phages Bacteriophages that can infect bacteria and establish a lysogenic relationship rather than immediately lysing their hosts.

Temperate virus A virus that can reproduce without killing the host.

Temperature A measure of the intensity of heat in degrees, reflecting the average kinetic energy of the molecules.

Temperature-compensated rhythm The relative insensitivity of biological rhythms to the effects of temperature; this contrasts with the fact that many chemical reactions double in rate for every 10C increase in temperature.

Temperate rain forest A biome dominated by coniferous trees, high rainfall, and high humidity; moist coniferous forest.

Template An outline form which can be used to make many identical copies without being used itself. Metal templates can be used placed over a piece of clothing material, which is then cut according to the shape of the template. Many pieces can be made from the same template, and they will all be the right shape for that part of the garment. Molecular templates can guide the formation of crystals by providing a shape which is characteristic of, for example anapatite crystal. The role of templates in crystal formation is calledepitaxy.

Template strand The strand of DNA that is transcribed to make RNA.

Temporal lobe The lobe of the cerebral cortex that is responsible for processing auditory signals.

Temporary parasite Parasite that contacts its host only to feed and then leaves. Also called an intermittent parasite or micropredator.

Temporomandibular joint The joint between the condyle of the mandible and the glenoid fossa of the temporal bone. The joint is divided into an upper and lower compartment by a fibrous disc and surrounded by a capsule.

Tenascin An adhesive molecule of connective tissue related to fibronectin andlaminin .

Tendons Bundles of connective tissue that link muscle to bone. Fibrous connective tissue that connects muscle to bone.

Tendril A modified leaf or stem in which only a slender strand of tissue constitutes the entire structure.

Teneral Newly emerged adult arthropod that is soft and weak.

Tension wood The reaction wood produced along the upper side of leaning woody trees, straightening the trunk by contracting and "pulling" the tree upright.

Tentacles Appendages which are flexible, because they have no rigid skeleton. Cnidarians and molluscs are two kinds of orgnaisms which may have tentacles.

Tentacle-tube-foot suspension feeder Suspension feeder that traps particles on distinct tentacles or tube feet (in echinoderms)

Tentaculata The class of ctenophorans with tentacles that may or may not be associated with sheaths into which tentacles can be retracted. Pleurobranchia.

Tentaculocyst One of the sense organs along the margin of medusae; a rhopalium.

Tentorium An extension of the parietal bone between the cerebral and cerebellar hemispheres; actually an ossification of the meninges.

Tepal When the sepals and petals of a flower are indistinguishable, they are referred to as tepals. Tepals are common in many groups of monocots.

Teratogens Chemicals or other factors that specifically cause abnormalities during embryonic growth and development.

Terebra Functional unit of a hymenopteran ovipositor, formed from first and second valvulae.

Tergite The primary plate or sclerite forming the dorsal surface of any body segment.

Tergum The dorsal surface of any body segment.

Terminal anchor In hydraulically burrowing organisms: any device used to anchor the leading portion of the burrower, permitting muscular contraction to drag the rest of the body into the sediment

Terminal bud scale scars The scars left by the reduced leaves that enclose the apical meristem of a twig. The distance between successive terminal bud scale scars represents one year's growth.

Termination codon One of three three-base sequences that initiate termination of the protein synthesis process.

Terminator A sequence *downstream* from the 3' end of an *open reading frame* that serves to halt *transcription* by the RNA polymerase. In bacteria these are commonly sequences that are *palindromic* and thus capable of forming *hairpins*. Sometimes termination requires the action of a protein, such as Rho factor in *E. coli.*

Terpenes A group of secondary compounds composed of two to many isoprene units in a chain or ring; sometimes categorized as hydrocarbons only, sometimes to include terpenoids.

Terpenoids A term referring to all compounds composed of isoprene units.

Terracing Shaping the land to create level shelves of earth to hold water and soil; requires extensive hand labor or expensive machinery, but it enables farmers to farm very steep hillsides.

Terrane A general term used to refer to a piece of the crust that is usually smaller than a continent but larger than an island; exotic terrane- terrane that has an unknown origin or a different origin than its surrounding rocks.

Terrarium A closed biological system in which plants and animals coexist without external inputs or discharges; H_2O, CO_2, O_2, and nutrients cycle in the closed system.

Terrestrial Living on land, as opposed to marine or aquatic.

Territoriality An intense form of intraspecific competition in which organisms define an area surrounding their home site or nesting site and defend it, primarily against other members of their own species.

Territory An area or space occupied and defended by an individual or a group; trespassers are attacked (and usually defeated); may be the site of breeding, nesting, food gathering, or any combination thereof.

Tertian malaria Malaria in which fevers recur every 48 hours. Caused by Plasmodium vivax, P. ovale, and P. falciparum.

Tertiary consumer A member of a trophic level of an ecosystem consisting of carnivores that eat mainly other carnivores.

Tertiary production The production of living material per unit area (or volume) per unit time by organisms consuming the herbivores. Usually expressed as grams carbon per meter square per year

Tertiary structure Refers to higher ordered structures conferred on proteins or nucleic acids by interactions between *amino acid residues* or *nucleotides* which are not closely positioned within the sequence (primary structure) of the molecule.

Tertiary treatment The removal from sewage of inorganic nutrients, heavy metals, viruses, etc., by chemical and biological means after microorganisms have degraded dissolved organic material during secondary sewage treatment.

Test A hard shell produced by some unicellular protists; may be made of calcium carbonate, silica, or sand grains.

Test cross Genetic crossing of an organism with known genotype (one that exhibits the recessive phenotype) with an individual expressing the dominant phenotype but of unknown heritage.

Testes The male gonad; produce spermatozoa and male sex hormones. Male gonads in mammals. Singular, testis. Paired organs that contain seminiferous tubules in which sperm are produced.

Testicular cyst Found in the testicular lobes of grasshoppers (Insecta, Orthoptera), a membrane bound structure containing a group of germ cells that are all at the same stage of spermatogenesis.

Testis Primary reproductive organ of a male; a sperm-cell producing organ.

Testosterone Male sex hormone that stimulates sperm formation, promotes the development of the male duct system in the fetus, and is responsible for secondary sex characteristics such as facial hair growth.

Tetanospasmin The neurotoxic component of the tetanus toxin, which causes the muscle spasms of tetanus. Tetanospasmin production is under the control of a plasmid gene.

Tetanus An often fatal disease caused by the anaerobic, spore-forming bacillus Clostridium tetani, and characterized by muscle spasms and convulsions.

Tethys ocean A small ocean that existed from the Triassic to the Jurassic; as Pangea was split into Gondwana and Laurasia in the Jurassic, an arm developed westward called the Tethys Seaway or Tethys Sea.

Tethys Sea A shallow sea that once separated the Eurasian and African sections of the supercontinent Pangaea. It eventually gave rise to the modern Mediterranean Sea.

Tetracotyle Strigeoid metacercaria in the family Strigeidae.

Tetracyclines A family of antibiotics with a common four-ring structure, which are isolated from the genus Streptomyces or produced semisynthetically; all are related to chlortetracycline or oxytetracycline.

Tetrad A pair of homologous chromosomes during

synapsis (prophase I of meiosis). A tetrad consists of four chromatids.

Tetrapartite associations A mutualistic association of the same plant with three different types of microorganisms.

Tetrapod A vertebrate possessing two pairs of limbs, such as amphibians, reptiles, birds, and mammals.

Tetrathyridium Only metacestode form known in the tapeworm cyclophyllidean genus Mesocestoides. A large, solid-bodied cysticercoid.

Thalamus An oval mass of gray matter within the diencephalon that serves as a sensory relay area. The brain region that serves as a switching center for sensory signals passing from the brain stem to other brain regions; part of the diencephalon. The major co-ordinating centre or sensory information in the brain.

Thaliacea A class of urochordates whose members are planktonic. Adults are tailless and barrel shaped. Oral and atrial openings are at opposite ends of the tunicate. Water currents are produced by muscular contractions of the body wall and result in a weak form of jet propulsion.

Thalidomide Drug which is a powerful human teratogen. First synthesized in Germany in 1954 as a new antihistamine, the drug was found to be a safe and effective sedative. It was widely used beginning in 1958 to relieve symptoms of morning sickness and nausea in pregnancy. By the 1960's the teratogenic effects of the drug had become apparent; the most common birth defects included severe limb abnormalities and malformed internal organs.

Thalloid Plants which have no roots, stems, or leaves are called thalloid, such as liverworts and hornworts.

Thallus A simple plant or algal body without true roots, leaves, or stems.

Theca General term for any stiff outer covering of a unicellular protist, and usually made up of interlocking plates. dinoflagellates and diatoms are examples of protists with thecae.

Theca externa A layer of fibrous connective tissue situated on the outer surface of a maturing ovarian follicle, covering the theca interna.

Theca interna A layer of vascularized connective tissue situated between the stratum granulosum and the theca externa in maturing ovarian follicles.

Thecodonts An informal term for a variety of Permian and Triassic reptiles that had teeth set in individual sockets. Small, bipedal thecodontians are the probable ancestors of dinosaurs.

Theileriosis Disease of cattle and other ruminants, caused by Theileria parva. Also called East Coast fever.

T-helper (TH) cell A cell that is needed for T-cell-dependent antigens to be effectively presented to B cells. It also promotes cell-mediated immune responses.

Thelyotoky Type of parthenogenesis in which all individuals are uniparental and essentially no males are produced.

Theory A hypothesis that has withstood extensive testing by a variety of methods, and in which a higher degree of certainty may be placed. A theory is never a fact, but instead is an attempt to explain one or more facts.

Theory of inheritance of acquired characteristics The mistaken idea that organisms develop new organs, or modify existing organs as environmental problems present themselves, and that these traits are passed on to offspring.

Therapsids Extinct Mesozoic mammal-like reptiles from which true mammals evolved.

Thermacidophiles A group of archaebacteria that are able to tolerate high temperatures and acidic pH.

Thermal death time (TDT) The shortest period of time needed to kill all the organisms in a microbial population at a specified temperature and under defined conditions.

Thermal plume A plume of hot water discharged into a stream or lake by a heat source, such as a power plant.

Thermal pollution Pollution by heated water.

Thermal stratification The layering of different temperatures of water or air caused by different densities, less dense floating on more dense layers.

Thermiogenesis The generation of heat by raising the body's metabolic rate; controlled by the hypothalamus.

Thermoacidophiles A group of bacteria that grow best at acid pHs and high temperatures; they are members of the Archaea.

Thermocline Layer of water separating upper warmer and lighter water from lower colder and heavier water in a lake or sea; a stratum of abrupt change in water temperature.

Thermoconformer To conform to the temperature of one's external environment.

Thermodynamics The study of transformations of energy. The first law of thermodynamics states that, in all processes, the total energy of a system plus its surroundings remains constant. The second law states that all natural processes tend to proceed in such a direction that the disorder or randomness of the system increases.

Thermohaline circulation Movement of seawater that is controlled by density differences that are largely explained in terms of temperature and salinity

Thermoplastics Soft plastics composed of single-chain, unlinked polymers, such as polyethylene, polypropylene, polyvinylchloride, polystyrene, and polyester, that can be remelted and reformed to make useful products.

Thermoreceptor A sensory receptor that is sensitive to changes in temperature; a heat receptor.

Thermoregulation The maintenance of internal temperature within a tolerable range.

Thermoset polymers Hard plastics composed of cross-linked molecular networks, such as acrylic, phenolic, or epoxy resins, that cannot be remelted or recycled.

Thick filament The thick filament is 130 nm long and 12 to 15 nm in diameter.

It consists mainly of a bundle of myosin molecule. Within each bundle, the myosin molecule are arranged in a bipolar fashion with their protruding club-like heads directed towards the ends and their shafts directed towards the middle. A myosin bundle, therefore, has a central, bare zone and an array of protruding heads of opposite polarity at each end.

Bipolar myosin-II filaments (12-14nm diameter, 1.6 mm long) are found in striated muscle. Myosin filaments elsewhere are often referred to as 'thick filaments', although their length may be considerably less.

The myosin club-like heads are directed towards the ends and their shafts are directed towards the middle. A myosin bundle, therefore, has a central, bare zone without projecting heads and an array of protruding heads of opposite polarity at each end.

Thick filaments will self-assemble in vitro under the right ionic conditions.

Thigmomorphogenesis A response in plants to chronic mechanical stimulation, resulting from increased ethylene production; an example is thickening stems in response to strong winds.

Thigmotropism Plants' response to contact with a solid object; e.g., tendrils' twining around a pole. Plant response to touch.

Thin filament The thin filament, is 7-9 nm in diameter

The major component of the thin filaments is F-actin, a filamentous protein composed of helical polymeric chains of the globular protein, G-actin.

The double helix of the actin filament has 13 G-actin molecules (diameter about 5.6 nm) per turn.

Third world Less-developed countries that are not capitalistic and industrialized (First World) or centrally-planned socialist economies (Second World); not intended to be derogatory.

Thoracic Pertaining to the thorax or chest.

Thoracic cavity The chest cavity in which the heart and lungs are located.

Thorax In insects, the second body region, between the head and thorax. It is the region where the legs and wings are attached.In many arthropods, one of three regions formed by the fusion of the segments (others are the head and abdomen).The insect body region behind the head which bears the legs and wings.The middle of the three major divisions of the insect body. The legs and wings (if present) are always attached to the thorax.

Thorn A modified stem terminating in a sharp point.

Thorn forest The dry end of a savanna community; hotter and drier than a typical savanna, these areas are characterized by dense, thorny small trees.

Thoroughfare channels Shortcuts within the capillary network that allow blood to bypass a capillary bed.

Threatened species While still abundant in parts of its territorial range, this species has declined significantly in total numbers and may be on the verge of extinction in certain regions or localities.

Threshold The minimum level of a signal(sound. pressure, pain) which is detectable.

Thrombin The final chain in the series of blood clotting forms fibrin from fibrinogen. Thrombin is formed from prothrombin by a prothrombin activator, itself the end of a series of reactions. This cascade of events may begin two ways. One, is the release of tissue factors from damaged vessels. The other, is the activation of factors in blood platelets which are altered by coming into contact with collagen or an artificial surface.

Thromboxanes Concerned with platelet clotting and a member of the eicosanoid family of hormones.

Thrombus Blood clot in a blood vessel or in one of the cavities of the heart.

Thrush Infection of the oral mucous membrane by the fungus Candida albicans; also known as oral candidiasis.

Thylakoid A flattened sac in the chloroplast stroma that contains photosynthetic pigments and the photosynthetic electron transport chain; light energy is trapped and used to form ATP and NAD(P)H in the thylakoid membrane.

Thymine (T) A nitrogenous base, one member of the *base pair* A- T (*adenine*- thymine). The pyrimidine 5-methyluracil that is found in nucleosides, nucleotides, and DNA.A nitrogen-containing base found in DNA molecules. Pairs with adenine.

Thymus/ Thymus gland An endocrine gland in the neck region of mammals that is active in establishing the immune system; secretes several messengers, including thymosin, that stimulate T cells.

Thyroid gland An endocrine gland that secretes iodine-containing hormones (T_3 and T_4), which stimulate metabolism and influence development and maturation in vertebrates, and cacitonin, which lowers blood calcium levels in mammals.

Thyroid-stimulating hormone A hormone produced by the anterior pituitary that stimulates the production and release of thyroid hormones.

Ti plasmid A plasmid of a tumor-inducing bacterium that integrates a segment of its DNA into the host chromosome of a plant; frequently used as a carrier for genetic engineering in plants.

Tibia Podomere of an insect or acarine leg that articulates proximally with the femur in insects and patella in acarines and distally with the tarsus in insects or with the metatarsus or tarsus in acarines.

Tidal bore A steep wave generated as high tides move up some estuaries and rivers.

Tidal current A current that is generated by tides.

Tidal energy Energy that can be harnessed as a result of the movement of tides.

Tidal range The difference in water level between successive high and low tides.

Tidal station A dam built across a narrow bay or estuary traps tide water flowing both in and out of the bay. Water flowing through the dam spins turbines attached to electric generators.

Tide The periodic, rhythmic rise and fall of the sea surface.Periodic movement of water resulting from gravitational attraction between the earth, sun, and moon

Tide table A table that gives the predicted time and height of tides for particular points along a coast.

Tied ridges Series of ridges running at right angles to each other so that runoff is blocked in all directions and water is allowed to soak into the soil.

Tiedemann's bodies Four or five pairs of pouchlike bodies attached to the ring canal of sea stars, apparently functioning in production of coelomocytes.

Tight junction A type of intercellular junction in animal cells that prevents the leakage of material between cells.

Timberline In mountains, the highest-altitude edge of forest that marks the beginning of the treeless alpine tundra.

T-independent antigen An antigen that triggers a B cell into immunoglobulin production without T-cell cooperation.

Tinea A name applied to many different kinds of superficial fungal infections of the skin, nails, and hair, the specific type (depending on characteristic appearance, etiologic agent, and site) usually designated by a modifying term.

Tinea capitis A fungal infection of scalp hair caused by species of Trichophyton or Microsporum.

Tinea corporis A fungal infection of the smooth parts of the skin caused by either Trichophyton rubrum, T. mentagrophytes, or Microsporum canis.

Tinea manuum A fungal infection of the hand caused by Trichophyton rubrum, T. mentagrophytes, or E. floccosum.

Tinea pedis A fungal infection of the foot caused by Trichophyton rubrum, T. mentagrophytes, or E. floccosum; also known as athlete's foot.

Tinea unguium A fungal infection of the nail bed caused by either Trichophyton rubrum or T. mentagrophytes.

Tinea versicolor A fungal infection caused by the yeast, Malassezia furfur, that forms brownish-red scales on the skin of the trunk, neck, face, and arms.

Tintinnids Ciliates that secrete vase-like cases, or loricas.

Tissue A group of cells with a specific function in the body of an organism. Lung tissue, vascular tissues, and muscle tissue are all kinds of tissues found in some animals. Tissues are usually composed of nearly identical cells, and are often organized into larger units called organs.

Tissue-specific expression Gene function which is restricted to a particular tissue or cell type. For example, the glycoprotein hormone alpha subunit is produced only in certain cell types of the anterior pituitary and placenta, not in lungs or skin; thus expression of the glycoprotein hormone alpha-chain gene is said to be tissue-specific. Tissue specific expression is usually the result of an enhancer which is activated only in the proper cell type.

Titer Reciprocal of the highest dilution of an antiserum that gives a positive reaction in the test being used.

Tm The melting point for a double-stranded nucleic acid. Technically, this is defined as the temperature at which 50% of the strands are in double-stranded form and 50% are single-stranded, i.e. midway in the melting curve. A primer has a specific Tm because it is assumed that it will find an opposite strand of appropriate character. The midpoint of the temperature range over which DNA is melted or denatured by heat; the temperature at which a *duplex* nucleic acid molecule is 50% *melted* into single strands, it is dependent upon the number and proportion of G-C *base pairs* as well as the ionic conditions. Often referred to as a measure of the thermal stability of a nucleic acid *probe*:target sequence hybrid.

Tolerance The ability of a plant to withstand injury by pests.

Tolerance range The range of variation in an environmental parameter that is compatible with the life of an organism.

Tonicity The state of tissue tone or tension; in body fluid physiology, the effective osmotic pressure equivalent.

Tonoplast A membrane that encloses the central vacuole in a plant cell, separating the cytosol from the cell sap.

Tonsillitis Inflammation of the tonsils, especially the palatine tonsils often due to S. pyogenes infection.

Tool-making revolution The invention of tools—probably first of stone, wood, or bone_ sometime in the early Paleolithic period. Knowledge of tool-making and tool-using technologies may have predated modern humans, as evidenced by tool utilization by other primates.

Top carnivore The consumer at the end of a food chain or web; a carnivore that ordinarily has no predator under those ecological conditions.

Top-down control Refers to food webs where control of a population is mainly explained by consumption by a species or group of species at higher levels of the food chain (e.g., population change of population of mussels controled by seastar predation).

Topical In a local area. e.g application of medication to the affected part only.

Topography The relief features of the Earth's surface, above and below sea level; the set of landforms in a region.

Topsoil The first true layer of soil; layer in which organic material is mixed with mineral particles; thickness ranges from a meter or more under virgin prairie to zero in some deserts.

Tornado A violent storm characterized by strong swirling winds and updrafts; tornadoes form when a strong cold front pushes under a warm, moist air mass over the land.

Tornaria A free-swimming larva of enteropneusts that rotates as it swims; resembles somewhat the bipinnaria larva of echinoderms.

Torpid Characterized by a dormant, inactive condition or state.

Torpor A time of decreased metabolism and lowered body temperature that occurs in daily activity cycles.

Torsion A developmental twisting of the visceral mass of a gastropod mollusc that results in an anterior opening of the mantle cavity and a twisting of nerve cords and the digestive tract.

Tortoiseshell The polished shell of hawksbill turtles.

Total fertility rate The number of children born to an average woman in a population during her entire reproductive life.

Total growth rate The net rate of population growth resulting from births, deaths, immigration, and emigration.

Totipotency The potential, throughout life, to express the full behavioral repertoire of the population (even if never actually expressed), and the ability to produce offspring like oneself, exhibiting the full behavioral repertoire of the population, without help.

Toxemia The condition caused by toxins in the blood of the host.

Toxic colonialism Shipping toxic wastes to a weaker or poorer nation.

Toxic shock-like syndrome (TSLS) A disease caused by an invasive group A streptococcus infection that is characterized by a rapid drop in blood pressure, failure of many organs, and a very high fever. It probably results from the release of one or more streptococcal pyrogenic exotoxins.

Toxicyst Structures possessed by predatory ciliate protozoa, which on stimulation expel a poison to subdue the prey.

Toxigenicity The capacity of an organism to produce a toxin.

Toxin A microbial product or component that can injure another cell or organism at low concentrations. Often the term refers to a poisonous protein, but toxins may be lipids and other substances.

Toxin neutralization The inactivation of toxins by specific antibodies, called antitoxins, that react with them.

Toxoid A bacterial exotoxin that has been modified so that it is no longer toxic but will still stimulate antitoxin formation when injected into a person or animal.

Toxoplasmosis A disease of animals and humans caused by the parasitic protozoan, Toxoplasma gondii.

Trabecula In general anatomical usage, a septum extending from an envelope through an enclosed substance, which, together with other trabeculae, forms part of the framework of various organs; here referring specifically to the cell processes connecting the perikarya of cestode and trematode tegumental cells with the distal cytoplasm. Also called internuncial process.

Trabecula-bone A description of the radiographic appearance of spongy bone. Radiographs provide an unusual opportunity to see condensations within spongy bone. These condensations form lines, or beams with are orientated so as to give the best support to loads tending to crush or fracture the bone.(the Latin word for a wooden beam was trabes, a small beam was a trabecula)

Trabecular net Network of living tissue formed by pseudopodia of amebocytes in Hexactinellida (phylum Porifera).

Trace element an element needed for normal metabolism but available only in minute amounts from the environment

Trace fossil Any indication of prehistoric organic activity, such as tracks, trails, burrows, or nests.

Trachea In insects and spiders, a series of tubes that carry air directly to cells for gas exchange; in humans, the air-conducting duct that leads from the pharynx to the lungs.Internal tubes through which air is taken for respiration. Vertebrates with lungs have a single trachea carrying air to the lungs, while insects and some other land-living arthropods have a complex network of tracheae carrying air from the spiracles to all parts of the body.

Tracheids Long, tapered cells with pitted walls that form a system of tubes in the xylem and carry water and solutes from the roots to the rest of the plant. One type of xylem cells. Tracheids are long and relatively narrow, and transport materials from the roots upward. Tracheids are dead at maturity and have lignin in their secondary walls.

Tracheophyte Any member of the clade of plants possessing vascular tissue; a vascular plant.

Trachoma A chronic infectious disease of the conjunctiva and cornea, producing pain, inflammation and sometimes blindness. It is caused by Chlamydia trachomatis serotypes A-C.

Tract A group or bundle of nerve fibers with accompanying connective tissue, located within the central nervous system.

Trade winds Persistent winds at low latitudes in both the Northern and Southern hemispheres, blowing toward the west and the equator

Tradition A behavior pattern that is passed from one generation to the next by learning.

Tragedy of the commons An inexorable process of degradation of communal resources due to selfish self-interest of "free riders" who use or destroy more than their fair share of common property.

Tragus The cartilaginous projection in front of the external auditory meatus.

Trans As used in molecular biology, an interaction that involves two sites which are located on separate molecules.

Transamination The removal of amino acid's amino group by transferring it to an a-keto acid acceptor.

Transcriptase/ Transcription An enzyme that catalyzes transcription; in viruses with RNA genomes, this enzyme is an RNA-dependent RNA polymerase that is used to make RNA copies of the RNA genomes.A process which leads to the copying of a gene's code, from a section of DNA, onto a strand of messenger RNA and which eventually leads to the synthesis of the peptide or protein which that gene codes for.

Transcription factor A protein which is involved in the transcription of genes. These usually bind to DNA as part of their function (but not necessarily). A transcription factor may be general (i.e. acting on many or all genes in all tissues), or tissue-specific (i.e. present only in a particular cell type, and activating the genes restricted to that cell type). Its activity may be constitutive, or may depend on the presence of some stimulus; for example, the glucocorticoid receptor is a transcription factor which is active only when glucocorticoids are present.

Transcription terminator A short base sequence found at the 3' end of a gene which causes the RNA polymerase to stop transcription.

Transcription/translation reaction An in vitro series of reactions, involving the synthesis (*transcription*) of an mRNA from a *plasmid* (usually with *T7* or *SP6 RNA polymerase*), followed by use of the mRNA to program *translation* in a cell-free system such as a rabbit *reticulocyte lysate*. The *polypeptide* product of translation in usually labelled with [35S]-methionine, and examined in an *SDS-PAGE* gel with or without prior *immunoprecipitation*. This series of reactions permits the synthesis of a polypeptide from DNA in vitro.

Transcriptional start site The nucleotide of a gene or cistron at which *transcription* (RNA synthesis) starts; the most common triplet at which transcription begins in *E. coli* is CAT. *Primer extension* identifies the transcriptional start site.

Transcytosis Process of transport of substance across an epithelium by uptake into and release from coated vesicles

Transducer A receptor that converts one form of energy into another.

Transduction The transfer of genes between bacteria by bacteriophages. Condition in which bacterial DNA (and the genetic characteristics it bears) is transferred from one bacterium to another by the agent of viral infection.

Transect A straight line placed on the ground along which ecological measurements are taken. If an ecologist wanted to sample the diversity of intertidal organisms in the intertidal, he/she would place a number of transects perpendicular to the shore and take samples at predetermined interval lengths.

Transfection A method by which experimental DNA may be put into a cultured mammalian cell. Such experiments are usually performed using cloned DNA containing coding sequences and control regions (promoters, etc) in order to test whether the DNA will be expressed. Since the cloned DNA may have been extensively modified (for example, protein binding sites on the promoter may have been altered or removed), this procedure is often used to test whether a particular modification affects the function of a gene.

Transfection cassette The carrier of the DNA used in transfection. Usually the virus itself.

Transfer host A host that is not necessary for the completion of a parasite's life cycle, but is used as a vehicle for reaching a final host.

Transfer RNA (tRNA) A class of *RNA* having structures with triplet *nucleotide* sequences that are *complementary* to the triplet nucleotide coding sequences of *mRNA*. The role of tRNAs in protein synthesis is to bond with *amino acids* and transfer them to the ribosomes, where proteins are assembled according to the genetic code carried by mRNA.

Transform fault A large horizontal displacement in the mid-ocean ridge.

Transformation A change in cell morphology and behavior which is generally related to carcinogenesis. Transformed cells tend to exhibit characteristics known collectively as the "transformed phenotype" (rounded cell bodies, reduced attachment dependence, increased growth rate, loss of contact inhibition, etc). There are different "degrees" of transformation, and cells may exhibit only a subset of these characteristics. Not well understood, the process of transformation is the subject of intense research. (1) The conversion of a normal animal cell to a cancerous cell.

(2) A phenomenon in which external DNA is assimilated by a cell.

Transformed Transfer and expression of a gene (e.g., for Bt toxin) into another organism.

Transforming factor A substance that can be passed from one cell to another and cause a permanent change in heredity.

Transforming growth factor TGF a superfamily of cytokines secreted by a variety of cells (monocytes, T cells, platelets, fibroblasts). The family include bone morphogenic proteins, which stimulate angiogenesis, fibroblast proliferation and inhibit T cell proliferation.

Transgenic Having artificially altered genetic material. A transgenic organism is one that has had its genotype altered by the introduction of a gene or DNA sequence into its genome by genetic manipulation; the introduced gene or DNA segment is called a transgene.

Transgenic animal or plant An animal or plant that has gained new genetic information from the insertion of foreign DNA. It may be produced by such techniques as injecting DNA into animal eggs, electroporation of mammalian cells and plant cell protoplasts, or shooting DNA into plant cells with a gene gun.

Transgenic mouse A mouse which carries experimentally introduced DNA. The procedure by which one makes a transgenic mouse involves the injection of DNA into a fertilized embryo at the pro-nuclear stage. The DNA is generally cloned, and may be experimentally altered. It will become incorporated into the genome of the embryo. That embryo is implanted into a foster mother, who gives birth to an animal carrying the new gene. Various experiments are then carried out to test the functionality of the inserted DNA.

Transgression A rise in sea level relative to the land.

Transient Short-lived or temporary

Transient transfection When DNA is transfected into cultured cells, it is able to stay in those cells for about 2-3 days, but then will be lost (unless steps are taken to ensure that it is retained). During those 2-3 days, the DNA is functional, and any functional genes it contains will be expressed. Investigators take advantage of this transient expression period to test gene function.

Transition A nucleotide substitution in which one pyrimidine is replaced by the other pyrimidine, or one purine replaced by the other purine (e.g., A is changed to G, or C is changed to T).

Transition mutations Mutations that involve the substitution of a different purine base for the purine present at the site of the mutation or the substitution of a different pyrimidine for the normal pyrimidine.

Transition reaction Biochemical process of converting 3-carbon pyruvate into 2-carbon acetyl and attaching it to coenzyme A (CoA) so it can enter Kreb's cycle. Carbon dioxide is also released and NADH is formed (from NAD and H) in this process.

Transitional mutants A type of mutation in which a single purine-pyrimidine base pair is replaced by another.

Transitional zone A zone in which populations from two or more adjacent communities meet and overlap.

Translocation (1) An aberration in chromosome structure resulting from an error in meiosis or from mutagens; attachment of a chromosomal fragment to a nonhomologous chromosome.

(2) During protein synthesis, the third stage in the elongation cycle when the RNA carrying the growing polypeptide moves from the A site to the P site on the ribosome.

(3) The transport via phloem of food in a plant.

Transmission electron microscope A microscope in which an image is formed by passing an electron beam through a specimen and focusing the scattered electrons with magnetic lenses.

Transovarian passage The passage of a microorganism such as a rickettsia from generation to generation of hosts through tick eggs. (No humans or other mammals are needed as reservoirs for continued propagation of the rickettsias.)

Transpeptidation 1. The reaction that forms the peptide cross-links during peptidoglycan synthesis.

2. The reaction that forms a peptide bond during the elongation cycle of protein synthesis.

Transpiration The loss of water molecules from the leaves of a plant; creates an osmotic gradient; producing tension that pulls water upward from the roots.

Transpirational pull Water molecules being "pulled" up the xylem are triggered by transpirational water loss at the leaf surface.

Transpiration-Cohesion Theory The theory that explains water movement in the xylem; the driving force is the pull of transpiration and the cohesion of water molecules.

Transplantation The movement of neural or hormonal tissue from one area of an animal to another or from one animal to another. Also, the movement of individuals or colonies from one location to another.

Transport host Paratenic host.

Transposable element Section of DNA which can copy itself and insert the copy into another position in the chromosome or another chromosome. Free transposons are not found; they spread via transposition into mobile DNA molecules such as plasmids. Bacterial transposonsmay carry, in addition to genes encoding enzymes needed for transposition, genes for antibiotic resistance or other functions.

Transposition The movement of DNA from one location to another location on the same molecule, or a different molecule within a cell.

Transposon A DNA segment that carries the genes required for transposition and moves about the chromosome; if it contains genes other than those required for transposition, it may be called a composite transposon. Often the name is reserved only for transposable elements that also contain genes unrelated to transposition.A transposable genetic element; certain sequence elements which are capable of moving from one site to another in a DNA molecule without any requirement for sequence relatedness at the donor and acceptor sites. Many transposons carry antibiotic resistance determinants and have *insertion sequences* at both ends, and thus have two sets of *inverted repeats*.

Transverse cut A section cut at right angles to the long axis. Also termed a cross section.

Transverse plane A plane or section that lies or passes across a body or structure.

Transverse suture A suture running across the thorax of many flies and dividing the mesonotum into a scutum and a prescutum.

Transversion A nucleotide substitution in which a purine replaces a pyrimidine, or vice versa (e.g., A is changed to T, or T is changed to G).

Transversion mutations Mutations that result from the substitution of a purine base for the normal pyrimidine or a pyrimidine for the normal purine.

Trap crop A small area of a crop used to divert pests from a larger area of the same or another crop. The pests, once diverted to the trap crop, may be treated with an insecticide.

Trauma Injury caused by accident or violence.

Traveler's diarrhea A type of diarrhea resulting from ingestion of certain viruses, bacteria, or protozoa normally absent from the traveler's environment. One of the major pathogens is enterotoxigenic Escherichia coli.

Tree Any tall plant, including many conifers and flowering plants, as well as extinct lycophytes and sphenophytes.

Trematoda The class of platyhelminthes that has members that are all parasitic; several holdfast devices present; have complicated life cycles involving both sexual and asexual reproduction.

Trench Deep and sinuous depression in the ocean floor, usually seaward of a continental margin or an arcuate group of volcanic islands

Triactinomyxon Stage in the life cycle of a myxozoan, formerly assigned to a separate class.

Triangle A triangular region near the base of the dragonfly wing, often divided into smaller cells.

Triassic period The first period of the Mesozoic Era between 225 and 185 million years ago. Pangaea began to breakup during this time. The ancestors of dinosaurs were present, as were early mammals and mammal-like reptiles.

Tribocytic organ Glandular, padlike organ behind the acetabulum of a strigeoid trematode.

Tricarboxylic acid cycle The cycle that oxidizes acetyl coenzyme A to CO2 and generates NADH and FADH2 for oxidation in the electron transport chain; the cycle also supplies carbon skeletons for biosynthesis.

Trichinosis A disease resulting from infection by Trichinella spiralis (Nematoda) larvae by eating undercooked meat; characterized by muscular pain, fever, edema, and other symptoms.

Trichocyst Organelle in ciliates and dinoflagellates which releases long filamentous proteins when the cell is disturbed. Used as a defense against would-be predators.

Trichogon Spiny male larva of a rhizocephalan cirripede that comes to lie within a special receptacle in the female.

Trichomes Extensions from the epidermis of the plant that provide shade and protection for the plant.

Trichomoniasis A sexually transmitted disease caused by the parasitic protozoan Trichomonas vaginalis.

Trichoptera The caddis flies, and order of insects with two pairs of hairy, scaled wings and long antennae and legs; the wormlike aquatic larvae build cases of sand and other particles glued with secretions.

Trickling filter A bed of rocks covered with a microbial film that aerobically degrades organic waste during secondary sewage treatment.

Triglyceride A type of lipid formed from three fatty acids bonded to a molecule of glycerol; a fat or oil.

Trihalomethanes Halogenated one-carbon compounds formed during water disinfection; many of these compounds are potential carcinogens.

Trilobitamorpha The subphylum of arthropods whose members had bodies divided into three longitudinal lobes. Head, thorax, and abdomen present. One pair of antennae and biramous appendages. Entirely extinct.

Trilobites A group of benthonic, detritus-feeding, extinct marine invertebrate animals (phylum Arthropoda), having skeletons of an organic compound called chitin.

Trimerophytophyta An extinct division of early vascular plants characterized by a monopodial axis, dichotomously branched lateral axes bearing terminal sporangia. Believed to have been the ancestor of the ferns and progymnosperms.

Trimerous Body in three main divisions, as in lophophorates and some deuterostomes.

Tripartite associations A mutualistic association of the same plant with two types of microorganisms.

Triplet A group of three nucleotides on a nucleic acid that codes for a particular amino acid.

Triplet code A group of three nucleotides in a DNA molecule that codes for a specific amino acid in the protein product of that DNA. Also called a codon.

Triploblastic Animals whose body parts are organized into layers that are derived embryologically from three tissue layers: ectoderm, mesoderm, and endoderm. Platyhelminthes and all coelomate animals are triploblastic.

Triploid A cell or nucleus that contains three sets of chromosomes; common in endosperm.

Triquetrum A carpal bone.

Trisomy A condition where a cell has an extra chromosome.

Tritonymph Third nymphal stage in most acarines.

Tritosternum Ventral, bristlelike sensory organ just behind the gnathosoma of a mesostigmatid mite.

Triungulin First instar larva of some parasitic, hypermetamorphic Neuroptera and Coleoptera and of the Strepsiptera, which is an active, campodeiform oligopod.

TRNA Small, tightly folded RNA molecules which act to bring specific amino acids into *translationally* active *ribosomes* in a fashion which is dependent upon the *mRNA* sequence. One end of the tRNA molecule recognizes the nucleotide triplet which is the *codon* of the mRNA, while the other end (when activated) is covalently linked to the relevant *amino acid*.

Trochanter The second segment of the leg, between coxa and femur: often very small and easily overlooked.

Trochophore A free-swimming ciliated marine larva characteristic of most molluscs and certain ectoprocts, brachiopods, and marine worms; an ovoid or pyriform body with preoral circlet of cilia and sometimes a secondary circlet behind the mouth.

Trombe wall An interior, heat-absorbing wall; may be water-filled glass tubes that absorb heat rays and let light into interior rooms.

Trophallaxis Exchange of food between young and adults, especially certain social insects.

Trophi Jaw-like structures in the mastax of rotifers.

Trophic hormones Hormonal or neurosecretory products from endocrine glands or neurosecretory cells that influence the production and release of other hormone products from endocrine glands.

Trophic levels The feeding level of an organism in an ecosystem; green plants and other autotrophs function at producer trophic levels; animals function at the consumer trophic levels.

Trophic structure The different feeding relationships in an ecosystem that determine the route of energy flow and the pattern of chemical cycling.

Trophoblast In a mammalian blastocyst, the outer superficial layer of cells surrounding the inner cell mass. This specialized tissue forms a trophic interface between the embryo and mother during cleavage, and subsequently forms the embryonic portion of the placenta upon implantation into the endometrial epithelium of the uterine wall. A portion of cytotrophoblast cells (near the inner cell mass) form the highly invasive syncytiotrophoblast which penetrates into the endometrial stroma.

Trophont Stage in the life cycle of gregarines.

Trophosome Organ in poganophorans bearing mutualistic bacteria, derived from midgut.

Trophozoite The active, motile feeding stage of a protozoan organism; in the malarial parasite, the stage of schizogony between the ring stage and the schizont.

Tropic Pertaining to behavior or action brought about by specific stimuli, for example, phototropic ("light-oriented") motion, gonadotropic ("stimulating the gonads") hormone.

Tropic hormone Hormone made by one gland that causes another gland to secrete a hormone.

Tropical rain forest An ecosystem characterized by very high rainfall and temperatures between 20 and 25C; very diverse life-forms; broad-leaved, nondeciduous trees; highly stratified forest.

Tropical seasonal forest Semievergreen or partly deciduous forests tending toward open woodlands and grassy savannas dotted with scattered, drought-resistant tree species; distinct wet and dry seasons, hot year-round.

Tropism The movement of plant parts toward or away from a stimulus in the plant's environment. Plant movement in response to an environmental stimulus.

Tropocollagen The precursor to the collagen molecule secreted by the cell. The removal of terminal peptides on the tropocollagen allows each molecule to join end to end with another to make a collagen fibril.

Tropomyosin Low-molecular weight protein surrounding the actin filaments of striated muscle.

Troponin Complex of globular proteins positioned at intervals along the actin filament of skeletal muscle; thought to serve as a calcium-dependent switch in muscle contraction.

Tropopause The boundary between the troposphere and the stratosphere.

Troposphere The layer of air nearest to the earth's surface; both temperature and pressure usually decrease with increasing altitude.

True language Communication that includes the use of symbols to represent abstract objects or ideas, and syntax, where those symbols convey different messages depending on their sequence.

True navigation The ability to maintain or establish reference to a goal without the use of landmarks.

True-breeding Occurs when self-fertilization gives rise to the same traits in all offspring, generation after generation. Now interpreted as equivalent to homozygous.

Truncate Ending abruptly: squared off.

Trypanosome A protozoan of the genus Trypanosoma. Trypanosomes are parasitic flagellate protozoa that often live in the blood of humans and other vertebrates and are transmitted by insect bites.

Trypanosomiasis An infection with trypanosomes that live in the blood and lymph of the infected host.

Trypomastigote Form of Trypanosomatidae with an undulating membrane and the kinetoplast located posterior to the nucleus. An example is Trypanosoma.

Tsetse fly Bloodsucking fly of the genus Glossina.

Tsunami Giant seismic sea swells that move rapidly from the center of an earthquake; they can be 10 to 20 meters high when they reach shorelines hundreds or even thousands of kilometers from their source.

Trochanter Podomere of an insect or acarine leg that articulates basally with the coxa and distally with the femur; usually fixed to the femur in insects.

Tubal ligation A contraceptive procedure in women in which the oviducts are cut, preventing the ova from reaching the uterus.

Tubal pregnancy Occurs when the morula remains in the oviduct and does not descend into the uterus.

Tube feet Extensions of the water-vascular system of echinoderms, protruding from the body and often ending in suckers. May be used for locomotion and/or for maintaining a tight grip on prey or on the bottom.

Tube nucleus One division of the microscope nucleus in a pollen grain that is responsible for the formation of a pollen tube from the stigma through the style to the ovule.

Tuber An enlarged, fleshy, underground stem tip, such as the potato.An underground stem which has been modified for storage of nutrients, such as a potato.

Tubercle A small knob like or rounded protuberance. A small, rounded, discrete hump or bump in skin.Any small rounded protrusion. In pycnogonids and some cheliceramorph arthropods, the central eyes are carried on a tubercle.

Tuberculoid (neural) leprosy A mild, nonprogressive form of leprosy that is associated with delayed-type hypersensitivity to antigens on the surface of Mycobacterium leprae. It is characterized by early nerve damage and regions of the skin that have lost sensation and are surrounded by a border of nodules.

Tuberculosis An infectious disease of humans and other animals resulting from an infection by a species of Mycobacterium and characterized by the formation of tubercles and tissue necrosis, primarily as a result of host hypersensitivity and inflammation. Infection is usually by inhalation, and the disease commonly affects the lungs (pulmonary tuberculosis), although it may occur in any part of the body.

Tuberculous cavity An air-filled cavity that results from a tubercle lesion caused by M. tuberculosis.

Tuberculum Process on some bones.

Tuberosity A normal raised area on a bone, such as the deltoid tuberosity, which is the site of insertion of the deltoid muscle.

Tuberous roots Modified fibrous roots that have become fleshy and enlarged with food reserves.

Tube-within-a-tube system A type of body plan in animals. The organism has two openings&emdash;one for food and one for the elimination of waste&emdash;and a specialized digestive system.

Tubular eyes Specialized eyes of many midwater animals that allow upward or downward vision.

Tubular nerve cord A hollow nerve cord that runs middorsally along the back of chordates; one of four unique chordate characteristics; also called the neural tube and, in vertebrates, the spinal cord.

Tubular secretion The process in which ions and other waste products are transported into the distal tubules of the nephron.

Tubulin Abundant cytoplasmic protein (55 kD), found mainly in two forms, a and b. A tubulin heterodimer (one a, one b), constitutes the protomer for microtubule assembly. Multiple copies of tubulin genes are present (and are expressed) in most eukaryotic cells studied so far. The different tubulin isoforms seem, however, to be functionally equivalent. G-tubulin is localized in the centrosome and is involved in nucleation of microtubule assembly during the cell cycle. Highly conserved from yeast to mammals.

Tuff A general term for consolidated rocks made of material ejected from volcanic explosions.

Tularemia A disease of rodents, lagomorphs, certain birds and sometimes humans, due to infection caused by the microorganism *Pasteurella tularensis* and transmitted by fleas and ticks; characterized by fever, headache, muscle pain, and nodule formations in the liver, spleen, and lymph nodes.

Tumble Random turning or tumbling movements made by bacteria when they stop moving in a straight line.

Tumor A spherical mass of cells in which cell divisions occur at random and often in an uncontrolled fashion.

Tumor necrosis factor. A cytokine, the most important source of which is macrophages, that is a major mediator of inflammation.

Tumor suppressor A gene that inhibits progression towards neoplastic transformation. The best-known examples of tumor suppressors are the proteins p53 and Rb. Genes that normally keep cell division under control, preventing the cell from responding to internal and external commands to divide.

Tundra/ Tundra biome A vast, nearly level, barren, treeless region located in the Arctic; characterized by very low winter temperatures, short, cool summers, and vegetation consisting of various grasses, rushes, perennial herbs, lichens, and dwarf woody plants. Extensive treeless plain across northern Europe, Asia, and North American between the taiga to the south and the permanent ice to the north. Much of the soil remains frozen in permafrost, and grasses and other vegetation support herds of large grazing mammals.

Tunic In tunicates, a cuticular, cellulose-containing covering of the body secreted by the underlying body wall.

Tunicates Chordates that show the three basic chordate characteristics only in the larva.

Turbellaria The class of Platyhelminthes that has members that are mostly free living and aquatic; external surface usually ciliated; predaceous; possess rhabdites; protrusable proboscis; mostly hermaphroditic. Examples: Convoluta, Notoplana, Dugesia.

Turbidite The sediments or rocks that formed as a result of a turbidity flow.

Turbidity current A bottom fast-flowing current that moves down a slope, depositing suspended sediments over the floor of a body of water

Turbidostat A continuous culture system equipped with a photocell that adjusts the flow of medium through the culture vessel so as to maintain a constant cell density or turbidity.

Turbinate Folded parts of the ethmoid bone.

Turbulence random, nonlaminar flow of a fluid

Turf The intertwined fibrous roots of grasses forming a mass with the soil just below ground level.

Turgid Firm; walled cells become turgid as a result of the entry of water from a hypotonic environment.

Turgor pressure Force exerted outward on a cell wall by the water contained in the cell. This force gives the plant rigidity, and may help to keep it erect.

Turner syndrome In humans, a genetically determined condition in which an individual has only one sex chromosome (an X). Affected individuals are always female and are typically short and infertile.

Turning over The mixing of thermal zones in a standing body of freshwater due to cold, more dense water "sinking" from the top through the other layers.

Turnover The replacement of cells by mitosis which keeps pace with cell loss, as in epithelia and blood cells. Also refers to the continual replacement of connective tissues like bone and fibrous tissue.

Turpentine A solvent that includes two terpenes—camphor and pinene.

Tusk shells Molluscs that have an elongate, tapered shell that is open at both ends.

Tussock A hummock of grasses or sedges bound together by their roots

Two-component phosphorelay system A signal transduction regulatory system that uses the transfer of phosphoryl groups to control gene transcription and protein activity. It has two major components: a sensor kinase and a response regulator.

Tymbal The sound-producing 'drum-skin' of a cicada.

Tympanal organs Auditory receptors present on the abdomen or legs of some insects.

Tympanic bulla bony case in the middle ear that encloses the sound-processing structures of mammals

Tympanum The membrane covering the external ear opening of the middle ear chamber or vestibule.

Type The type specimen of a species is the actual insect from which the original description of that species was produced. If several specimens were used for this purpose, one of them should have been designated as the type. Because the type can be of only one sex, it is usual to designate a certain individual of the opposite sex as the allotype. The original type specimen is then called the holotype. These 'type specimens' are very important in taxonomy and classification.

Type I hypersensitivity A form of immediate hypersensitivity arising from the binding of antigen to IgE attached to mast cells, which then release anaphylaxis mediators such as histamine. Examples: hay fever, asthma, and food allergies.

Type II hypersensitivity A form of immediate hypersensitivity involving the binding of antibodies to antigens on cell surfaces followed by destruction of the target cells (e.g., through complement attack, phagocytosis, or agglutination).

Type III hypersensitivity A form of immediate hypersensitivity resulting from the exposure to excessive amounts of antigens in which antibodies bind to the antigens and produce antibody-antigen complexes. These activate complement and trigger an acute inflammatory response with subsequent tissue damage. Examples: poststreptococcal glomerulonephritis, serum sickness, and farmer's lung disease.

Type IV hypersensitivity A delayed hypersensitivity response (it appears 24 to 48 hours after antigen exposure). It results from the binding of antigen to activated T lymphocytes, which then release cytokines and trigger inflammation and macrophage attacks that damage tissue. Type IV hypersensitivity is seen in contact dermititis from poison ivy, leprosy, and tertiary syphilis.

Type specimen. A specimen deposited in a museum that formally defines the name of the species that it represents.

Typhlosole A longitudinal fold projecting into the intestine in certain invertebrates such as the earthworm.

Typhoid fever A bacterial infection transmitted by contaminated food, water, milk, or shellfish. The causative organism is Salmonella typhi, which is present in human feces.

Typology A classification of organisms in which members of a taxon are perceived to share intrinsic, essential properties, and variation among organisms is regarded as uninteresting and unimportant.

Tyrosine kinase An enzyme that catalyzes the transfer of phosphate groups from ATP to the amino acid tyrosine in a substrate protein.

Tyrosine kinase receptor A receptor protein in the plasma membrane that responds to the binding of a signal molecule by catalyzing the transfer of phosphate groups from ATP to tyrosines on the cytoplasmic side of the receptor. The phosphorylated tyrosines activate other signal-transduction proteins within the cell.

U

Ulcer Area of inflammation that opens out to the skin or a mucous surface.

Ultimate causation The hypothetical evolutionary explanation for the existence of a certain pattern of animal behavior.

Ultimate cause The evolutionary factors responsible for the origin, state of being, or purpose of a biological system.

Ultimate phalange The last digit.

Ultimobranchial The last embryonic gill slit [often a pouch].

Ultradian rhythm A cyclical rhythm of less than 24 hours.

Ultramicrobacteria Bacteria that can exist normally in a miniaturized form or which are capable of miniaturization under low-nutrient conditions. They may be 0.2 mm or smaller in diameter.

Ultraplankton phytoplankton with cells between 2 and 5 micrometers in width

Ultrastructure The detailed structure of a specimen, such as a cell, tissue, or organ, that can be observed only by electron microscopy. Also called fine structure.

Ultraviolet (UV) radiation/ Ultraviolet Radiation of fairly short wavelength, about 10 to 400 nm, and high energy.That portion of the sun's total range of radiation having wavelengths immediately shorter than the shortest of the visible spectrum (purple); between approximately 380 and 100 nm.

Umbel An inflorescence in which the individual pedicels all arise from the apex of the peduncle.

Umbilical cord The structure that connects the placenta and the embryo; contains the umbilical arteries and the umbilical vein.

Umbo One of the prominences on either side of the hinge region in a bivalve mollusc shell. Also, the "beak" of a brachiopod shell.

Umwelt The sensory and perceptual world of the animal, dependent upon the types of sensory receptors that it possesses and the internal nervous system processes for receiving and interpreting those stimuli.

Unconditioned response (UCR) In classical conditioning, the animal's response that is initially given to the unconditioned stimulus.

Unconditioned stimulus (US) In classical conditioning, the stimulus that produces what is initially the unconditioned response.

Unconformity Any interruption of the continuity of a depositional sequence.

Unconventional air pollutants Toxic or hazardous substances, such as asbestos, benzene, beryllium, mercury, polychlorinated biphenyls, and vinyl chloride; also called noncriteria pollutants.

Unconventional oil Resources such as shale oil and tar sands that can be liquefied and used like oil.

Undernourished Those who receive less than 90 percent of the minimum dietary intake over a long-term time period; they lack energy for an active, productive life and are more susceptible to infectious diseases.

Understory That vegetation which characterizes the lower level of plants in a forest; the vegetation below the canopy.

Undifferentiated Unable to distinguish between. Undifferentiated rocks: rocks for which it is not possible to specify finer age divisions.

Undiscovered resources Potential supplies of a mineral or other useful material believed to exist based on history, scientific theory, or general knowledge of geology, biology, or geography of an unexplored area.

Undulating membrane Name applied to two quite different structures in protozoa. In some Mastigophora it is a finlike ridge across the surface of a cell, with the axoneme of a flagellum near its surface. In some ciliates it is a line of cilia that are fused at their bases, usually beating to force food particles toward the gullet.

Undulating ridges Undulatory waves in the surface of some protozoa, probably aided by subpellicular microtubules; the means of locomotion in some species.

Undulipodium Another term for a eukaryotic flagellum.

Ungulate A term that refers to hoofed animals. This includes Family Cervidae (deer) and Bovidae (sheep and goats) as well as the pronghorn. They typically have antlers or horns and are herbivores.

Unicellular (cytoplasmic) organization The life-form in which all functions are carried out within the confines of a single plasma membrane; members of the kingdom Protista display unicellular organization; also called cytoplasmic organization.

Uniform spatial distribution Situation in which individuals are more evenly spread in space than would be expected on the basis of chance alone

Uniformitarianism The idea that geological processes have remained uniform over time and that slight changes over long periods can have large-scale consequences; proposed by James Hutton in 1795 and refined by Charles Lyell during the 1800s. The principle on which modern geology was founded: processes operating today on the earth operated in much the same way in the geologic past. Sometimes expressed as "the present is the key to the past".

Uninucleate Term applied to cells having only a single nucleus.

Uniramia The subphylum of arthropods whose members are characterized by a head with one pair of antennae and one pair of mandibles. All appendages are uniramous. Among arthropods, uniramous refers to appendages that have only one branch. Insects, centipedes and millipedes, and their relatives are uniramous arthropods; land-living chelicerates such as scorpions, spiders,and mites are also uniramous but probably descended from ancestors with biramous appendages. Contrast with biramous.

Uniramous appendage Arthropod appendage that is unbranched, characteristic of living arthropods other than Crustacea, although some crustacean appendages are uniramous.

Unisexual The condition of having either stamens or pistil, not both; an imperfect flower.

Unit membrane The transmission electron microscopic interpretation of a biological membrane, consisting of two electron-dense lines separated by a translucent space.

Universal veil A membrane that totally encloses some young mushrooms; after it breaks, its remnants appear as a volva at the base and scales on the cap.

Universalists Those who believe that some fundamental ethical principles are universal and unchanging. In this vision, these principles are valid regardless of the context or situation.

Univoltine Having only one brood or generation per season.

Unsaturated fat A triglyceride with double coavent bonds between some carbon atoms.

Unsaturated fatty acid A fatty acid possessing one or more double bonds between the carbons in the hydrocarbon tail. Such bonding reduces the number of hydrogen atoms attached to the carbon skeleton.

Upland An area that is higher relative to the surrounding areas, but not mountainous; highland– syn.; *lowland*– ant.

Uplift The process or result of raising a portion of the Earth's crust through different tectonic mechanisms.

Upstream/downstream In an RNA, anything towards the 5' end of a reference point is "upstream" of that point. This orientation reflects the direction of both the synthesis of mRNA, and its translation - from the 5' end to the 3' end. In DNA, the situation is a bit more complicated. In the vicinity of a gene (or in a cDNA), the DNA has two strands, but one strand is virtually a duplicate of the RNA, so it's 5' and 3' ends determine upstream and downstream, respectively. Note that in genomic DNA, two adjacent genes may be on different strands and thus oriented in opposite directions. Upstream or downstream is only used on conjunction with a given gene.

Upstream activator sequence A binding site for transcription factors, generally part of a promoter region. A UAS may be found upstream of the TATA sequence (if there is one), and its function is (like an enhancer) to increase transcription. Unlike an enhancer, it can not be positioned just anywhere or in any orientation.

Upwelling The raising of benthic nutrients to the surface waters. This occurs in regions where the flow of water brings currents of differing temperatures together, and increases productivity of the ecosystem.

Uracil A nitrogenous base normally found in RNA but not DNA; uracil is capable of forming a *base pair* with *adenine*.

Urban area An area in which a majority of the people are not directly dependent on natural resource-based occupations.

Urban renewal Programs to revitalize old, blighted sections of inner cities.

Urbanization An increasing concentration of the population in cities and a transformation of land use to an urban pattern of organization.

Urea A nitrogen-containing waste product excreted in the urine of many vertebrates

Uredium The structure that produces uredospores in rust fungi; sometimes called a uredinium.

Uredospore A reddish, binucleate spore formed by rust fungi; often forms the repeating stage of the rust; also called a urediniospore.

Ureotelic excretion Having urea as the chief excretory product of nitrogen metabolism; occurs in mammals.

Ureter A muscular tube that transports urine by peristaltic contractions from the kidney to the bladder.

Urethra A narrow tube that transports urine from the bladder to the outside of the body. In males, it also conducts sperm and semen to the outside.

Uric acid the main nitrogenous excretory product in birds, reptiles, some invertebrates, and insects

Uricotelic excretion Having uric acid as the chief excretory product of nitrogen metabolism; occurs in reptiles and birds.

Urinary bladder The storage organ for urine.

Urine Fluid containing various wastes that is produced in the kidney and excreted from the bladder.

Urn Region near the center of an infusoriform larva of a dicyemid mesozoan.

Urochordata The subphylum of chordates whose members have all four chordate characteristics as larvae. Adults are sessile or planktonic and enclosed in tunic that usually contains cellulose. Sea squirts or tunicates.

Urogenital opening The common opening for urine and gametes in bony fishes and other animals.

Urophysis A discrete structure in the spinal cord of the fish tail that produces neuropeptides that help control water and ion balance, blood pressure, and smooth muscle contractions.

Uropod Posteriormost appendage of many crustaceans.

Uropolar cells Somatoderm cells at the posterior end of the trunk of a dicyemid mesozoan.

Urosome Portion of the body posterior to the major point of body flexion in many copepods; usually includes one or more free thoracic segments and abdomen.

Urostyle A rod-like bone composed of fused tail vertebrae, present in frogs and toads.

Urstigmata Sense organs between the coxae of the first and second pairs of legs on some mites. Apparently they are humidity receptors. Also called Claparedé organs.

Uterine bell Structure in female acanthocephalans that allows fully developed, shelled embryos to pass out of the body and that retains undeveloped ones.

Uterine tube The tube that leads from the ovary to the uterus; also called fallopian tube.

Uterus The hollow muscular organ in female mammals in which the fertilized ovum usually becomes embedded, and in which the developing embryo and fetus are nourished.

Utilitarian conservation A philosophy that resources should be used for the greatest good for the greatest number for the longest time.

Utricle That part of the inner ear containing the receptors for dynamic body balance; the semicircular canals lead from and to the utricle.

Ultimobranchial gland In jawed fishes, primitive tetrapods, and birds, the small gland(s) that forms ventral to the esophagus; produces the hormone calcitonin that helps regulate calcium concentrations.

V

Vaccination The process of protecting against infectious disease by introducing into the body a vaccine that stimulates a primary immune response and the production of memory cells against the disease-causing agent.

Vaccine A planned exposure to an antigen in order that memory B lymphocytes can retain a memory for it. In practice the organism carrying the antigen is either killed or modified so that it does not cause the disease. When encountered again, the antigen is recognised and there will be a rapid production of antibodies. For example smallpox, polio, measles. Influenza vaccines are less effective as new strains of the virus are always occurring which do not have recognisable antigens.

Vacuole Membrane-bound fluid-filled space within a cell. In most plant cells, there is a single large vacuole filling most of the cell's volume. Some bacterial cells contain gas vacuoles.

Vagabond's disease Darkened, thickened skin caused by years of infestation with body lice, Pediculus humanus humanus.

Vagina Part of the female reproductive system between the uterus and the outside opening; the birth canal in mammals; also accommodates the male's penis and receives sperm during copulation.

Valence Degree of combining power of an element as expressed by the number of atoms of hydrogen (or its equivalent) that the element can hold (if negative) or displace in a reaction (if positive). The oxidation state of an element in a compound. The number of electrons gained, shared, or lost by an atom when forming a bond with one or more other atoms.

Valence shell The outermost energy shell of an atom, containing the valence electrons involved in the chemical reactions of that atom.

Valve One of the two shells of a typical bivalve mollusc or brachiopod.

Valvifers Basal portions of the ovipositor in Hymenoptera, derived from coxae of segmental appendages.

Valvulae Processes from the valvifers to form the body of the ovipositor (terebra) and the ovipositor sheath (third valvulae) in Hymenoptera.

Variability Generally refers to genetic differences between all organisms that results in uniqueness.

Variable region The region at the N-terminal end of immunoglobulin heavy and light chains whose amino acid sequence varies between antibodies of different specificity. Variable regions form the antigen binding site.

Variance A statistical measure of the amount of variation in a series of measurements.

Variant antigen type (VAT) Applied to certain trypanosomes, any one of numerous antigenic types expressed on the surface of the organisms and "seen" by the immune system of the host.

Variant-specific surface glycoprotein (VSG) Glycoprotein on the surface of certain trypanosomes recognized by the host's immune system. Each VSG is responsible for one VAT.

Variation Differences among individuals of a group or species that cannot be ascribed to age, sex, or position in the life cycle.

Vas deferens vasa deferentia The tube in the male reproductive system in which sperm travel from the epididymis to the urethra.

Vascular Containing or concerning vessels that conduct fluid.

Vascular bundle Groups of xylem, phloem and cambium cells in stems of plants descended from the procambium embryonic tissue layer.

Vascular cambium A layer of lateral meristematic tissue between the xylem and phloem in the stems of woody plants. Lateral meristem tissue in plants that produces secondary growth.

Vascular cylinder A central column formed by the vascular tissue of a plant root; surrounded by parenchymal ground tissue.

Vascular parenchyma Specialized parenchyma cells in the phloem of plants.

Vascular plant A general name for any plant that has xylem and phloem.

Vascular ray Sheet of parenchyma that extends radially through the wood, across the cambium and into the secondary phloem; rays are produced by the vascular cambium and function in lateral transport.

Vascular Refers to a network of tubes which distribute nutrients and remove wates from the tissues of the body. Large multicellular animals must rely on a vascular system to keep their cells nourished and alive.

Vascular system Specialized tissues for transporting fluids and nutrients in plants; also plays a role in supporting the plant; one of the four main tissue systems in plants.

Vascular tissue Plant tissue consisting of cells joined into tubes that transport water and nutrients throughout the plant body.

Vasectomy A contraceptive procedure in men in which the vas deferens is cut and the cut ends are sealed to prevent the transportation of sperm. Surgical separation of the vas deferens so that sperm, while still produced, do not leave the body.

Vasoconstriction A reduction in the diameter of small arteries (arterioles) which is caused by constriction of the smooth muscle fibres in the wall of the arteriole. Vasoconstriction is an important method of increasing the blood pressure. In local areas of damage it prevents blood loss by haemostasis. Local vasoconstriction can be caused by nerve impulses to the smooth muscle from the sympathetic nervous system, by locally released prostaglandins, serotonin and Epinephrine.

Vasodilation An increase reduction in the diameter of small arteries (arterioles) which is caused by relaxation of the smooth muscle fibres in the wall of the arteriole. While vasoconstriction prevents blood loss in damaged tissues, vasodilation follows in order to allow the blood flow to slow down and clotting factors and leucotrines to seep into the damaged tissue. Local vasodilation can be caused by nerve impulses to the smooth muscle from the parasympathetic nervous system, and by locally released bradykinins.

Vector 1. In genetic engineering, another name for a cloning vector. A DNA molecule that can replicate (a replicon) and is used to transport a piece of inserted foreign DNA, such as a gene, into a recipient cell. It may be a plasmid, phage, cosmid or artificial chromosome.

2. In epidemiology, it is a living organism, usually an arthropod or other animal, that transfers an infective agent from one host to another.

Vector-borne transmission The transmission of an infectious pathogen between hosts by means of a vector.

Vegan A pure vegetarian consuming no animal products at all.

Vegetal hemisphere The half of an oocyte with more yolk or the corresponding half of an embryo with the slower dividing yolky cells.

Vegetal pole The lower pole of an egg; usually more dense than the animal pole because it contains more yolk.

Vegetarian A person who does not consume animal flesh; some consume dairy products and eggs, while others are vegans.

Vegetative growth Growth of a plant by division of cells, without sexual reproduction.

Vegetative reproduction Cloning of plants by asexual means.

Vehicle An inanimate substance or medium involved in the transmission of a pathogen.

Veins Blood vessels conducting blood toward the heart in any animal. Also more heavily sclerotized portions of wings of insects, which are remains of lacunae.

Velarium Shelf-like extension of the subumbrella edge in cubozoans (phylum Cnidaria).

Veliger larva The second free-swimming larval stage of many molluscs; develops from the trochophore and forms rudiments of the shell, visceral mass, and head-foot before settling to the substrate and undergoing metamorphosis.

Veliger Larval form of certain molluscs; develops from the trochophore and has the beginning of a foot, mantle, shell, and so on.

Velum A membrane on the subumbrella surface of jellyfish of class Hydrozoa. Also, a ciliated swimming organ of the veliger larva.

Vena cava A large vein that brings blood from the tissues to the right atrium of the four-chambered mammalian heart. The superior vena cava collects blood from the forelimbs, head, and anterior or upper trunk; the inferior vena cava collects blood from the posterior body region.

Venation The arrangement and pattern of veins in a leaf.

Veneer A thin sheet of wood, often with attractive grain, used to cover less expensive wood.

Venereal syphilis A contagious, sexually transmitted disease caused by the spirochete Treponema pallidum.

Ventilation The mechanics of breathing in and out through the use of the diaphragm and muscles in the wall of the thoracic cavity.

Ventral Pertaining to the undersurface of an animal that holds its body in a horizontal position; to the front surface of an animal that holds its body erect.

Ventricle The chamber of the heart that pumps the blood into the blood vessels that carry it away from the heart. The lower chamber of the heart through which blood leaves the heart.

Ventriculi ventricul = belly.

Venules The smallest veins. Blood flows into them from the capillary beds. Small veins that connect a vein with capillaries.

Vermicle Infective stage of Babesia in a tick.

Vermiculite Expanded mica used as a sterile medium for the rooting of cuttings; high water-holding capacity and relatively inert.

Vermiform Adjective to describe any wormlike organism; an adult (nematogen) rhombozoan (phylum Mesozoa).

Vernalization Artificial exposure of seeds or seedlings to cold to enable the plant to flower.

Verrucae vulgaris The common wart; a raised, epidermal lesion with horny surface caused by an infection with a human papillomavirus.

Verruga peruana Clinical form of Carrion's disease, caused by the bacterium Bartonella bacilliformis and transmitted by sand flies.

Vertebrae The segments of the spinal column;

separated by disks made of connective tissue (sing.: vertebra).

Vertebral column The backbone; in nearly all vertebrates, it forms the supporting axis of the body and protects the spinal cord.

Vertebrata The subphylum of chordates whose members are characterized by cartilaginous or bony vertebrae surrounding a nerve cord. The skeleton is modified anteriorly into a skull for protection of the brain.

Vertebrate A chordate animal with a backbone: the mammals, birds, reptiles, amphibians, and various classes of fishes.

Vertical migration daily or seasonal movement of small marine animals between the photic zone and midwater depths

Vertical stratification The vertical distribution of specific subcommunities within a community.

Vertically homogeneous estuary An estuary in which, at any given location, wind or tidal mixing homogenizes salinity throughout the water column

Vesicles Small membrane-bound spaces in most plant and animal cells that transport macromolecules into and out of the cell and carry materials between organelles in the cell.

Vesicular disease Any disease of the urinary bladder, such as vesicular schistosomiasis.

Vessel A tubelike column of vessel elements that are connected by open end walls and are specialized for the conduction of water and minerals.

Vessel elements Short, wide cells arranged end to end, forming a system of tubes in the xylem that moves water and solutes from the roots to the rest of the plant. Large diameter cells of the xylem that are extremely specialized and efficient at conduction. An evolutionary advance over tracheids. Most angiosperms have vessels.

Vessel member One of the cells comprising a vessel, characterized by the possession of a lumen and perforation plate shared with other vessel members above and below in the series. Also termed vessel elements.

Vestibulum Cavity leading into another cavity or passage, such as in the ciliate order Vestibulifera.

Vestige A rudimentary organ that may have been well developed in some ancestor or in the embryo.

Vestigial/ Vestigial organ Any structure or organ that is reduced in size or function when compared to a previous embryonic stage or similar structure in related organisms or in the fossil record.

Vestimentiferans Pogonophorans that are common at hydrothermal vents.

Viability The period of time an organism remains alive; often used to describe the length of time before a seed will fail to germinate.

Vibrio A rod-shaped bacterial cell that is curved to form a comma or an incomplete spiral.

Vibrissae 1. Any of the erectile, tactile hairs found on the face of most mammals except humans.

2. any of the feathers near the mouth of many birds that may help in keeping insects caught as food from escaping.

Vicariance Geographical separation of populations, especially as imposed by discontinuities in the physical environment that fragmented populations that were formerly geographically continuous.

Village A collection of rural households linked by culture, custom, and association with the land.

Villi Finger-like projections of the lining of the small intestine that increase the surface area available for absorption. Also, projections of the chorion that extend into cavities filled with maternal blood and allow the exchange of nutrients between the maternal and embryonic circulations. Projections of the inner layer of the small intestine that increase the surface area for absorbtion of food.

Viral hemagglutination The clumping or agglutination of red blood cells caused by some viruses.

Viral neutralization An antibody-mediated process in which IgA, IgM, and IgA antibodies bind to some viruses during their extracellular phase and inactivate or neutralize them.

Viremia The presence of viruses in the blood stream.

Viricide An agent that inactivates viruses so that they cannot reproduce within host cells.

Virion A complete virus particle that represents the extracellular phase of the virus life cycle; at the simplest, it consists of a protein capsid surrounding a single nucleic acid molecule.

Virioplankton Viruses that occur in waters; high levels are found in marine and freshwater environments.

Viroids Infective forms of nucleic acid without a protective coat of protein; unencapsulated single-stranded RNA molecules. Naked RNA, possibly of degenerated virus, that infects plants.

Virology The branch of microbiology that is concerned with viruses and viral diseases.

Virulence The degree or intensity of pathogenicity of an organism as indicated by case fatality rates and/or ability to invade host tissues and cause disease.

Virulence factor A bacterial product, usually a protein or carbohydrate, that contributes to virulence or pathogenicity.

Virulent bacteriophages Bacteriophages that lyse their host cells during the reproductive cycle.

Virus A noncellular biological entity that can reproduce only within a host cell. Viruses consist of *nucleic acid* covered by *protein*; some animal viruses are also surrounded by membrane. Inside the infected cell, the virus uses the synthetic capability of the host to produce progeny virus.

Viscera Internal organs in the body cavity.

Visceral mass The region of a mollusc's body that contains visceral organs.

Visceral muscle Smooth muscle found in the walls of the digestive tract, bladder, arteries, and other internal organs.

Visco-elastic A property of a material which combines elasticity and viscosity. The suspension of a car and the periodontal ligament of the tooth are examples of visco-elastic support. Elasticity refers to the return of a material to its original shape after being stretched or compressed. Viscosity is the resistance of a fluid to flowing fast.

Viscosity the resistance of water molecules to external forces that would separate them

Viscous A liquid which has a high viscosity, or resistance to flow.

Visible light A portion of the electromagnetic spectrum that includes the wavelengths used for photosynthesis.

Visible light That portion of the electromagnetic spectrum detected as various colors by the human eye, ranging in wavelength from about 400 nm to about 700 nm.

Vitalism The view that natural processes are controlled by supernatural forces and cannot be explained through the laws of physics and chemistry alone, as opposed to mechanism.

Vitamin An organic compound required by organisms in minute quantities for growth and reproduction because it cannot be synthesized by the organism; vitamins often serve as enzyme cofactors or parts of cofactors.

Vitellaria Structures in many flatworms that produce vitelline cells, that is, cells that provide eggshell material and nutrient for the embryo.

Vitelline Referring to the yolk of an egg or the ovum.

Vitelline circulation The embryonic circulatory system supplying blood to the yolk sac via vitelline arteries and its return to embryo through the vitelline veins.

Vitelline envelope/membrane Membrane outside the plasma membrane forming a fibrous mat over the sea urchin egg; becomes the fertilization membrane via the cortical reaction.

Viviparity Reproduction in which eggs develop within the female body, with nutritional aid of maternal parent as in therian mammals, many reptiles, and some fishes; offspring are born as juveniles. Adj., viviparous.

Viviparous An animal whose eggs develop inside the female while the embryo derives nutrition from the mother.

Viviporous Characterized by beginning embryo growth while still attached to the mother plant.

VNTR Variable number tandem repeat A type of DNA sequence found in eukaryotic genomes, in which a short sequence is repeated. The number of repeats varies between individuals and is used as a basis for genetic identification.

Volatile oils Terpenes composed of two to four isoprene units; also known as essential oils, such as lemon and peppermint.

Volatile organic compounds Organic chemicals that evaporate readily and exist as gases in the air.

Volatilization The process of evaporation; to pass into the atmosphere as a gas.

Volcanic Describes the action or process of magma and gases rising to the crust and being extruded onto the surface and into the atmosphere; also applies to the resulting igneous rocks that cool on the surface of the Earth, including beneath water, which typically have small crystals due to the rapidity of cooling.

Volcanic arc A curved chain of islands that rise from the sea floor, usually near a continent. The convex side usually faces the open ocean, while the concave side usually faces the continent; island arc - syn.

Volcanism The process by which magma and associated gases rise to the Earth's crust and are extruded, or expelled, onto the surface and into the atmosphere.

Voltage-gated channel Ion channel in a membrane that opens and closes in response to changes in membrane potential (voltage); the sodium and potassium channels of neurons are examples.

Voluntary (somatic) nervous system That part of the nervous system that relays commands to skeletal muscles.

Volvocaceans Group of colonial eukaryotic organisms, including Volvox.

Vomer The narrow bone forming the lower and posterior half of the nasal septum.

Vomerine teeth Teeth lying on the vomer, in the palate of amphibians.

Vulnerable A species of special concern because of characteristics that make it particularly sensitive to human activities or natural events. Compare *endangered*, *extinct*, *extirpated*, *threatened.*

Vulva A collective term for the external genitals in women.

W

Warm front A long, wedge-shaped boundary caused when a warmer advancing air mass slides over neighboring cooler air parcels.

Warning coloration Coloration that allows organisms to escape from predators by advertising something harmful or distasteful.

Wash zone The depth zone in which sediments are disturbed by wave action near the shoreline

Waste stream The steady flow of varied wastes, from domestic garbage and yard wastes to industrial, commercial, and construction refuse.

Wastewater treatment The use of physical and biological processes to remove particulate and dissolved material from sewage and to control pathogens.

Water activity A quantitative measure of water availability in the habitat; the water activity of a solution is one-hundredth its relative humidity.

Water column The vertical column of seawater that extends from the surface to the bottom.

Water cycle Worldwide circulation of water molecules, powered by the sun. Water evaporates from oceans, lakes, rivers, and, in smaller amounts, soil surfaces and bodies of organisms; water returns to the Earth in the form of rain and snow. Of the water falling on land, some flows into rivers that pour water back into the oceans and some percolates down through the soil until it reaches a zone where all pores and cracks in the rock are filled with water (groundwater); the deep groundwater eventually reaches the oceans, completing the cycle.

Water droplet coalescence A mechanism of condensation that occurs in clouds too warm for ice crystal formation.

Water mass A body of water that maintains its identity and can be characterized by such parameters as temperature and salinity

Water mold A common term for a member of the division Oomycota.

Water potential The physical property predicting the direction in which water will flow, governed by solute concentration and applied pressure.

Water stress A shortage of water in plants resulting in decreased ability to function properly, because water affects all biochemical controls and is the medium for all chemical reactions.

Water table The top layer of the zone of saturation; undulates according to the surface topography and subsurface structure.

Water vascular system A network of hydraulic canals unique to echinoderms that branches into extensions called tube feet, which function in locomotion, feeding, and gas exchange.

Waterlogging Water saturation of soil that fills all air spaces and causes plant roots to die from lack of oxygen; a result of overirrigation.

Watershed The land area that is drained by a river or estuary and its tributaries

Water soluble vitamin A vitamin that is not readily stored in the body, with excess eliminated in the urine; includes B vitamins and vitamin C.

Water vascular system. System of fluid-filled closed tubes and ducts peculiar to echinoderms; used to move tentacles and tube feet that serve variously for clinging, food handling, locomotion, and respiration.

Wave The undulation that forms as a disturbance moves along the surface of the water. Waves can be described by their height (the vertical distance between crest and trough), wavelength (the horizontal distance between adjacent crests), and period (the time the wave takes to move past a given point).

Wave crest The highest part of a wave.

Wave reinforcement The process by which two waves collide and add together to create a higher wave.

Wave shock The intensity of the impact of a wave.

Wave trough The lowest part of a wave.

Wavelength The distance between crests of waves, such as those of the electromagnetic spectrum.

Wax A lipid material with considerable oxygen inserted in the molecule; high melting point and relatively impermeable to water.

Wean To accustom a child or young animal to food other than the mother's milk.

Weather Description of the physical conditions of the atmosphere (moisture, temperature, pressure, and wind).

Weathering The physical, chemical, and biological processes by which rock is broken down into smaller pieces.

Weed A species that volunteers in artificially modified habitats and is considered undesirable by people. The same species may occur elsewhere in a wild state, or even in cultivation. *Weedy habit* refers to the propensity of certain species to disperse easily and widely and to colonize disturbed habitats.

Weil-Felix reaction A test for the diagnosis of typhus and certain other rickettsial diseases. In this test, the blood serum of a patient with suspected rickettsial disease is tested against certain strains of Proteus vulgaris (OX-2, OX-19, OX-K). The agglutination reactions, based on antigens common to both organisms, determine the presence and type of rickettsial infection.

Weir Filtration apparatus in flame-cell protonephridia of flatworms.

Westerlies Persistent eastward-equatorward winds in midlatitudes in both the Northern and Southern Hemispheres

Western blot A technique for analyzing mixtures of proteins to show the presence, size and abundance of one particular type of protein. Similar to Southern or Northern blotting, except that (1) a protein mixture is electrophoresed in an acrylamide gel, and

(2) the "probe" is an antibody which recognizes the protein of interest, followed by a radioactive secondary probe (such as 125I-protein A).

Wetlands Ecosystems of several types in which rooted vegetation is surrounded by standing water during part of the year.

Whirling disease Disease of fishes, caused by the protozoan Myxobolus cerebralis.

White blood cell Component of the blood that functions in the immune system. Also known as a leukocyte.

White piedra A fungal infection caused by the yeast Trichosporon beigelii that forms light-colored nodules on the beard and mustache.

Whole-genome shotgun sequencing An approach to genome sequencing in which the complete genome is broken into random fragments, which are then individually sequenced. Finally the fragments are placed in the proper order using sophisticated computer programs.

Whole-organism vaccine A vaccine made from complete pathogens, which can be of four types: inactivated viruses; attenuated viruses; killed microorganisms; and live, attenuated microorganisms.

Whorl An arrangement of appendages, such as branches or leaves, such that all are equally spaced around the stem at the same point, much like the spokes of a wheel or the ribs of an umbrella

Widal test A test involving agglutination of typhoid bacilli when they are mixed with serum containing typhoid antibodies from an individual having typhoid fever; used to detect the presence of Salmonella typhi and S. paratyphi.

Wild type An individual with the normal phenotype.

Wilderness An area of undeveloped land affected primarily by the forces of nature; an area where humans are visitors who do not remain.

Wildlife Plants, animals, and microbes that live independently of humans; plants, animals, and microbes that are not domesticated.

Wildlife refuges Areas set aside to shelter, feed, and protect wildlife; due to political and economic pressures, refuges often allow hunting, trapping, mineral exploitation, and other activities that threaten wildlife.

Wildtype The native or predominant genetic constitution before mutations, usually referring to the genetic consitution normally existing in nature.

Wind farms Large numbers of windmills concentrated in a single area; usually owned by a utility or large-scale energy producer.

Windbreak Rows of trees or shrubs planted to block wind flow, reduce soil erosion, and protect sensitive crops from high winds.

Windward The side toward which the wind is blowing.

Winnowing The process that separates the grain from the fragments of chaff.

Winogradsky column A glass column with an anaerobic lower zone and an aerobic upper zone, which allows growth of microorganisms under conditions similar to those found in a nutrient-rich lake.

Winter sleep A period of inactivity in which a mammal's body temperature remains near normal and the mammal is easily aroused.

Winterbottom's sign Swollen lymph nodes at the base of the skull, symptomatic of African sleeping sickness.

Wisconsin general test apparatus (WGTA) An apparatus designed originally at the University of Wisconsin Primate Center for use with various species of primates in a series of learning problems. Modifications of this apparatus have been used for a variety of animal species.

Wise use movement A coalition of ranchers, loggers, miners, industrialists, hunters, off-road vehicle users, land developers, and others who call for unrestricted access to natural resources and public lands.

Withdrawal A description of the total amount of water taken from a lake, river, or aquifer.

Within-habitat comparison A contrast of diversity between two localities of similar habitat type

Wobble A violation of the base-pairing rules in that third nucleotide (5' end) of a tRNA anticodon can form hydrogen bonds with more than one kind of base in the third position (3' end) of a codon.

Wood A secondary tissue found in seed plants which consists largely of xylem tissue

Wood pulp A watery suspension of pulverized wood used in the production of paper, cardboard, fiberboard, rayon, cellophane, and other products.

Woodland A forest where tree crowns cover less than 20 percent of the ground; also called open canopy.

Work The application of force through a distance; requires energy input.

Worker A member of the nonreproductive laboring caste of social insects.

World conservation strategy A system of maintaining essential ecological processes, preserving genetic diversity, and ensuring that utilization of species and ecosystems is sustainable.

World Ocean The interconnected world seas and oceans.

Wort The filtrate of malted grains used as the substrate for the production of beer and ale by fermentation.

Wrack zone A bank of accumulated litter at the strandline

X

X ray Very short wavelength in the electromagnetic spectrum; can penetrate soft tissue; although it is useful in medical diagnosis, it also damages tissue and causes mutations.

Xanthophore A chromatophore containing yellow pigment.

Xanthophyll a group of yellow or golden photosynthetic pigments

X-chromosome One of the sex chromosomes.

Xenodiagnosis Diagnosis of a disease by infecting a test animal.

Xenograft Graft of a piece of tissue or organ from one individual to another of a different species.

Xenosomes Body or organelle living within a cell that contains its own DNA and is capable of reproducing itself, once having functioned as a free-living organism. Examples are zooxanthellae and zoochlorellae.

Xeric 1.Describing a location or habitat with very little moisture. 2. describing an organism that lives in such an environment.

Xeric Characterized by a very dry environment.

Xerophilic microorganisms Microorganisms that grow best under low aw conditions, and may not be able to grow at high aw values.

Xerophyte A plant well adapted to withstand prolonged drought. The typical xerophyte is a deciduous shrub with tiny leaves and a shallow root system that etends well beyond the crown of the shrub.

Xerophytic leaves The leaves of plants that grow under arid conditions with low levels of soil and water. Usually characterized by water-conserving features such as thick cuticle and sunken stomatal pits.

Xiphidiocercaria Cercaria with a stylet in the anterior rim of its oral sucker.

X-organ. Neurosecretory organ in eyestalk of crustaceans that secretes molt-inhibiting hormone.

X-ray crystallography A technique which allows elucidation of the structure of macromolecules. In this technique, an X-ray beam is passed through a crystal of a particular substance. The atoms which make up the crystal cause the X-rays to be deflected. This deflection pattern is directly dependent on the structure of the molecule being analyzed, and can thus be used to help determine the molecular structure.

X-ray diffraction Technique utilized to study atomic structure of crystalline substances by noting the patterns produced by x-rays shot through the crystal.

Xylem Tissue in the vascular system of plants that moves water and dissolved nutrients from the roots to the leaves; composed of various cell types including tracheids and vessel elements. Plant tissue type that conducts water and nutrients from the roots to the leaves.

Xylem ray That part of a vascular ray which is located in the wood or secondary xylem. Consists of a radial file of parenchyma cells, Also termed wood ray.

Y

Yard Lowland area with dense coniferous cover in areas.

Ya.·ding Herding up in lowland areas with dense coniferous cover in areas called "yards."

Yaws Bacterial disease caused by the spirochete Treponema pertenue, often transmitted by flies.

Year-class effect The common domination of a species population by individuals recruited in one reproductive season

Yearling An animal that is between one and two years old.

Yeast A unicellular fungus that lives in liquid or moist habitats, primarily reproducing asexually by simple cell division or by budding of a parent cell.

Yeast artificial chromosome (YAC) A stretch of DNA that contains all the elements required to propagate a chromosome in yeast and which is used to clone foreign DNA fragments in yeast cells.

Yeast cultures Yeast cells growing ***in vitro*** in a solution containing all the necessary nutrients and trace elements necessary for normal growth.

Yellow fever An acute infectious disease caused by a flavivirus, which is transmitted to humans by mosquitoes. The liver is affected and the skin turns yellow in this disease.

Yellowcake The concentrate of 70 to 90 percent uranium oxide extracted from crushed ore.

YM shift The change in shape by dimorphic fungi when they shift from the yeast (Y) form in the animal body to the mold or mycelial form (M) in the environment.

Yolk Nutritive stores found within the ovum cytoplasm consisting of proteins and fats.

Yolk platelets Membrane-bound discs containing high concentrations of yolk found in eggs.

Yolk plug The large yolk-filled cells that protrude from beneath the ectoderm at the blastopore of the amphibian gastrula.

Yolk sac The first of four extraembryonic membranes of amniotes to form during embryogenesis. Like the allantois, it arises from the splanchnopleure (endoderm and splanchnic mesoderm) to surround the mass of yolk in reptile and bird eggs. It is connected to the midgut by the yolk stalk. The yolk sac is also formed in mammals, despite the absence of yolk.

Y-organ. Gland in the antennal or maxillary segment of some crustaceans that secretes molting hormone.

Z

Z lines Dense areas in myofibrils that mark the beginning of the sarcomeres. The actin filaments of the sarcomeres are anchored in the Z lines.

Z value The increase in temperature required to reduce the decimal reduction time to one-tenth of its initial value.

Zebroid A hybrid animal that results from breeding zebras and horses.

Zeitgeber Any entraining agent that plays a role in setting or resetting an internal biological clock. Examples include sunrise or sunset.

Zero population growth (ZPG) The number of births at which people are just replacing themselves; also called the replacement level of fertility.

Zinc finger A protein structural motif common in DNA binding proteins. Four Cys residues are found for each "finger" and one finger can bind a molecule of zinc. A typical configuration is: CysXxxXxxCys—(intervening 12 or so aa's)—CysXxxXxxCys.

Zoea an early larval stage of many crustaceans

Zoecium , zooecium Cuticular sheath or shell of Ectoprocta.

Zoid Member of a colonial organism.

Zoitocyst Tissue phase in some of the coccidia of the Isospora group. They usually have internal septae and contain thousands of bradyzoites. Also called a sarcocyst or Miescher's tubule.

Zona pellucida A translucent non-cellular membrane surrounding a mammalian egg through which sperm must penetrate in order for fertilization to occur.

Zona reaction In mammals, modification of the zona pellucida that blocks polyspermy; enzymes released by cortical granules digest sperm receptor proteins ZP2 and ZP3 so that they can no longer bind sperm.

Zonation Occurrence of single species or groups of species in recognizable bands that might delineate a range of water depth or a range of height in the intertidal zone

Zone of aeration Upper soil layers that hold both air and water.

Zone of differentiation An area within the plant, proximal to the apical meristems, where cells begin to assume their srtuctural and functional characters.

Zone of elongation Area in plant roots where recently produced cells grow and elongate prior to differentiation.

Zone of intolerance The area outside the geographic range where a population is absent; grades into the zone of physiological stress.

Zone of leaching The layer of soil just beneath the topsoil where water percolates, removing soluble nutrients that accumulate in the subsoil; may be very different in appearance and composition from the layers above and below it.

Zone of physiological stress The area in a population's geographic range where members of population are rare due to physical and biological limiting factors.

Zone of saturation Lower soil layers where all spaces are filled with water.

Zoned reserve systems Habitat areas that are protected from human alteration and surrounded by lands that are used and more extensively altered by human activity.

Zonite The individual body unit of a member of the phylum Kinorhyncha.309

Zoochlorella Any of various minute green algae (usually Chlorella) that live symbiotically within the cytoplasm of some protozoa and other invertebrates.

Zooflagellates Members of the Zoomastigophora, the animal-like flagellates (phylum Sarcomastigophora).

Zooid An individual member of a colony of animals, such as colonial cnidarians and ectoprocts, produced by incomplete budding.

Zoology The study of animals.

Zoomastigophorea The protozoan class where members lack chloroplasts; heterotrophic or saprozoic. Examples: Trypanosoma, Trichonympha, Trichomonas, Giardia.

Zoonosis Disease of animals that is transmissible to humans. Some authors subdivide the concept into zooanthroponosis, an infection humans can acquire from animals, and anthropozoonosis, a disease of humans transmissible to other animals.

Zooplankton Minute aquatic animals that drift freely in the plankton, feeding mainly on phytoplankton (plant-like organisms) and having no locomotory structures.

Zoospore A motile asexual spore.

Zooxanthellae Symbiotic dinoflagellates in the genus *Symbiodinium* that live in the tissues of a number of marine invertebrates and protists, notably in many foraminiferans, cnidarians, and some mollusks.

Zosterophyllophyta An extinct division of early vascular plants characterized by lateral sporangia and commonly spine-like emergences. Believed by evolutionary biologists to be the ancestor of the lycopods.

Z-scheme A diagrammatic representation of electron flow through PS II and PS I.

Zugunruhe Restlessness that an animal displays at the time of its usual migration.

Zygomatic arch The arch formed by the articulation of the broad temporal process of the zygomatic bone and the slender zygomatic process of the temporal bone.

Zygomorphic flower A bilaterally symmetrical flower, capable of being divided into two symmetrical halves only by a single longitudinal plane passing through the axis.

Zygomycetes One of the division of the fungi, characterized by the production of zygospores; includes the bread molds.

Zygosporangium (or zygospore) The thick-walled sexual spore formed by members of the zygomycetes.

Zygospore In fungi, a structure that forms from the diploid zygote created by the fusion of haploid hyphae of different mating types. After a period of dormancy, the zygospore forms sporangia, where meiosis occurs and spores form.

Zygote A fertilized egg. Or the diploid cell that results from the joining of two haploid gametes (sperm and egg) during sexual reproduction, that will cleave to form an embryo.

Zygotene The second stage in prophase 1 of meiosis, characterized by the synapsis, or pairing, of homologous chromosomes to form a bivalent.

Zygotic meiosis Meiosis that takes place within the first few divisions after zygote formation; thus all stages in the life cycle other than the zygote are haploid.